Principles of Electric Machines with Power Electronic Applications

M. E. El-Hawary

Electrical Engineering Department
Technical University of Nova Scotia
Halifax, Nova Scotia, Canada

A Reston Book, Prentice-Hall
Englewood Cliffs, NJ 07632

Library of Congress Cataloging-in-Publication Data

El-Hawary, M. E.
    Principles of electric machines with power
electronic applications.

    "A Reston book."
    1. Electric machinery.   2. Power electronics.
I. Title.
TK2000.E38115   1986   621.31'042   85-30069
ISBN 0-8359-5785-3

O 8 1 8 8 7

Editorial/production supervision and
interior design by Camelia Townsend

10   9   8   7   6   5   4   3   2   1

PRINTED IN THE UNITED STATES OF AMERICA

FOR BOB: GLORY OF FAITH

You are the light of the world. A city that is set on an hill cannot be hid.

Neither do men light a candle, and put it under a bushel, but on a candlestick; and it giveth light unto all that are in the house.

Let your light so shine before men, that they may see your good works, and glorify your Father which is in heaven

Matthew, 5:14–16

Contents

Preface

This book is designed mainly as an introduction to principles of electric machines and the closely related area of power electronics and adjustable speed drives. It is intended for students in electrical and other engineering disciplines as well as being useful as a reference and self-study guide for the professional dealing with this important area. The coverage of the book is intended to enable its use in a number of ways including service courses taught to non-electrical majors. The organization and details of the material in this book allows a maximum flexibility for the instructor to select topics for inclusion in courses in the modern engineering curriculum.

The book does not require a level of mathematical sophistication beyond that given in undergraduate courses in basic physics and introductory electric circuits. The emphasis in our coverage is given to an improved understanding of the operational characteristics of the electric apparatus discussed, on the basis of linear mathematical models. Almost every key concept is illustrated through the use of in-text examples, that are worked out in detail to enforce the reader's understanding. Many practical problems in electric machines operation involve the use of known performance variables under a given operational condition to predict the same variables under different operating conditions. These problems can be easily dealt with using the basic performance characteristics to obtain some corollary results that are useful for this purpose. On many occasions in this text we take the time to derive some of these useful relations to allow the student to deal with these common and important problems.

The first chapter in this book provides a historical perspective on the development of electromechanical energy conversion devices and starts by tracing the origins of electricity leading up to the fundamental discoveries of the not too distant past. While this topic is not an integral part of the conventional coverage in texts and courses in this area, this chapter should provide interesting insights into the influence of these developments on present day's civilization. It is through an appreciation of the past developments and achievements that we can understand our present and forge ahead with future advances.

Chapters two to eight deal with the conventional topics covered in present courses in electric machines and transformers. Emphasis is given to practical aspects such as dealing with matching motors to loads, speed control, starting, and in general to the main performance characteristics of the devices discussed.

The advent of the SCR and subsequent developments in solid state and power electronics technology introduced new elements in the practice of motor speed control. This important area is now of sufficient maturity that it should form an integral part of any comprehensive treatment of electric motors. This text recognizes this need by devoting three chapters to this important area. In Chapter 9 we explore the characteristics of thyristors and discuss a number of important issues related to its use. Chapter 10 deals with power electronic systems such as rectifiers, inverters, choppers, controllers and converters. In Chapter 11 we discuss adjustable speed drives with special emphasis on dc and induction motor control.

I have attempted to make this book as self-containing as much as possible. As a result the reader will find that many background topics such as magnetic circuits, the per unit system and three phase circuits are included in the text's main body as opposed to the recent trend towards including many appendices dealing with these topics. In studying and teaching electrical machines it has been my experience that a problem solving approach is most effective in exploring this rich area. As a result, the reader will find many solved problems at the end of each chapter that are intended to reinforce the concepts learned in the chapter. Readers familiar with the author's other Prentice-Hall-Reston texts on Electric Power Systems and Control Systems Engineering will note the similarity in the formats of the treatment in these and the present book. The positive comments about these features encouraged me to retain this problem-solving approach.

A textbook such as this could not have been written without the continuing input of the many students who have gone through many versions of its material as it was developed. My sincere thanks to the members of the many classes whom I was privileged to teach this fascinating area. I wish to acknowledge the able work of Wanda Roy and Verilea Ellis of TUNS in putting this manuscript in a better shape than I was able to produce. My continuing association with the Prentice-Hall and Reston staff has been valuable through out the many stages of preparing this text. I wish to express my appreciation for the work of Camelia Townsend in the production of this text. Thanks to Ben Wentzel, Greg Michael, Elio Ennamorati, and Tim Bozik for their work and support.

I owe a debt of gratitude to many of my colleagues who reviewed this manuscript and provided many valuable comments that improved this work considerably: Dr. A. Bose, Arizona State University; Dr. G. S. Christensen, University of Alberta; Dr. B. T. Ooi, McGill University in Montreal; Dr. A. Sharaf, University of New Brunswick.

It is always a great pleasure to acknowledge with thanks the continuing support of President J. C. Callaghan and Dean D. A. Roy of Technical University of Nova Scotia during the course of preparing this text. As always have been the case, my wife Dr. Ferial El-Hawary's patience and understanding made this project another joy to look forward to completing. It goes without saying that our three children deserve a greater share of my appreciation for their continuing understanding.

M. E. EL-HAWARY

Introduction

Electrical engineering is a wide and diversified area of human activity that plays a very important role in our present-day civilization. If one attempts to single out a field of electrical engineering that heralded the dawn of the Electrical Age, the field of electromechanical energy conversion emerges as a strong contender. It is the main purpose of this chapter to discuss the origins and historical developments that led to present-day electromechanical energy conversion engineering. The treatment is brief but is intended to provide a historical perspective that highlights advances gained over the years. The brief history is followed by an outline of the text.

1.1 ELECTRIC MACHINES

An electric machine may be defined as an electric apparatus that has one or more components capable of rotary or linear motion. The operation of the machine depends on electromagnetic induction, which is defined as the production of an electromotive force in an electric circuit by a change in the magnetic flux linking with the circuit.

Electric machines are encountered in almost all areas of human activity and it would be difficult to attempt to list such areas. It is worth noting, however, that present high-tech applications such as robotics and computers depend, in no small measure, on advanced, precise, and energy-efficient electric machines operating as motors. Disk and tape drives, printer mechanisms, and manipulator arms rely on the electric motor in their performance.

The definitions offered at the beginning of this section include key concepts that are traced back in history in the following sections.

1.2 ROOTS IN OBSERVATION

For many centuries, human beings have observed natural electric and magnetic effects. Atmospheric electrical phenomena such as lightning have been known to the human race since antiquity. Italian sailors navigating the Mediterranean discovered emissions of light from the mastheads of their ships on nights of dry stormy weather. This light is known as the St. Elmo's fire and has been referred to in the records of the second voyage of Columbus in 1493. Polar lights in the north (aurora borealis) and in the south (aurora australis) have been observed for many centuries.

Biological electric field effects displayed by certain creatures of the sea have been observed and reported as far back as more than 2000 years ago in the writings of Greek philosophers, including Aristotle. A human being could receive a shock by touching an electric torpedo with a spear. The connection between the torpedo's electric property and that of atmospheric electric effects is recognized in the Arabic name for the torpedo, *ráad*, which is also the word for *thunder*. The electric torpedo was used for medical treatment of headaches at the time of Anthony and Cleopatra.

Lightning, St. Elmo's fire, polar lights, and animal electricity are phenomena that people observe but in whose generation they do not take an active part. Two other phenomena, however, were observed and required human action. The property of amber to acquire the power of attracting light objects through friction has been known for at least 2500 years, as evidenced by the writings of Thales, one of the seven wise men of Greece. An ancient reference to the electric property of amber is found in the Middle Eastern romance "The Loves of Majnoon and Leila." In the text, Majnoon says of his beloved, "She was an amber, and I but as straw; she touched me and I shall ever be attracted to her." The second phenomenon involves the emission of sparks from the human body due to friction. It was known, for example, that Servius Tullius, the sixth King of Rome, gave off sparks as his locks were combed.

The observations of the ancients led to the invention in 1660 of the first frictional machine by Otto Von Guericke of Magdeburg. The machine consisted of a sulfur globe mounted on an axis which when rotated against a cloth pressed to its surface emitted sparks and attracted light pieces of straws.

Protection of buildings against damage due to lightning strokes is believed to date back several thousand years. The nature of construction of famous buildings such as the Temple of Juno and Solomon's Temple was such that their roofs were covered by metallic points. The concept of deliberate protection of buildings using lightning conductors was introduced by Benjamin Franklin following the famous (but extremely dangerous) kite experiments in 1750.

Magnetic field effects appearing in nature have been observed since antiquity. Ancient civilizations made use of magnetism in direction finding as early as 2637 B.C., one of the earlier gadgets being that attributed to the time of the Chinese emperor Hoang-Ti. In that invention, a prominent female figure was mounted on a chariot and always pointed to the south regardless of the Chariot's direction of motion. Chinese south-seeking instruments were again reported 16 centuries later for use as navigational aids on land and sea. By the twelfth century, knowledge of the magnetic compass and its application had spread to many countries. It is well known that Christopher Columbus employed the magnetic compass on his voyages (1492).

The attraction of iron to lodestone has been known for countless years. Attempts to explain this phenomenon date back to the poetic work of Lucretius in 55 B.C. The early Christian writer Saint Augustine described in his *De Civitate Dei* (A.D. 428) magnetic experiments involving the attraction of a piece of iron lying on a silver dish by a lodestone underneath. The phenomenon of magnetic attraction lead to the story of sympathetic needles that prevailed in the sixteenth to eighteenth centuries. William Gilbert of Colchester, physician to Queen Elizabeth I, published a book entitled *De Magnete* dealing with observed phenomena of magnetism and electricity. The publication of the book helped to focus the attention of investigators on new directions that resulted in significant discoveries.

1.3 BEGINNINGS

Further discoveries and applications of electricity resulted from the availability of the frictional machine in the eighteenth century. A notable discovery occurred in 1720, with Stephen Grey announcing the principle of insulation and conduction of electricity. Electric charges were transmitted for hundreds of feet through a hempen cord suspended by silk threads. Transmission of electric charge for several miles was made possible using metallic wire instead of the hempen cord. Grey's discoveries were a prelude to the discovery of charge storage in a Leyden jar that took place a quarter of a century later.

It has been observed that electrified bodies lost their charge quickly and it was believed that air abstracted the charge from the conductor. It thus seemed that if an electrified body were surrounded by an insulating material, contact with air, and hence loss of charge, would be reduced.

Water appeared to be a convenient recipient of the electric charge and glass seemed to be the most effective insulator. This provided the basis for experiments carried out by many investigators. It is believed that Cunaeus working with Musschenbröek of Leyden on this concept discovered in January 1746 what is now known as the Leyden jar for storing electric charge. Cunaeus held the glass bottle in his right hand; the prime conductor of a powerful frictional machine was connected by a wire to the water inside the bottle. Using his left hand, Cunaeus attempted to disengage the wire from the conductor, believing that enough charge had been accumulated in the water. Needless to say, our friend received quite a convulsive shock, which made him drop the bottle. Musschenbröek repeated the experiment and felt struck in the arms and shoulders and lost his breath. It took him two days to recover and he stated later that he would not repeat the experiment for the entire kingdom of France.

Improvements on the Leyden experiments followed in many parts of Europe. The object of the experiments was to transmit shock for large distances using chains of Leyden jars. William Watson, a London physician, established the idea of using two coatings separated by a dielectric to improve the Leyden jar. Watson also established the idea of positive and negative electrical charge.

The discovery of steady electric current is due to the observation of muscle contractions in the legs of dead frogs when in contact with two different metals. In 1678, the Dutch scientist Swammerdam produced convulsions in the muscle of a frog by holding it against a brass ring from which it hung by a silver wire. This experiment clearly resembles that by which Luigi Galvani became famous over 100 years later. The Swiss philosopher Sulzer described an experiment carried out in 1762. In his experiment, Sulzer placed two pieces of different metals (silver and zinc) above and below his tongue. When the metals were brought in contact, Sulzer reported an itching sensation and a taste of sulfate of iron. Obviously, both Swammerdam and Sulzer had the opportunity to discover galvanism but did not.

Luigi Galvani, born in Bologna in September 1737, was destined to discover the significance of the Swammerdam and Sulzer experiments following his own famous experiments. On November 6, 1780, Galvani was preparing a frog for dissection in the vicinity of an electrical machine and observed that the animal's body suddenly convulsed. After a number of similar experiments, he resolved in 1781 to try the effect of atmospheric electricity. Galvani erected a lightning conductor on the roof of his house and connected it to his laboratory. Nerves of frogs and other animals were connected to the lightning conductor and muscles were observed to convulse whenever lightning appeared. In 1786, Luigi Galvani resumed the experiments with the help of his nephew Camillo Galvani. On September 20, 1786, Camillo had some frogs prepared for the experiments and hung them by an iron hook from the top of an iron rail of the balcony. Soon, he noted convulsions when a frog was pressed against a rail. Luigi Galvani experimented at length and concluded that the convulsions were the result of the simultaneous contact of iron with nerves and muscles. It is interesting to note that Galvani attributed the convulsions to animal electricity, that is, electricity excited by the animal organs themselves.

Alessandro Volta, then a professor of physics at Pavia, was interested in the results of Galvani's experiments. He concluded that the exciting cause of convulsions was nothing but ordinary electricity produced by the contact of two metals. His theory of contact was derived from the Sulzer experiments. In August 1796, Volta took disks, one of copper and the other of zinc, brought them into contact, and suddenly separated them without friction. The usual indications of electricity were found. On June 26, 1800, he made his formal announcement to the Royal Society describing the now famous voltaic pile. In his setup, Volta placed a disk of silver on a table; on top of it, he placed a disk of zinc, followed by a disk of spongy matter (cardboard) impregnated with a saline solution. The arrangements were repeated thirty to forty times. Shocks and sparks were found to occur, but it was remarkable that the pile recharged itself after a shock had been given. Thus was "born" one of mankind's greatest boons, the electric current.

The thermal effects of electricity were known as early as 1772, when the lightning conductors at St. Paul's Cathedral were observed to become red hot during thunderstorms. The idea of electric conduction through a liquid was established, and decomposition of water into its constituent gases by the use of electricity from a Leyden jar was accomplished by the English physician George Pearson in 1797. Following the invention of the voltaic pile, Nicholson and Carlisle established that water can be decomposed using voltaic current. The similarity between the chemical effects of static and voltaic electricity was thus established.

At the forefront of contributors to electrochemistry in the ninteenth century is Humphry Davy, who was born in Penzance on December 17, 1778, and became professor of chemistry at the Royal Institution in 1801. His discoveries advanced the understanding and appreciation of the newly found steady electric current. In 1808, Davy made the first public display of the electric arc between two carbon electrodes connected to a voltaic pile. His most brilliant success was in the decomposition of fixed alkalies by electricity, in 1807.

1.4   FOUNDATIONS OF ELECTROMAGNETISM

The strong resemblance between electricity and magnetism had been observed for many years. These observations were supported by many facts, discovered as early as 1630, when Gassendi observed that magnetism was imparted to iron by lightning. In 1675, English navigators reported that the compass needles of their ships had their poles weakened and even reversed by lightning strokes. By the middle of the eighteenth century, speculation about the nature of the link between electricity and magnetism was a favorite pursuit.

Franklin noted that electricity is capable of reversing the polarity of magnetic needles. He also showed that the discharge of Leyden jars through a common sewing needle resulted in its magnetization. Many other investigators attempted at the turn of the nineteenth century to establish the connection between electricity and magnetism using the then available voltaic pile. It appears, however, that many reports of the effect of the voltaic pile on a magnetic needle were done with the pile open-circuited.

The main breakthrough is credited to a discovery by Hans Christian Oersted (1777–1851), born in Rudköbing on the island of Langeland in the Baltic. He was appointed in 1806 to the Chair of Physics at the University of Copenhagen and was destined to produce the answer to the mystery of electromagnetism. It is reported that in a private lecture to advanced students in the winter of 1819–1820, Oersted was demonstrating the then familiar experiment with a magnetic needle in the vicinity of a voltaic pile. Previously, the battery circuit had always been left open, as mentioned earlier. It is not clear whether in making the demonstration, Oersted closed the circuit by a lucky chance or on purpose. In any event, to his delight, he saw the needle move from its position of rest. Oersted pointed to the apparatus with trembling hands and invited his students to repeat the experiment. An account of this significant discovery was published in July 1820.

The enormous effect of Oersted's discovery on the direction of investigations in electricity and magnetism is evidenced by the quantum leap in related reports during that period. Sir Humphry Davy repeated Oersted's experiment and noted that iron filings on a paper near the wire were immediately attracted. Davy also anticipated lines of magnetic force by observing the orientation of steel needles in the neighborhood of the current-carrying conductors. Davy was not alone in immediately taking up Oersted's discoveries; the list includes Ampére and Arago in France, Faraday and Sturgeon in England, Schweiger of Germany, and Joseph Henry of New York.

One of those who followed up on Oersted's discoveries is André Marie Ampère (1775–1836), who was then a professor at the College of France and at the Ecole Polytechnique in Paris. Ampère assumed the first and highest place by his research results, reported on September 18, 1820, to the Academy of Sciences—within two months of Oersted's formal announcement. In his paper, he explained the law that determines the position of the magnetic needle relative to the electric current. To illustrate the concept, Ampère proposed a little bon homme (good man) swimming in the direction of the current with the positive pole at the feet and the negative at the head. The needle will be aligned at right angles to the current with the north pole pointing toward the man's left and the south pole pointing toward his right. Ampère established the concept of a complete electric circuit by discussing the effect on the magnetic needles of the current in the battery. He also showed that current-carrying conductors have a magnetic effect on one another.

Yet another important step in the development of electrodynamics was taken during the same period. Schweigger (1779–1857), a German chemist in Halle, announced on September 16, 1820, his famous "multiplier principle." The principle states that a conducting wire twisted upon itself and forming N turns produces a magnetic effect N times greater than a wire with a single turn. Schweigger's discovery resulted in the invention of the galvanometer.

Arago (1786–1853) was another French physicist who took an interest in Oersted's discovery and in Ampère's theory of magnetism. He placed an ordinary sewing needle in a glass tube and wound a copper wire around the tube. The application of a current through the wire resulted in the magnetization of the needle. Arago's observations bore fruit in the first electromagnet, constructed by William Sturgeon (1783–1850) and displayed in 1825. Sturgeon used a bar of soft iron bent in the form of a horseshoe and coated with insulating varnish. Sixteen turns of bare cooper wire were wound around the bar while keeping the turns separate. Upon the application of electric current, the electromagnet lifted a weight of 9 pounds.

The concept of Sturgeon's electromagnet was further improved by Joseph Henry (1797–1878) of New York. Henry, brought the turns of wire closer together in a circumferential formation around the rod. To do this, Henry insulated the wires with silk covering. He was thus able to improve on the lifting power by increasing the number of turns at nearly right angles to the axis of the rod. Henry's first electromagnet was exhibited in June 1828 at the Albany Institute in New York, at which he was a professor. Henry continued with his experiments but encountered a problem in attempting to increase the lifting power by adding more turns. He discovered that turns added beyond a certain point were less effective in improving the lifting power. It is interesting to note that in 1827, Georg Simon Ohm, a professor of physics at Munich, published his now celebrated law. Ohm's law can be used to explain Henry's observation. It is unfortunate, however, that Ohm's law failed to attract attention in that era and was totally unknown to Henry or even to Wheatstone in 1837. Nonetheless, Henry worked out a practical solution and by doing so discovered some practical implications of Ohm's law.

Henry's solution resulted in two kinds of electromagnets. The first, which he called a quantity magnet, had numerous short coils (low resistance). The second, which he referred to as an intensity magnet, had one very long coil (high resistance). The coils of the quantity magnet were connected (in parallel) to a galvanic cell, which Henry called a quantity battery.

To drive a single long coil of the intensity magnet, Henry used a number of galvanic cells (in series), which he called an intensity battery. By so doing, Henry arrived at some very practical corollaries to Ohm's law without even being aware of it. As a professor of natural philosophy at Princeton, Joseph Henry was able in 1833 to demonstrate electromagnets that were capable of lifting up to 3500 pounds.

We have seen that Arago's work inspired the invention of the electromagnet. Another experiment that he carried out provided the inspiration for major breakthroughs by Michael Faraday. In the now famous Arago's experiment, a copper disk was rotated about an axis. Just above the disk, a magnetic needle was suspended such that its point of suspension was exactly above the center of the disk. A screen of glass separated the needle and disk so that air circulation due to the motion of the disk was not imparted to the needle. The needle deviated from its rest position as the disk rotated. With a high rotational speed, the needle ended up rotating with the disk.

1.5 THE DAWN OF ELECTRODYNAMICS

Oersted's discovery was followed by a wave of new experiments and discoveries that provided the foundation for linking electricity and magnetism, as we have seen in the preceding section. Up to that point, we know that electrical investigators depended entirely on the voltaic current and the magnetic needle for their work. From our present knowledge, we can claim that the stage was set for the arrival of a person who changed the course of electrical engineering history by discovering a new source of electric current. The human race is indebted to Michael Faraday for his series of experimental researches.

Michael Faraday was born in Surrey, outside London, on September 22, 1791. At the age of 13, Faraday became an apprentice bookbinder and developed an appetite for reading on scientific subjects. At age 21, he had an opportunity to attend a course of four public lectures by Sir Humphry Davy (the tickets were purchased by one of the shop's customers). The young enthusiast took full notes of the lectures and copied them neatly. A bound volume of the lectures was produced by Faraday and presented to Sir Humphry. It is evident that Davy was impressed by Faraday's account of the lecture, for in 1813 he recommended that Faraday be appointed his assistant in the laboratory at the Royal Institution. Thus started an association that was to last for nearly 50 years.

Faraday's scientific activities encompassed many areas, with emphasis on chemistry. His investigations led to discoveries of new compounds, such as alloy steels and optical glass. The subject of electromagnetism occupied him upon the heels of Oersted's discovery in 1820–1821 and resulted in the development of what is now known as electromagnetic rotation. Ten years later, Faraday's interest in the area resulted in the development of electromagnetic induction.

Faraday repeated Oersted's experiment and saw the significance of the mutual effect of the current and the magnet. On September 3 and 4, 1821, he set up an experiment whereby a free-moving current-carrying wire was made to rotate around the pole of a fixed magnet and established that a free-moving magnet can be made to rotate around a fixed current-carrying wire. Faraday designed a device to show the two effects, which he had made for him by an instrument maker. A sketch of the device is shown in Figure 1.1. Faraday's experimental success demonstrated for the first time the possibility of converting electric current into continuous mechanical motion.

During the period 1821–1831, Faraday's interests were primarily in the chemical field, although from time to time, he returned to consider electromagnetic effects. He read published accounts of experiments and theories proposed in the area, but continued to be puzzled by one facet of Arago's experiment. Iron was the only substance attracted by a magnet, yet copper appeared to be magnetic when rotated near a pivoted magnet. It was also known that a copper current-carrying wire attracted iron filings. He concluded after much thought that the key lay in investigating the inductive

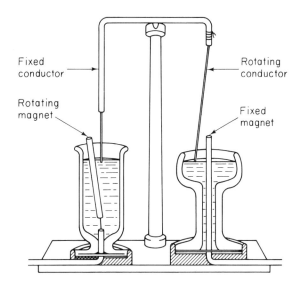

Fixed conductor

Rotating conductor

Rotating magnet

Fixed magnet

Rotation of magnetic pole around a current-carrying conductor

Rotation of a current-carrying conductor around a fixed magnetic pole

FIGURE 1.1   Device showing Faraday's electromagnetic rotations.

effects of electric currents. He was rewarded by obtaining a full explanation of Arago's magnetic phenomenon.

On August 29, 1831, Faraday achieved success by carrying out his famous ring experiment. The experiment involved an iron ring and two coils, *A* and *B*. Coil *A* was made of three sections wound on the left-hand side of the ring. Coil *B* was made of two sections wound on the right-hand side. It is of interest to note that this arrangement constituted the first transformer. The terminals of coil *B* were connected by a long wire passing above a magnetic needle. One section of coil *A* was connected to a battery. On making the connection, Faraday observed the needle to move, oscillate, and then settle to its original rest position. On breaking the connection of the *A* coil with the battery, the needle again moved. Faraday repeated the experiment but with three sections of coil *A* connected to the battery and observed that the effect on the needle was much stronger then before.

The results of the transformer experiment were not satisfying to Faraday, for apparently, he was looking for a steady deflection of the needle. The intermittent nature of the secondary current puzzled him. He continued with his experiments and discovered that the same effect, but weaker, could be obtained when the iron ring was replaced by a wooden core. In one of his experiments, he moved a flat wire helix between opposite magnetic poles. The effect on the needle was so clear that he claimed to have discovered the long-awaited conversion of magnetism to electricity.

On October 17, 1831, a second major discovery was achieved. A helix consisting of eight coils of copper wire, connected in parallel to a galvanometer and wound around a hollow paper cylinder, was used. He thrust the end of a cylindrical bar magnet quickly into the cylinder. There was a sudden deflection of the needle which then died down. On withdrawing the magnet, the same effect took place. Although the transformer experiment and the magnet plunged into a coil were successful to show the effects of induction, Faraday was still interested in following up on Arago's experiment and was thus led to his third experiment.

The third experiment was carried out on October 28, 1831, and was modified to its now well-known arrangement on November 4, 1831. A revolving copper plate was mounted on a horizontal brass axle. By means of small magnets fixed to the pole pieces of a powerful magnet, Faraday produced a short gap in which the plate revolved. One contact was made on the disk's periphery and the second one on the axle. This arrangement produced significant currents lasting for as long as the plate was rotated.

Faraday's most outstanding discoveries were thus made during a very short period toward the end of 1831 following much study and thought. These discoveries clearly provide the origins of electrodynamic devices, including the transformer, the alternator, and the direct-current generator. Faraday summarized all these results in a famous paper read before the Royal Society on November 24, 1831.

1.6 EARLY ELECTRIC GENERATORS

The news of Faraday's experiments sparked an unprecedented interest in developing machines that generated electricity from magnetism, as the process was known at the time. In July 1832, an oscillating setup was made by dal Negro of Padua in Italy. The arrangements included four coils on a table and four small magnets on a carriage that could be advanced up to the coils and then withdrawn mechanically. Clearly, this setup was not practical, but it served to demonstrate the phenomenon.

The Pixii Machine

The first electromagnetic generator employing rotation was invented by M. H. Pixii of Paris in 1832. In this first generator, a permanent magnet of the horseshoe type was rotated about a vertical axis by means of a hand crank and gearing. Immediately above the magnet were the poles of a wire-wound soft-iron core fixed in a suspending frame. The terminals of the winding were connected to a commutator. The machine produced a rapid succession of sparks when cranked. The first public display of this machine was on September 3, 1832, in Paris.

The Saxton Machine

A magnetoelectric machine made by Saxton was displayed at the June 1833 meeting of the British Association for the Advancement of Science in Cambridge. The machine had a horizontal horseshoe magnet. The armature consisted of a four-armed soft-iron cross. A soft-iron core carrying a coil was attached to each arm. The four-pole armature rotated on a horizontal spindle opposite the end of the magnet. The rotation was provided by a belt from a vertical pulley and hand crank.

Further Improvements

Sturgeon, the English magnet maker, introduced another variation in the construction. He made a machine with a coil that rotated between the poles of the magnet. He also produced an all-metallic commutator made of four semicylindrical pieces attached to the rotating spindle. A few years later, Stöhrer of Leipzig made a six-pole machine by setting up three horseshoe magnets with their poles forming a circle. Above these, he rotated a frame carrying six coils and a commutator. Woolrich produced a continuous (non-intermittent) unidirectional current by using more coils than magnets and taking the current only from coils that are active at the moment of contact. The use of commutators allowed a number of machines to be used for electroplating.

Machines for Lighthouse Illumination

By 1850, the development of dynamos depended on their perceived application as a source of electricity for illuminating lighthouses. The most important designs of that era were due to Holmes. The first of his designs was a $2\frac{1}{2}$-hp machine that was tried at Blackwall in 1857 under the direction of Faraday. The rotor consisted of six wheels each carrying six compound permanent magnets, for a total of 36. The frame consisted of five rings each carrying 24 coils. A direct-current output was obtained through a large commutator.

The success of the trial resulted in Holmes receiving an order for two machines with a stipulation that the driving speed must not exceed 90 rpm. Holmes complied by modifying the design so that the magnets were stationary. He used 60 magnets that were carried in three vertical planes. Two coil-carrying wheels each with 80 coils rotated at 90 rpm. Each of the $2\frac{1}{4}$-hp machines were driven by a steam engine through a belt drive. On December 8, 1858, at South Foreland High Lighthouse, the electric light illuminated the sea for the first time, with Faraday in attendance.

An additional Holmes machine was placed in use in 1862 and five years later, two more were ordered. The 1867 design ran at 400 rpm and had fixed magnets and rotating coils. In all these machines, it should be remembered, permanent magnets were employed.

The Electromagnet and Self-Excitation

The idea of replacing permanent magnets by electromagnets in the design of electric machines dates to observations by the Abbés Moigno and Raillard in 1838. Starting with a small machine whose permanent magnet was capable of lifting only a few grams, the electricity generated enabled a larger electromagnet to lift a load of 600 kg. About the same time, Charles G. Page (1812–1868) found that by adding a current-carrying coil around a permanent magnet, the effect of the original magnet was greatly increased.

In February 1867, William Siemens (1823–83) communicated a paper to the Royal Society asserting that permanent magnets were not necessary to convert mechanical into electrical energy. Within a few years, he developed a practical machine in which the field magnet was in two parts and the poles faced one another on opposite sides of the armature. The design employed a fundamental invention made in 1856 by Werner Von Siemens (William's brother) in which a cylindrical armature was made to rotate in a closely encircling tunnel in the magnet pole pieces. Experiments similar to those of Seimens were communicated almost simultaneously by Wheatstone and Varley. Another inventor, Wilde, could also lay claim to the invention. Wilde's experiment involved a small magneto whose electricity output was fed to the field magnet of a second machine. The output of the second machine was led to the field of a third and larger one. Using this cascading approach, Wilde observed the melting of pieces of iron rod 15 in. in length and ¼ in. in diameter.

Wilde effected self-excitation in some of his machines in 1867 by diverting some of the armature current through the field winding. His machines were used in electroplating and as the first electric searchlights on battleships. Within a year, the principles of "shunt" and "series" field windings were well recognized. The machines constructed in this era all suffered from current fluctuations. A solution was proposed by Pacinotti in 1860, but unfortunately, this went unnoticed until it was reinvented by Gramme in 1870.

The Pacinotti machine consisted of a toothed iron wheel rotating on a vertical axis. Sixteen coils connected in series were wound between the teeth of the wheel. Each junction between adjacent coils were connected to one commutator bar. The wheel rotated inside a circular magnet. Much steadier currents were obtained through the coil arrangement.

The Gramme design employed a ring armature that rotated on a horizontal spindle between vertical field magnets. The armature of the Gramme machine was thicker than Pacinotti's design and had many more sections. As a result, more uniform current was obtained. The design features of the Gramme machines opened up a new era in this new technology. Practical dynamos could then be produced on a commercial scale.

Design modifications soon followed the introduction of the Gramme machine. The armature design was soon modified to the drum type to eliminate idle copper and enabled coils to be preformed and applied to the armature as a whole. Needless to say, this was followed immediately by the more efficient design of a slotted drum armature. This enabled the laying of coils in longitudinal slots on the periphery of the drum. This design is used in present-day machines.

Improvements to design of dynamos continued, aided by theoretical foundations provided by the publication of Maxwell's treatise on electricty and magnetism published in 1872. It is worth noting here that Maxwell's work influenced Hopkinson of King's College, Cambridge, who established the concept of the magnetic circuit, with its associated magnetomotive force, magnetic flux, and reluctance. These advances had a significant impact on the design of more economic and advanced machines.

1.7 EARLY ELECTRIC MOTORS

The concept of the electric motor emerged at almost the same time as that of the electric generator. Dal Negro demonstrated in 1830 that a rotary motion can be obtained from current supplied by a voltaic battery. The experiment showed that a pivoted magnet could be kept in continuous oscillation by current in an adjacent coil. In the United States, Joseph Henry discussed in 1831 the principles of operation of one of the first proposed motors. The proposal for a simple electromagnetic motor based on a rocking energized iron bar is contained in Henry's paper "On a Reciprocating Motion Produced by Magnetic Attraction and Repulsion" published in *Silliman's Journal*, Vol. 20, pp. 340–343. In 1839, a Russian, M. H. Von Jacobi, described to M. Faraday in a private letter how an electric motor that he built was used to propel a boat. Von Jacobi did not pursue the development, as he foresaw a prohibitive expenditure to refine the idea.

The Vermont inventor Thomas Davenport experimented with some early motor designs and by 1840 had a workable motor. Simultaneously, Page experimented with an electric motor that emulated beam engines, which were well known at the time. Page replaced the steam cylinder of a beam engine by a solenoid and the piston by an iron core. With the aid of a congressional grant, he built a double-acting beam-type electric motor with a flywheel and connecting rods that produced 1 hp.

The fact that a dynamo can be used as well as a motor had been discovered by Lenz as early as 1838. This simple fact went unnoticed until 1860, when Pacinnoti announced a motor concept which resulted in a great advance. The basic idea was to create magnetic poles in the armature that were controlled by the commutator such that they did not move with the armature rotation but remained stationary as the iron itself moved.

The Pacinnoti motor consisted of a vertical armature that rotated vertically between the pole pieces of a stationary field magnet. A vertical spindle with a four-armed brass spider carried the iron-ring armature. The iron ring had 16 teeth projecting outward. The armature winding consisted of 16 sections of coil connected through wiring to a 16-bar commutator. The field magnet consisted of two vertically wound coils. Pole pieces were fitted to the upper ends of an iron bar, thus enclosing the greater part of the armature with two narrow gaps. The width of each gap was approximately equal to the width of two armature coils. Pacinnoti's interpretation of the working of the motor is enlighting. The armature is simply an electromagnet that is repelled from the one pole piece it is leaving and is attracted by the pole piece it is approaching.

It was not until the time of the Vienna Exhibition in 1873 that a public display of two Gramme machines, one generating current and the other receiving it and reconverting the power back into mechanical energy, took place. From that time on, many firms started manufacturing motors that were intended for traction purposes.

Edison designed a miniature motor for use in his famous electric pen. The pen motor may be claimed as the first electric motor in history to be produced and sold commercially in large quantities. The motor had a two-pole electromagnet and a flywheel armature carrying a short steel bar magnet. A small commutator reversed the current in the electromagnet coils as the magnet poles passed its cores. The motor's dimensions were 1 inch by 1 inch and ½ inch. What was the electric pen intended for? The pen was designed to produce punctured copying stencils. The motor ran at 4000 rpm and by means of a short crank caused a long needle in the pen holder to vibrate so that the paper was pierced with a large number of holes. Edison also designed a motor for running sewing machines. The motor's no-load current was made small by the design of a small air gap and a good cross section of iron. The superior qualities of Edison's motors were due primarily to the application of knowledge gained in designing dynamos.

A motor specially designed for traction purposes is due to Immisch (1886). It featured a double horizontal field magnet and a long Gramme armature. In an attempt to reduce magnetic field distortion and consequent sparking, he employed a particular commutator design. The commutator had two sections side by side with segments staggered to short-circuit the coils as they passed through the neutral position.

In 1887, Frank Julian Sprague built a 12-mile street railway in Richmond, Virginia. The sprague motor design incorporated effective solutions to problems of speed control, housing, suspension, geared drive, and the proper form of trolley contact. The Sprague system can be argued to have innagurated the golden age of electric motors for traction purposes, for railroad electrification grew vigorously beginning in the early 1890s.

1.8 ALTERNATING CURRENT

Our discussion so far has focused on the history of generation and utilization of electric power using direct-current apparatus. Although it is true that most of the early developments were directed at emulating the steady or continuous current possible using voltaic sources, alternating current emerged as a challenger during the last two decades of the nineteenth century. In that period, the field of electrical engineering experienced phenomenal growth following the introduction of the incadescent lamp. Alternating current was used for early arc lamps, as it ensured equal burning of the two carbon electrodes.

Direct current (dc) held the field for a number of years for a number of reasons. One of the strongest arguments is that dc permitted the use of standby batteries, thus ensuring reliability of customer service and providing nighttime supply when generators were shut down. Efficient and reliable dc motors were developed out of the fundamental work related to the dynamo. However, the scales were tipped in favor of alternating current with the emergence of two inventions, the transformer and the induction motor. Both devices were developed in the last decade of nineteenth century.

The Transformer

We have seen that the concept of induction between closely coupled coils has been known since Faraday's famous ring transformer experiments of 1831. The idea was developed further in Henry's experiments (1832) and was closely followed in 1836 by C. G. Page's work on what he termed a "dynamic multiplier." Page correlated the phenomena of self-induction and induction between two discrete conductors. From the prototype of the autotransformer, Page evolved a design that featured a separate primary and a secondary.

In 1856, C. F. Varley devised a construction in which the advantages of a subdivided iron core to secure minimum eddy-current loss were combined with a simple construction. The core in Varley's construction was a bundle of iron wires. The primary and secondary windings were wound over the center one-third of the core length. The ends of core wires were turned back over the windings to complete the magnetic circuit.

Close to 30 years later, Lucien Gaulard and John D. Gibbs introduced a system of single-phase 2000-V ac distribution. The system's backbone is a transformer with a core of soft-iron wire with a primary of insulated wire coil that was surrounded by six equal coils. The secondaries were brought out to separate terminals on the side so that six sections could be used if required.

In 1885, Westinghouse read of the use of ac in Europe in conjunction with transformers displayed in England by L. Gaulard and J. D. Gibbs. The transformer configuration patented by Gaulard and Gibbs utilized multiple one-to-one-turns-ratio transformers with the primary windings connected in series across the high-voltage primary circuit. The secondary circuit then supplied an individual low-voltage secondary. Westinghouse bought the American rights to Gaulard–Gibbs patents and authorized the development of equipment for an experimental power plant at Great Barrington, Massachusetts. Under the direction of William Stanley, transformers designed in 1886 were such that the primary windings for the individual units were in parallel across the high-voltage primary circuit rather than in series. Tests being successful, Westinghouse marketed the first commercial ac system at the end of 1886.

Ferranti followed the pattern of Gaulard and Gibbs, but employed iron strips for the core. One of Ferranti's early transformers for 15-kW ratings used as a core six bundles of hoop iron laid side by side. Over the center of the core of the low voltage, heavy current winding was wound. The secondary was made in sections carried by light frames and slipped over the primary. The ends of the core strips were bent over one by one to overlap and the whole assembly was clamped in a cast-iron frame.

In 1891, Mordey invented the construction used up to the present. The magnet core was made of soft-iron stampings. Mordey employed the shell-type construction with the iron sheets assembled around the primary and secondary coils so that the windings were completely surrounded by iron.

The importance of the transformer to the advance of alternating current is due to the growth of electric power demand. With a large demand in major centers, the need for producing cheap power at remote centers was an obvious solution. The transformer used in a step-up configuration enabled the transmission of power at high voltages with lower transmission losses. Once there, the transformer is used to step-down the voltage for further utilization. Parallel to the development of the transformer were inroads made to produce an efficient ac motor. We review this presently.

The Induction Motor

In 1879, Walter Bailey demonstrated the possibility of producing a magnetic field that rotated without the aid of mechanical motion at an exhibition before the Physical Society. Bailey constructed a four-pole electromagnet. The poles were arranged in a vertical configuration from a common yoke. Designating the poles A, B, A', and B', two pulsating currents were sent through the coils, the first current through the opposite coils A and A' and the other current through coils B and B'. The currents were supplied from a commutator which reversed the currents periodically. In addition, the current through B and B' was reversed 90° later than the current AA'. The result is a rotating magnetic field in the horizontal plane. A copper disk was suspended over the four pole pieces and was shown to rotate along with the magnetic field.

The principles of design and operation of alternating current motors were patented by Nikola Tesla on May 1, 1888. Details were announced on May 16 of the same year in a paper to the American Institute of Electrical Engineers. Tesla's paper described three forms of his invention. The common factor between the three forms is the ring-wound stator with four salient poles. The first form had a rotor with four salient poles, forming a reluctance motor that would run at synchronous speed. This form is not self-starting. The second form had a wound rotor and ran at slightly less than synchronous speed. The third form was a synchronous motor, as direct current was supplied to the rotor windings. Tesla's motors were designed for two-phase current taken from a dynamo with a drum armature through two coils at right angles connected to two pairs of slip rings instead of the normal commutator.

George Westinghouse bought Tesla'a patents and employed him to develop them. The Westinghouse organization produced a practical induction motor in 1892. At the Chicago World Fair in 1893, Westinghouse displayed a 220-V 300-hp two-phase induction motor. The power supply consisted of two single-phase 60-Hz 50-hp alternators on the same engine shaft but displaced 90°. The motor had 12 poles with a distributed two-phase primary winding of cable threaded through rotor slots.

The Thomson-Houston Company entered the field of designing and producing induction motors and soon both Westinghouse and Thompson-Houston adopted three-phase induction motors for commercial purposes. The invention of the T-connection of two transformers to provide three-phase supply from two phase systems is due to C. F. Scott at Westinghouse. The T or Scott connection was a determining factor in the transition from two-phase to three-phase designs.

Single-phase motors require a rotating magnetic field. Tesla invented the split-phase winding to achieve this from a single-phase supply. Two coils displaced by 90° were used, one winding having a much higher resistance than the other, so that the currents were displaced in phase. The high-resistance winding had to be opened when the motor approached full speed. This type of motor has been largely replaced by the appearance of the capacitor motor in 1925.

While developments were taking place in North America as described above, parallel developments in the same direction in Britain and the Continent added to the rapid progress in motor technology. Galileo Ferrari of Turin produced a prototype two-phase induction motor that used high inductance in one coil to produce the necessary phase shift in 1888. W. Langdon Davies produced an experimental single-phase induction motor in 1891 and in 1897, a more advanced version. The Davies design featured slotted laminated sheets for stator construction and squirrel-cage conductors in the rotor. At the Frankfurt Exhibition in 1890, a three-phase 100-hp induction motor was displayed.

1.9 OUTLINE OF THE TEXT

The importance of the experimental and theoretical foundations of the area of electrical and electromechanical energy conversion engineering is evident from the preceding brief historical review. It is on this basis that Chapters 2 and 4 are founded. In Chapter 2, basic laws, definitions, and relevant phenomena of electromagnetism are discussed. Emphasis is placed on the magnetic circuits of devices treated in this text. Ampère's circuital law is used to establish engineering-oriented procedures for analyzing the electromagnetic behavior of some simple structures. Properties of ferromagnetic materials, including eddy current and hysterisis loss, are discussed in this chapter. Chapter 3 deals with transformers and offers the student an opportunity to appreciate the significance of Faraday's law and principles discussed in Chapter 2 to the development of this important device. Analysis of the performance of transformers is the central topic of this chapter. The multitude of uses and connections of transformers are discussed from an application point of view.

In Chapter 4, theoretical foundations of electromechanical energy conversion devices are treated. The object here is the development of a simple but general approach to the energy conversion process in rotating electric machines. The discussion in this chapter leads to an appreciation of the differences between the various types of electrical machines. Of special note here is the discussion on reluctance and stepper motors, which are fundamental importance in robotics, numerical machine control, and computer hardware.

Chapter 5 is dedicated to the induction motor, which is the versatile workhorse of today's industrial complex. Our treatment relies on principles discussed in Chapter 4 as well as the treatment of transformers in Chapter 3. Modeling the performance of the induction motor is dealt with, together with important characteristics. Motor starting and control issues are discussed in this chapter. In Chapter 6, the synchronous machine is treated from a systems application point of view. Again, performance characteristics for round-rotor and salient-pole machines are dealt with. The applications of synchronous motors are presented in this chapter.

In Chapter 7 the direct-current machine as generator and motor is considered. The discussion of dc generators is brief by necessity, but various connections and performance characteristics are given. The discussion of dc motors is more detailed and incorporates such important issues as motor matching to load, starting, speed control, and applications. A number of fractional-horsepower machines are discussed in Chapter 8 to highlight their applications and performance characteristics.

Chapters 9–10 and 11 are devoted to the important area of solid-state power conversion and applications in adjustable-speed drives. This exciting and relatively new area of electric machine control is reviewed from a practical point of view.

The text is organized such that additional worked-out problems are given under the heading "Some Solved Problems" following the text material. This should prove useful in gaining more appreciation of the methods treated in each chapter. An ample number of drill problems at the end of each chapter are designed to follow the natural development of the text material.

The journey of exploration of this wonderful area of electrical engineering is continued in the following chapters.

Principles of
Electromagnetism

2.1 INTRODUCTION

The operation of electromechanical energy conversion devices depends on a number of fundamental principles that originate in basic experimental work. In this chapter we discuss some concepts and terminology used in this text to study electric machines. The chapter begins with a brief review of the Lorentz force law, Biot–Savart law, and Ampère's circuital law. Magnetic circuit concepts are discussed and properties of magnetic materials are covered. Faraday's law, inductance, and energy relations are treated here. It should be noted that chapter 4 is a natural extension of the fundamentals treated in this chapter.

2.2 SOME MAGNETIC FIELD LAWS

It is known that electric fields are associated with stationary electric charges. If the charges are moving with uniform velocity, a second effect, known as magnetism takes place. It is thus clear that a magnetic field is associated with moving charges and as a result, electric currents are sources of magnetic fields.

A magnetic field is identified by a vector $\mathbf{B}$ called the magnetic flux density. In the SI system of units, the unit of $\mathbf{B}$ is the tesla (T). The magnetic flux $\Phi = \mathbf{B}.\mathbf{A}$. The unit of magnetic flux Φ in the SI system of units is the weber (Wb). It is clear that

$$1 \text{ T} = 1 \text{ Wb/m}^2$$

The tesla is a large unit and we may use the smaller unit of flux density in the cgs system denoted by the gauss, where

$$1 \text{ tesla } = 10^4 \text{ gauss}$$

It is appropriate at this point to review a useful concept from vector analysis.

The Cross (Vector) Product

The cross product (sometimes called the vector product) of two vectors **U** and **V** is another vector, **W**. This product is expressed by inserting a cross (an ×) between **U** and **V**, thus:

$$\mathbf{W} = \mathbf{U} \times \mathbf{V}$$

The cross product **W** has a magnitude that is equal to the product of the magnitudes of **U** and **V** and the sine of the angle θ between **U** and **V**.

$$|\mathbf{W}| = |\mathbf{U} \times \mathbf{V}| = |\mathbf{U}|\,|\mathbf{V}|\sin\theta$$

Geometrically, the magnitude of the cross product is the area of the parallelogram formed with **U** and **V** as adjacent sides. The direction of the cross product **W** is perpendicular to the plane of **U** and **V** and follows the right-hand rule as shown in Figure 2.1.

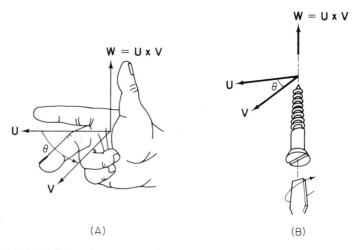

FIGURE 2.1 Defining the cross-product **W** of two vectors **U** and **V**, using: (a) the right-hand rule;   (b) the right-hand screw rule.

The concept of cross product of two vectors is useful in formulating magnetic field laws based on the experimental work of Oersted, Ampère, and Biot-Savart. We start with the Lorentz force law, which deals with the force on a moving charge in a magnetic field.

The Lorentz Force Law

A charged particle q in motion at a velocity $\mathbf{V}$ in a magnetic field of flux density $\mathbf{B}$ is found experimentally to experience a force whose magnitude is proportional to the product of the magnitude of the charge q, its velocity, and the flux density $\mathbf{B}$ and to the sine of the angle between the vectors $\mathbf{V}$ and $\mathbf{B}$ and is given by a vector in the direction of the $\mathbf{V} \times \mathbf{B}$. Thus we write

$$\mathbf{F} = q\mathbf{V} \times \mathbf{B} \tag{2.1}$$

Equation (2.1) is known as the Lorentz force equation. An interpretation of Eq. (2.1) is given in Figure 2.2.

On the basis of Eq. (2.1), one can define the tesla as the magnetic flux density that exists when a charge q of 1 coulomb, moving normal to the field at a velocity of 1 m/s, experiences a force of 1 newton.

A distribution of charge experiences a differential force $d\mathbf{F}$ on each moving incremental charge element dq given by

$$d\mathbf{F} = dq(\mathbf{V} \times \mathbf{B})$$

Moving charges over a line constitute a line current and thus we have

$$d\mathbf{F} = (I \times \mathbf{B})\,\mathbf{dl} \tag{2.2}$$

Equation (2.2) simply states that a current element $I\,\mathbf{dl}$ in a magnetic field

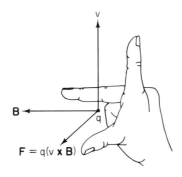

FIGURE 2.2   Lorentz force law.

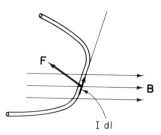

FIGURE 2.3   The force on the current element I **dl** subject to the magnetic field **B** is perpendicular to the plane of **B** and I **dl** according to Eq. (2.2)

B will experience a force $d\mathbf{F}$ given by the cross product of I **dl** and **B**. A pictorial presentation of Eq. (2.2) is given in Figure 2–3.

The differential form (2.2) is of mathematical significance only, as the current element I **dl** cannot exist by itself and must be a part of a complete circuit. The force on an entire loop can be obtained by integrating the current element

$$\mathbf{F} = \oint I \, \mathbf{dl} \times \mathbf{B} \tag{2.3}$$

Equations (2.2) and (2.3) are fundamental in the analysis and design of electric motors, as will be seen later. Another important law in magnetic field theory is the Biot–Savart law, which is based on Ampère's work showing that electric currents exert forces on each other and that a magnet could be replaced by an equivalent current.

The Biot–Savart Law

A magnetic field results from the motion of a charged particle q at a velocity **V**. The flux density **B** at a point P which is at a distance r from q in free space is given by the Biot–Savart equation:

$$\mathbf{B} = \frac{\mu_0}{4\pi} q \, \frac{\mathbf{V} \times \mathbf{a}_r}{r^2} \qquad \text{T} \tag{2.4}$$

The constant μ_0 is called the permeability of free space and in SI units is given by

$$\mu_0 = 4\pi \times 10^{-7}$$

The vector $\mathbf{a}_r$ is the unit vector originating at the charge q and directed toward the point P. A distribution of charge causes differential magnetic flux density $d\mathbf{B}$ due to each incremental charge element dq given by

$$d\mathbf{B} = \frac{\mu_0}{4\pi} \, dq \, \frac{\mathbf{V} \times \mathbf{a}_r}{r^2}$$

Recall that for line currents

$$(dq)\mathbf{V} = I \, \mathbf{dl}$$

As a result, we have

$$d\mathbf{B} = \frac{\mu_0}{4\pi} \frac{I \, \mathbf{dl} \times \mathbf{a}_r}{r^2} \tag{2.5}$$

The interpretation of Eq. (2.5) is shown in Figure 2.4. The total magnetic field is obtained by integrating the contributions from all the incremental elements:

$$\mathbf{B} = \frac{\mu_0}{4\pi} \oint \frac{I \, \mathbf{dl} \times \mathbf{a}_r}{r^2} \tag{2.6}$$

As an application of the Biot–Savart law, let us consider the magnetic field of a long straight wire carrying a current I. The current element $I \, \mathbf{dl}$ is shown in Figure 2.5 directed upward. The point P experiences an incremental magnetic flux density $d\mathbf{B}$ directed into the plane of the paper as shown. The magnitude of $d\mathbf{B}$ is given by

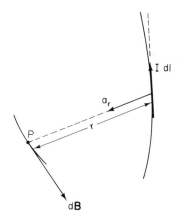

FIGURE 2.4   Interpreting the Biot–Savart law.

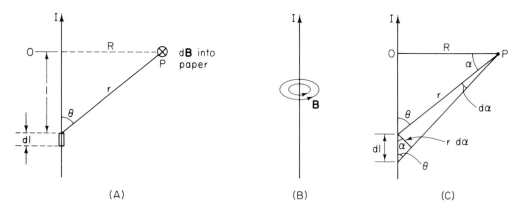

FIGURE 2.5 Finding the field due to an infinite wire carrying a current I: (a) geometry of wire used to apply Biot–Savart law; (b) showing the magnetic field; (c) elements of geometry to evaluate overall field.

$$d\mathbf{B} = \frac{\mu_0}{4\pi} \frac{I\ d\mathbf{l}\ \sin\theta}{r^2}$$

The total field at P is obtained by integrating over the entire length of the wire.

From Figure 2.5(c) we can use the angle α defined by

$$\alpha = \frac{\pi}{2} - \theta$$

We thus have

$$d\mathbf{B} = \frac{\mu_0}{4\pi} I \frac{\cos\alpha\ d\mathbf{l}}{\mathbf{r}^2}$$

Now from the figure

$$r\ d\alpha = \cos\alpha\ d\mathbf{l}$$

Thus the magnitude of $d\mathbf{B}$ is obtained as

$$d\mathrm{B} = \frac{\mu_0}{4\pi} I \frac{d\alpha}{r}$$

However, we have

$$r = \frac{R}{\cos\alpha}$$

As a result,

$$dB = \frac{\mu_0 I}{4\pi R} \cos \alpha \, d\alpha$$

The total field from an infinitely long conductor is obtained by integration from $-\pi/2$ to $+\pi/2$, with the result

$$B = \frac{\mu_0 I}{4\pi R} \int_{-\pi/2}^{\pi/2} \cos \alpha \, d\alpha$$

The total flux density at P is thus given by

$$B = \frac{\mu_0 I}{2\pi R} \tag{2.7}$$

We can thus conclude that the magnetic field is in the form of concentric circles about the wire, with a magnitude that increases in proportion to the current I and decreases as the distance from the wire is increased.

The Biot–Savart law provides us with a relation between current and the resulting magnetic flux density **B**. An alternative to this relation is Ampère's circuital law, which we discuss next.

Ampère's Circuital Law

Ampère's circuital law states that the line integral of **B** about any closed path in free space is exactly equal to the current enclosed by that path times μ_0.

$$\oint_c \mathbf{B} \cdot \mathbf{dl} = \begin{cases} \mu_0 I & \text{path } c \text{ encloses } I \\ 0 & \text{path } c \text{ does not enclose } I \end{cases} \tag{2.8}$$

It should be noted that the path c can be arbitrarily shaped closed loop about the net current I.

To illustrate Ampère's circuital law, let us consider the infinite wire carrying a current I treated earlier using the Biot–Savart law. We have shown that the field is concentric circles about the wire with flux density magnitude given by Eq. (2.7) as

$$B = \frac{\mu_0 I}{2\pi R}$$

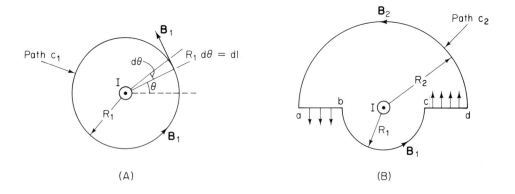

(A)   (B)

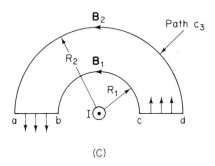

(C)

FIGURE 2.6 Illustrating Ampère's circuital law: (a) path c_1 is a circle enclosing current I;   (b) path c_2 is not a circle but encloses current I;   (c) path c_3 does not enclose current I.

where R is the shortest distance between P and the wire. As shown in Figure 2.6(a), the flux density at points on the circle with radius R_1 is constant at

$$\mathbf{B}_1 = \frac{\mu_0 I}{2\pi R_1}$$

We can choose this circle to be the closed path c_1 indicated in Eq. (2.8). The incremental length dl is given by

$$\mathbf{dl} = R_1\, d\theta$$

The flux density is in the same direction as dl and thus we have

$$\mathbf{B}_1 \cdot \mathbf{dl} = \frac{\mu_0 I}{2\pi R_1}\, R_1\, d\theta$$

As a result,

$$\oint_{c_1} \mathbf{B}_1 \cdot \mathbf{dl} = \int_0^{2\pi} \frac{\mu_0 I}{2\pi} d\theta = \mu_0 I$$

This confirms the first statement of Ampère's law.

The closed path need not be circular, and to illustrate this, we consider the path c_2 shown in Figure 2.6(b). c_2 consists of the line segment ab and the semicircle bc with radius R_1, the line segment cd, and the semicircle da with radius R_2. The left-hand side of Eq. (2.8) is thus

$$\oint_{c_2} \mathbf{B} \cdot \mathbf{dl} = \int_a^b \mathbf{B} \cdot \mathbf{dl} + \int_b^c \mathbf{B} \cdot \mathbf{dl} + \int_c^d \mathbf{B} \cdot \mathbf{dl} + \int_d^a \mathbf{B} \cdot \mathbf{dl}$$

We note that on the line segments ab and dc, the flux is perpendicular to dl and the inner (scalar) product is zero. Thus we have

$$\oint_{c_2} \mathbf{B} \cdot \mathbf{dl} = \int_b^c \mathbf{B} \cdot \mathbf{dl} + \int_d^a \mathbf{B} \cdot \mathbf{dl}$$

The first integral is given by

$$\int_b^c \mathbf{B} \cdot \mathbf{dl} = \int_{-\pi}^0 \frac{\mu_0 I}{2\pi R_1} R_1 \, d\theta = \frac{\mu_0 I}{2}$$

Similarly, the second integral is

$$\int_d^a \mathbf{B} \cdot \mathbf{dl} = \int_0^{\pi} \frac{\mu_0 I}{2\pi R_2} R_2 \, d\theta = \frac{\mu_0 I}{2}$$

We can therefore conclude that

$$\oint_{c_2} \mathbf{B} \cdot \mathbf{dl} = \mu_0 I$$

The closed path c_2 encloses the current I and the statement of Ampère's law is confirmed for this path.

We will consider path c_3 of Figure 2.6(c), which does not enclose the current I. Ampère's circuital law asserts that the line integral is zero in this case. This is confirmed as follows:

$$\oint_{c_3} \mathbf{B} \cdot \mathbf{dl} = \int_a^b \mathbf{B} \cdot \mathbf{dl} + \int_b^c \mathbf{B} \cdot \mathbf{dl} + \int_c^d \mathbf{B} \cdot \mathbf{dl} + \int_d^a \mathbf{B} \cdot \mathbf{dl}$$

Again, the integrands from a to b and c to d are zero since $\mathbf{B}$ and dl are orthogonal. As a result,

$$\oint_{c_3} \mathbf{B} \cdot \mathbf{dl} = \int_b^c \mathbf{B} \cdot \mathbf{dl} + \int_d^a \mathbf{B} \cdot \mathbf{dl}$$

Here we have

$$\int_b^c \mathbf{B} \cdot \mathbf{dl} = \int_{-\pi}^0 \frac{-\mu_0 I}{2\pi R_1} R_1 \, d\theta = \frac{-\mu_0 I}{2}$$

$$\int_d^a \mathbf{B} \cdot \mathbf{dl} = \int_0^\pi \frac{\mu_0 I}{2\pi R_2} R_2 \, d\theta = \frac{\mu_0 I}{2}$$

As a result, we conclude that, as expected, we have

$$\oint_{c_3} \mathbf{B} \cdot \mathbf{dl} = 0$$

2.3 PERMEABILITY AND MAGNETIC FIELD INTENSITY

The discussion of Section 2.2 is valid in free space, as we have consistently used the free-space permeability μ_0 in all expressions of magnetic field laws. To extend the validity of these laws to situations where material other than free space is encountered, one must clearly deal with the concept of permeability. The Biot–Savart law given by Eq. (2.5) shows that μ_0 is essentially a ratio of $\mathbf{B}$ to the current I that produces it. For free space $\mathbf{B}$ varies linearly with I and μ_0 is therefore a constant. We can conclude that for materials that exhibit a linear variation of $\mathbf{B}$ with I, all expressions discussed in Section 2.2 are valid provided that μ_0 is replaced by the permeability μ corresponding to the material considered.

Material can be classified from a $\mathbf{B}$–I-variation point of view into two classes:

1. Nonmagnetic material such as all dielectrics and metals with permeability equal to μ_0 for all practical purposes.

2. Magnetic material that belongs to the iron group known as ferromagnetic material. For this class, a given current produces a much larger $\mathbf{B}$ field than in free space. Permeability of ferromagnetic material is much higher than that of free space and is nonlinear since it varies over a wide range with variations in current.

Ferromagnetic material can be further categorized into two classes:

1. Soft ferromagnetic material for which a linearization of the **B**–I variation in a region is possible.
2. Hard ferromagnetic material for which it is difficult to give a meaning to the term *permeability*. Material in this group is suitable for permanent magnets.

The source of **B** in the case of soft ferromagnetic material can be modeled as due to the current I. For hard ferromagnetic material, the source of **B** is a combined effect of current I and material magnetization M, which originates entirely in the medium. In order to separate the two sources of the magnetic **B** field, the concept of magnetic field intensity **H** is introduced.

Magnetic Field Intensity

The magnetic field intensity (or strength) denoted by **H** is a vector defined by the relation

$$\mathbf{B} = \mu\mathbf{H} \tag{2.9}$$

For isotropic media (have the same properties in all directions), μ is a scalar and thus **B** and **H** are in the same direction. On the basis of Eq. (2.9) we can write the statements of Biot–Savart law given by Eqs. (2.5) and (2.6) as

$$d\mathbf{H} = \frac{1}{4\pi}\frac{I\mathbf{dl} \times \mathbf{a}_r}{r^2} \tag{2.10}$$

$$\mathbf{H} = \frac{1}{4\pi}\oint\frac{I\,\mathbf{dl} \times \mathbf{a}_r}{r^2} \tag{2.11}$$

Moreover, Ampère's circuital law given by Eq. (2.8) is written as

$$\oint \mathbf{H}\cdot\mathbf{dl} = \begin{cases} I & \text{path } c \text{ encloses } I \\ 0 & \text{path } c \text{ does not enclose } I \end{cases} \tag{2.12}$$

Expressions (2.10), (2.11), and (2.12) are independent of the medium and relate the magnetic field intensity **H** to the current causing it, I.

Equation (2.9) requires further examination, as the permeability μ is not a constant in general but is dependent on **H** and, strictly speaking, we should state this dependence in the form

$$\mu = \mu(\mathbf{H}) \tag{2.13}$$

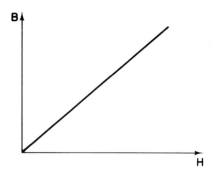

FIGURE 2.7 B–H characteristic for nonmagnetic material.

For nonmagnetic material μ is constant at a value equal to $\mu_0 = 4\pi \times 10^{-7}$ for all practical purposes. The **B–H** characteristic of nonmagnetic materials is shown in Figure 2.7.

The **B–H** characteristics of soft ferromagnetic material, often called the magnetization curve, follow the typical pattern displayed in Figure 2.8. The permeability of the material in accordance with Eq. (2.9) is given as the ratio of **B** to **H** and is clearly a function of **H**, as indicated by Eq. (2.13).

$$\mu = \frac{\mathbf{B}}{\mathbf{H}} \tag{2.14}$$

The permeability at low values of **H** is called the initial permeability and is much lower than the permeability at higher values of **H**. The maximum

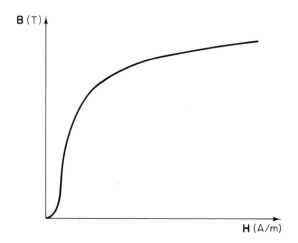

FIGURE 2.8 B–H characteristic for a typical ferromagnetic material.

value of μ occurs at the knee of the **B–H** characteristic. The permeability of soft ferromagnetic material μ is much larger than μ_0 and it is convenient to define the relative permeability μ_r by

$$\mu_r = \frac{\mu}{\mu_0} \tag{2.15}$$

A typical variation of μ_r with **H** for a ferromagnetic material is shown in Figure 2.9.

For electromechanical energy conversion devices treated in this text, a linear approximation to the magnetization curve provides satisfactory answers in the normal region of operation. The main idea is illustrated in Figure 2.10, where a straight line passing through the origin of the **B–H** curve that best fits the data points is drawn and taken to represent the characteristics of the material considered. Within the acceptable range of **H** values, one may then use the following relation to model the ferromagnetic material:

$$\mathbf{B} = \mu_0\mu_r\mathbf{H} \tag{2.16}$$

It should be noted that μ_r is in the order of thousands for magnetic materials used in electromechanical energy conversion devices (2000 to 80,000, typically). Properties of magnetic materials are discussed further in the following sections. For the present, we assume that μ_r is constant.

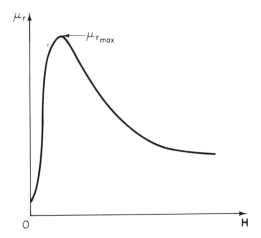

FIGURE 2.9   Typical variation of μ_r with **H** for a ferromagnetic material.

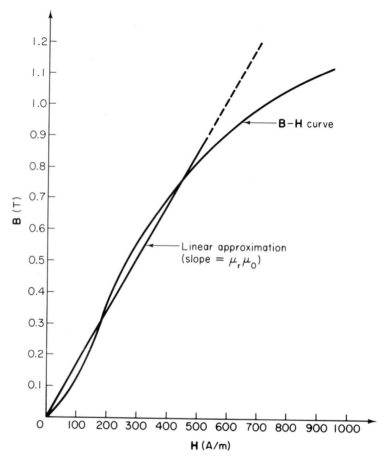

FIGURE 2.10 Linear approximation of the magnetization curve for a soft ferrogmagnetic material.

2.4 MAGNETIC CIRCUITS

The statement of Ampère's circuital law of Eq. (2.12) is based on the assumption that the path C encloses the current one time. If a path encloses the current N times, the right-hand side of Eq. (2.12) becomes simply NI. Thus, in this case, we write

$$NI = \oint \mathbf{H} \cdot \mathbf{dl} \tag{2.17}$$

The left-hand side of Eq. (2.17) is commonly referred to as the magneto-motive force (MMF) $\mathscr{F}$, and hence we have

$$\mathscr{F} = NI = \oint \mathbf{H} \cdot \mathbf{dl} \qquad (2.18)$$

A magnetic circuit is a structure made for the most part of high-permeability magnetic material. The magnetic flux is confined to paths defined by the structure due to the high permeability of the magnetic material. An elementary example of a magnetic circuit is shown in Figure 2.11. Here, a toroidal ferromagnetic core carrying an N-turn coil is considered. The torus has a cross-sectional area A, an inner radius a, and an outer radius b. Ampère's circuital law tells us that for a circular path with radius $r < a$, there is no magnetic field, since no current is enclosed by path.

$$H = 0 \qquad r < a$$

For a circular path $r \geq b$, we have

$$\oint \mathbf{H} \cdot \mathbf{dl} = N(I - I) = 0$$

Thus we have

$$H = 0 \qquad r \geq b$$

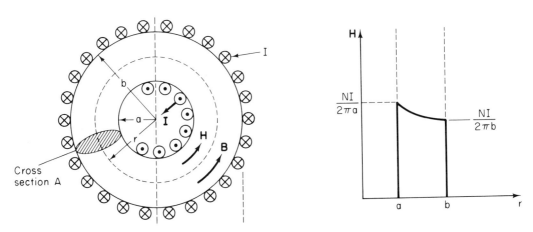

FIGURE 2.11   Toroidal coil and magnetic field intensity distribution.

Inside the torus, the magnetic field is constant on a circular path of radius r, due to symmetry. The field $\mathbf{H}$ is tangential to the circle, and thus

$$\mathbf{H}\cdot\mathbf{dl} = \mathrm{H} r \, d\theta$$

As a result, the magnitude of $\mathbf{H}$ is obtained as

$$\oint \mathbf{H}\cdot\mathbf{dl} = \int_0^{2\pi} \mathrm{H} r \, d\theta = 2\pi r \mathrm{H}$$

We can conclude that

$$2\pi r \mathrm{H} = NI$$

The path clearly encloses the current N times. We can assert that the magnetic field intensity H inside the structure is given by

$$\mathrm{H} = \frac{NI}{2\pi r} \qquad a < r < b$$

The field inside the toroid varies inversely with the radius r of the path. The flux density $\mathbf{B}$ inside the core is obtained by multiplying $\mathbf{H}$ by the permeability of the core material.

$$\mathrm{B} = \frac{\mu NI}{2\pi r}$$

The flux Φ is obtained using the expression

$$\Phi = \int \mathbf{B}\cdot\mathbf{dA}$$

As an approximation, we take the flux as the product of B at the average radius r_{av} and the uniform cross-sectional area A. Therefore, we have

$$\Phi = \frac{\mu NIA}{2\pi r_{\mathrm{av}}}$$

where

$$r_{\mathrm{av}} = a + \frac{b - a}{2}$$

As we have already defined the magnetomotive force by $\mathcal{F} = NI$, we can write

$$\Phi = \mathcal{F} \, \frac{\mu A}{2\pi r_{av}}$$

The length of the circular path at the average radius is given by

$$l_{av} = 2\pi r_{av}$$

As a result, we conclude that

$$\Phi = \mathcal{F} \left[\frac{\mu A}{l_{av}} \right] \qquad (2.19)$$

Let us examine Eq. (2.19) from an input–output (cause–effect) point of view. If we consider the magnetomotive force $\mathcal{F}$ as an input (cause), the flux Φ set up by $\mathcal{F}$ is an output (effect). This relation is depicted in Figure 2.12(a) in block diagram form. Here the output Φ is obtained by multiplying the input $\mathcal{F}$ by the quantity in brackets in Eq. (2.19). We are already familiar with Ohm's law for dc circuits as interpreted in Figure 2.12(b). Inspection of the figure shows us that an analogy between magnetic variables and electric circuit variables is possible. We can thus establish a correspondence between electric voltage V (electromotive force) and the magnetomotive force $\mathcal{F}$. Next, we have a correspondence of the electric current I and the flux Φ. As a consequence, we can claim a correspondence between the quantity $\mu A/l_{av}$ and the inverse of the resistance to complete the picture. Indeed, we now define the reluctance of the magnetic circuit denoted by $\mathcal{R}$ as

$$\mathcal{R} = \frac{l_{av}}{\mu A} \qquad (2.20)$$

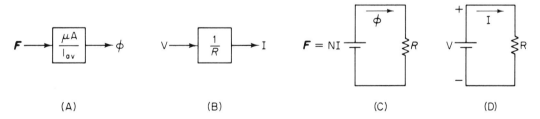

(A) (B) (C) (D)

FIGURE 2.12 Establishing the magnetic circuit concept.

As a result, we write Eq. (2.19) as

$$\Phi = \frac{\mathscr{F}}{\mathscr{R}} \tag{2.21}$$

The units of reluctance are thus seen to be amperes per weber.

Let us observe that the magnetic reluctance as defined in Eq. (2.20) decreases as the permeability is increased and in view of Eq. (2.21), a higher value of flux is obtained for the same mmf by using a material of lower reluctance. The advantages of the analogy just established include the ability to utilize our circuit analysis tools in dealing with magnetic field problems.

The development of the magnetic circuit concept in this section relies on the analysis of the toroid of Figure 2.11, where a uniformly wound coil of N turns is the source of mmf. In practice, coils occupying only a portion of the toroid as shown in Figure 2.13 are frequently encountered. Our results are valid from a practical engineering point of view as long as the toroid thickness $(b - a)$ is much smaller than the average length l_m. It should also be noted that analysis results apply to structures that are not circular. It is appropriate now to take a look at a simple example.

Example 2.1   The mean radius of a toroid is 25 cm and its cross-sectional area is 3 cm². The toroid is wound with a coil of 600 turns and a direct current of 1.5 A is passed through the coil. Assume that the relative permeability of the toroid is 1500. It is required to calculate:

(a) The reluctance of the circuit.
(b) The magnetomotive force $\mathscr{F}$ and magnetic field intensity H.
(c) The flux and flux density inside the toroid.

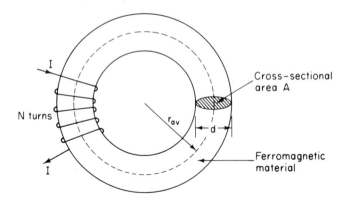

FIGURE 2.13   Partially wound toroid.

Solution

(a) the reluctance is calculated using

$$\mathcal{R} = \frac{l_{av}}{\mu A}$$

We have

$$l_{av} = 2\pi(0.25) \text{ m}$$

$$\mu = 1500(4\pi \times 10^{-7})$$

$$A = 3 \times 10^{-4} \text{ m}^2$$

As a result,

$$\mathcal{R} = \frac{2\pi(0.25)}{1500(4\pi \times 10^{-7})(3 \times 10^{-4})} = 2.78 \times 10^6 \text{ At/W}$$

(b) The magnetomotive force is

$$\mathcal{F} = NI = 600(1.5) = 900 \text{ At}$$

The magnetic field intensity is

$$H = \frac{\mathcal{F}}{l_{av}} = \frac{900}{2\pi(0.25)} = 572.96 \text{ At/m}$$

(c) The flux Φ is obtained as

$$\Phi = \frac{\mathcal{F}}{\mathcal{R}} = \frac{900}{2.78 \times 10^6}$$

$$= 3.24 \times 10^{-4} \text{ Wb}$$

The flux density is

$$B = \frac{\Phi}{A} = \frac{3.24 \times 10^{-4}}{3 \times 10^{-4}} = 1.08 \text{ T}$$

Of course, we could use H directly from part (b) to obtain

$$B = \mu H = 1500(4\pi \times 10^{-7})(572.96)$$

$$= 1.08$$

This would provide

$$\Phi = \mathbf{B} \cdot \mathbf{A} = 1.08(3 \times 10^{-4}) = 3.24 \times 10^{-4} \text{ Wb}$$

Naturally, we obtain the same results.

Assuming uniform fields, the magnetic circuit concept can be applied to structures composed of different materials, such as that shown in Figure 2.14. The path of integration in Ampère's law is shown as the closed contour *abcdefgha*. The length of the path *abcde* is denoted by l_1, the length of the path *ef* is denoted by l_2, the length of the path *fgh* is denoted by l_3, and that of *ha* is denoted by l_4. The application of Ampère's law results in

$$\mathscr{F} = NI = H_1 l_1 + H_2 l_2 + H_3 l_3 + H_4 l_4$$

The flux Φ is the same everywhere in this single-loop structure, (neglecting leakage.) and we thus have the following flux densities:

$$B_1 = \frac{\Phi}{A_1} \qquad B_2 = \frac{\Phi}{A_2}$$

$$B_3 = \frac{\Phi}{A_3} \qquad B_4 = \frac{\Phi}{A_4}$$

The areas A_1, A_2, A_3, and A_4 correspond to the sections indicated in the structure. Recall that

$$B_i = \mu_i H_i \qquad i = 1, 2, 3, 4$$

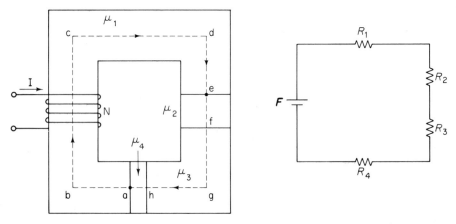

FIGURE 2.14 Magnetic structure exhibiting reluctances in series.

Thus

$$H_i = \frac{\Phi}{\mu_i A_i}$$

As a result, we have

$$\mathcal{F} = NI = \Phi \left(\frac{l_1}{\mu_1 A_1} + \frac{l_2}{\mu_2 A_2} + \frac{l_3}{\mu_3 A_3} + \frac{l_4}{\mu_4 A_4} \right)$$

Recall the definition of the reluctance:

$$\mathcal{R}_i = \frac{l_i}{\mu_i A_i}$$

We can thus assert that

$$\mathcal{F} = (\mathcal{R}_1 + \mathcal{R}_2 + \mathcal{R}_3 + \mathcal{R}_4)\Phi$$

It is clear that the reluctances are in series, as shown in the magnetic equivalent circuit of Figure 2.14.

The following example deals with a single-loop structure with two materials, The unknown is the MMF, and this problem is the inverse of that given in Example 2.1.

Example 2.2   Consider the magnetic structure shown in Figure 2.15. Find the current I in the 500-turn winding to set up a flux density of 1 T in the structure. Assume the relative permeability of the structure material to be 3980.

Solution

The length of the magnetic material path is denoted by l_i and is found from the geometry of Figure 2.15 as

$$l_i = 2(16 + 14) = 60 \text{ cm} = 0.6\text{m}$$

The air-gap length is denoted by l_g and is

$$l_g = 0.5 \text{ mm} = 0.5 \times 10^{-3} \text{ m}$$

The cross-sectional area of the structure is

$$A_i = A_g = 16 \times 10^{-4} \text{ m}^2$$

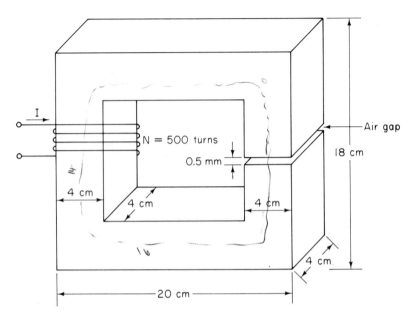

FIGURE 2.15   Magnetic structure for Example 2.2.

The reluctance of the air gap is thus obtained as

$$\mathcal{R}_g = \frac{l_g}{\mu_0 A_g} = \frac{0.5 \times 10^{-3}}{(4\pi \times 10^{-7})(16 \times 10^{-4})} = 248.68 \times 10^3$$

The reluctance of the iron path is

$$\mathcal{R}_i = \frac{l_i}{\mu_0 \mu_r A_i} = \frac{0.6}{(4\pi \times 10^{-7})(3980)(16 \times 10^{-4})} = 74.979 \times 10^3$$

Since a flux density of 1 T is required, the flux is

$$\Phi = \mathbf{B} \cdot \mathbf{A} = (1)(16 \times 10^{-4}) = 1.6 \times 10^{-3} \text{ Wb}$$

The MMF required is thus given by

$$\mathcal{F} = (\mathcal{R}_g + \mathcal{R}_i)\, \phi$$
$$= 1.6 \times 10^{-3}(248.68 \times 10^3 + 74.979 \times 10^3)$$
$$= 517.85 \text{ At}$$

But we need I, so we use

$$I = \frac{\mathcal{F}}{N} = \frac{517.85}{500} = 1.0357 \text{ A}$$

Our discussion so far has been limited to single-loop structures. In Figure 2.16, we have a structure with more than one loop. We reason the equivalent circuit shown in the figure by appealing again to Ampère's circuital law. The contour composed of l_a, l_c, and l_g provides us with

$$\mathcal{F} = \mathrm{H}_a l_a + \mathrm{H}_c l_c + \mathrm{H}_g l_g$$

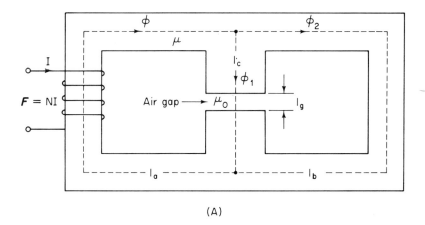

(A)

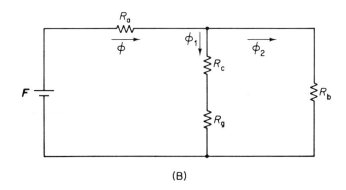

(B)

FIGURE 2.16 (a) Multiloop magnetic structure and (b) its equivalent circuit.

Now

$$H_a = \frac{\Phi}{A_a\mu}$$

$$H_c = \frac{\Phi_1}{A_c\mu}$$

$$H_g = \frac{\Phi_1}{A_g\mu_0}$$

By definition of the reluctances, we obtain the following relation, which is satisfied by the left-hand loop of the equivalent circuit:

$$\mathscr{F} = \Phi\,\mathscr{R}_a + \Phi_1\,(\mathscr{R}_c + \mathscr{R}_g)$$

where

$$\mathscr{R}_a = \frac{l_a}{\mu A_a}$$

$$\mathscr{R}_c = \frac{l_c}{\mu A_c}$$

$$\mathscr{R}_g = \frac{l_g}{\mu_0 A_g}$$

By continuity of flux, we have

$$\Phi = \Phi_1 + \Phi_2$$

For the contour l_c, l_g, and l_b, we have

$$0 = H_g l_g + H_c l_c - H_b l_b$$

Thus we have

$$0 = \Phi_1\left(\frac{l_g}{\mu_0 A_g} + \frac{l_c}{\mu A_c}\right) - \Phi_2\left(\frac{l_b}{\mu A_b}\right)$$

As a result,

$$0 = \Phi_1(\mathscr{R}_g + \mathscr{R}_c) - \Phi_2\mathscr{R}_b$$

where

$$\mathcal{R}_b = \frac{l_b}{\mu A_b}$$

The relation obtained above is satisfied by the right-hand loop of the equivalent circuit. It is clear that once we have the equivalent circuit, familiar techniques of circuit analysis can be applied. It is clear that $\mathcal{R}_b$ is in parallel with the series combination of $\mathcal{R}_c$ and $\mathcal{R}_g$.

From a problem-solving point of view, once the geometric dimension and permeability values are available, two classes of problems arise. In the first, the MMF is given and the flux (or flux densities) values in various sections of the structure are to be found. The second class involves the inverse problem of finding $\mathcal{F}$ given a flux (flux density) value in the structure.

Example 2.3   For the magnetic structure of Figure 2.16, assume that the soft ferromagnetic structure has a constant relative permeability $\mu_r = 4100$. Assume the following dimensions:

$$l_a = l_b = 0.85 \text{ m}$$

$$l_c = 0.36 \text{ m}$$

$$l_g = 0.8 \text{ mm}$$

The structure has a constant cross section of area $A = 8 \times 10^{-3} \text{ m}^2$. Find the flux density in the air gap for an MMF of 180 At.

Solution

We first compute the reluctances involved:

$$\mathcal{R}_a = \frac{l_a}{\mu_0 \mu_r A} = \frac{0.85}{(4\pi \times 10^{-7})(4100)(8 \times 10^{-3})} = 20.622 \times 10^3$$

Note that

$$\mathcal{R}_b = \mathcal{R}_a$$

Now

$$\mathcal{R}_c = \frac{l_c}{\mu_0 \mu_r A} = \frac{0.36}{(4\pi \times 10^{-7})(4100)(8 \times 10^{-3})} = 8.734 \times 10^3$$

$$\mathcal{R}_g = \frac{l_g}{\mu_0 A} = \frac{0.8 \times 10^{-3}}{(4\pi \times 10^{-3})(8 \times 10^{-3})} = 79.577 \times 10^3$$

We have $\mathcal{R}_c$ and $\mathcal{R}_g$ in series and thus

$$\mathcal{R}_c + \mathcal{R}_g = (8.734 + 79.577)10^3 = 88.311 \times 10^3$$

Now $\mathcal{R}_b$ is in parallel with $(\mathcal{R}_c + \mathcal{R}_g)$ and the combination is in series with $\mathcal{R}_a$. The equivalent reluctance is thus

$$
\begin{aligned}
\mathcal{R}_{eq} &= \mathcal{R}_a + \frac{\mathcal{R}_b(\mathcal{R}_c + \mathcal{R}_g)}{\mathcal{R}_b + \mathcal{R}_c + \mathcal{R}_g} \\[2mm]
&= \left[20.622 + \frac{20.622(88.311)}{20.622 + 88.311} \right] 10^3 \\[2mm]
&= 37.34 \times 10^3
\end{aligned}
$$

since $\mathcal{F} = 180$, we obtain

$$\Phi = \frac{180}{37.34 \times 10^3} = 4.8205 \times 10^{-3}$$

Now

$$
\begin{aligned}
\Phi_1 &= \frac{\mathcal{R}_b}{\mathcal{R}_b + \mathcal{R}_c + \mathcal{R}_g} \Phi \\[2mm]
&= 4.8205 \times 10^{-3} \left(\frac{20.622}{20.622 + 88.311} \right) \\[2mm]
&= 912.57 \times 10^{-6}
\end{aligned}
$$

As a result, the flux density B_1 is

$$B_1 = \frac{\Phi_1}{A} = \frac{912.57 \times 10^{-6}}{8 \times 10^{-3}} = 0.114 \text{ T}$$

Magnetic structures may have more than one MMF source. An example structure is shown in Figure 2.17 where two coils set up the flux in the structure. The concept of an equivalent magnetic circuit is useful in solving such problems, as indicated in the next example.

Example 2.4   The soft ferromagnetic material of the structure shown in Figure 2.17 has a relative permeability of 10,000. The thickness of the core is 2 cm. Find the flux in the core given that $N = 400$, $I_1 = 1$ A, and $I_2 = 1.2$ A.

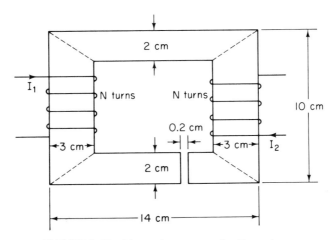

FIGURE 2.17   Magnetic structure for Example 2.4

Solution

We compute the reluctances in the structure by first finding relevant lengths and areas. It should be noted that the structure is nonuniform. The mean length of the path in the vertical sections is

$$l_1 = 8 \times 10^{-2} \text{ m}$$

The mean length of the path in the horizontal sections is

$$l_2 = 11 \times 10^{-2} \text{ m}$$

The air-gap length is

$$l_g = 0.2 \times 10^{-2} \text{ m}$$

The cross-sectional areas are

$$A_1 = 6 \times 10^{-4} \text{ m}^2$$
$$A_2 = 4 \times 10^{-4} \text{ m}^2$$
$$A_g = 4 \times 10^{-4} \text{ m}^2$$

The reluctance values are therefore given by

$$\mathcal{R}_1 = \frac{l_1}{\mu A_1} = \frac{8 \times 10^{-2}}{10^4 (4\pi \times 10^{-7})(6 \times 10^{-4})} = 10.61 \times 10^3 \text{ A/W}$$

$$\mathscr{R}_2 = \frac{l_2}{\mu A_2} = \frac{11 \times 10^{-2}}{10^4(4\pi \times 10^{-7})(4 \times 10^{-4})} = 21.884 \times 10^3 \text{ A/W}$$

$$\mathscr{R}_g = \frac{l_g}{\mu_0 A_g} = \frac{0.2 \times 10^{-2}}{(4\pi \times 10^{-7})(4 \times 10^{-4})} = 3.9789 \times 10^6 \text{ A/W}$$

The equivalent magnetic circuit is shown in Figure 2.18. As a result, we conclude that

$$\Phi = \frac{N(I_1 + I_2)}{2(\mathscr{R}_1 + \mathscr{R}_2) + \mathscr{R}_g}$$

$$= \frac{400(1 + 1.2)}{[2(10.61 + 21.884) + 3.9789 \times 10^3]10^3}$$

$$= 0.2176 \times 10^{-3} \text{ Wb}$$

It should be noted that the two coils aid in setting up the flux. The polarity of the sources is determined by the right-hand rule.

Effect of Air Gaps

Two points are worth discussing in connection with the presence of air gaps in a magnetic structure. The first has to do with the value of the reluctance of an air gap relative to that of a path in the magnetic material. In Example 2.4 we found that $\mathscr{R}_g$ is a little over 180 times $\mathscr{R}_2$. We can conclude that the air-gap reluctance is much higher than the reluctance of a path in iron. This can be seen from

$$\mathscr{R}_i = \frac{l_i}{\mu_0 \mu_r A}$$

$$\mathscr{R}_g = \frac{l_g}{\mu_0 A}$$

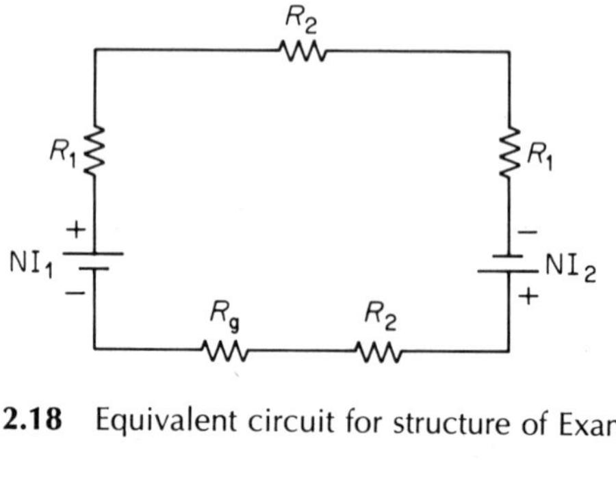

FIGURE 2.18   Equivalent circuit for structure of Example 2.4.

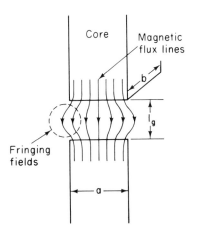

FIGURE 2.19 Fringing effect in an air gap.

Thus

$$\mathcal{R}_g = \mathcal{R}_i \frac{\mu_r}{l_i/l_g}$$

It is clear that if $l_i/l_g \ll \mu_r$, then $\mathcal{R}_g/\mathcal{R}_i \gg 1$. In this case, the magnetic path can be approximated by a short circuit (zero reluctance) in the equivalent magnetic circuit.

   The second point is that in practice, the magnetic field lines extend outward somewhat as they cross the air gap, as shown in Figure 2.19. This phenomenon is called fringing and is accounted for by taking a larger air gap cross-sectional area. The common practice is to use an air-gap area made up by adding the air-gap length to each of the two dimensions making the cross-sectional area. As a result, with reference to Figure 2.19, accounting for fringing,

$$A_g = (a + l_g)(b + l_g)$$

This approximation is valid for short air gaps. The following example illustrates this point.

Example 2.5   Allow for fringing effects in solving for the required current in the magnetic structure of Example 2.2

Solution

The air-gap area allowing for fringing is

$$A_g = (4 + 0.05)(4 + 0.05) = 16.4 \text{ cm}^2$$
$$= 16.4 \times 10^{-4} \text{ m}^2$$

As a result, the reluctance of the air gap is recalculated as

$$\mathcal{R}_g = \frac{0.5 \times 10^{-3}}{(4\pi \times 10^{-7})(16.403 \times 10^{-4})} = 242.58 \times 10^3$$

Note that accounting for fringing reduces the air-gap reluctance. Now, we have from Example 2.2,

$$\Phi = 1.6 \times 10^{-3}$$
$$\mathcal{R}_i = 74.979 \times 10^3$$

Thus

$$\mathcal{F} = 1.6 \times 10^{-3}(242.58 + 74.979) \times 10^3$$
$$= 508.09 \text{ At}$$

As a result,

$$I = \frac{508.09}{500} = 1.016 \text{ A}$$

Less current is required in this case than the value obtained neglecting fringing.

2.5 NONLINEAR MAGNETIC CIRCUITS

The development of solution techniques for magnetic structure problems using the equivalent-circuit concept of Section 2.4 is based on the assumption of constant permeabilities for the materials involved. Realistic problems involve materials for which μ is not constant, as indicated in Section 2.3. The reader should refer to Figures 2.8 to 2.10 for details. The concept of an equivalent magnetic circuit including reluctances to model the material behavior can still be used to solve problems where the **B–H** curve of the material is nonlinear. In this case, the reluctances are functions of the magnetic loading **H**.

In this section we illustrate the problem for structures consisting of a single loop and subject to one source MMF. Two types of problems arise. In the first, the flux (or flux density) in the structure is specified and the problem is to find the required source MMF. The second type is the inverse problem where the flux is the unknown. Let us illustrate the first problem type by way of an example.

Example 2.6   The magnetic structure of a relay is shown in Figure 2.20. Calculate the magnetomotive force required to set up a flux of 50×10^{-6} Wb in the structure. Assume that the core material is cast steel with the magnetization curve shown in Figure 2.21. Allow for fringing.

Solution

We first calculate the length of the magnetic path in iron:

$$l_i = 2(3 + 7.5) + 2(0.5) = 22 \text{ cm}$$
$$= 0.22 \text{ m}$$

The air-gap length for our calculation is

$$l_g = 2 \times 0.5 \times 10^{-3} = 10^{-3} \text{ m}$$

The area of the cross section of the iron is

$$A_i = 1 \times 10^{-4} \text{ m}^2$$

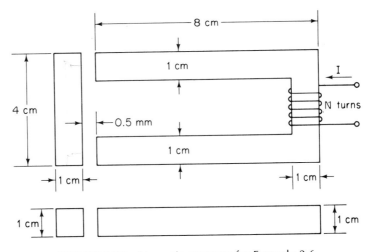

FIGURE 2.20   Magnetic structure for Example 2.6.

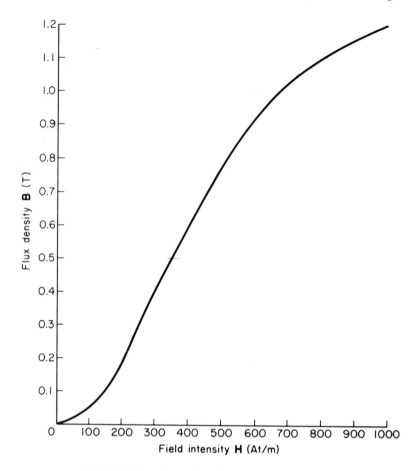

FIGURE 2.21   Magnetization curve for cast steel.

The area of cross section of the air gap is

$$A_g = (0.5 \times 10^{-3} + 10^{-2})^2 = 110.25 \times 10^{-6} \text{ m}^2$$

The flux density in the air gap is

$$B_g = \frac{\phi}{A_g} = \frac{50 \times 10^{-6}}{110.25 \times 10^{-6}} = 0.4535 \text{ T}$$

As a result, we find the magnetic field intensity in the gaps:

$$H_g = \frac{B_g}{\mu_0} = \frac{0.4535}{4\pi \times 10^{-7}} = 360.9 \times 10^3$$

As a consequence, the MMF required by the gaps is

$$\mathscr{F}_g = H_g l_g = 360.9 \text{ At}$$

The solution so far has been fairly along familiar lines. The departure comes with the computation for the iron path. Hence we have

$$B_i = \frac{\phi}{A_i} = \frac{50 \times 10^{-6}}{10^{-4}} = 0.5 \text{ T}$$

Having obtained B_i, we use the magnetization curve (we do not have the relative‘permeability specified as a constant) to find the corresponding H as

$$H_i = 350 \text{ At/m}$$

The required MMF for the iron path is thus

$$\mathscr{F}_i = H_i l_i = 350(0.22) = 77 \text{ At}$$

The total MMF required is thus given by

$$\begin{aligned} \mathscr{F}_t &= H_g l_g + H_i l_i \\ &= 360.9 + 77 = 437.9 \text{ At} \end{aligned}$$

It is clear that the problem type treated in Example 2.6 is straightforward to solve. We summarize our procedure for solving the problem by reference to the magnetic structure and the equivalent circuit of Figure 2.22.

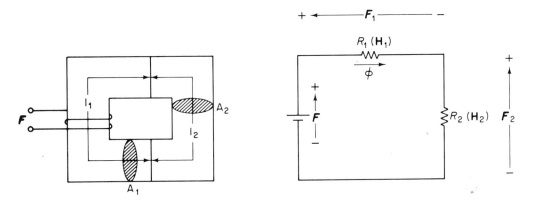

FIGURE 2.22   A Single-loop magnetic structure and its equivalent circuit.

Procedure

1. Given the flux ϕ, calculate the corresponding flux densities:

$$B_1 = \frac{\phi}{A_1} \qquad B_2 = \frac{\phi}{A_2}$$

2. Obtain H_1 corresponding to B_1 from the magnetization curve of material 1. Similarly, obtain H_2 for material 2.
3. Calculate the MMFs required for each section:

$$\mathscr{F}_1 = H_1 l_1$$
$$\mathscr{F}_2 = H_2 l_2$$

4. The required result is the source MMF given by

$$\mathscr{F} = \mathscr{F}_1 + \mathscr{F}_2$$

It is clear that this procedure is applicable to structures with more than two sections of different materials. The difference of course is that we will have to make more than two visits to the magnetization curves.

The inverse problem where $\mathscr{F}$ is specified and the resulting flux ϕ is to be calculated is not as straightforward. The main hurdle is to find the proper division of $\mathscr{F}$ into the two MMFs $\mathscr{F}_1$ and $\mathscr{F}_2$ such that the resulting flux in both sections is the same. A simple trial-and error scheme may be adopted in this case. Here, a trial value of the flux is used in conjunction with the procedure detailed earlier. The resulting MMF is then compared with the MMF specified. If the two values agree, the problem is solved. If the two values do not agree, we modify our trial value of the flux and repeat the procedure until reasonable agreement is reached.

There is an intelligent graphical procedure to solve this problem. Here, we first convert the available **B–H** curves into Φ–$\mathscr{F}$ curves, as shown in Figure 2.23 for material 1. This is done by applying the relations $\Phi = BA$, and $\mathscr{F} = Hl$ to a few points on the **B–H** curve. It can be seen that this process is simply a change of scale and the curve retains its initial shape. The Φ–$\mathscr{F}$ curve for material 2 is reversed as shown in the upper right-hand corner of Figure 2.24. The point P is located on the $\mathscr{F}_1$ axis such that $O_1 P = \mathscr{F}$. The origin O_2 of the Φ_2–$\mathscr{F}_2$ curve is then established at P as shown in the bottom curve of Figure 2.24. The intersection of the two curves provides us with the value of the flux as well as the division of $\mathscr{F}$ into $\mathscr{F}_1$ and $\mathscr{F}_2$. Care must be taken to draw both Φ–$\mathscr{F}$ curves to the same scale to obtain meaningful results.

Magnetization curves for cast iron, cast steel, and sheet steel are

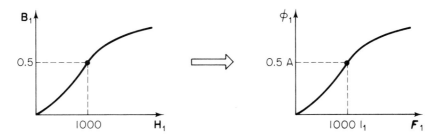

FIGURE 2.23 Converting the **B–H** curve to a Φ–$\mathcal{F}$ curve for material 1.

shown in Figure 2.25. The same curves are shown reversed in Figure 2.26. Let us note here that for the special case $A_1 = A_2$, the conversion of B into a Φ is not necessary and we can work just as well with B–$\mathcal{F}$ characteristics. We also note that if $l_1 = l_2$, we can also work with H values rather than $\mathcal{F}$. The following example is directed toward solving this special case.

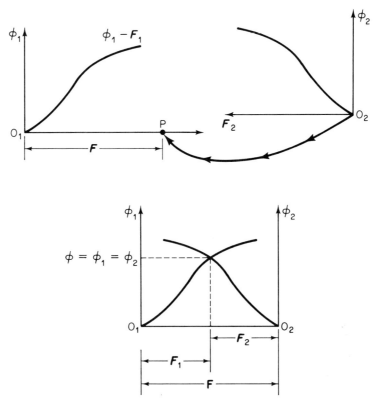

FIGURE 2.24 Graphical procedure to solve for the division of $\mathcal{F}$ between sections 1 and 2 for structure of Figure 2.22.

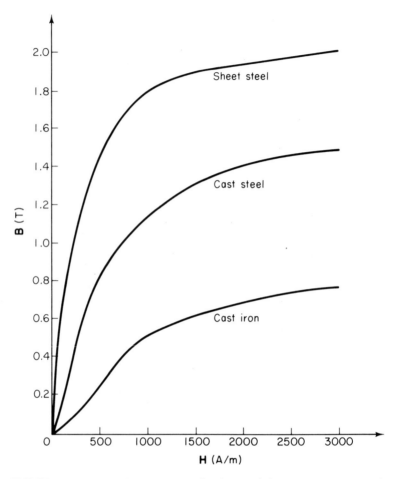

FIGURE 2.25   Magnetization curves for three soft ferromagnetic materials.

Example 2.7   Consider the magnetic structure shown in Figure 2.22. Assume that section 1 is made of sheet steel and section 2 is made of cast steel. Let

$$l_1 = l_2 = 0.4 \text{ m}$$
$$A_1 = A_2 = 8 \times 10^{-4} \text{ m}^2$$

Find the flux, flux densities B_1 and B_2, and magnetic field intensities H_1 and H_2, for the following MMFs.

(a) $\mathcal{F} = 800$ At.
(b) $\mathcal{F} = 1200$ At.
(c) $\mathcal{F} = 1400$ At.

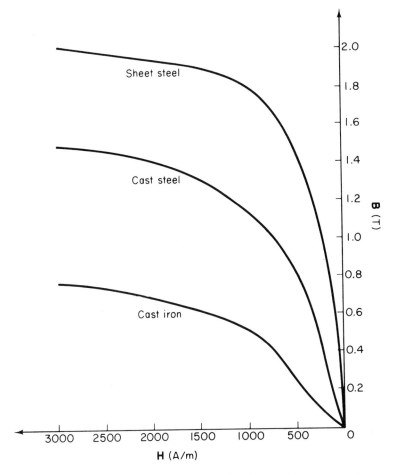

FIGURE 2.26 Reversed magnetization curves for three soft ferromagnetic materials.

Solution

It will not be necessary for us to convert the **B–H** curves for cast steel and sheet steel, as the two sections have the same length and area. The MMF equation is

$$\mathcal{F} = H_1 l_1 + H_2 l_2$$

Thus since $l_1 = l_2 = l$,

$$\frac{\mathcal{F}}{l} = H_1 + H_2$$

As a result, for part (a):

$$\frac{800}{0.4} = H_1 + H_2$$

$$2000 = H_1 + H_2$$

We thus have the construction of Figure 2.27. From the intersection of the curves, we obtain

$$H_1 = 400 \text{ At/M} \qquad \mathcal{F}_1 = 160 \text{ At}$$

$$H_2 = 1600 \text{ At/m} \qquad \mathcal{F}_2 = 640 \text{ At}$$

$$B_1 = B_2 = 1.35 \text{ T}$$

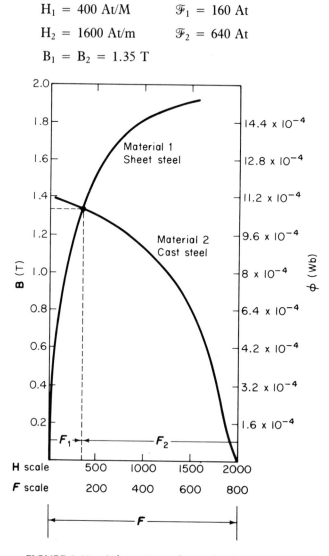

FIGURE 2.27   Solving Example 2.7 for $\mathcal{F} = 800$ At.

The flux is therefore given by

$$\Phi = 1.35 \times 8 \times 10^{-4} = 10.80 \times 10^{-4} \text{ Wb}$$

The construction for part (b) is shown in Figure 2.28. From the intersection of the two curves, we get

$$H_1 = 450 \text{ At/m} \qquad \mathscr{F}_1 = 180 \text{ At}$$
$$H_2 = 1550 \text{ At/m} \qquad \mathscr{F}_2 = 1020 \text{ At}$$
$$B_1 = B_2 \simeq 1.48 \text{ T}$$

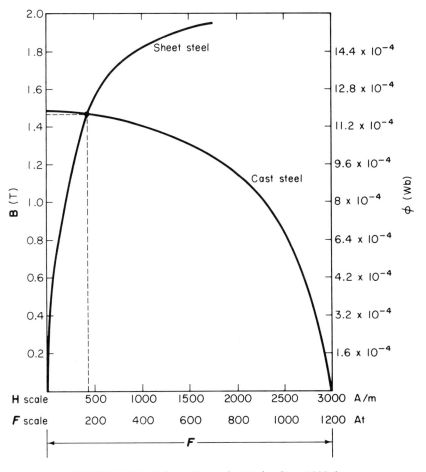

FIGURE 2.28   Solving Example 2.7 for $\mathscr{F} = 1200$ At.

The flux is therefore given by

$$\Phi = 1.48 \times 8 \times 10^{-4} = 11.84 \times 10^{-4}\ \text{Wb}$$

For part (c), we have Figure 2.29, giving

$$H_1 = 550\ \text{At/m} \qquad \mathscr{F}_1 = 220\ \text{At}$$
$$H_2 = 1950\ \text{At/m} \qquad \mathscr{F}_2 = 1180\ \text{At}$$
$$B_1 = B_2 \simeq 1.49\ \text{T}$$

The flux is therefore given by

$$\Phi = 1.49 \times 8 \times 10^{-4} = 11.92 \times 10^{-4}\ \text{Wb}$$

The reader should note that the results obtained here are approximate.

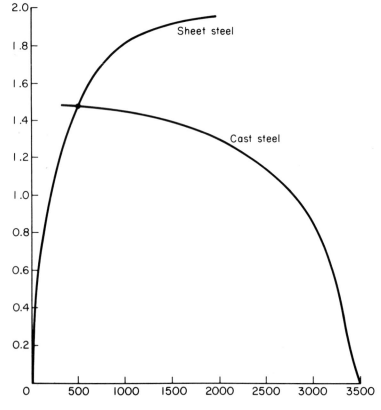

FIGURE 2.29 Solving Example 2.7 for $\mathscr{F} = 1400$ At.

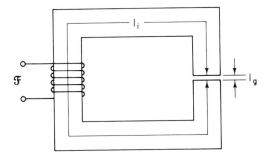

FIGURE 2.30 Magnetic structure for Example 2.8.

Our next example deals with the case when section 2 is an air gap. In this case, the construction is easier than the case with two magnetic materials, as the **B–H** characteristic for an air gap is a straight line.

Example 2.8 Find the flux density in the structure shown in Figure 2.30 given that the source MMF is 2400 At. The material of section 1 is cast steel, with a length of 80 cm, while the air-gap length is 0.4π mm.

Solution

We will need the air-gap characteristic. Since the area of the structure is uniform ($A_i = A_g$), we need only work with flux densities.

$$\mathscr{F}_g = H_g l_g = \frac{B_g l_g}{\mu_0}$$

$$= \frac{B_g(0.4\pi \times 10^{-3})}{4\pi \times 10^{-7}}$$

Thus

$$\mathscr{F}_g = 1000 B_g$$

As a result, the air-gap line has a slope of 1000 At/T, or alternatively for $B_g = 1$ T, we have $\mathscr{F}_g = 1000$ At.

The H scale of the **B–H** characteristic of cast steel is converted to $\mathscr{F}$ scale by multiplying by the iron path length, which is 0.8 m. Thus a 3000 At/m is converted to 2400 At on the $\mathscr{F}$ scale as shown in Figure 2.31. From the construction, we find the flux density as

$$B \cong 1.27 \text{ T}$$

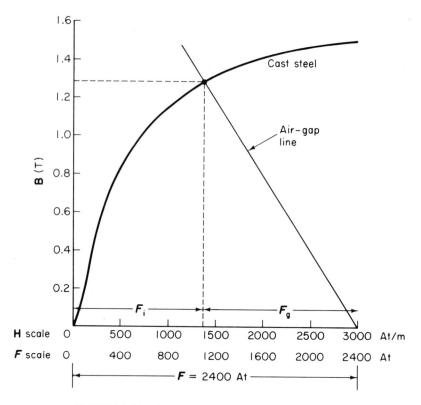

FIGURE 2.31 Construction for solving Example 2.8.

2.6 FLUX LINKAGES, INDUCED VOLTAGES, INDUCTANCE, AND ENERGY

The discussion of the previous sections is limited to situations where the variables of interest such as current and flux do not change with time. It is our intent now to consider the effects of variations with time of the magnetic system. A change in the magnetic field is associated with the establishment of an electric field that is manifested as an induced voltage. This basic fact is due to Faraday's experiments and is expressed by Faraday's law of electromagnetic induction.

Consider a toroidal coil with N turns through which a current i flows producing a total flux Φ. Each turn encloses or links the total flux and we also note that the total flux links each of the N turns. In this situation we define the flux linkages λ as the product of the number of turns N and the flux Φ linking each turn.

$$\lambda = N\Phi \tag{2.22}$$

In terms of flux linkages, Faraday's law is stated as

$$e = \frac{d\lambda}{dt} = N\frac{d\Phi}{dt} \tag{2.23}$$

The electromotive force (EMF) or induced voltage is thus equal to the rate of change of flux linkages in the structure.

The flux linkages λ can be related to the current i in the coil by the definition of inductance L through the relation

$$\lambda = Li \tag{2.24}$$

From this point of view, inductance is the ratio of total flux linkages to the current which they link. Inductance is a passive circuit element that is related to the geometry and material properties of the structure. In the case of the toroid, we have

$$L = \frac{N\Phi}{i} = \frac{NBA}{\mathrm{H}l/\mathrm{N}}$$

$$= \frac{\mathrm{N}^2\mathrm{A}}{l}\frac{\mathrm{B}}{\mathrm{H}}$$

For the case of a linear **B–H** curve, we get

$$L = \frac{N^2A}{l}\mu_0\mu_r \tag{2.25}$$

There is no single definition of inductance which is useful in all cases for which the medium is not linear. The unit of inductance is the henry or weber-turns per ampere.

The inductance L is related to the reluctance $\mathcal{R}$ of the magnetic structure of a single-loop structure, since we have

$$L = \frac{N\Phi}{i} = \frac{N\mathcal{F}}{i\mathcal{R}}$$

As a result, we get

$$L = \frac{N^2}{\mathcal{R}} \tag{2.26}$$

Faraday's law (2.23) can be written in the form

$$e = \frac{d}{dt} Li \tag{2.27}$$

In electromechanical energy conversion devices, the reluctance varies with time and thus L also varies with time. In this case

$$e = L \frac{di}{dt} + i \frac{dL}{dt} \tag{2.28}$$

Note that if L is constant, we get the familiar equation for modeling an inductor in elementary circuit analysis.

Power and energy relationships in a magnetic circuit are important in evaluating performance of electromechanical energy conversion devices treated in this book. We presently explore some basic relationships, starting with the fundamental definition of power $p(t)$ given by

$$p(t) = e(t)i(t) \tag{2.29}$$

The power into a component (the coil in the case of toroid) is given as the product of the voltage across its terminals $e(t)$ and the current through $i(t)$. Using Faraday's law (2.23), we can write

$$p(t) = i(t) \frac{d\lambda}{dt} \tag{2.30}$$

The units of power are watts (or joules per second).

Let us recall the basic relation stating that power $p(t)$ is the rate of change of energy $W(t)$:

$$p(t) = \frac{dW}{dt} \tag{2.31}$$

Combining Eqs. (2.30) and (2.31), we obtain

$$dW = i \, d\lambda \tag{2.32}$$

Using Eq. (2.22), we can also write the alternative expression

$$dW = Ni \, d\Phi \tag{2.33}$$

Yet another expression of dW can be obtained in terms of B and H. This is done by recalling that

$$Ni = Hl \qquad \text{and} \qquad \Phi = \mathbf{B}A$$

As a result, we have the expression

$$dW = (lA)\text{H } d\text{B} \tag{2.34}$$

Equations (2.32) to (2.34) provide us with three related interpretations of the energy taken by the magnetic structure in terms of its properties as shown in Figure 2.32. Part (a) depicts Eq. (2.32) assuming that the λ–i characteristic is available. In part (b) it is assumed that the Φ–$\mathscr{F}$ characteristic is available, and thus Eq. (2.33) is applicable. In part (c) we assume that the **B–H** curve is available and we can thus obtain the incremental energy per unit volume denoted $d\tilde{w}$ as an application of Eq. (2.34).

Consider the case of a magnetic structure that experiences a change in state between the time instants t_1 and t_2. The change in energy into the system is denoted by ΔW and is given by

$$\Delta W = W(t_2) - W(t_1) \tag{2.35}$$

Using Eq. (2.32), we get

$$\Delta W = \int_{\lambda_1}^{\lambda_2} i \, d\lambda \tag{2.36}$$

Using Eq. (2.33), we get

$$\Delta W = \int_{\Phi_1}^{\Phi_2} Ni \, d\Phi \tag{2.37}$$

Finally, using Eq. (2.34), we get

$$\Delta W = lA \int_{\mathbf{B}_1}^{\mathbf{B}_2} \text{H } d\text{B} \tag{2.38}$$

Equations (2.36) to (2.38) are interpreted in Figure (2.32) (d) to (f). It is clear that the energy per unit volume expended between t_1 and t_2 is the area between the **B–H** curve and the B axis between B_1 and B_2.

It is important to realize that the energy relations obtained thus far do

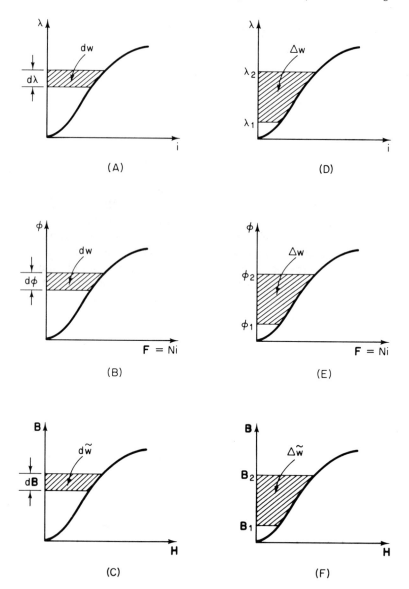

FIGURE 2.32 Energy in magnetic structures.

not require linearity of the characteristics. For a linear structure, we can develop these relations further. Using Eq. (2.24) and substituting into Eq. (2.36), we get

$$\Delta W = \int_{\lambda_1}^{\lambda_2} \frac{\lambda}{L} \, d\lambda$$

Thus

$$\Delta W = \frac{1}{2L} (\lambda_2^2 - \lambda_1^2) \qquad (2.39)$$

or

$$\Delta W = \frac{1}{2} L(i_2^2 - i_1^2) \qquad (2.40)$$

The energy expressions obtained in this section provide us with measures of energy stored in the magnetic field treated. This information is useful in many ways, as will be seen in this text.

2.7 HYSTERESIS LOOP

Ferromagnetic materials are characterized by a **B–H** characteristic that is both nonlinear and multivalued. This is generally referred to as a hysteresis characteristic. To illustrate this phenomenon, we use the sequence of portraits of Figure 2.33 showing the evolution of a hysteresis loop for a toroid with virgin ferromagnetic core. Assume that the MMF and hence **H** is a slowly varying sinusoidal waveform with period T as shown in the lower portion graphs of Figure 2.33. We will discuss the evolution of the **B–H** hysteresis loop in the following intervals.

Interval I: Between $t = 0$ and $T/4$, the magnetic field intensity H is positive and increasing. The flux density increases along the initial curve (*oa*) up to the saturation value B_s. Increasing H beyond saturation level does not result in an increase in B.

Interval II: Between $t = T/4$ and $T/2$, the magnetic field intensity is positive but decreasing. The flux density B is observed to decrease along the segment *ab*. Note that *ab* is above *oa* and thus for the same value of H, we get different value of B. This is true at *b*, where there is a value for $B = B_r$ different from zero even though H is zero at that point in time $t = T/2$. The value of B_r is referred to as the residual field, remanence, or retentivity. If we leave the coil unenergized, the core will still be magnetized.

Interval III: Between $t = T/2$ and $3T/4$, the magnetic field intensity H is reversed and increases in magnitude. B decreases to zero at point *c*. The value of H at which magnetization is zero is called the coercive force H_c. Further decrease in H results in reversal of B up to point *d*, corresponding to $t = 3T/4$.

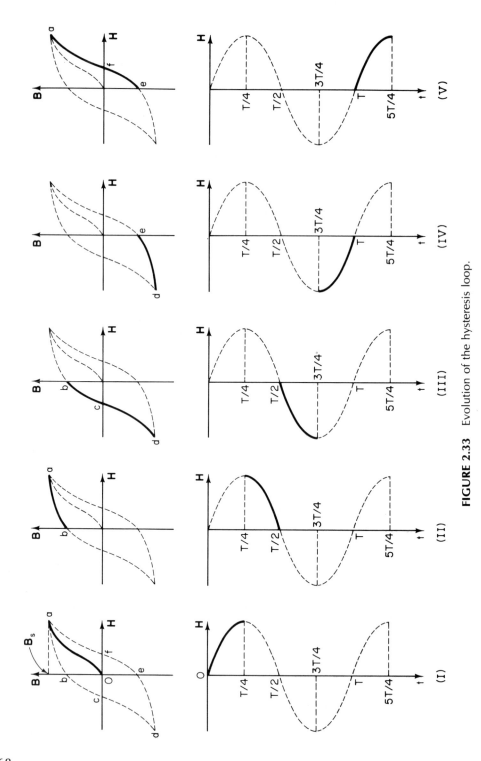

FIGURE 2.33 Evolution of the hysteresis loop.

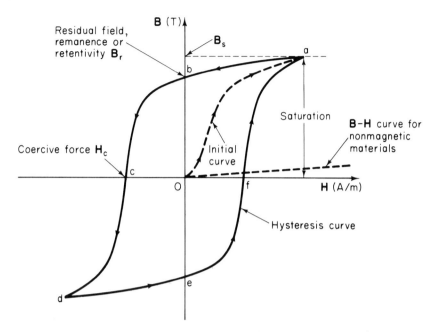

FIGURE 2.34 Hysteresis loop for a ferromagnetic material.

Interval IV: Between $t = 3T/4$ and T, the value of H is negative but increasing. The flux density B is negative and increases from d to e. Residual field is observed at e with $H = 0$.

Interval V: Between $t = T$ and $5T/4$, H is increased from 0, and the flux density is negative but increasing up to f, where the material is demagnetized. Beyond f, we find that B increases up to a again.

A typical hysteresis loop is shown in Figure (2.34). On the same graph, the **B–H** characteristic for nonmagnetic material is shown to show the relative magnitudes involved. It should be noted that for each maximum value of the ac magnetic field intensity cycle, there is a steady-state loop, as shown in Figure 2.35. The dashed curve connecting the tips of the loops in the figure is the dc magnetization curve for the material. The distinction between hard and soft magnetic material on the basis of their hysteresis loops is shown in Figure 2.36. It is evident from the figure that the coercive force H_c for the soft magnetic material is much lower than that for a hard magnetic material. Table 2.1 lists some typical values for H_c, B_r, and B_s for common magnetic materials.

In Section 2.6, we have shown that the energy supplied by the source per unit volume of the magnetic structure is given by

$$d\widetilde{W} = H\, dB$$

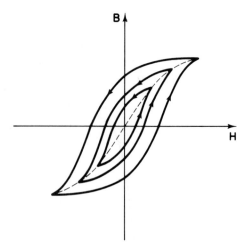

FIGURE 2.35 Family of hysteresis loops.

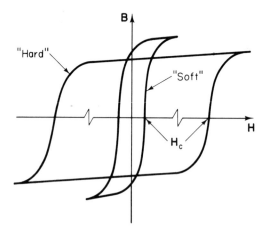

FIGURE 2.36 Hysteresis loops for soft and hard magnetic materials.

and

$$\Delta \widetilde{W} = \int_{\mathbf{B}_1}^{\mathbf{B}} \mathbf{H} \, d\mathbf{B}$$

The energy supplied by the source in moving from a to b in the graph of Figure 2.37(a) is negative since H is positive but B is decreasing. If we continue on from b to d through c, the energy is positive as H is negative but B is decreasing [see Figure 2.37(b)].

The second half of the loop is treated in Figure 2.37(b) and is self-explanatory. Superimposing both halves of the loop, we obtain Figure 2.37(c), which clearly shows that the net energy per unit volume supplied by the source is the area enclosed by the hysteresis loop. This energy is

TABLE 2.1

Properties of Magnetic Materials and Magnetic Alloys

Material (Composition)	Initial Relative Permeability, μ_r/μ_0	Maximum Relative permeability, μ_{max}/μ_0	Coercive Force H_r (A/m)	Residual Field B_r (Wb/m^2)	Saturation Field B_s (Wb/m^2)
Commercial iron (0.2 imp)	250	9,000	≈ 80	0.77	2.15
Silicon-iron (4 Si)	1,500	7,000	20	0.5	1.95
Silicon-iron (3 Si)	7,500	55,000	8	0.95	2.00
Mu metal (5 Cu, 2 Cr, 77 Ni)	20,000	100,000	4	0.23	0.65
78 Permalloy (78.5 Ni)	8,000	100,000	4	0.6	1.08
Supermalloy (79 Ni, 5 Mo)	100,000	1,000,000	0.16	0.5	0.79

expended in the magnetization-demagnetization process and is dissipated as a heat loss. Note that the loop is described in one cycle and as a result, the hysteresis loss per second is equal to the product of the loop area and the frequency f of the waveform applied. The area of the loop depends on the maximum flux density, and as a result, we assert that the power dissipated through hysteresis P_h is given by

$$P_h = k_h f (B_m)^n$$

where k_h is a constant, f is the frequency, and B_m is the maximum flux density. The exponent n is determined from experimental results and ranges between 1.5 and 2.5.

2.8   EDDY CURRENT AND CORE LOSSES

In Section 2.7 we saw that if the core is subject to a time-varying magnetic field (sinusoidal input was assumed), energy is extracted from the source in the form of hysteresis losses. There is another loss mechanism that arises in connection with the application of time-varying magnetic field, called eddy-current loss. A rigorous analysis of the eddy-current phenomenon is a complex process but the basic model can be explained in simple terms on the basis of Faraday's law, Eq. (2.23).

The change in flux will induce voltages in the core material which will result in currents circulating in the core. The induced currents tend to establish a flux that opposes the original change imposed by the source. The induced currents which are essentially the eddy currents will result in power loss due to heating of the core material. To minimize eddy current losses, the magnetic core is made of stackings of sheet steel laminations, ideally separated by highly resistive material. It is clear that this effectively results

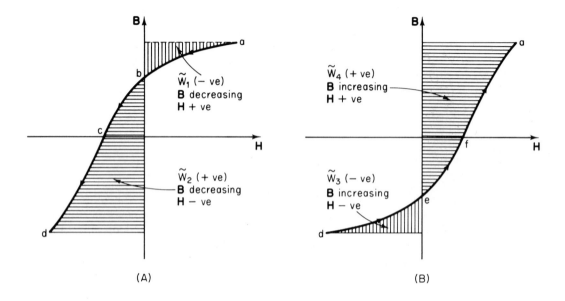

(A)

(B)

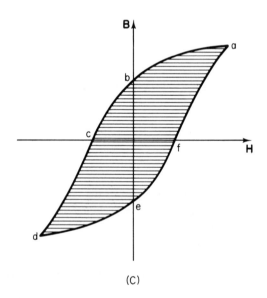

(C)

FIGURE 2.37 Illustrating the concept of energy loss in the hysteresis process.

in the actual area of the magnetic material being less than the gross area presented by the stack. To account for this, a stacking factor is employed for practical circuit calculations.

$$\text{Stacking factor} = \frac{\text{actual magnetic cross-sectional area}}{\text{gross cross-sectional area}}$$

Typically, lamination thickness ranges from 0.01 mm to 0.35 m with associated stacking factors ranging between 0.5 to 0.95. The eddy–current power loss per unit volume can be expressed by the empirical formula

$$P_e = K_e(f B_m t_1)^2 \quad \text{W/m}^3$$

The formula above shows that the eddy-current power loss per unit volume varies with the square of frequency f, maximum flux density B_m, and the lamination thickness t_1. Of course, K_e is a proportionality constant.

The term *core loss* is used to denote the combination of eddy-current and hysteresis power losses in the material. In practice, manufacturer-supplied data are used to estimate the core loss P_c for given frequencies and flux densities for a particular type of material.

SOME SOLVED PROBLEMS

Problem 2.A.1

A magnetic structure is made of two parts, as shown in Figure 2.38. The flux is 0.5×10^{-3} Wb in the structure. The cast steel portion is square in cross section and operates at a flux density of 1 T. The sheet steel portion operates at a flux density of 1.5 T. Determine the required dimensions of the structure and find the required MMF.

Solution

The area of the cast steel portion is

$$A_c = d^2$$

From the flux and flux density specifications, we have

$$A_c = \frac{0.5 \times 10^{-3}}{1} \text{ m}^2$$

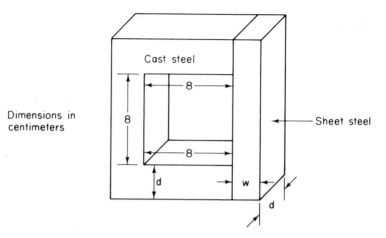

FIGURE 2.38   Magnetic structure for Problem 2.A.1.

As a result,

$$d = 2.236 \times 10^{-2} \text{ m}$$

The area of the sheet steel portion is

$$A_s = W \cdot d$$

From the flux and flux density specifications, we have

$$A_s = \frac{0.5 \times 10^{-3}}{1.5}$$

As a result,

$$W = \frac{0.5 \times 10^{-3}}{1.5(2.236 \times 10^{-2})} = 1.491 \times 10^{-2} \text{ m}$$

From Figure 2.25, we obtain the values of H corresponding to the specified flux densities:

$$H_c = 750$$

$$H_s = 550$$

The length of the cast steel portion is

$$l_c = 24 \times 10^{-2} + 2(2.236 \times 10^{-2}) = 28.472 \times 10^{-2}$$

The length of the sheet steel portion is

$$l_s = 8 \times 10^{-2} + 1.491 \times 10^{-2} + 2.236 \times 10^{-2} = 11.73 \times 10^{-2}$$

As a result,

$$\mathscr{F} = H_c l_c + H_s l_s = 278.06 \text{ At}$$

Problem 2.A.2

Find the current I required to set up a flux of 2×10^{-3} Wb in the air gap of the magnetic structure shown in Figure 2.39. Assume that the coil has 500 turns and use the following dimensions:

$l_1 = 40$ cm

$A_1 = 40$ cm^2 $l_2 = 24$ cm $l_3 = l_4 = 26$ cm

$l_g = 25 \times 10^{-3}$ cm $A_2 = 12$ cm^2 $A_3 = A_4 = 25$ cm^2

$A_g = 26$ cm^2

The core material is silicon sheet steel with the magnetization curve shown in Figure 2.40.

Solution

To set up a flux of 2×10^{-3} Wb in the air gap, the flux density in the gap is

$$B_g = \frac{\Phi_3}{A_g} = \frac{2 \times 10^{-3}}{26 \times 10^{-4}} = 0.769 \text{ T}$$

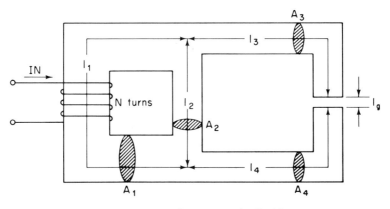

FIGURE 2.39 Magnetic structure for Problem 2.A.2

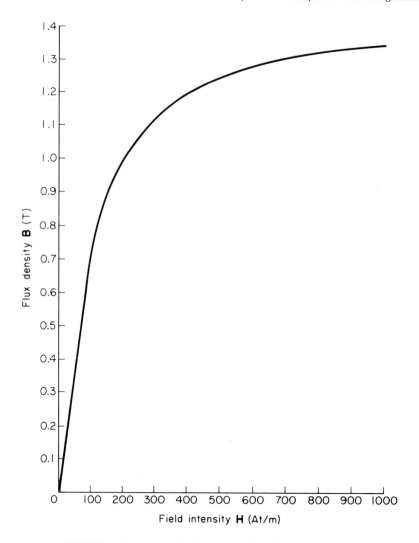

FIGURE 2.40   Magnetization curve for silicon sheet steel.

Thus

$$H_g = \frac{\mathbf{B}_g}{\mu_0}$$

$$= 612.13 \times 10^3$$

The MMF required for the gap is

$$\mathscr{F}_g = H_g l_g = 15.303 \times 10^3 \times 10^{-2} = 153.03 \text{ At}$$

The flux density in the paths l_3 and l_4 is

$$B_3 = B_4 = \frac{\Phi_3}{A_3} = \frac{2 \times 10^{-3}}{25 \times 10^{-4}} = 0.8 \text{ T}$$

From the magnetization curve of Figure 2.40, we obtain

$$H_3 = H_4 = 120 \text{ At/m}$$

As a result,

$$\mathcal{F}_3 = H_3 l_3 = 120(0.26) = 31.2 \text{ At}$$

The MMF around l_3, l_g, and l_4 is thus

$$
\begin{aligned}
\mathcal{F}_2 &= \mathcal{F}_3 + \mathcal{F}_4 + \mathcal{F}_g \\
&= 2\mathcal{F}_3 + \mathcal{F}_g \\
&= 62.4 + 153.03 = 215.43 \text{ At}
\end{aligned}
$$

At this point, the reader may refer to the equivalent circuit of Figure 2.41, as the last statement is based on Kirchhoff's voltage law (Ampère's circuital law).

We can now obtain H_2 as

$$H_2 = \frac{\mathcal{F}_2}{l_2} = \frac{215.43}{0.24} = 897.63 \text{ At/m}$$

From the magnetization curve

$$B_2 = 1.34 \text{ T}$$

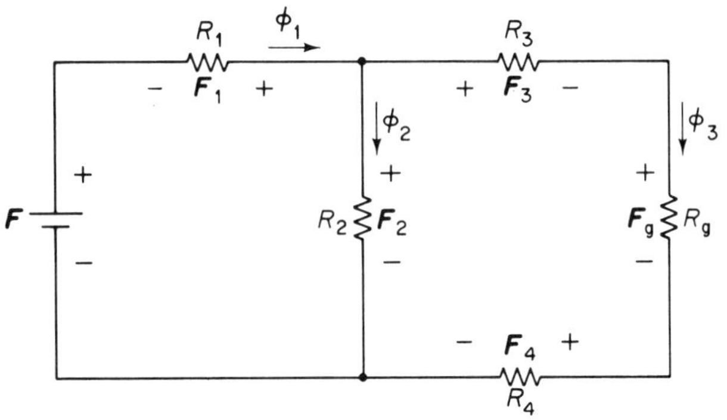

FIGURE 2.41 Equivalent circuit for Problem 2.A.2

We note that the middle section is operating very close to saturation. The flux Φ_2 is now obtained as

$$\Phi_2 = B_2A_2 = (1.34)(12 \times 10^{-4}) = 1.608 \times 10^{-3} \text{ Wb}$$

The flux Φ_1 in the left-hand section is now obtained as

$$\Phi_1 = \Phi_2 + \Phi_3$$
$$= 3.608 \times 10^{-3} \text{ Wb}$$

As a result, we obtain the flux density B_1 as

$$B_1 = \frac{\Phi_1}{A_1} = \frac{3.608 \times 10^{-3}}{40 \times 10^{-4}} = 0.902 \text{ T}$$

From the magnetization curve, we get

$$H_1 = 150 \text{ At/m}$$

Thus the MMF for the left-hand section is

$$\mathscr{F}_1 = H_1l_1 = (150)(0.4) = 60 \text{ At}$$

The source MMF is thus obtained as

$$\mathscr{F} = \mathscr{F}_1 + \mathscr{F}_2$$
$$= 60 + 215.43$$
$$= 275.43 \text{ At}$$

Since the coil is a 500-turn coil, we obtain

$$I = \frac{\mathscr{F}}{N} = \frac{275.43}{500} = 0.551 \text{ A}$$

Problem 2.A.3

A cross section of the magnetic structure of a dc machine is shown in Figure 2.42. There are four identical poles carrying identical coils with the same number of turns and current. The stator poles and rotor are made of sheet steel with relative permeability of 3800. The stator is made of cast steel with relative permeability of 1500. The following dimensions are available:

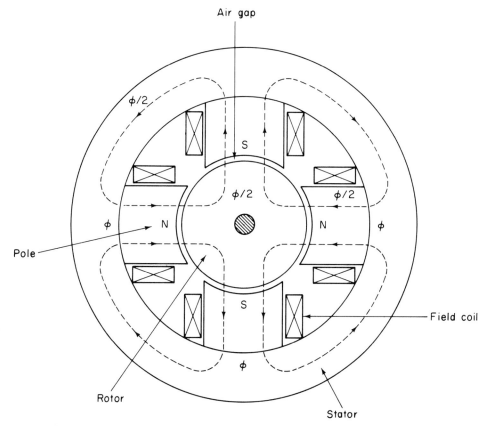

FIGURE 2.42 Direct-current machine magnetic structure for Problem 2.A.3.

Rotor diameter	= 18 cm
Rotor axial length	= 10 cm
Pole axial length	= 10 cm
Pole circumferential width	= 8.5 cm
Pole height	= 9 cm
Stator yoke mean diameter	= 42 cm
Stator yoke cross section	= 14 × 6 cm
Air-gap length	= 1.5 mm

It is required to:

(a) Draw an equivalent magnetic circuit,
(b) Determine the MMF per pole required to produce a flux density of 0.9 T in the air gap.

Solution

(a) The magnetic equivalent circuit of the structure is shown in Figure 2.43. The following subscripts are employed: r for rotor, g for airgap, p for stator pole, and s for stator yoke. The structure is symmetric and therefore analysis of one quarter is sufficient.

  (b) The solution for the MMF per pole can be obtained from consideration of Ampère's law:

$$2\mathscr{F} = [2(\mathscr{R}_p + \mathscr{R}_g) + \tfrac{1}{2}(\mathscr{R}_r + \mathscr{R}_s)]\Phi$$

Thus

$$\mathscr{F} = \Phi[\mathscr{R}_p + \mathscr{R}_g + 0.25(\mathscr{R}_r + \mathscr{R}_s)]$$

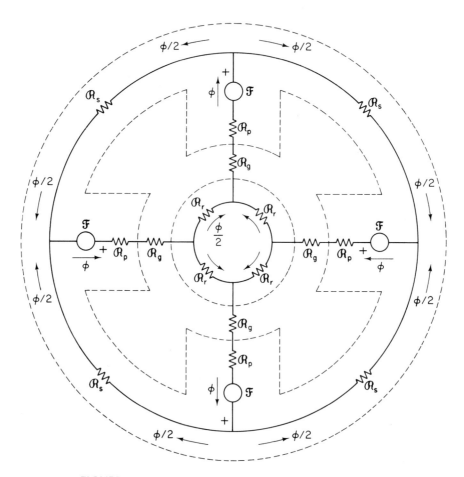

FIGURE 2.43 Magnetic equivalent circuit for the dc machine of Figure 2.42.

The flux Φ is obtained from the specification of flux density of the air gap and the air-gap area. Here we have

$$A_g = 0.085 \times 0.1 = 8.5 \times 10^{-3} \text{ m}^2$$

The air gap is the product of the pole circumferential width and its axial length. As a result,

$$\Phi = B_g A_g = 0.9 \times 8.5 \times 10^{-3}$$
$$= 7.65 \times 10^{-3} \text{ Wb}$$

The reluctances are obtained on the basis of the formula

$$\mathscr{R} = \frac{l}{\mu_0 \mu_r A}$$

For the air gap, we have

$$\mathscr{R}_g = \frac{l_g}{\mu_0 A_g} = \frac{1.5 \times 10^{-3}}{4\pi \times 10^{-7} \times 8.5 \times 10^{-3}} = 1.4043 \times 10^5$$

For the pole, we have

$$\mathscr{R}_p = \frac{l_p}{\mu_0 \mu_p A_p}$$

Here we have

$$l_p = 0.09 \text{ m}$$
$$A_p = 8.5 \times 10^{-2} \times 10 \times 10^{-2} = 8.5 \times 10^{-3} \text{ m}^2$$
$$\mu_p = 3800$$

Thus

$$\mathscr{R}_p = \frac{9 \times 10^{-2}}{4\pi \times 10^{-7} \times 3800 \times 8.5 \times 10^{-3}} = 2.2173 \times 10^3$$

For the rotor, we note that four reluctances represent the rotor; thus we take the length l_r to be 0.25 of the rotor circumference

$$l_r = \frac{\pi D_r}{4} = \frac{\pi(0.18)}{4} = 0.14137 \text{ m}$$

The area of the rotor subject to flux is taken as the product of rotor radius times the length of the stator pole:

$$A_r = 0.09 \times 0.1 = 9 \times 10^{-3} \text{ m}^2$$

The permeability of the rotor is

$$\mu_r = 3800$$

As a result,

$$\mathcal{R}_r = \frac{0.14137}{(4\pi \times 10^{-7})(3800)(9 \times 10^{-3})} = 3.2895 \times 10^3$$

The stator is treated in a similar manner:

$$l_s = \frac{\pi D_s}{4} = \frac{\pi(0.42)}{4} = 3.2987 \times 10^{-1}$$

$$A_s = 0.14 \times 0.06 = 8.4 \times 10^{-3}$$

$$\mu_s = 1500$$

As a result,

$$\mathcal{R}_s = \frac{3.2987 \times 10^{-1}}{(4\pi \times 10^{-7})(1500)(8.4 \times 10^{-3})} = 2.0833 \times 10^4$$

We can now obtain $\mathcal{F}$ as

$$\mathcal{F} = 7.65 \times 10^{-3} \left[2.2173 \times 10^3 + 1.4043 \times 10^5 \right.$$
$$\left. + \frac{(3.2985 + 20.833)10^3}{4} \right]$$
$$= 1137.4 \text{ At}$$

Problem 2.A.4

The λ–i characteristic of a magnetic structure is shown in Figure 2.44. Find the energy stored in the magnetic field for a current of 1.75 A.

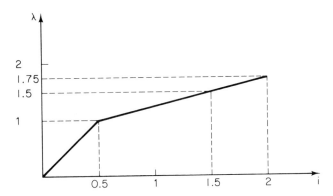

FIGURE 2.44 λ–i characteristic for Problem 2.A.4.

Solution

The energy stored in the magnetic field is given by formula (2.36), with integration carried out between zero and the present value of λ corresponding to the given current. From Figure 2.44, we get

$$\lambda_f = 1.625 \text{ weber-turns}$$

We thus have

$$\Delta W = \int_0^{\lambda_f} i \, d\lambda$$

We can do the integrations graphically as areas between the characteristic lines and the λ axis.

$$\Delta W = \Delta W_1 + \Delta W_2 + \Delta W_3$$

Here we have, as shown in Figure 2.45,

$$\Delta W_1 = \frac{(1)(0.5)}{2} = 0.25$$

$$\Delta W_2 = 0.5(0.5) + \frac{(1)(0.5)}{2} = 0.5$$

$$\Delta W_3 = 0.125(1.5) + \frac{0.125(0.25)}{2} = 0.20313$$

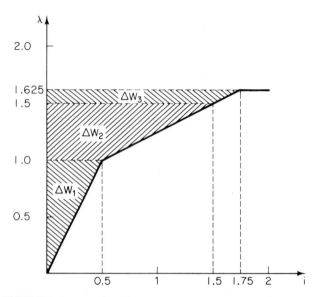

FIGURE 2.45   Finding the stored energy for Problem 2.A.4.'

Thus

$$\Delta W = 0.95313$$

Problem 2.A.5

Determine the inductance of the coil for Example 2.3 given that it consists of 100 turns.

Solution

In Example 2.3, we obtained the equivalent reluctance of the magnetic circuit as

$$\mathscr{R}_{eq} = 37.34 \times 10^3$$

According to Eq. (2.26), we have

$$L = \frac{N^2}{\mathscr{R}}$$

As a result,

$$L = \frac{10^4}{37.34 \times 10^3} = 0.26781 \text{ H}$$

PROBLEMS

Problem 2.B.1

A small toroidal coil is shown in Figure 2.46. Assume that the relative permeability of the core material is 1500 and that the outside radius of the toroid is 0.2 cm. The area of cross section is circular with diameter 0.05 cm. Find the magnetomotive force required to set up a flux density of 1 millitesla in the core. Find the flux in the core.

Problem 2.B.2

For the toroidal coil of Problem 2.B.1, find the flux density in the core for an MMF of 200 At.

Problem 2.B.3

Consider the toroid shown in Figure 2.47, which is wound with a coil of 200 turns. Assume that the current is 40 A and that the core is nonmagnetic. It is required to calculate:

  a. The reluctance of the circuit.
  b. The flux and flux density inside the core.

Problem 2.B.4

Consider the magnetic structure shown in Figure 2.48. Find the current in the 50-turn winding to set up a flux density of 0.8 T in the structure, assuming that the relative permeability of the core material is 2000. Use the concept of reluctance in your solution.

Problem 2.B.5

For the magnetic structure shown in Figure 2.49, find the current I required to set up a flux density of 1.5 T in the air gap. Assume that the relative permeability of the core material is 1800. Use the concepts of magnetic circuit and reluctance in your solution.

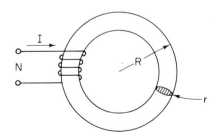

FIGURE 2.46 Toroidal coil for Problem 2.B.1

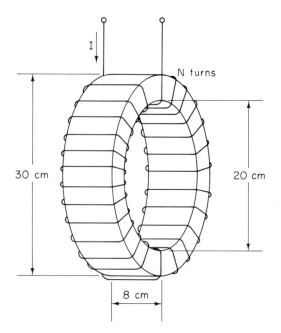

FIGURE 2.47   Toroid for Problem 2.B.3.

Problem 2.B.6

For the magnetic structure of Problem 2.B.4, find the flux density in the air gap for a current of 45 A.

Problem 2.B.7

For the magnetic structure of Problem 2.B.5, find the flux density in the air gap for a current of 20 A.

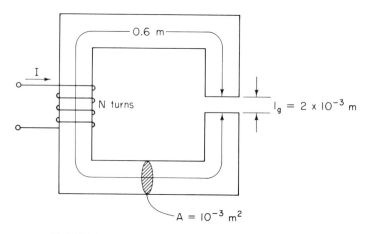

FIGURE 2.48   Magnetic structure for Problem 2.B.4.

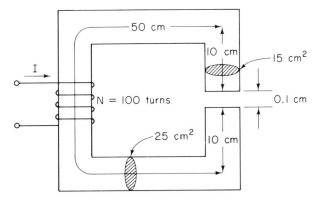

FIGURE 2.49   Magnetic structure for Problem 2.B.5.

Problem 2.B.8

The material of the magnetic structure of Figure 2.50 has a relative permeability of 2500. The coil current is 0.1 A and the corresponding flux is 10^{-4} Wb. Find the depth d of the structure.

Problem 2.B.9

The source MMF in the magnetic structure of Figure 2.51 is fixed at 200 At. The two rotors are made of a material with relative permeability μ_r while the core material has a permeability μ_i. With the rotors absent, the flux is found to be 0.5236×10^{-5} Wb. With the rotors in the circuit, the flux is found to be 3.122×10^{-5} Wb. Find the relative permeabilities μ_r and μ_i given that the cross section of the structure is uniform with area 4×10^{-4} m^2.

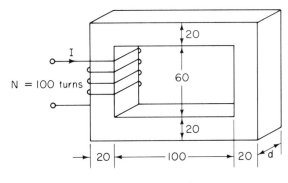

Dimensions in mm

FIGURE 2.50   Magnetic structure for Problem 2.B.8.

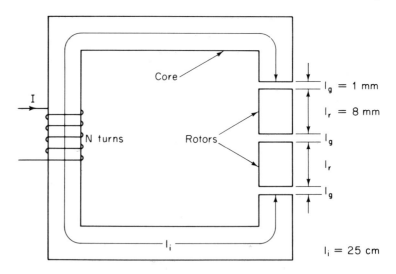

FIGURE 2.51   Magnetic structure for Problem 2.B.9.

Problem 2.B.10

The relative permeability of the core of the magnetic structure of Figure 2.52 is 2000. The MMF is 500 At and the flux is 7×10^{-4} Wb. Find the length l_0.

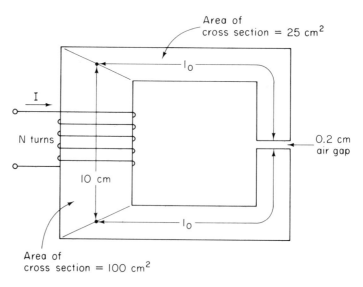

FIGURE 2.52   Magnetic structure for Problem 2.B.10.

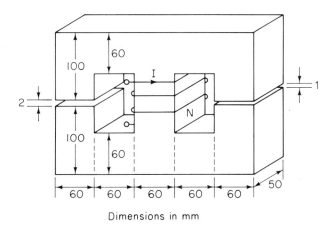

FIGURE 2.53   Magnetic structure for Problem 2.B.11.

Problem 2.B.11

Consider the magnetic structure shown in Figure 2.53. Draw an equivalent magnetic circuit for the system. Assume that the relative permeability of the core is 2500. Calculate the flux in the right-hand air gap for an MMF in the coil of 800 At.

Problem 2.B.12

Consider the magnetic structure shown in Figure 2.54. The flux density in the cast iron portion is 0.6 T and the flux density in the sheet steel portion is 1 T. This corresponds to an MMF of 410 At and a flux of 0.35 mWb.

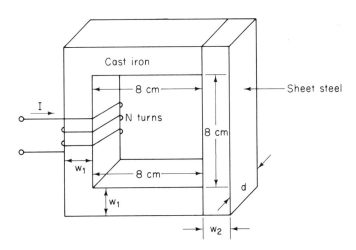

FIGURE 2.54   Magnetic structure for Problem 2.B.12.

Assume that the permeability of cast iron is 4×10^{-4} and that the permeability of sheet steel is 4.348×10^{-3}. Find the dimensions w_1, w_2, and d for this structure.

Problem 2.B.13

For the magnetic structure of Figure 2.16, assume that the ferromagnetic material has a relative permeability $\mu_r = 3000$. Assume the following dimensions:

$$l_a = 0.9 \text{ m} \qquad l_b = 0.8 \text{ m}$$
$$l_a = 0.3 \text{ m} \qquad l_g = 1 \text{ mm}$$

The structure has a cross-sectional area $A = 6 \times 10^{-3}$ m. Find the flux in the air gap for an MMF of 240 At.

Problem 2.B.14

The magnetic structure of a synchronous machine is shown in Figure 2.55. Assume that the rotor and stator iron have relative permeabilities of 3500 and 1800, respectively. The outside and inside diameters of the stator yoke are 1 m and 0.72 m, respectively. The rotor length is 0.7 m and is 0.1 m wide. The air gaps are 1 cm long each. The axial length of the machine is

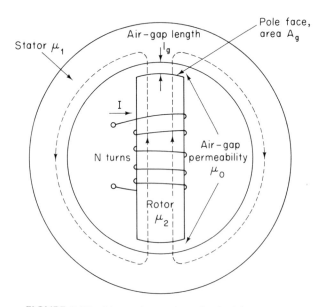

FIGURE 2.55   Magnetic structure for Problem 2.B.14.

1 m. Draw an equivalent magnetic circuit of the machine and find the flux density in the air gap for an MMF of 10,000 At.

Problem 2.B.15

Consider the magnetic structure of Problem 2.B.13 and assume that the air-gap length is 1.5 mm with all other data unchanged. Find the required MMF to set up a flux of 0.7×10^{-3} Wb in the center section.

Problem 2.B.16

The magnetic structure of Figure 2.56 is made of material with relative permeability 2000. The air-gap length is 2 mm. Find the flux in the air gap given that

$$\mathscr{F}_1 = \mathscr{F}_2 = 4000 \text{ At}$$

Problem 2.B.17

For the magnetic structure of Problem 2.B.16, find $\mathscr{F}_2$ given that $\mathscr{F}_1 = 4000$ At and that the flux in the air gap is 0.5×10^{-3} Wb.

Problem 2.B.18

Repeat problem 2.B.4 accounting for fringing of flux in the air gap using 1.1 as a fringe factor:

$$\text{fringe factor} = \frac{\text{cross-sectional area of flux lines}}{\text{cross section of air gap}}$$

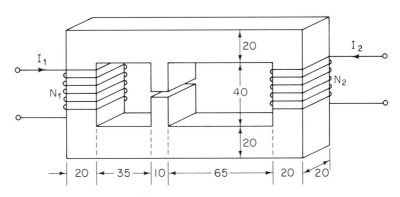

Dimensions in mm

FIGURE 2.56 Magnetic structure for Problem 2.B.16.

Problem 2.B.19

Repeat Problem 2.B.5 accounting for fringing effects assuming that the air gap is square in cross section.

Problem 2.B.20

Repeat Problem 2.B.4 assuming that the iron path is cast steel with a magnetization curve as shown in Figure 2.21.

Problem 2.B.21

Repeat Problem 2.B.17 assuming that the magnetization curve of the core is given in Figure 2.21. Assume that the air-gap flux is 2×10^{-4} Wb and that $\mathscr{F}_1$ is 1800 At.

Problem 2.B.22

Repeat Example 2.7 assuming that section 1 is made of cast steel and that section 2 is made of cast iron.

Problem 2.B.23

Repeat Example 2.8 assuming that the air-gap length is 0.48π mm.

Problem 2.B.24

Find the inductance of the coil of Problem 2.B.1 assuming that $N = 10$ turns.

Problem 2.B.25

Find the inductance of the coil of Problem 2.B.4.

Problem 2.B.26

The flux linkages in a coil are related to the current by the relation

$$\lambda = 1 - 0.5e^{-1.25i}$$

Find the energy stored for a current of 0.8 A.

Problem 2.B.27

Consider the coil of Problem 2.B.4. Find the energy stored in the air gap and the magnetic core.

Problem 2.B.28

The eddy-current and hysteresis losses in a transformer are 500 and 600 W, respectively, when operating from a 60-Hz supply. Find the eddy-current and hysteresis losses when the transformer is operated from a 50-Hz supply with an increase of 10% in flux density. Find the change in core losses.

Transformers

3.1 INTRODUCTION

One of the most valuable apparatus in electric systems is the transformer, for it enables us to utilize different voltage levels across the power system for the most economical value. Generation of power at the synchronous machine level is normally at a relatively low voltage, which is most desirable economically. Stepping up of this generated voltage to high voltage, extrahigh voltage, or even to ultrahigh voltage is done through power transformers to suit the power transmission requirement to minimize losses and increase the transmission capacity of the lines. This transmission voltage level is then stepped down in many stages for distribution and utilization purposes. Transformers are used at all levels of the system. An electronic device designed to operate on normal household voltage and frequency has a transformer to supply suitable voltage for the various components of the device. Audio-frequency (up to 20 kHz) transformers are used for impedance matching at the input or output of audio frequency amplifiers or between amplifiers. Control-type transformers are used to provide desired modifications to magnitude and phase of voltages at various points in the electric system. Instrument-type transformers are used to interface the high-energy side of a system to instruments and devices that monitor the state of the system.

Transformer action requires the existence of the varying flux that links the two windings. This will be obtained more effectively if an iron core is used because an iron core confines the flux to a definite path linking both windings. A magnetic material such as iron undergoes a loss of energy due to the application of alternating voltage to its B–H loop. The losses are

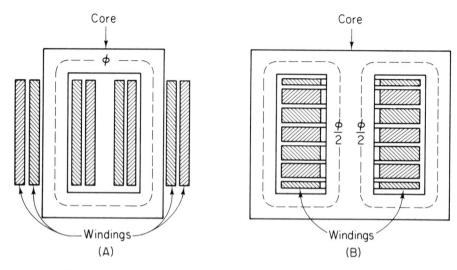

FIGURE 3.1 (a) Core-type and (b) shell-type transformer construction.

composed of two parts. The first is called the eddy-current loss, and the second is the hysteresis loss. Eddy-current loss is basically an I^2R loss due to the induced currents in the magnetic material. To reduce these losses, the magnetic circuit is usually made of a stack of thin laminations. For high-frequency applications, the core is made of fine-particle magnetic material. Hysteresis loss is caused by the energy used in orienting the magnetic domains of the material along the field. The loss depends on the material used.

Two types of construction are used, as shown in Figure 3.1. The first is denoted the core type, which is a single ring encircled by one or more groups of windings. The mean length of the magnetic circuit for this type is long, whereas the mean length of windings is short. The reverse is true for the shell type, where the magnetic circuit encloses the windings.

The present discussion is mainly oriented toward power system transformers operating at power-frequency (50 or 60 Hz) levels. In this case, capacitive effects are negligible. For high-frequency applications, the effects of capacitance must be included. The first order of business to understand transformer operation is to study the ideal transformer. Although we will never encounter an ideal transformer in real life, the concept is extremely useful in modeling and analysis studies.

3.2 IDEAL TRANSFORMERS

A transformer consists of two or more windings, as shown in Figure 3.2, linked by a mutual field. The primary winding is connected to an alternating voltage source, which results in an alternating flux whose magnitude

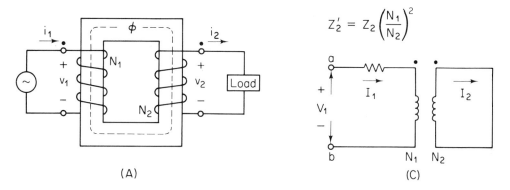

FIGURE 3.2 Ideal transformer and a load impedance.

depends on the voltage, frequency, and number of turns of the primary winding. The alternating flux links the secondary winding and induces a voltage in it with a value that depends on the number of turns of the secondary winding. If the primary voltage is v_1, the core flux ϕ is established such that the counter EMF e equals the impressed voltage (neglecting winding resistance). Thus

$$v_1 = e_1 = N_1 \frac{d\phi}{dt} \tag{3.1}$$

Here N_1 denotes the number of turns of the primary winding. The EMF e_2 is induced in the secondary by the alternating core flux:

$$v_2 = e_2 = N_2 \frac{d\phi}{dt} \tag{3.2}$$

Taking the ratio of Eq. (3.1) to Eq. (3.2), we see that

$$\frac{v_1}{v_2} = \frac{N_1}{N_2} \tag{3.3}$$

Neglecting losses, the instantaneous power is equal on both sides of the transformer:

$$v_1 i_1 = v_2 i_2 \tag{3.4}$$

Combining Eqs. (3.3) and (3.4), we get

$$\frac{i_1}{i_2} = \frac{N_2}{N_1} \tag{3.5}$$

Thus, in an ideal transformer, the current ratio is the inverse of the voltage ratio. We can conclude that almost any desired voltage ratio of transformation can be obtained by adjusting the number of turns of the transformer.

Transformers can be broadly classified as either step-up or step-down types. In a step-up transformer, the ratio N_1/N_2 is less than 1 and hence v_2 is higher than v_1. A step-down transformer will have a ratio (N_1/N_2) higher than 1 and in this case, v_2 is lower than v_1.

Let us assume that the flux is given by the sinusoidal waveform

$$\phi(t) = \phi_m \sin \omega t \tag{3.6}$$

As a result of Eq. (3.1), we see that

$$e_1(t) = N_1 \phi_m \omega \cos \omega t$$

Since $\omega = 2\pi f$, with f being the supply frequency, we see that

$$e_1(t) = 2\pi N_1 \phi_m f \sin (\omega t + 90°)$$

The primary EMF $e_1(t)$ is written as

$$e_1(t) = \sqrt{2} E_1 \sin (\omega t + 90°) \tag{3.7}$$

where E_1 is the effective (or root mean square) value of the primary EMF. It is thus clear that

$$E_1 = \frac{2\pi}{\sqrt{2}} N_1 \phi_m f$$

or

$$E_1 = 4.44 N_1 \phi_m f \tag{3.8}$$

The voltage is thus proportional to the number of turns, flux magnitude, and frequency. It should also be noted that according to Eqs. (3.6) and (3.7), the voltage phasor E_1 leads the flux phasor by 90°.

Consider an ideal transformer (with negligible winding resistances and leakage reactances and no exciting losses) connected to a load as shown in Figure 3.2. Clearly Eqs. (3.1) to (3.5) apply. The dot markings indicate terminals of corresponding polarity in the sense that both windings encircle the core in the same direction if we begin at the dots. Thus comparing the voltages of the two windings shows that the voltages from a dot-marked terminal to an unmarked terminal will be of the same polarity for the primary and secondary windings (i.e., v_1 and v_2 are in phase). From Eqs. (3.3) and (3.5), we can write for sinusoidal steady-state operation

$$\frac{V_1}{I_1} = \left(\frac{N_1}{N_2}\right)^2 \frac{V_2}{I_2} \tag{3.9}$$

But the load impedance Z_2 is

$$\frac{V_2}{I_2} = Z_2 \tag{3.10}$$

Thus

$$\frac{V_1}{I_1} = \left(\frac{N_1}{N_2}\right)^2 Z_2 \tag{3.11}$$

The result is that as far as its effect is concerned, Z_2 can be replaced by an equivalent impedance Z_2' in the primary circuit. Thus

$$Z_2' = \left(\frac{N_1}{N_2}\right)^2 Z_2 \tag{3.12}$$

The equivalence is shown in Figure 3.3.

The impedance Z_2' is simply the load impedance Z_2, referred to the primary side. From Eq. (3.12) it is clear that impedance ratios vary as the square of the turns ratio. An example is appropriate at this point.

Example 3.1 A 440/110-V single-phase 5-kVA 60-Hz transformer delivers a secondary current of 40 A at 0.8 power factor leading at rated secondary voltage. Assume that the transformer is ideal.

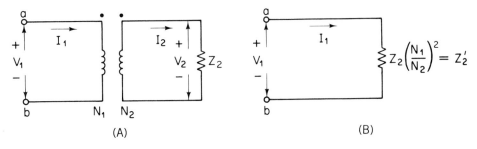

(A)   (B)

FIGURE 3.3 Equivalent representations of impedance loads on the secondary of an ideal transformer.

(a) Find the primary voltage and current.

(b) Find the load impedance.

(c) Find the impedance of the load referred to the primary side.

Solution

From the specifications available we have

$$V_2 = 110 \; \underline{/0}$$
$$I_2 = 40 \; \underline{/\cos^{-1}0.8}$$
$$= 40 \; \underline{/36.87°}$$

(a) The transformer ratio is

$$\frac{V_1}{V_2} = \frac{N_1}{N_2} = \frac{440}{110} = 4$$

Thus

$$V_1 = 440V$$

$$I_1 = \frac{N_2}{N_1}I_2 = \tfrac{1}{4}(40 \; \underline{/36.87°})$$

$$= 10 \; \underline{/36.87°} \; A$$

(b) The load impedance is

$$Z_2 = \frac{V_2}{I_2} = \frac{110 \; \underline{/0}}{40 \; \underline{/36.87°}} = 2.75 \; \underline{/-36.87°} \; \Omega$$

(c) The impedance of the load referred to the primary side is

$$Z_2' = \left(\frac{N_1}{N_2}\right)^2 Z_2$$

$$= (4)^2(2.75 \; \underline{/-36.87°})$$

$$= 44 \; \underline{/-36.87°}$$

Note that we can get this value by simply using the basic definition of impedance:

$$Z_2' = \frac{V_1}{I_1} = \frac{440}{10 \underline{/36.87°}} = 44 \underline{/-36.87°} \; \Omega$$

With this background on the performance of an ideal transformer, we can now proceed to deal with a more realistic view of the transformer.

3.3 TRANSFORMER MODELS

The ideal transformer is a very simple mathematical model that accounts only for the voltage, current, and impedance transformations in a transformer. It is our intention presently to discuss more realistic models of a transformer in the form of equivalent circuits.

Let us start by assuming that the primary winding is connected to a source with voltage $v_1(t)$ and that the current in the primary winding is $i_1(t)$. We will assume that the resistance of the primary winding is R_1 and as a result, a net voltage $\tilde{v}_1(t)$ is produced, as shown in Figure 3.4.

$$v_1(t) = \tilde{v}_1(t) + R_1 i_1(t) \tag{3.13}$$

According to Faraday's law, we have

$$\tilde{v}_1(t) = N_1 \frac{d\phi_1}{dt} \tag{3.14}$$

The flux ϕ_1 links the primary winding. If the permeability of the core is infinite, that is, zero reluctance, then the flow of ϕ_1 should be confined to the iron path. To account for this less than ideal situation, we assume that a portion of ϕ_1 (however small) is leaked, ϕ_{l1}. The remainder, denoted by

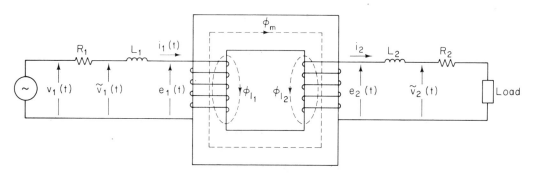

FIGURE 3.4 Modeling a transformer.

ϕ_m, is the flux in the core structure and links both primary and secondary windings.

$$\phi_1 = \phi_{l1} + \phi_m \tag{3.15}$$

An equivalent magnetic circuit of the transformer is shown in Figure 3.5.

As a result of Eq. (3.15), we can combine Eqs. (3.13) and (3.14) to obtain

$$v_1(t) = R_1 i_1(t) + N_1 \frac{d\phi_{l1}}{dt} + N_1 \frac{d\phi_m}{dt} \tag{3.16}$$

The leakage inductance of the primary winding is now defined by

$$L_1 \frac{di_1}{dt} = N_1 \frac{d\phi_{l1}}{dt} \tag{3.17}$$

We also define the EMF induced in the primary e_1 as

$$e_1 = N_1 \frac{d\phi_m}{dt} \tag{3.18}$$

As a result, Kirchhoff's voltage law for the primary loop given by Eq. (3.16) reduces to

$$v_1(t) = R_1 i_1(t) + L_1 \frac{di_1}{dt} + e_1(t) \tag{3.19}$$

Let us take a look at the equivalent magnetic circuit of Figure 3.5. It is clear that the flux ϕ_m is set up as a result of the difference of $\mathcal{F}_1$ and $\mathcal{F}_2$. Thus we have

$$\phi_m \mathcal{R}_m = N_1 i_1 - N_2 i_2 \tag{3.20}$$

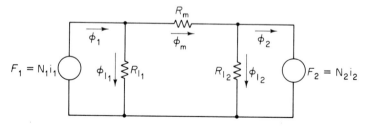

FIGURE 3.5 Equivalent magnetic circuit of a transformer.

The expression above assumes a linear B–H characteristic. Alternatively, we define an exciting current i_ϕ that is required of the primary to set up the flux ϕ_m.

$$N_1 i_\phi = N_1 i_1 - N_2 i_2 \tag{3.21}$$

Note that by doing this we are no longer bound by the linearity assumption. From the above, we can write the primary current as the sum:

$$i_1 = i_\phi + \frac{N_2}{N_1} i_2 \tag{3.22}$$

At this point we assume that the source voltage $v_1(t)$ is sinusoidal and that the core material is represented by a hysteresis loop to model the relation between ϕ_m and the exciting current i_ϕ. Assuming that $\phi_m(t)$ is sinusoidal, the construction of Figure 3.6 reveals that the resulting $i_\phi(t)$ is not sinusoidal but is periodic. The implications of this discovery are that i_1 and e_1 are not sinusoidal which creates a paradox. To get around this difficulty, we recall that a periodic function can be resolved using Fourier series into the sum of sinusoids with frequencies that are integer multiples of the main function frequency. Essentially what we are saying is that

$$i_\phi = i_{\phi 1} + i_{\phi 3} + \cdots \tag{3.23}$$

The current $i_{\phi 1}$ is sinusoidal with a frequency equal to that of the source. The current $i_{\phi 3}$ is a third-harmonic current with frequency that is three times the source frequency, and so on. Figure 3.7 illustrates this concept. Note that, in practice, the third harmonic of the exciting current is of a much smaller magnitude than the fundamental's $i_{\phi 1}$. By assuming that $i_{\phi 3}$ is negligible, we get around the difficulty of the distortion in i_ϕ.

Let us proceed with the development of the transformer model assuming that the exciting current $i_\phi(t)$ is sinusoidal as indicated by the first harmonic $i_{\phi 1}$. Comparing the $\phi_m(t)$ and $i_{\phi 1}(t)$ waveforms, we discover that there is a phase shift between the two waves. Note that if the λ–i or ϕ–i characteristic were linear, we would expect no phase shift between ϕ and i, and thus we have an inductance to model the exciting current process. The phasor diagram of Figure 3.8 shows that to account for the phase shift, we can resolve i_ϕ into two components $i_m(t)$ and $i_c(t)$. The current i_m is referred to as the magnetizing current and is in phase with the flux. The current $i_c(t)$ represents the component that goes to cover the core losses and is in phase with the voltage E_1. The exciting current phenomenon discussed here can be modeled by a parallel combination of a resistor whose conductance is G_c and an inductor whose susceptance (inverse of reactance) is B_m.

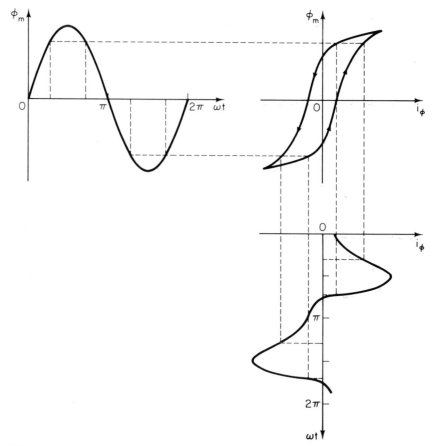

FIGURE 3.6 Developing the exciting current waveform from a sinusoidal flux waveform.

The voltage e_1 is transformed into e_2 through the action of an ideal transformer of turns ratio N_1/N_2, and the current in the secondary winding is i_2. Through an argument similar to that used with the primary winding, we can conclude that a leakage inductance L_2 appears on the secondary side as well as a resistance R_2 representing the ohmic voltage drop in the secondary winding. This concludes the process of modeling a realistic transformer and we can therefore reason that the equivalent circuit of Figure 3.9 is a realistic representation of a practical transformer under steady-state sinusoidal operating conditions.

The equivalent circuit of Figure 3.9(a) can be reduced to that of Figure 3.9(b) by simply referring the secondary side to the primary side using the following:

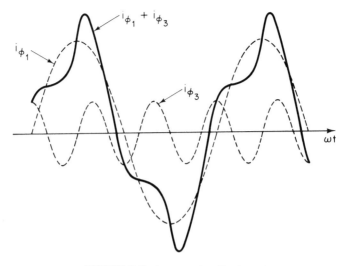

FIGURE 3.7 Interpreting Eq. (3.23).

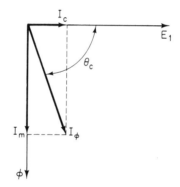

FIGURE 3.8 Phasor representation of exciting current.

$$V_2' = \frac{N_1}{N_2} V_2 \qquad (3.24)$$

$$I_2' = \frac{N_2}{N_1} I_2 \qquad (3.25)$$

$$R_2' = R_2 \left(\frac{N_1}{N_2}\right)^2 \qquad (3.26)$$

$$X_2' = X_2 \left(\frac{N_1}{N_2}\right)^2 \qquad (3.27)$$

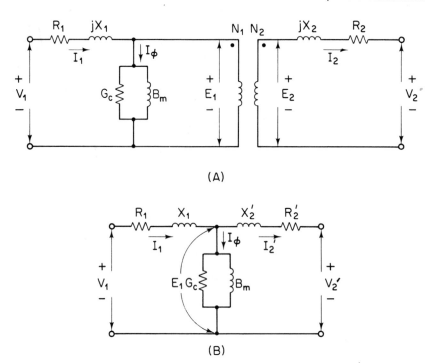

FIGURE 3.9 Equivalent circuits of transformer.

Although the equivalent circuit illustrated above is simply a T-network, it is customary to use approximate circuits such as shown in Figure 3.10. In the first two circuits, we move the shunt branch either to the secondary or primary sides to form inverted L circuits. Further simplifications are shown where the shunt branch is neglected in Figure 3.10(c) and finally with the resistances neglected in Figure 3.10(d). These last two circuits are of sufficient accuracy in most power system applications. In Figure 3.10 note that

$$R_{eq} = R_1 + R_2' \tag{3.28}$$

$$X_{eq} = X_1 + X_2' \tag{3.29}$$

A practical example will illustrate the principles and orders of approximations involved.

Example 3.2   A 100-kVA 400/2000-V single-phase transformer has the following parameters:

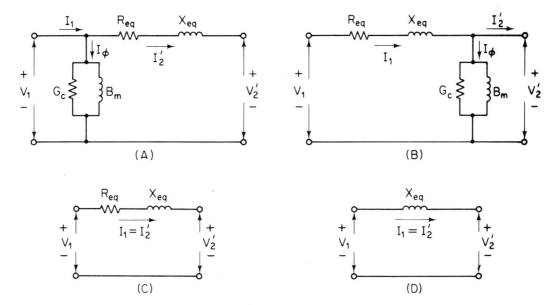

FIGURE 3.10 Approximate equivalent circuits for the transformer, all referred to primary side.

$$R_1 = 0.01 \ \Omega \quad R_2 = 0.25 \ \Omega$$

$$X_1 = 0.03 \ \Omega \quad X_2 = 0.75 \ \Omega$$

$$G_c = 2.2 \ \text{mS} \quad B_m = -6.7 \ \text{mS}$$

Note that G_c and B_m are given in terms of primary reference. The transformer supplies a load of 90 kVA at 2000 V and 0.8 PF lagging. Calculate the primary voltage and current using the equivalent circuits shown in Figure 3.10. Use the equivalent circuit of Figure 3.9(b) to establish the orders of approximation involved.

Solution

Let us refer all the data to the primary (400 V) side:

$$R_1 = 0.01 \ \Omega \qquad\qquad X_1 = 0.03 \ \Omega$$

$$R_2' = 0.25 \left(\frac{400}{2000}\right)^2 \quad X_2' = 0.75 \left(\frac{400}{2000}\right)^2$$

$$= 0.01 \ \Omega \qquad\qquad = 0.03 \ \Omega$$

Thus

$$R_{eq} = R_1 + R_2' \quad X_{eq} = X_1 + X_2'$$
$$= 0.02 \ \Omega \qquad\quad = 0.06 \ \Omega$$

The voltage $V_2 = 2000$ V; thus

$$V_2' = 2000\left(\frac{400}{2000}\right) = 400 \text{ V}$$

The current I_2' is thus

$$|I_2'| = \frac{90 \times 10^3}{400} = 225 \text{ A}$$

The power factor of 0.8 lagging implies that

$$I_2' = 225 \ \underline{/-36.87°} \text{ A}$$

For ease of computation, we start with the simplest circuit of Figure 3.10(d). Let us denote the primary voltage calculated through this circuit by V_{1d}. It is clear that

$$V_{1d} = V_2' + jI_2'(X_{eq})$$
$$= 400 \ \underline{/0} + j \ (0.06 \ \underline{/-36.87°}) \ (225)$$
$$= 408.243 \ \underline{/1.516°} \text{ V}$$
$$I_{1d} = 225 \ \underline{/-36.87°} \text{ A}$$

Comparing circuits (c) and (d) in Figure 3.10, we deduce that

$$V_{1c} = V_2' + I_2'(R_{eq} + jX_{eq}) = V_{1d} + I_2'(R_{eq})$$

Thus

$$V_{1c} = 408.243 \ \underline{/1.516°} + (225 \ \underline{/-36.87°})(0.02)$$
$$= 411.7797 \ \underline{/1.127°} \text{ V}$$
$$I_{1c} = I_2' = 225 \ \underline{/-36.87°} \text{ A}$$

Let us consider circuit (a) in Figure 3.10, where we see that

$$V_{1a} = V_{1c} = 411.7797 \; \underline{/1.127°}$$

But

$$
\begin{aligned}
I_{1a} &= I_2' + (G_c + jB_m)V_{1a} \\
&= 225 \; \underline{/-36.87°} + (411.7797 \; \underline{/1.127°}) \\
&\quad (2.2 - j6.7)10^{-3} \\
&= 227.418 \; \underline{/-37.277°} \; \text{A}
\end{aligned}
$$

Circuit (b) is a bit different since we start with V_2' impressed on the shunt branch. Thus

$$
\begin{aligned}
I_{1b} &= I_2' + (G_c + jB_m)V_2' \\
&= 225 \; \underline{/-36.87°} \\
&\quad + (2.2 - j6.7)10^{-3}(400) \\
&= 227.3177 \; \underline{/-37.277°}
\end{aligned}
$$

Now

$$
\begin{aligned}
V_{1b} &= V_2' + I_{1b}(R_{eq} + jX_{eq}) \\
&= 400 \; \underline{/0} + (227.3177 \; \underline{/-37.277°}) \\
&\quad \times (0.02 + j0.06)) \\
&= 411.958 \; \underline{/1.12652°}
\end{aligned}
$$

The exact equivalent circuit of Figure 3.9(b) is now considered. We first calculate E_1:

$$
\begin{aligned}
E_1 &= V_2' + I_2'(R_2' + jX_2') \\
&= 400 \; \underline{/0} + (225 \; \underline{/-36.87°})(0.01 + j0.03) \\
&= 405.87 \; \underline{/0.57174°}
\end{aligned}
$$

Next, we calculate I_1 as

$$
\begin{aligned}
I_1 &= I_2' + E_1(G_c + jB_m) \\
&= (225 \; \underline{/-36.87°}) + (405.87 \; \underline{/0.57174°})(2.2 - j6.7)10^{-3} \\
&= 227.37 \; \underline{/-37.277°}
\end{aligned}
$$

As a result, we obtain the primary voltage as

$$V_1 = E_1 + I_1(R_1 + jX_1)$$

$$= 405.87 \ \underline{/0.57174°} + (227.37 \ \underline{/-37.277°})(0.01 + j0.03)$$

$$= 411.87 \ \underline{/1.127°} \ V$$

The computed values of primary voltage and current from the exact equivalent circuit are used as a benchmark to evaluate the approximate circuits. We start by the simplest form of Figure 3.10(d). The error in voltage magnitude computation is 3.627 V or 0.88%, while the error in voltage phase angle is $-0.389°$ or -34%. In the current magnitude, we encounter a 1.04% error, while for the current phase angle, the error is -1.09%.

Including the resistance in the form of Figure 3.10(c) results in a considerable improvement in the error in voltage magnitude and phase angle. The errors in current are the same as those for the circuit (d). The circuit (b) has improved current errors and reasonably close voltage errors.

Depending on the purpose of the study, one of the simplified circuits can provide satisfactory results. Of course, one would expect circuit (c) to be a favorite in large-scale system applications.

3.4 TRANSFORMER PERFORMANCE MEASURES

The selection of a proper transformer for a given application involves evaluating certain important performance measures. Two important measures are the voltage regulation and efficiency of the device.

The voltage regulation is a measure of the variation in the secondary voltage when the load is varied from zero to rated value at a constant power factor. The percentage voltage regulation (PVR) is thus given by

$$PVR = 100 \frac{|V_{2(no \ load)}| - |V_{2rated}|}{|V_{2rated}|} \tag{3.30}$$

If we neglect the exciting current and refer the equivalent circuit to the secondary side, we have by inspection of Figure 3.11

$$PVR = 100 \frac{|V_1/a| - |V_2|}{|V_2|}$$

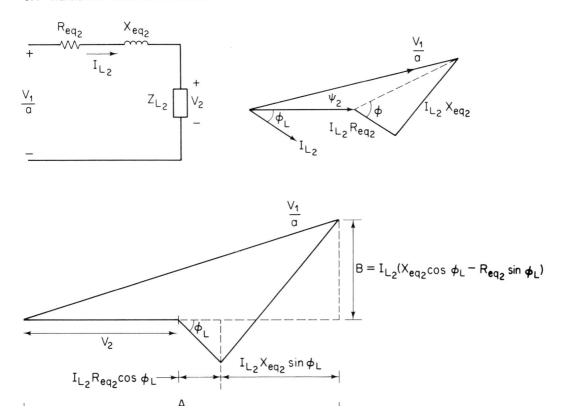

FIGURE 3.11 Transformer approximate equivalent circuit and associated phasor diagrams for voltage regulation derivation.

where a is the transformer ratio. Thus

$$a = \frac{N_1}{N_2}$$

From the phasor diagram of Figure 3.11, we have

$$\left| \frac{V_1}{a} \right| = \sqrt{A^2 + B^2} = A\left(1 + \frac{B^2}{A^2}\right)^{1/2}$$

We use the following approximation:

$$\frac{V_1}{a} = A\left(1 + \frac{B^2}{2A^2} + \cdots\right) \simeq A + \frac{B^2}{2A}$$

Hence the percentage voltage regulation is

$$\text{PVR} = \left(\frac{A - V_2 + B^2/2A}{V_2}\right) 100 \simeq \left(\frac{A - V_2 + B^2/2V_2}{V_2}\right) 100$$

In terms of transformer constants, we get

$$\text{PVR} \simeq 100 \left\{ \frac{I_{L2}(R_{eq2} \cos \phi_L + X_{eq2} \sin \phi_L)}{V_2} \\ + \frac{1}{2}\left[\frac{I_{L2}(X_{eq2} \cos \phi_L - R_{eq2} \sin \phi_L)}{V_2}\right]^2 \right\} \quad (3.31)$$

The efficiency of the transformer is the ratio of output (secondary) power to the input (primary) power. Formally, the efficiency is η:

$$\eta = \frac{P_2}{P_1} \quad (3.32)$$

If we deal with the transformer as referred to the secondary side, we have

$$P_2 = |V_2| \, |I_L| \cos \phi_L \quad (3.33)$$

where I_L is the load current. The input power P_1 is the sum of the output power and power loss in the transformer. Thus

$$P_1 = P_2 + P_l \quad (3.34)$$

The power loss in the transformer is made of two parts: the I^2R loss and the core loss P_c. Thus

$$P_l = P_c + |I_L|^2(R_{eq}) \quad (3.35)$$

As a result, the efficiency is given by

$$\eta = \frac{|V_2| \, |I_L| \cos \phi_L}{|V_2| \, |I_L| \cos \phi_L + P_c + |I_L|^2(R_{eq})} \quad (3.36)$$

The efficiency of a transformer varies with the load current I_L. It attains a maximum when

$$\frac{\partial \eta}{\partial I_L} = 0 \quad (3.37)$$

Using Eq. (3.32), the derivative is

$$\frac{\partial \eta}{\partial I_L} = \frac{P_1(\partial P_2/\partial |I_L|) - P_2 (\partial P_1/\partial |I_L|)}{P_1^2} \tag{3.38}$$

Thus the condition for maximum power is

$$\frac{P_1}{P_2} = \frac{\partial P_1/\partial |I_L|}{\partial P_2/\partial |I_L|} \tag{3.39}$$

Using Eqs. (3.33) to (3.35), we get

$$\frac{P_1}{P_2} = \frac{|V_2|\cos \phi_L + 2|I_L|R_{eq}}{|V_2| \cos \phi_L} \tag{3.40}$$

This reduces to

$$P_1 = P_2 + 2|I_L|^2 R_{eq} \tag{3.41}$$

Thus, for maximum efficiency, we have

$$P_1 = P_c + |I_L|^2 R_{eq} \tag{3.42}$$

As a result, the maximum efficiency occurs for

$$P_c = |I_L|^2 R_{eq} \tag{3.43}$$

That is, when the I^2R losses equal the core losses, maximum efficiency is attained.

The following example utilizes results of Example 3.2 to illustrate the computations involved.

Example 3.3   Find the percentage voltage regulation and efficiency of the transformer of Example 3.2.

Solution

Let us refer quantities to the secondary side:

$$V_2 = 2000 \text{ V}$$

$$I_{L_2} = \frac{90 \times 10^3}{2000} = 45 \text{ A}$$

$$R_{eq2} = 0.02(5)^2 = 0.5 \ \Omega$$

$$X_{eq2} = 0.06(5)^2 = 1.5 \ \Omega$$

Thus, substituting in Eq. (3.31), we get

$$PVR = 100 \left\{ \frac{45[0.5(0.8) + 1.5(0.6)]}{2000} + \frac{1}{2} \left[\frac{45[1.5(0.8) - 0.5(0.6)]}{2000} \right]^2 \right\}$$

$$= 2.9455\%$$

Let us compare this with the results of applying Eq. (3.30) with no approximations but for those of circuit (c). Here we have

$$V_1 = 411.7797$$

$$V_2' = 400 \ V$$

Referred to the secondary, we have

$$V_1' = 411.7797(5)$$

On no-load, the current is zero and hence

$$V_{2(no \ load)} = V_1'$$

The rated secondary voltage is

$$V_{2rated} = 2000$$

As a result,

$$PVR = \frac{411.7797 - 400}{400}$$

$$= 2.9449\%$$

To calculate the efficiency, we apply the basic definition, for which we need P_1 and P_2. We use the exact circuit results, for which

$$P_2 = V_2 I_2 \cos \phi_2$$

$$= 400(225)(0.8) = 72,000 \ W$$

$$P_1 = V_1 I_1 \cos \phi_1$$

The angle ϕ_1 is that between the voltage V_1 and current I_1 and is given by

$$\phi_1 = 1.127 - (-37.277)$$
$$= 38.404$$

We also have

$$V_1 = 411.87 \text{ V}$$
$$I_1 = 227.37 \text{ A}$$

As a result,

$$P_1 = 411.87(227.37) \cos 38.404$$
$$= 73{,}386 \text{ W}$$

The efficiency is now found as

$$\eta = \frac{72{,}000}{73{,}386} = 0.9811$$

Example 3.4   Find the maximum efficiency of the transformer of Example 3.2 under the same power factor and voltage conditions.

Solution

We need first the core losses. These are obtained from the exact equivalent circuit as

$$P_c = |E_1|^2(G_c)$$
$$= (405.87)^2 (2.2 \times 10^{-3})$$
$$= 362.41 \text{ W}$$

For maximum efficiency,

$$P_c = I_L^2(R_{eq})$$

Referred to the primary, we thus have

$$362.41 = I_L^2(0.02)$$

Thus for maximum efficiency, we get

$$I_L = 134.61 \text{ A}$$

As a result,

$$\eta_{max} = \frac{V_2'|I_L| \cos \phi_L}{V_2'|I_L| \cos \phi_L + 2P_c}$$

$$= \frac{400(134.61)(0.8)}{400(134.61)(0.8) + 2(362.41)}$$

$$= 0.98345$$

This concludes this example.

3.5 EQUIVALENT-CIRCUIT PARAMETER EVALUATION

The models of the transformer discussed in Section 3.3 include six parameters: R_1 and X_1 representing the primary resistance and leakage reactance, R_2 and X_2 representing the secondary resistance and leakage reactance, and G_c and B_m representing the core loss conductance and magnetizing current susceptance, respectively. It is logical to assume that the turns ratio N_1/N_2 is available in our present discussion. It is reasonable to expect that in most instances, the values of R_1 and R_2' are equal and also that the values of X_1 and X_2' are equal. We can prove the first statement by noting that

$$R_1 = \frac{\rho l_1}{A_1}$$

$$R_2 = \frac{\rho l_2}{A_2}$$

where ρ is the resistivity of the winding material, l_1 and A_1 are the length and cross-sectional area of the primary, and l_2 and A_2 are the length and cross-sectional area of the secondary. Thus

$$\frac{R_1}{R_2} = \frac{l_1 A_2}{l_2 A_1}$$

We assume that the winding length is proportional to the number of turns and thus

$$\frac{l_1}{l_2} = \frac{N_1}{N_2}$$

The cross-sectional area is proportional to the current carried:

$$\frac{A_2}{A_1} = \frac{I_2}{I_1} = \frac{N_1}{N_2}$$

Combining the relations stated above leads us to conclude that

$$R_1 = R_2 \left(\frac{N_1}{N_2}\right)^2$$

As a result,

$$R_1 = R_2'$$

Using (3.28), and (3.29), we conclude that

$$R_{eq} = 2R_1 = 2R_2' \qquad (3.44)$$

$$X_{eq} = 2X_1 = 2X_2' \qquad (3.45)$$

It is clear that we need to evaluate only four parameters under the foregoing assumptions. These are R_{eq}, X_{eq}, G_c, and B_m.

From a purely theoretical point of view, all that we need to evaluate the four parameters is a set of four equations or more based on the results of operating data combined with the equivalent-circuit model of our choice. It is customary, however, to employ a wonderfully simple procedure based on practical approximations and the results of two simple tests, called the open-circuit and the short-circuit tests.

Open-Circuit Test

A voltage equal to the rated primary voltage at rated frequency is applied to the primary winding with the secondary winding open (no load connected). The open-circuit voltage V_{oc}, current I_{oc}, and active power at the primary P_{oc} are measured. To obtain values of G_c, and B_m, in terms of the measured values, we use the approximate circuit of Figure 3.10(a) with $I_2' = 0$ to obtain

$$P_{oc} = V_{oc}^2 G_c \qquad (3.46)$$

$$Y_{oc} = \frac{I_{oc}}{V_{oc}} = (G_c^2 + B_m^2)^{1/2} \qquad (3.47)$$

Equation (3.46) provides us immediately with G_c, which is then used in Eq. (3.47) to obtain B_m.

Short-Circuit Test

A voltage V_{sc}, which is a fraction of rated voltage, is applied to the primary to obtain rated current in the primary while the secondary winding is short-circuited. The values of I_{sc} and P_{sc} are measured along with V_{sc}. The approximate circuit of Figure 3.10(b) is now used with $V_2' = 0$ to obtain

$$P_{sc} = I_{sc}^2 R_{eq} \qquad (3.48)$$

$$Z_{sc} = \frac{V_{sc}}{I_{sc}} = (R_{eq}^2 + X_{eq}^2)^{1/2} \qquad (3.49)$$

Clearly, Eq. (3.48) provides us with R_{eq} and hence X_{eq} is obtained using Eq. (3.49).

It is about time to take in an example to illustrate the procedure.

Example 3.5 A single-phase 2300/230-V 500-kVA distribution transformer is tested to find the equivalent circuit parameters. The following information is available:

Open-circuit test: $V_{oc} = 2300$ V, $I_{oc} = 9.4$ A, $P_{oc} = 2250$ W.
Short-circuit test: $V_{sc} = 94.5$ V, $I_{sc} = I_{rated}$, $P_{sc} = 8.22 \times 10^3$ W.

Find the parameters of the equivalent circuit.

Solution

Substituting in Eq. (3.46), we get

$$2250 = (2300)^2 G_c$$

Thus

$$G_c = 4.253 \times 10^{-4} \text{ S}$$

Using Eq. (3.47), we have

$$\left(\frac{9.4}{2300} \right)^2 = (4.253 \times 10^{-4})^2 + B_m^2$$

Thus

$$B_m = 4.0648 \times 10^{-3} \text{ S}$$

The short-circuit test is conducted with current equal to rated value

$$I_{sc} = \frac{500 \times 10^3}{2300} = 217.39 \text{ A}$$

Using Eq. (3.48), we get

$$R_{eq} = \frac{8.22 \times 10^3}{(217.39)^2} = 0.1739 \ \Omega$$

Using Eq. (3.49), we now obtain

$$\left(\frac{94.5}{217.39}\right)^2 = (0.1739)^2 + X_{eq}^2$$

As a result, we get

$$X_{eq} = 0.3984 \ \Omega$$

3.6   PER UNIT SYSTEMS

The per unit (p.u.) value representation of electrical variables in power system problems is favored by the electric power systems engineer. The numerical per unit value of any quantity is its ratio to a chosen base quantity of the same dimension. Thus a per unit quantity is a normalized quantity with respect to the chosen base value. The per unit value of a quantity is thus defined as

$$\text{p.u. value} = \frac{\text{actual value}}{\text{reference or base value of the same dimension}} \quad (3.50)$$

In an electrical network, five quantities are usually involved in the calculations. These are the current I, and the voltage V, the complex power S, the impedance Z, and the phase angles. The angles are dimensionless; the other four quantities are completely described by knowledge of only two of them. It is thus clear that an arbitrary choice of two base quantities will fix the other base quantities. Let $|I_b|$ and $|V_b|$ represent the base current and base voltage expressed in amperes and volts, respectively. The product of the two gives the base complex power in volt-amperes (VA):

$$|S_b| = |V_b| |I_b| \quad \text{VA} \quad (3.51)$$

The base impedance will also be given by

$$|Z_b| = \frac{|V_b|}{|I_b|} = \frac{|V_b|^2}{|S_b|} \quad \Omega \tag{3.52}$$

The base admittance will naturally be the inverse of the base impedance. Thus

$$|Y_b| = \frac{1}{|Z_b|} = \left|\frac{I_b}{V_b}\right|$$

$$= \frac{|S_b|}{|V_b|^2} \quad \text{siemens} \tag{3.53}$$

The nominal voltage of lines and equipment is almost always known as well as the apparent (complex) power in volt-amperes, so these two quantities are usually chosen for base-value calculation. The same volt-ampere base is used in all parts of a given system. One base voltage is chosen; all other base voltages must then be related to the one chosen by the turns ratio of the connecting transformers.

From the definition of per unit impedance, we can express the ohmic impedance Z_Ω in the per unit value $Z_{\text{p.u.}}$ as

$$Z_{\text{p.u.}} \overset{\Delta}{=} \frac{Z_\Omega}{|Z_b|}$$

$$= \frac{Z_\Omega |I_b|}{|V_b|} \tag{3.54}$$

Thus

$$Z_{\text{p.u.}} = \frac{Z_\Omega |S_b|}{|V_b|^2} \quad \text{p.u.} \tag{3.55}$$

As for admittances, we have

$$Y_{\text{p.u.}} \overset{\Delta}{=} \frac{1}{Z_{\text{p.u.}}} = \frac{|V_b|^2}{Z_\Omega |S_b|} = Y_s \frac{|V_b|^2}{|S_b|} \quad \text{p.u.} \tag{3.56}$$

It is interesting to note that $Z_{\text{p.u.}}$ can be interpreted as the ratio of the voltage drop across Z with base current injected to the base voltage. This can be verified by inspection of the expression

$$Z_{\text{p.u.}} = \frac{Z_\Omega I_b}{V_b} \tag{3.57}$$

An example will illustrate the procedure.

Example 3.6   Consider the transformer of Example 3.5. Find the value of the parameters obtained in per unit terms based on

$$V_b = 2300 \text{ V}$$

$$S_b = 500 \times 10^3 \text{ VA}$$

Solution

We first obtain

$$Z_b = \frac{(2300)^2}{500 \times 10^3} = 10.58 \ \Omega$$

As a result,

$$R_{\text{eqp.u.}} = \frac{0.1739}{10.58} = 16.44 \times 10^{-3} \text{ p.u.}$$

$$X_{\text{eqp.u.}} = \frac{0.3984}{10.58} = 37.437 \times 10^{-3} \text{ p.u.}$$

The base admittance is

$$Y_b = \frac{1}{Z_b} = 94.518 \times 10^{-3}$$

As a result,

$$G_{c\text{p.u.}} = \frac{4.253 \times 10^{-4}}{94.518 \times 10^{-3}} = 4.4997 \times 10^{-3} \text{ p.u.}$$

$$B_{m\text{p.u.}} = \frac{4.0648 \times 10^{-3}}{94.518 \times 10^{-3}} = 43.006 \times 10^{-3} \text{ p.u.}$$

Given an impedance in per unit on a given base S_{b_o} and V_{b_o}, it is sometimes required to obtain the per unit value referred to a new base set S_{b_n} and V_{b_n}. The conversion expression is obtained as follows:

$$Z_\Omega = Z_{\text{p.u.o}}\left(\frac{|V_{b_o}|^2}{|S_{b_o}|}\right) \tag{3.58}$$

Also, the same impedance Z in ohms is given referred to the new base by

$$Z_\Omega = Z_{\text{p.u.}_n}\left(\frac{|V_{b_n}|^2}{|S_{b_n}|}\right)$$

Thus, equating the two expressions above, we get

$$Z_{\text{p.u.}_n} = Z_{\text{p.u.}_o}\frac{|S_{b_n}|}{|S_{b_o}|}\frac{|V_{b_o}|^2}{|V_{b_n}|^2} \tag{3.60}$$

which is our required conversion formula. The admittance case simply follows the inverse rule. Thus

$$Y_{p.u._n} = Y_{\text{p.u.}_o}\frac{|S_{b_o}|}{|S_{b_n}|}\frac{|V_{b_n}|^2}{|V_{b_o}|^2} \tag{3.61}$$

3.7 SINGLE-PHASE CONNECTIONS

Single-phase transformers can be connected in a variety of ways. To start with, consider two single-phase transformers A and B. They can be connected in four different combinations provided that the polarities are observed.

Series–Series Connections

The primaries of the two transformers are connected in series, whereas the secondaries are connnected in series. The connection is shown in Figure 3.12. The voltage V_1 is the sum of the voltages V_{1_A} and V_{1_B}, and similarly, V_2 is the sum of the secondary voltages V_{2A} and V_{2B}. The primary current of the combination passes through the primaries of transformers A and B and similarly, the secondary current of the combination passes through the secondaries of both transformers. It is clear that the allowed currents should not exceed the rated current of either transformer. This results in a derating of the combination as shown in the following example.

Example 3.7   Assume that transformer A is rated at 1000 VA and transformer B is rated at 600 VA. Both transformers have a voltage ratio of 240/120. It is clear that the rated currents on the primary side are

Transformer A

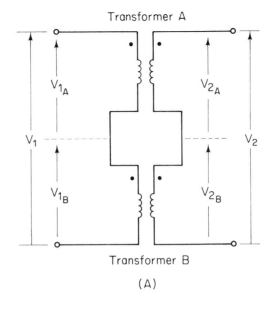

(A)

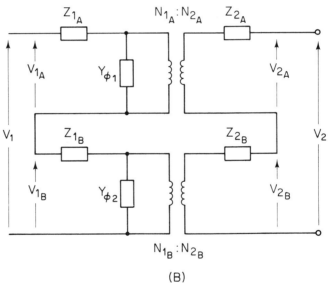

(B)

FIGURE 3.12   Connection diagram for two transformers with primaries connected in series and secondary connected in series.

$$I_{1A_r} = \frac{1000}{240} = 4.167 \text{ A}$$

$$I_{1B_r} = \frac{600}{240} = 2.5 \text{ A}$$

The allowable current for the primaries connected in series is the lowest of the two rated currents,

$$I_1 = 2.5 \text{ A}$$

On the secondary side, we should have

$$I_2 = 5 \text{ A}$$

The series combination has a voltage ratio of 480/240 V. The volt-ampere rating is now given by

$$\text{VA} = 480 \times 2.5 = 1200 \text{ VA}$$

Series–Parallel Connections

The primaries of transformers A and B are connected in series, whereas their secondaries are connected in parallel, as shown in Figure 3.13(a).

Parallel–Series Connections

The primaries of transformers A and B are connected in parallel, whereas their secondaries are connected in series, as shown in Figure 3.13(b). Note that when windings are connected in parallel, those having the same voltage and polarity are paralleled. When connected in series, windings of opposite polarity are joined in one junction. Coils of unequal voltage ratings may be series-connected either aiding or opposing.

Parallel Connections

Figure 3.14 shows two transformers A and B connected in parallel with their approximate equivalent circuit indicated as well. Assume that Z_A is the ohmic equivalent impedance of transformer A referred to its secondary side. Similarly, Z_B is the ohmic equivalent impedance of transformer B referred to its secondary side. Z_L is the load impedance. V_A is the primary voltage on both transformer primaries. The turns ratio for each of the transformers should be identical for the parallel combination to make sense

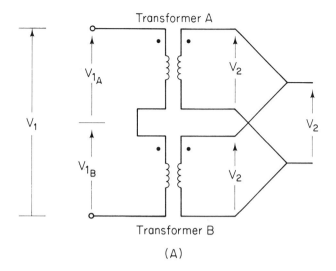

Transformer A

(A)

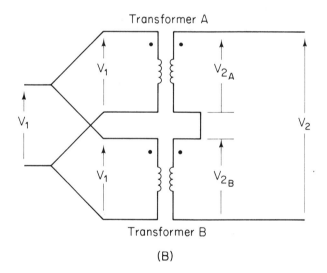

Transformer A

(B)

FIGURE 3.13   Series–parallel and parallel–series connections for single-phase transformers.

$$a = \frac{N_{1A}}{N_{2A}} = \frac{N_{1B}}{N_{2B}}$$

The current delivered by transformer A is I_A. Thus

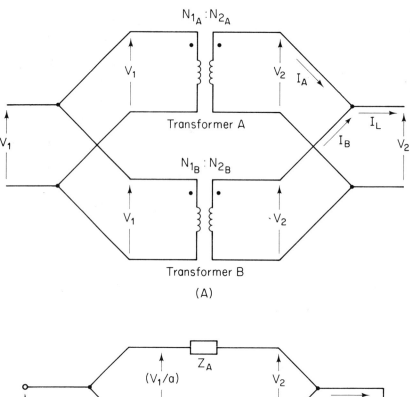

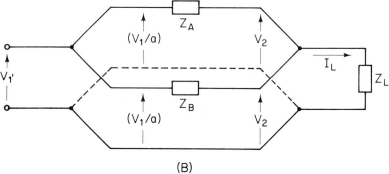

FIGURE 3.14 Parallel-connected single-phase transformers: (a) connection diagram; (b) equivalent circuit.

$$I_A = \frac{\Delta V}{Z_A} \tag{3.62}$$

Similarly,

$$I_B = \frac{\Delta V}{Z_B} \tag{3.63}$$

where

$$\Delta V = \frac{V_1}{a} - V_2 \qquad (3.64)$$

The load current is

$$I_L = I_A + I_B \qquad (3.65)$$

Thus using Eqs. (3.62) to (3.65), we get

$$I_L = \frac{\Delta V}{Z_{eq}} \qquad (3.66)$$

where Z_{eq} is the parallel combination of Z_A and Z_B given by

$$\frac{1}{Z_{eq}} = \frac{1}{Z_A} + \frac{1}{Z_B} \qquad (3.67)$$

Now

$$\Delta V = \frac{V_1}{a} - I_L Z_L \qquad (3.68)$$

Thus using Eq. (3.68) in Eq. (3.66), we conclude that

$$I_L = \frac{V_1/a}{Z_L + Z_{eq}} \qquad (3.69)$$

This result is expected, as Z_A and Z_B in parallel are in series with Z_L across the voltage (V_1/a). The current division between the two transformers is obtained from

$$I_A = \frac{I_L Z_{eq}}{Z_A} = \frac{\Delta V}{Z_A} \qquad (3.70)$$

$$I_B = \frac{I_L Z_{eq}}{Z_B} = \frac{\Delta V}{Z_B} \qquad (3.71)$$

3.8 THREE-WINDING TRANSFORMERS

The three-winding transformer is used in many parts of the power system for the economy achieved when using three windings on the one core. Figure 3.15 shows a three-winding transformer with a practical equivalent cir-

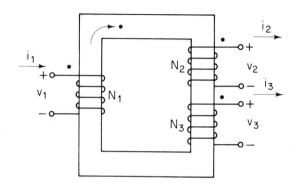

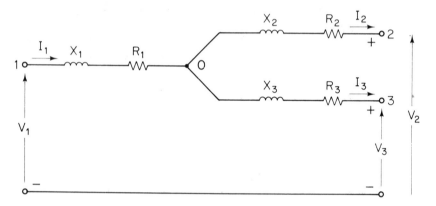

FIGURE 3.15   Three-winding transformer and its practical equivalent circuit.

cuit. The impedances Z_1, Z_2, and Z_3 are calculated from the three impedances obtained by considering each pair of windings separately with

$$Z_1 = \frac{Z_{12} + Z_{13} - Z_{23}}{2} \tag{3.72}$$

$$Z_2 = \frac{Z_{12} + Z_{23} - Z_{13}}{2} \tag{3.73}$$

$$Z_3 = \frac{Z_{13} + Z_{23} - Z_{12}}{2} \tag{3.74}$$

The I^2R loss for a three-winding transformer can be obtained from analysis
of the equivalent circuit shown.

Example 3.8   Consider a three-winding transformer with the particulars shown in the
equivalent circuit referred to the primary side given in Figure 3.16. Assuming that V_1 is the reference, calculate the following:

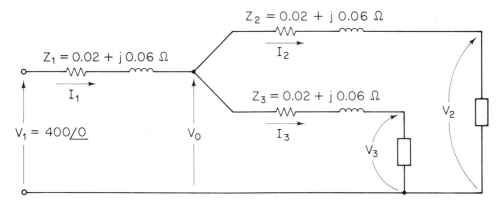

FIGURE 3.16 Circuit for Example 3.8.

(a) The secondary and tertiary voltages referred to the primary side.

(b) The apparent powers and power factors at the primary, secondary, and tertiary terminals.

(c) The transformer efficiency.

Assume that

$$I_2 = 50 \underline{/-30°}$$

$$I_3 = 50 \underline{/-35°}$$

Solution

(a) The primary current is

$$I_1 = I_2 + I_3$$

$$= 99.9048 \underline{/-32.5°}$$

Now the voltage at point 0 is

$$V_0 = V_1 - I_1 Z_1$$

$$= 400 - (99.9048 \underline{/-32.5°})(0.02 + j0.06)$$

$$= 395.114 \underline{/-0.577°} \text{ V}$$

The secondary voltage is obtained referred to the primary as

$$V_2 = V_0 - I_2 Z_2$$

$$= 395.114 \underline{/-0.577°} - (50 \underline{/-30°})(0.02 + j0.06)$$

$$= 392.775 \underline{/-0.887°} \text{ V}$$

The tertiary voltage is obtained referred to the primary as

$$V_3 = V_0 - I_3 Z_3$$
$$= 395.114 \underline{/-0.577°} - (50 \underline{/-35°})(0.02 + j0.06)$$
$$= 392.598 \underline{/-0.856°} \text{ V}$$

(b) The apparent power into the load connected to the secondary winding is thus

$$S_2 = V_2 I_2^*$$
$$= 19{,}638.75 \underline{/29.113°} \text{ VA}$$

As a result,

$$PF_2 = \cos(29.113°) = 0.87366$$

Similarly, for the tertiary winding, we get

$$S_3 = V_3 I_3^*$$
$$= 19{,}629.9 \underline{/34.144°} \text{ VA}$$

As a result,

$$PF_3 = \cos(34.144°) = 0.8276$$

The apparent power at the primary side is

$$S_1 = V_1 I_1^*$$
$$= 39{,}961.9 \underline{/32.5°} \text{ VA}$$

As a result,

$$PF_1 = \cos(32.5°) = 0.84339$$

(c) The active powers are

$$P_2 = 19{,}638.75 \cos 29.113°$$
$$= 17{,}157.6 \text{ W}$$
$$P_3 = 19{,}629.9 \cos 34.144°$$

$$= 16{,}246.28 \text{ W}$$

$$P_1 = 39{,}961.92 \cos 32.5°$$

$$= 33{,}703.54 \text{ W}$$

The efficiency is therefore

$$\eta = \frac{P_2 + R_3}{P_1}$$

$$= 0.991$$

3.9 THREE-PHASE SYSTEMS AND TRANSFORMER CONNECTIONS

The major portion of all the electric power presently used is generated, transmitted, and distributed using balanced three-phase voltage systems. A balanced three-phase voltage system is composed of three single-phase voltages having the same magnitude and frequency but time-displaced from one another by 120°. Figure 3.17(a) shows a schematic representation where the three single-phase voltage sources appear in a Y connection; a Δ configuration is also possible. A phasor diagram showing each of the phase voltages is also given in Figure 3.17(b). As the phasors revolve at the angular frequency with respect to the reference line in the counterclockwise (positive) direction, the positive maximum value first occurs for phase a and then in succession for phases b and c. Stated in a different way, to an observer in the phasor space, the voltage of phase a arrives first followed by that of b and then that of c. For this reason, the three-phase voltage of Figure 3.17 is

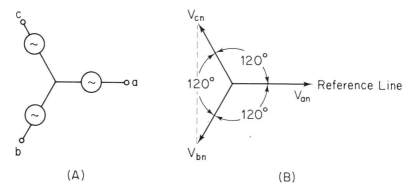

(A)   (B)

FIGURE 3.17 Y-connected three-phase system and the corresponding phasor diagram.

said to have the phase sequence *abc* (*order* or *phase sequence* or *rotation* are all synonymous terms). This is important for certain applications. For example, in three-phase induction motors, the phase sequence determines whether the motor turns clockwise or counterclockwise.

Balanced three-phase systems can be studied using techniques developed for single-phase circuits. The arrangement of the three single-phase voltages into a Y or a Δ configuration requires some modification in dealing with the overall system.

Y Connection

With reference to Figure 3.17, the common terminal *n* is called the neutral or star (Y) point. The voltages appearing between any two of the line terminals *a*, *b*, and *c* have different relationships in magnitude and phase to the voltages appearing between any one line terminal and the neutral point *n*. The set of voltages V_{ab}, V_{bc}, and V_{ca} are called the *line voltages*, and the set of voltages V_{an}, V_{bn}, and V_{cn} are referred to as the *phase voltages*. Analysis of phasor diagrams provides the required relationships.

The effective values of the phase voltages are shown in Figure 3.18 as V_{an}, V_{bn}, and V_{cn}. Each has the same magnitude, and each is displaced 120° from the other two phasors. To obtain the magnitude and phase angle of the line voltage from *a* to *b* (i.e., V_{ab}), we apply Kirchhoff's voltage law:

$$V_{ab} = V_{an} + V_{nb} \tag{3.75}$$

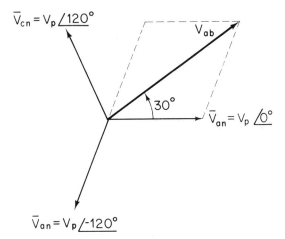

FIGURE 3.18 Illustrating the phase and magnitude relations between the phase and line voltage of a Y connection.

This equation states that the voltage existing from a to b is equal to the voltage from a to n (i.e., V_{an}) plus the voltage from n to b. Thus Eq. (3.75) can be rewritten as

$$V_{ab} = V_{an} - V_{bn} \tag{3.76}$$

Since for a balanced system, each phase voltage has the same magnitude, let us set

$$|V_{an}| = |V_{bn}| = |V_{cn}| = V_p \tag{3.77}$$

where V_p denotes the effective magnitude of the phase voltage. Accordingly, we may write

$$V_{an} = V_p \underline{/0} \tag{3.78}$$

$$V_{bn} = V_p \underline{/-120°} \tag{3.79}$$

$$V_{cn} = V_p \underline{/-240°} = V_p \underline{/120°} \tag{3.80}$$

Substituting Eqs. (3.78) and (3.79) in Eq. (3.76) yields

$$V_{ab} = V_p(1 - 1 \underline{/-120°})$$

or $\hspace{10cm}$ (3.81)

$$V_{ab} = \sqrt{3}\, V_p \underline{/30°}$$

Similarly, we obtain

$$V_{bc} = \sqrt{3}\, V_p \underline{/-90°} \tag{3.82}$$

$$V_{ca} = \sqrt{3}\, V_p \underline{/150°} \tag{3.83}$$

The expressions obtained above for the line voltages show that they constitute a balanced three-phase voltage system whose magnitudes are $\sqrt{3}$ times the phase voltages. Thus we write

$$V_L = \sqrt{3}\, V_p \tag{3.84}$$

A current flowing out of a line terminal a (or b or c) is the same as that flowing through the phase source voltage appearing between terminals n and a (or n and b, or n and c). We can thus conclude that for a Y-connected three-phase source, the line current equals the phase current. Thus

$$I_L = I_p \tag{3.85}$$

where I_L denotes the effective value of the line current and I_p denotes the effective value for the phase current.

Δ Connections

We consider now the case when the three single-phase source is rearranged to form a three-phase Δ connection, as shown in Figure 3.19. It is clear from inspection of the circuit shown that the line and phase voltages have the same magnitude:

$$|V_L| = |V_p| \tag{3.86}$$

The phase and line currents, however, are not identical, and the relationship between them can be obtained using Kirchhoff's current law at one of the line terminals.

In a manner similar to that adopted for the Y-connected source, let us consider the phasor diagram shown in Figure 3.20. Assume the phase currents to be

$$I_{ab} = I_p \underline{/0} \tag{3.87}$$

$$I_{bc} = I_p \underline{/-120°} \tag{3.88}$$

$$I_{ca} = I_p \underline{/120°} \tag{3.89}$$

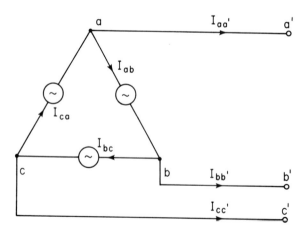

FIGURE 3.19 Δ-connected three-phase source.

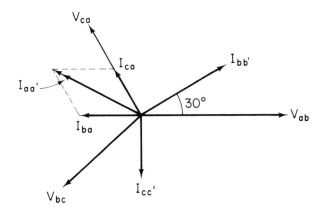

FIGURE 3.20 Illustrating relation between phase and line currents in Δ connection.

The current that flows in the line joining a to a' is denoted $I_{aa'}$ and is given by

$$I_{aa'} = I_{ca} - I_{ab}$$

As a result, we have

$$I_{aa'} = I_p(1\ \underline{/120°} - 1\ \underline{/0})$$

which simplifies to

$$I_{aa'} = \sqrt{3}\ I_p\ \underline{/150°} \tag{3.90}$$

Similarly,

$$I_{bb'} = \sqrt{3}\ I_p\ \underline{/30°} \tag{3.91}$$

$$I_{cc'} = \sqrt{3}\ I_p\ \underline{/-90°} \tag{3.92}$$

Note that a set of balanced three-phase currents yields a corresponding set of balanced line currents that are $\sqrt{3}$ times the phase values:

$$I_L = \sqrt{3}\ I_p \tag{3.93}$$

where I_L denotes the magnitude of any of the three line currents.

Power Relationships

Assume that the three-phase generator is supplying a balanced load with three sinusoidal phase voltages:

$$v_a(t) = \sqrt{2}\, V_p \sin \omega t$$

$$v_b(t) = \sqrt{2}\, V_p \sin (\omega t - 120°)$$

$$v_c(t) = \sqrt{2}\, V_p \sin (\omega t + 120°)$$

with the currents given by

$$i_a(t) = \sqrt{2}\, I_p \sin (\omega t - \phi)$$

$$i_b(t) = \sqrt{2}\, I_p \sin (\omega t - 120° - \phi)$$

$$i_c(t) = \sqrt{2}\, I_p \sin (\omega t + 120° - \phi)$$

where ϕ is the phase angle between the current and voltage in each phase. The total power in the load is

$$P_{3\phi}(t) = v_a(t)i_a(t) + v_b(t)i_b(t) + v_c(t)i_c(t)$$

This reduces to

$$P_{3\phi}(t) = 3V_p I_p \cos \phi$$

When referring to the voltage level of a three-phase system, one invariably understands the line voltages. From the discussion above, the relationship between the line and phase voltages in a Y-connected system is

$$|V_L| = \sqrt{3}\, |V| \tag{3.94}$$

The power equation thus reads in terms of line quantities:

$$P_{3\phi} = \sqrt{3}\, |V_L|\, |I_L| \cos \phi \tag{3.95}$$

We note that the total instantaneous power is constant, having a magnitude of three times the real power per phase. We may be tempted to assume that the reactive power is of no importance in a three-phase system since the Q terms cancel out. However, this situation is analogous to the summation of balanced three-phase currents and voltages that also cancel out. Although the sum cancels out, these quantities are still very much in evidence in each phase. We thus extend the concept of complex or apparent power (S) to three-phase systems by defining

$$S_{3\phi} = 3V_p I_p^{\star} \tag{3.96}$$

where the active power and reactive power are obtained from

$$S_{3\phi} = P_{3\phi} + jQ_{3\phi} \tag{3.97}$$

as

$$P_{3\phi} = 3 \, |V_p| \, |I_p| \cos \phi \tag{3.98}$$

$$Q_{3\phi} = 3 \, |V_p| \, |I_p| \sin \phi \tag{3.99}$$

In terms of line values, we can assert that

$$S_{3\phi} = \sqrt{3} \, V_L \, I_L^{\star} \tag{3.100}$$

and

$$P_{3\phi} = \sqrt{3}|V_L| \, |I_L| \cos \phi \tag{3.101}$$

$$Q_{3\phi} = \sqrt{3}|V_L| \, |I_L| \sin \phi \tag{3.102}$$

In specifying rated values for power system apparatus and equipment such as generators, transformers, circuit breakers, and so on, we use the magnitude of the apparent power $S_{3\phi}$ as well as line voltage for specification values. In specifying three-phase motor loads, we use the horsepower output rating and voltage.

Three-Phase Transformer Connections

For three-phase system applications, it is possible to install three-phase transformer units or banks made of three single-phase transformers connected in the desired three-phase configurations. The latter arrangement is advantageous from a reliability standpoint, since it is then possible to install a single standby single-phase transformer instead of a three-phase unit. This provides a considerable cost saving. We have seen that there are two possible three-phase connections; the Y connection and the Δ connection. We thus see that three-phase transformers can be connected in four different ways. In the Y/Y connection, both primary and secondary windings are connected in Y. In addition, we have Δ/Δ, Y/Δ, or Δ/Y connections. The Y-connected windings may or may not be grounded.

The Y/Δ configuration is used for stepping down from a high voltage to a medium or low voltage. This provides a grounding neutral on the high-voltage side. Conversely, the Δ/Y configuration is used in stepping up to a

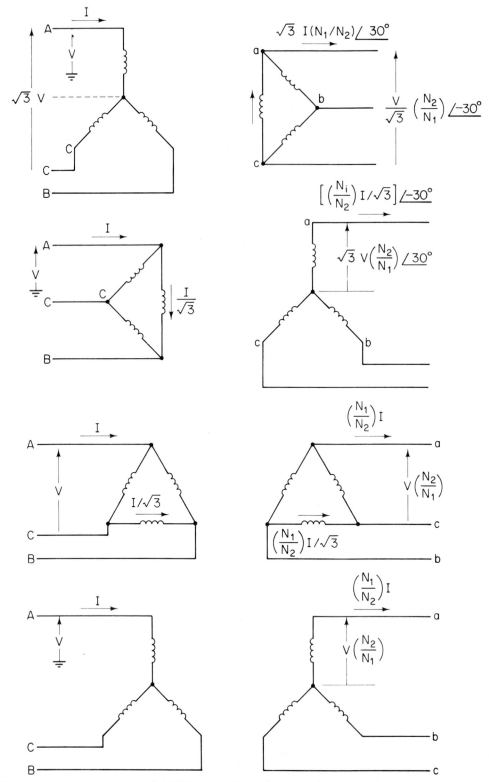

FIGURE 3.21 Three-phase transformer connections.

high voltage. The Δ/Δ connection enables one to remove one transformer for maintenance while the other two continue to function as a three-phase bank (with reduced rating) in an open-delta or V connection. The difficulties arising from the harmonic contents of the exciting current associated with the Y/Y connection make it seldom used.

In Figure 3.21, the four common three-phase transformer connections are shown along with the voltage and current relations associated with the transformation. It is important to realize that the line-to-ground voltages on the Δ side lead the corresponding Y-side values by 30° and that the line currents on the Δ side also lead the currents on the Y side by 30°. The proof of this statement is given now.

Consider the Y/Δ three-phase transformer shown in Figure 3.22. The secondary voltage E_s is given in terms of the line-to-ground voltages by

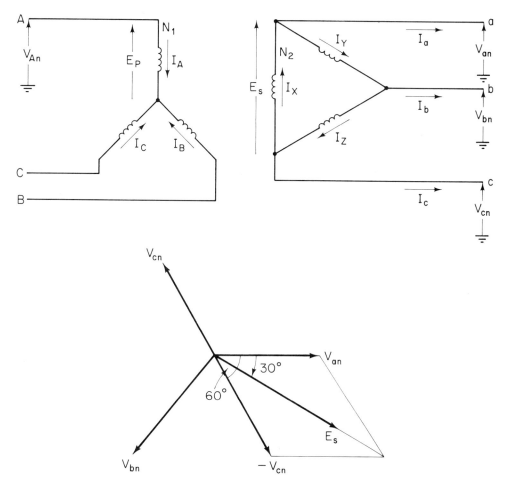

FIGURE 3.22 Y–Δ transformer and a phasor diagram.

$$E_s = V_{an} - V_{cn}$$

Assuming phase sequence a-b-c, taking V_{an} as the reference, we have

$$V_{cn} = V_{an} \, \underline{/120°}$$

As a result,

$$E_s = (1 - 1 \, \underline{/120°}) \, V_{an}$$

or   (3.103)

$$E_s = \sqrt{3} \, V_{an} \, \underline{/-30°}$$

This last result can be verified either analytically or by reference to the phasor diagram in Figure 3.22. Assuming that each winding of the primary has N_1 turns and that each secondary winding has a number of turns N_2, we have

$$E_p = \frac{N_1}{N_2} E_s$$

or

$$E_p = \frac{N_1}{N_2} \sqrt{3} \, V_{an} \, \underline{/-30°}$$

But the line-to-ground voltage on the Y side is

$$V_{An} = E_p$$

Thus we have

$$V_{An} = \frac{N_1}{N_2} \sqrt{3} \, V_{an} \, \underline{/-30°} \qquad (3.104)$$

We can conclude that the Δ-side line-to-ground secondary voltage V_{an} leads the Y-side line-to-ground primary voltage V_{an} by 30°.

Turning our attention now to the current relations, we start by

$$I_x = \frac{N_1}{N_2} I_A$$

$$I_y = \frac{N_1}{N_2} I_B = \frac{N_1}{N_2} I_A \, \underline{/-120°}$$

But

$$I_a = I_x - I_y$$

$$= \frac{N_1}{N_2} I_A (1 - 1 \underline{/-120°})$$

This reduces to

$$I_a = \frac{N_1}{N_2} \sqrt{3} I_A \underline{/30°} \tag{3.105}$$

Thus the secondary line current leads the primary current by 30°.

Example 3.9 A three-phase bank of three single-phase transformers steps up the three-phase generator voltage of 13.8 kV (line to line) to a transmission voltage of 138 kV (line to line). The generator rating is 41.5 MVA. Specify the voltage, current, and kVA ratings of each transformer for the following connections.

(a) Low-voltage windings Δ, high-voltage windings Y.
(b) Low-voltage windings Y, high-voltage windings Δ.
(c) Low-voltage windings Y, high-voltage windings Y.
(d) Low-voltage windings Δ, high-voltage windings Δ.

Solution

The low voltage is given by

$$V_1 = 13.8 \text{ kV (line to line)}$$

The high voltage is given by

$$V_2 = 138 \text{ kV (line to line)}$$

The apparent power is

$$|S| = 41.5 \text{ MVA}$$

(a) Consider the situation with the low-voltage windings connected in Δ, as shown in Figure 3.23. Each winding is subject to the full line-to-line voltage. Thus

$$E_p = 13.8 \text{ kV}$$

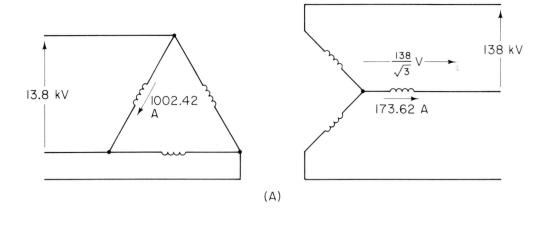

(A)

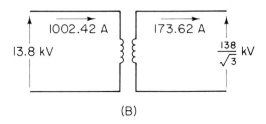

(B)

FIGURE 3.23 (a) Δ–Y transformer with variables indicated; (b) single transformer loading.

The power per winding is $|S|/3$; thus the current in each winding is

$$I_p = \frac{41.5 \times 10^6}{3(13.8 \times 10^3)} = 1002.42 \text{ A}$$

With the secondary connected in Y, the voltage on each winding is the line-to-ground value

$$E_s = \frac{138}{\sqrt{3}} = 79.67 \text{ kV}$$

The current in each winding is obtained as

$$I_s = \frac{41.5 \times 10^6}{3(79.67 \times 10^3)} = 173.62 \text{ A}$$

The kVA rating of each transformer is thus

$$|S_1| = E_p I_p = E_s I_s = 13.83 \text{ MVA}$$

(b) When the low-voltage windings are connected in Y, the voltage on each winding is the line-to-ground value

$$E_p = \frac{13.8}{\sqrt{3}} = 7.97 \text{ kV}$$

The current is

$$I_p = \frac{41.5 \times 10^6}{3(7.97)(10^3)} = 1736.23 \text{ A}$$

With the secondary windings connected in Δ, the voltage on each winding is

$$E_s = 138 \text{ kV}$$

The current is calculated as

$$I_s = \frac{41.5 \times 10^6}{3(138 \times 10^3)} = 100.24 \text{ A}$$

The kVA rating of each transformer is therefore

$$|S_1| = E_p I_p = E_s I_s = 13.83 \text{ MVA}$$

The arrangement is shown in Figure 3.24.

(c) With low-voltage windings connected in Y, from the solution to part (b), we have

$$E_p = 7.97 \text{ kV}$$

$$I_p = 1736.23 \text{ A}$$

With high-voltage windings connected in Y, from the solution to part (a) we have

$$E_s = 79.67 \text{ kV}$$

$$I_s = 173.62 \text{ A}$$

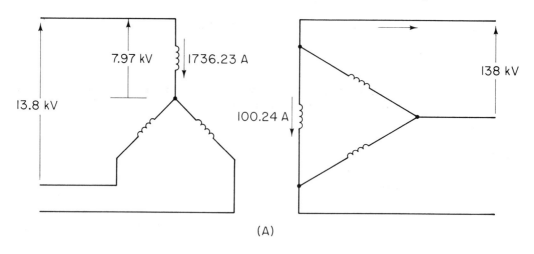

(A)

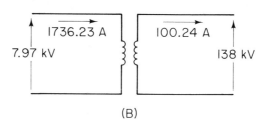

(B)

FIGURE 3.24   (a) Y–Δ transformer with variables indicated for Example 3.9;   (b) single transformer loading.

This arrangement is shown in Figure 3.25.

(d.) With low-voltage windings connected in Δ, from the solution to part (a), we get

$$E_p = 13.8 \text{ kV}$$

$$I_p = 1002.42 \text{ A}$$

With high-voltage windings connected in Δ, from the solution to part (b), we get

$$E_s = 138 \text{ kV}$$

$$I_s = 100.24 \text{ A}$$

The situation is shown in Figure 3.26. Table 3.1 summarizes the voltage and current ratings for the single-phase transformers associated with each transformer connection.

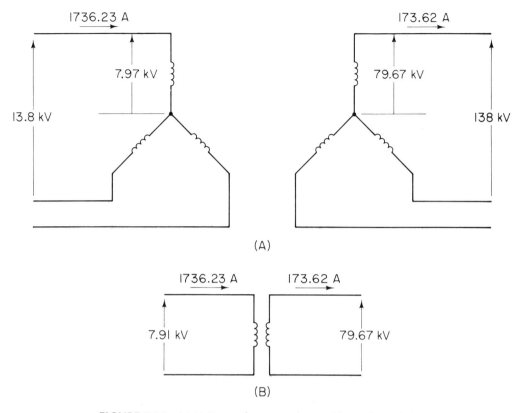

FIGURE 3.25 (a) Y–Y transformer with variables indicated for Example 3.9; (b) single transformer loading.

3.10 AUTOTRANSFORMERS

The basic idea of the autotransformer is permitting the interconnection of the windings electrically. Figure 3.27 shows a two-winding transformer con-

TABLE 3.1

Comparison of Single Transformer Ratings for Different Three-Phase Connections

	Δ–Y	Y–Δ	Y–Y	Δ–Δ
E_p (kV)	13.8	7.97	7.97	13.8
I_p (A)	1002.42	1736.23	1736.23	1002.42
E_s (kV)	79.67	138	79.67	138
I_s (A)	173.62	100.24	173.62	100.24

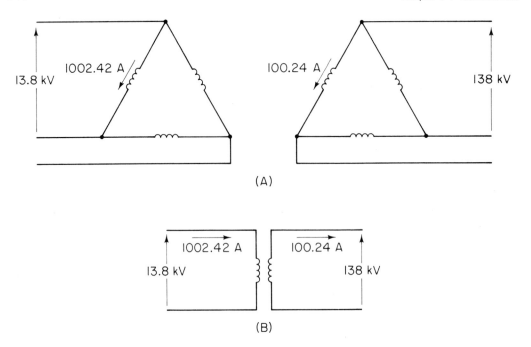

(A)

1002.42 A   100.24 A

13.8 kV   138 kV

(B)

FIGURE 3.26 (a) Δ–Δ transformer with variables indicated for Example 3.9; (b) single transformer loading.

nected in an autotransformer step-up configuration. We will assume the same voltage per turn; that is,

$$\frac{V_1}{N_1} = \frac{V_2}{N_2} \tag{3.106}$$

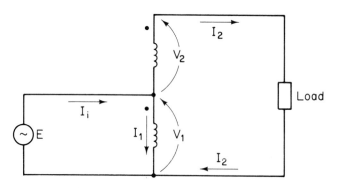

FIGURE 3.27 Step-up autotransformer.

The rating of the transformer when connected in a two-winding configuration is

$$S_{\text{rated}} = V_1 I_1 = V_2 I_2 \tag{3.107}$$

In the configuration chosen, the apparent power into the load is

$$S_0 = (V_1 + V_2)I_2 \tag{3.108}$$

$$= V_2 I_2 \left(1 + \frac{N_1}{N_2}\right)$$

The input apparent power is

$$S_i = V_1(I_1 + I_2) \tag{3.109}$$

$$= V_1 I_1 \left(1 + \frac{N_1}{N_2}\right)$$

Thus the rating of the autotransformer is higher than the original rating of the two-winding configuration. Note that each winding passes the same current in both configurations, and as a result, the losses remain the same. Due to the increased power rating, the efficiency is thus improved.

Autotransformers are generally used when the ratio is 3:1 or less. Two disadvantages are the lack of electric isolation between primary and secondary and the increased short-circuit current over that for the corresponding two-winding configuration.

Example 3.10 A 30-kVA 2.4/0.6-kV transformer is connected as a set-up autotransformer from a 2.4-kV supply. Calculate the currents in each part of the transformer and the load rating. Neglect losses.

Solution

With reference to Figure 3.27, the primary winding rated current is

$$I_1 = \frac{30}{2.4} = 12.5 \text{ A}$$

The secondary rated current is

$$I_2 = \frac{30}{0.6} = 50 \text{ A}$$

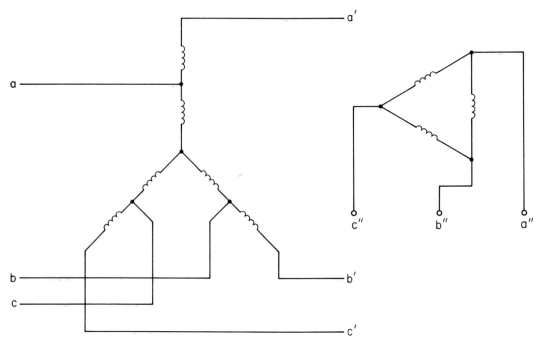

FIGURE 3.28   Schematic diagram of a three-winding autotransformer.

Thus the load current is

$$I_L = 50 \text{ A}$$

The load voltage is

$$V_L = V_1 + V_2 = 3 \text{ kV}$$

As a result, the load rating is

$$S_L = V_L I_L = 150 \text{ kVA}$$

Note that

$$I_i = I_1 + I_2$$
$$= 62.5 \text{ A}$$
$$V_i = V_1 = 2.4 \text{ kV}$$

Thus

$$S_i = 2.4(62.5) = 150 \text{ kVA}$$

Three-phase autotransformers are usually Y–Y connected with the neutral grounded. A third (tertiary) Δ-connected set of windings is included to carry the third harmonic component of the exciting current. A schematic diagram of a three-phase autotransformer with a Δ tertiary is shown in Figure 3.28.

3.11 V–V OR OPEN-DELTA CONNECTIONS

The V–V connection enables three-phase operation with two single-phase transformers as shown in Figure 3.29(b). It is clear that this configuration is identical with a Δ–Δ-connected three-phase transformer with one single transformer removed, hence the name *open delta*.

The phasor diagram of Figure 3.29(a) assumes that a balanced three-phase supply is connected across the two primaries. Note that the voltage V_{CA} is the resultant of V_{BC} and V_{AB}. The induced voltages in the secondaries V_{ab} and V_{bc} result in the balanced voltages V_{an}, V_{bn}, and V_{cn} at the load side as indicated in Figure 3.29(c). Let us assume that the load voltage and current phasors are as follows:

$$V_{an} = V_2 \underline{/0} \qquad V_{bn} = V_2 \underline{/-120°} \qquad V_{cn} = V_2 \underline{/+120°}$$
$$I_{an} = I_2 \underline{/-\phi} \qquad I_{bn} = I_2 \underline{/-120° - \phi} \qquad I_{cn} = I_2 \underline{/120° - \phi}$$

The secondary of transformer 1 supplies the load with current I_{an} and the voltage V_{ab}.

$$V_{ab} = \sqrt{3} \, V_2 \underline{/30°} \tag{3.110}$$

The apparent power supplied by the secondary of transformer 1 is thus

$$\begin{aligned} S_1 &= V_{ab} \, I_{an}^{\star} \\ &= \sqrt{3} \, V_2 I_2 \underline{/\phi + 30°} \end{aligned} \tag{3.111}$$

Thus transformer 1 operates at a power factor of $\cos(\phi + 30)$. The apparent power supplied by the secondary of transformer 2 is

$$\begin{aligned} S_2 &= V_{cb} \, I_{cn}^{\star} \\ &= (\sqrt{3} \, V_2 \underline{/90°})(I_2 \underline{/\phi - 120°}) \end{aligned}$$

Thus

$$S_2 = \sqrt{3} \, V_2 I_2 \underline{/\phi - 30°} \tag{3.112}$$

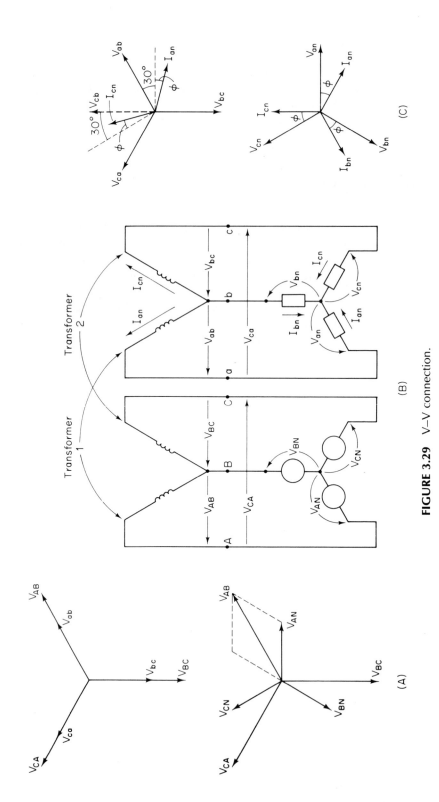

FIGURE 3.29 V–V connection.

Transformer 2 operates at a power factor of cos (ϕ − 30°). Of course, the total power supplied is

$$S_t = S_1 + S_2 = 3V_2I_2 \underline{/\phi} \qquad (3.113)$$

Each of the two transformers carries the same volt-amperes value of $\sqrt{3}\,V_2I_2$, which should be equal to the VA rating of its windings. Thus

$$VA \text{ rating } = \sqrt{3}\,V_2I_2$$

As a result, the load volt-amperes is related to the rating of each transformer by

$$|S_t| = \sqrt{3} \text{ VA rating} \qquad (3.114)$$

It should be noted now that if a Δ–Δ connection is used, then

$$|S_{t\Delta}| = 3 \text{ VA rating} \qquad (3.115)$$

We can thus assert that the V–V connection is capable of carrying ($1/\sqrt{3}$) of the apparent power that can be carried by a Δ–Δ connection with similarly rated transformers.

3.12 T–T CONNECTIONS

The T–T connection enables the use of only two transformers to transform three-phase power. The purpose of this connection is the same as that for the V–V connection. The first transformer is called the main transformer and is either a center-tapped transformer or a multiwinding transformer with two equal primary and two equal secondary windings. Assume that each of the primary windings has N_1 turns and that each of the secondary windings has N_2 turns. The second transformer is called the teaser transformer and has $\sqrt{3}\,N_1$ turns on its primary side and $\sqrt{3}\,N_2$ turns on its secondary side. Assuming equal voltage per turn, it is clear that the voltage ratings of the teaser transformer's primary and secondary are 0.866 ($\sqrt{3}/2$) of the main transformer's rating. The T–T connection is shown in Figure 3.30.

Assume that the primary three-phase voltages and currents are supplied by a Y-connected source with the following phase voltage values:

$$V_{AN} = \frac{V_1}{\sqrt{3}} \underline{/-30°}$$

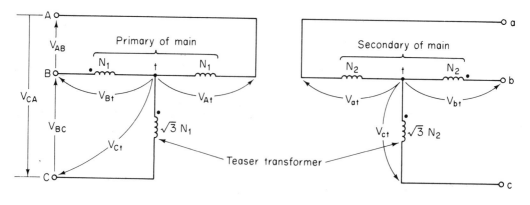

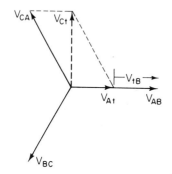

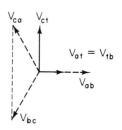

FIGURE 3.30   T–T connection.

$$V_{BN} = \frac{V_1}{\sqrt{3}} \underline{/-150°}$$

$$V_{CN} = \frac{V_1}{\sqrt{3}} \underline{/90°}$$

The phase current values are assumed to be in phase with the corresponding phase voltages:

$$I_{AN} = I_1 \underline{/-30°}$$

$$I_{BN} = I_1 \underline{/-150°}$$

$$I_{CN} = I_1 \underline{/90°}$$

The currents in the primaries are thus given by

$$I_{At} = I_1 \underline{/-30°}$$

$$I_{Bt} = I_1 \underline{/-150°}$$

$$I_{Ct} = I_1 \underline{/90°}$$

The voltages in the primary are thus balanced and given by

$$V_{AB} = V_1 \underline{/0}$$

$$V_{BC} = V_1 \underline{/-120°}$$

$$V_{CA} = V_1 \underline{/120°}$$

The voltage V_{AB} is shared equally by the windings A-t and B-t.

$$V_{AB} = V_{At} + V_{tB} = V_1 \underline{/0}$$

Therefore, as shown in Figure 3.30, we have

$$V_{At} = \frac{V_1}{2} \underline{/0}$$

$$V_{Bt} = \frac{V_1}{2} \underline{/180°}$$

The voltage V_{Ct} is obtained from

$$V_{Ct} = V_{CA} + V_{At}$$

$$= V_1 \underline{/120°} + \frac{V_1}{2} \underline{/0}$$

As a result,

$$V_{Ct} = \frac{\sqrt{3}}{2} V_1 \underline{/90°}$$

The secondary voltages can now be obtained as

$$V_{at} = \frac{N_2}{N_1} V_{At} = \frac{N_2}{N_1} \frac{V_1}{2} \underline{/0}$$

$$V_{bt} = \frac{N_2}{N_1} V_{Bt} = \frac{N_2}{N_1} \frac{V_1}{2} \underline{/180°}$$

$$V_{Ct} = \frac{\sqrt{3}N_2}{\sqrt{3}N_1} V_{Ct} = \frac{N_2}{N_1} \frac{\sqrt{3}}{2} V_1 \underline{/90°}$$

The secondary currents are obtained as

$$I_{ta} = \frac{N_1}{N_2} I_{At} = \frac{N_1}{N_2} I_1 \underline{/-30°}$$

$$I_{tb} = \frac{N_1}{N_2} I_{Bt} = \frac{N_1}{N_2} I_1 \underline{/-150°}$$

$$I_{tc} = \frac{\sqrt{3} N_1}{\sqrt{3} N_2} I_{Ct} = \frac{N_1}{N_2} I_1 \underline{/90°}$$

The apparent powers in the A and B parts of the primary of the main transformer, are given by

$$S_{At} = V_{At} I_{At}^{\star} = \frac{V_1 I_1}{2} \underline{/30°}$$

$$= \frac{V_1 I_1}{2} \left(\frac{\sqrt{3}}{2} + j \frac{1}{2} \right)$$

$$S_{Bt} = V_{Bt} I_{Bt}^{\star} = \frac{V_1 I_1}{2} \underline{/-30°}$$

$$= \frac{V_1 I_1}{2} \left(\frac{\sqrt{3}}{2} - j \frac{1}{2} \right)$$

We note that the A part carries an inductive reactive power while the B part carries a capacitive reactive power. The primary of the main transformer, however, carries in total power at unity power factor since

$$S_{\text{primary main}} = S_{At} + S_{Bt}$$

$$= V_1 I_1 \left(\frac{\sqrt{3}}{2} \right)$$

Note that the value obtained above is the rating of the primary of the main transformer in VA. There is a reduction from $V_1 I_1$ (when used as a single-phase transformer) to $0.866 V_1 I_1$. The apparent power in the teaser primary is given by

$$S_{ct} = V_{ct} I_{ct}^{\star}$$

$$= V_1 I_1 \frac{\sqrt{3}}{2}$$

Therefore, the primary and teaser share the VA rating equally.

In a similar fashion, the apparent power through the secondary of the main transformer is obtained as

$$S_{ta} + S_{tb} = \frac{V_1 I_1 \sqrt{3}}{2}$$

The apparent power through the secondary of the teaser is also given by

$$S_{tc} = \frac{V_1 I_1 \sqrt{3}}{2}$$

3.13 THREE-PHASE-TO-TWO-PHASE TRANSFORMATION: SCOTT CONNECTIONS

As can be seen from the analysis of the T–T connection, the induced secondary voltages V_{at} and V_{ct} are displaced by 90°, although not of equal magnitudes. The voltages V_{bt} and V_{ct} exhibit the same phase displacement. A two-phase system is characterized by voltage phasors that are equal in magnitude but 90° out of phase. It is clear that the T–T connection with some modification can provide a transformation from a set of balanced three-phase voltages to a two-phase voltage system. The modification results in the Scott connection.

The primary of the main transformer in the Scott connection is similar to that of the T–T connection in that it is made of two equal windings with N_1 turns each. The secondary, however, is not divided up into two windings and has a total number of turns of N_2. For analysis purposes, we may assume that the secondary of the main transformer is made of two windings of N_2 turns each but connected in series as shown in Figure 3.31. The primary of the teaser transformer consists of $\sqrt{3} \, N_1$ turns and is connected to the center of the primary of the main transformer. The secondary of the teaser has $2N_2$ turns and is connected to the terminals of the main transformer secondary.

Let us assume that the voltages on the primary terminals A, B, and C are balanced three-phase voltages with the same source as that assumed for the analysis of the T–T connection. We thus have

$$V_{At} = \frac{V_1}{2} \, \underline{/0}$$

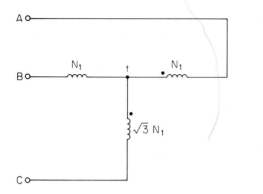
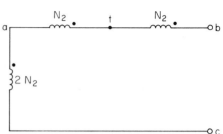

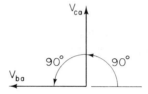

FIGURE 3.31   Scott connection.

$$V_{Bt} = \frac{V_1}{2} \angle 180°$$

$$V_{Ct} = \frac{\sqrt{3}}{2} V_1 \angle 90°$$

Using the assumed transformer ratios, we obtain for the secondary induced voltages

$$V_{at} = \frac{V_1}{2} \frac{N_2}{N_1} \angle 0$$

$$V_{bt} = \frac{V_1}{2} \frac{N_2}{N_1} \angle 180°$$

$$V_{ca} = \frac{\sqrt{3}}{2} V_1 \frac{2N_2}{\sqrt{3}N_1} \angle 90°$$

As a result, we conclude that the terminal voltages are

$$V_{ba} = V_{bt} + V_{ta}$$

$$= \frac{N_2}{N_1} V_1 \underline{/180°}$$

$$V_{ca} = \frac{N_2}{N_1} V_1 \underline{/90°}$$

The voltage of terminal b with respect to a leads that of terminal c with respect to a by 90° and the two voltage phasors are of equal magnitude.

SOME SOLVED PROBLEMS

Problem 3.A.1

A resistive load R_r is connected through an ideal transformer to a generator represented by a voltage source V_s in series with an impedance Z_s. Derive an expression for the transformer turns ratio so that maximum power is transferred to the load. Obtain the specifications of the required transformer for a load of 0.5 Ω to be supplied from a 240-V source with a series reactance of 32 Ω.

Solution

The load resistance referred to the primary side is denoted by R_r'. The current in the primary circuit is thus

$$I = \frac{V_s}{Z_s + R_r'}$$

The power in the load is therefore given by

$$P_r = \frac{V_s^2 R_r'}{(R_s + R_r')^2 + X_s^2}$$

For maximum power transfer, we set the derivative of P_r with respect to R_r' to zero to obtain

$$R_r' = (R_s^2 + X_s^2)^{1/2}$$

As a result,

$$\left(\frac{N_1}{N_2}\right)^2 = \frac{(R_s^2 + X_s^2)^{1/2}}{R_r}$$

Numerically, we have

$$V_s = 240 \text{ V} \qquad R_r = 0.5 \qquad R_s = 0 \qquad X_s = 32$$

As a result,

$$\left(\frac{N_1}{N_2}\right)^2 = \frac{32}{0.5} = 64$$

Thus the required turns ratio is

$$\frac{N_1}{N_2} = 8$$

The primary current is calculated as

$$I_1 = \frac{240}{32 + j32} = 5.303 \underline{/-45°} \text{ A}$$

The voltage V_1 on the primary of the transformer is

$$V_1 = V_s - IZ_s$$
$$= 240 - (5.303 \underline{/-45°})(j32)$$
$$= 169.707 \underline{/-45°}$$

The power into the transformer is

$$S_1 = V_1 I_1^*$$
$$= (169.707 \underline{/-45°})(5.303 \underline{/45°})$$
$$= 900 \text{ W}$$

The transformer is rated at about 170 V, 900 W, with a turns ratio of 8:1.

Problem 3.A.2

A two-winding transformer is rated at 50 kVA. The maximum efficiency of the transformer occurs when the output of the transformer is 35 kVA. Find the rated efficiency of the transformer at 0.8 PF lagging given that the no-load losses are 200 W.

Solution

At maximum efficiency, the copper losses are equal to the core losses; thus

$$I_A^2 R_{eq} = 200 \text{ W}$$

where I_A is the current at maximum efficiency. Let the current be I_B for 50-kVA loading. The copper losses at this loading are

$$I_B^2 R_{eq} = \left(\frac{I_B}{I_A}\right)^2 200$$

$$= \left(\frac{50}{35}\right)^2 200 = 408.16 \text{ W}$$

The efficiency at 50 kVA, 0.8 PF is thus

$$\eta = \frac{50 \times 10^3 \times 0.8}{50 \times 10^3 \times 0.8 + 408.16 + 200}$$

$$= 0.98502$$

Problem 3.A.3

The no-load input to a 5-kVA 500/100-V single-phase transformer is 100 W at 0.15 PF at rated voltage. The voltage drops due to resistance and leakage reactance are 0.01 and 0.02 times the rated voltage when the transformer operates at rated load. Calculate the input power and power factor when the load is 3 kW at 0.8 PF lagging at rated voltage.

Solution

The rated load current is

$$I_r = \frac{5 \times 10^3}{100} = 50 \text{ A}$$

The no-load current is

$$I_0 = \frac{100}{100 \times 0.15} = 6.667 \text{ A}$$

$$= 6.67 \, \underline{/-81.37°}$$

We can obtain the equivalent resistance and reactance as

$$0.01(100) = 50R_{eq}$$

$$0.02(100) = 50X_{eq}$$

Thus

$$R_{eq} = 0.02 \ \Omega$$

$$X_{eq} = 0.04 \ \Omega$$

For 3 kW, at 0.8 PF, we have

$$I = \frac{3 \times 10^3}{100 \times 0.8} \ \underline{/-\cos^{-1}0.8}$$

$$= 37.5 \ \underline{/-36.87°} \ A$$

Now

$$V_1' = V_2 + IZ_{eq}$$

$$= 100 + (37.5 \ \underline{/-36.87°})(0.02 + j0.04)$$

$$= 101.5028 \ \underline{/0.4234°} \ V$$

The primary current referred to secondary is thus

$$I_1' = I + I_0$$

$$= 37.5 \ \underline{/-36.87°} + 6.67 \ \underline{/-81.37°}$$

$$= 42.51 \ \underline{/-43.18°} \ A$$

Thus we have

$$\phi_1 = 43.18 + 0.4234 = 43.60°$$

$$\cos \phi_1 = 0.7241$$

As a result,

$$P_1 = V_1'I_1' \cos \phi_1$$

$$= 101.5028(42.51)(0.7241)$$

$$= 3124.68 \ W$$

Problem 3.A.4

Consider the three-winding transformer of Example 3.8. This time assume that the loads on the secondary and tertiary are specified by

$$|S_2| = |S_3| = 20 \text{ kVA}$$

The secondary winding's load has a power factor of 0.9 lagging. The tertiary windings load has a power factor of 0.8 lagging at a voltage of 400 V referred to the primary. Calculate the primary voltage and current for this loading condition as well as the voltage at the secondary terminals referred to the primary side.

Solution

The equivalent circuit of Figure 3.16 is utilized. We will work backward from the tertiary winding to the point 0 to obtain V_0. Knowledge of V_0 enables us to find V_2, but this turns out to be a somewhat involved process. Once this is done, the procedure is straightforward. From the specifications at the tertiary terminals, we have

$$I_3 = \frac{20 \times 10^3}{400} \underline{/-\cos^{-1}0.8} = 50 \underline{/-36.87°} \text{ A}$$

As a result, the voltage V_0 is obtained as

$$V_0 = V_3 + I_3 Z_3$$

$$= 400 \underline{/0} + (50 \underline{/-36.87°})(0.02 + j0.06)$$

$$= 402.604 \underline{/0.25616°} \text{ V}$$

Note that V_3 is taken as the reference voltage. The voltage V_2 is unknown at this stage, so we assume that

$$V_2 = |V_2| \underline{/\theta_2}$$

θ_2 is measured against the reference voltage. The current I_2 is therefore

$$I_2 = \frac{20 \times 10^3}{|V_2|} \underline{/\theta_2 - \cos^{-1}0.9}$$

As a result, we have

$$V_0 = V_2 + I_2 Z_2$$

$$402.604 \underline{/0.25616°} = |V_2| \underline{/\theta_2} + (0.02 + j0.06)$$

$$\times \frac{20 \times 10^3}{|V_2|} \underline{/\theta_2 - \cos^{-1}0.9}$$

$$= |V_2| \underline{/\theta_2} + \frac{1264.91}{|V_2|} \underline{/\theta_2 + 45.72°}$$

Separating real and imaginary parts of the relation above, we obtain the following two simultaneous equations in θ_2 and $|V_2|$:

$$|V_2| \cos \theta_2 + \frac{1264.91}{|V_2|} \cos (\theta_2 + 45.72) = 402.6$$

$$|V_2| \sin \theta_2 + \frac{1264.91}{|V_2|} \sin (\theta_2 + 45.72) = 1.79997$$

Expanding the trignometric functions and collecting terms, we obtain

$$\left(|V_2| + \frac{883.12}{|V_2|}\right) \cos \theta_2 - \frac{905.60}{|V_2|} \sin \theta_2 = 402.6$$

$$\left(|V_2| + \frac{883.12}{|V_2|}\right) \sin \theta_2 + \frac{905.60}{|V_2|} \cos \theta_2 = 1.79997$$

To simplify matters, let us substitute

$$a = |V_2| + \frac{883.12}{|V_2|}$$

$$b = \frac{905.60}{|V_2|}$$

$$c = 402.6$$

$$d = 1.79997$$

Thus we have

$$a \cos \theta_2 - b \sin \theta_2 = c$$

$$a \sin \theta_2 + b \cos \theta_2 - d$$

Dividing through by $\sqrt{a^2 + b^2}$, we get

$$\frac{a}{\sqrt{a^2 + b^2}} \cos \theta_2 - \frac{b}{\sqrt{a^2 + b^2}} \sin \theta_2 = \frac{c}{\sqrt{a^2 + b^2}}$$

$$\frac{a}{\sqrt{a^2 + b^2}} \sin \theta_2 + \frac{b}{\sqrt{a^2 + b^2}} \cos \theta_2 = \frac{d}{\sqrt{a^2 + b^2}}$$

Let

$$\cos \psi = \frac{a}{\sqrt{a^2 + b^2}} \qquad \sin \psi = \frac{b}{\sqrt{a^2 + b^2}}$$

Thus

$$\cos \psi \cos \theta_2 - \sin \psi \sin \theta_2 = \frac{c}{\sqrt{a^2 + b^2}}$$

$$\cos \psi \sin \theta_2 + \sin \psi \cos \theta_2 = \frac{d}{\sqrt{a^2 + b^2}}$$

or

$$\cos (\theta_2 + \psi) = \frac{c}{\sqrt{a^2 + b^2}}$$

$$\sin (\theta_2 + \psi) = \frac{d}{\sqrt{a^2 + b^2}}$$

Squaring and adding, we get

$$\frac{c^2}{a^2 + b^2} + \frac{d^2}{a^2 + b^2} = 1$$

or

$$c^2 + d^2 = a^2 + b^2$$

In terms of our physical variables, we thus have

$$(402.6)^2 + (1.7997)^2 = \left(|V_2| + \frac{883.12}{|V_2|}\right)^2 + \left(\frac{905.6}{|V_2|}\right)^2$$

or

$$|V_2|^4 - 160{,}323.76 \, |V_2|^2 + 1{,}600{,}012.29 = 0$$

$$|V_2|^2 = \frac{160{,}323.76 \pm \sqrt{(160{,}323.76)^2 - (4)\,(1{,}600{,}012.2)}}{2}$$

Take the positive sign to obtain

$$|V_2| = 400.39203 \text{ V}$$

Thus

$$a = |V_2| + \frac{883.12}{|V_2|} = 402.598$$

$$b = \frac{905.6}{|V_2|} = 2.26178$$

$$\cos \psi = \frac{402.598}{\sqrt{(402.598)^2 + (2.26178)^2}} = 0.999984$$

$$\psi = 0.32188°$$

$$\cos (\theta_2 + \psi) = \frac{402.6}{\sqrt{(402.598)^2 + (2.26178)^2}}$$

$$\theta_2 + \psi = 0.256094$$

$$\theta_2 = -0.06579°$$

Thus we have

$$V_2 = 400.39203 \, \underline{/-0.06579°}$$

The secondary current is now obtained as

$$I_2 = \frac{20 \times 10^3}{400.39203} \, \underline{/-0.06579° - 25.84°}$$

$$= 49.951 \, \underline{/-25.9077°}$$

The primary current is

$$I_1 = I_2 + I_3$$

$$= (49.951 \, \underline{/-25.9077°}) + (50 \, \underline{/-36.87°})$$

$$= 99.4940 \, \underline{/-31.3915°}$$

Finally, the primary voltage is obtained from

$$V_1 = V_0 + I_1 Z_1$$

$$= 402.604 \,\underline{/0.25616°} + (99.4940 \,\underline{/-31.3915°}) \,(0.02 + j0.06)$$

$$= 407.4502 \,\underline{/0.824°}$$

This completely solves the problem.

PROBLEMS

Problem 3.B.1

A single-phase transformer has a turns ratio of 4:1. The transformer may be considered ideal. A load impedance of $10\underline{/30°}$ is connected across the secondary terminals and the secondary voltage is 120 V and is taken as the reference. Find the current in the primary and secondary windings, the primary voltage, and the load impedance referred to the primary.

Problem 3.B.2

A 50-kVA, 400/2000-V, single-phase transformer delivers a load of 40 kVA at 2000 V and 0.8 PF lagging. Assume that the transformer is ideal.

  a. Find the load impedance.
  b. Find the load impedance referred to the primary side.

Problem 3.B.3

The load connected to the secondary of an ideal single-phase transformer is 10 kVA and its impedance is $Z_2 = 2\underline{/-32}$. The load impedance referred to the primary is $32 \,\underline{/-32°}$. Find the turns ratio and the primary and secondary currents and voltages.

Problem 3.B.4

A single-phase 10:1 ideal transformer has a primary voltage of 39.8 kV and a primary apparent power of 1000 kVA at 0.8 PF lagging. Find the load impedance connected to the secondary winding.

Problem 3.B.5

Calculate the turns ratio, voltage, and volt-ampere ratings of an ideal transformer so that maximum power is transferred to a resistive load of 1 Ω from a source with internal voltage of 250 and series reactance of 36 Ω (see Problem 3.A.1).

Problem 3.B.6

Having designed the transformer of Problem 3.B.5, it was decided that the source's internal resistance of 4 Ω should be included in the design. Instead of changing the transformer, a capacitor with reactance X_c is inserted in series with the source to maintain maximum power transfer to the load. Calculate the required reactance.

Problem 3.B.7

Consider the circuit shown in Figure 3.32 and assume that both transformers are ideal. Impedance values are given on the basis of the voltage level in the corresponding portion of the circuit. Obtain an equivalent representation of the circuit referred to the 33-kV side and hence find V_1 and I_1 taking the load voltage as reference.

$$Z_1 = 10.5 + j35 \ \Omega$$

$$Z_2 = 0.5 + j0.64 \ \Omega$$

$$Z_3 = (4 + j6)10^{-3} \ \Omega$$

Problem 3.B.8

Consider the circuit of Problem 3.B.7. Find the current I_2 and voltage V_2 by referring the circuit to the 2400-V side.

Problem 3.B.9

Find the current I and the source voltage V_1 in the circuit shown in Figure 3.33.

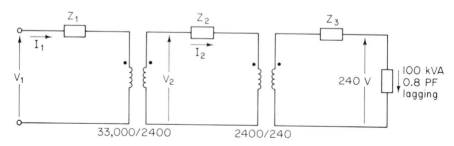

FIGURE 3.32 Circuit for Problem 3.B.7.

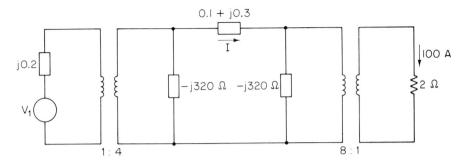

FIGURE 3.33 Circuit for Problem 3.B.9.

Problem 3.B.10

Find the current I in the circuit shown in Figure 3.34, given that

$$V_1 = 13.8 \qquad V_2 = 13.8 \underline{/15°}$$

$$Z_1 = Z_4 = 0.01 + j0.03$$

$$Z_2 = Z_3 = 1 + j3$$

$$Z_L = 0.06 + j0.08$$

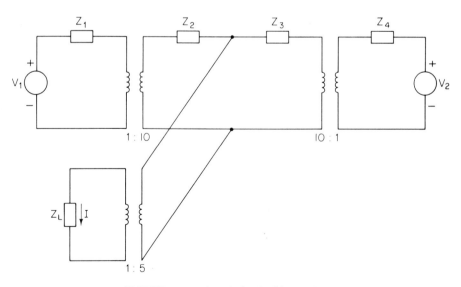

FIGURE 3.34 Circuit for Problem 3.B.10.

Problem 3.B.11

A 10-kVA, 480/120-V, single-phase transformer has the following parameters:

$$R_1 = 0.6 \, \Omega \qquad\qquad X_1 = 1 \, \Omega$$
$$R_2 = 37.5 \times 10^{-3} \, \Omega \qquad X_2 = 62.5 \times 10^{-3} \, \Omega$$
$$G_c = 0.333 \times 10^{-3} \, S \qquad B_m = 2 \times 10^{-3} \, S$$

Use the equivalent circuit of Figure 3.9(b) to calculate the voltage and current at the primary terminals for a load of 10 kVA at 120 V and a power factor of 0.85 lagging.

Problem 3.B.12

Calculate the power losses in the secondary winding, the primary winding, and in the core as well as the efficiency of the transformer of Problem 3.B.11.

Problem 3.B.13

Repeat Problem 3.B.11 using the equivalent circuit of Figure 3.10(a) and find the power factor at the primary and the efficiency of the transformer.

Problem 3.B.14

A 50-kVA 400/2000-V single-phase transformer has the following parameters:

$$R_1 = 0.02 \, \Omega \qquad R_2 = 0.5 \, \Omega$$
$$X_1 = 0.06 \, \Omega \qquad X_2 = 1.5 \, \Omega$$
$$G_c = 2 \, mS \qquad B_m = -6 \, mS$$

Note that G_c and B_m are given in terms of primary reference. The transformer supplies a load of 40 kVA at 2000 V and 0.8 PF lagging. Calculate the primary voltage and current using the equivalent circuits shown in Figures 3.9 and 3.10.

Problem 3.B.15

Find the voltage regulation and efficiency for the transformer of Problem 3.B.14.

Problem 3.B.16

Find the maximum efficiency of the transformer of Problem 3.B.14 under the same conditions.

Problem 3.B.17

The equivalent impedance referred to the primary of a 2300/230-V 500-kVA, single-phase transformer is

$$Z = 0.2 + j0.6 \ \Omega$$

Calculate the percentage voltage regulation (PVR) when the transformer delivers rated capacity at 0.8 PF lagging at rated secondary voltage. Find the efficiency of the transformer at this condition given that core losses at rated voltage are 2 kW.

Problem 3.B.18

A 500/100-V two-winding transformer is rated at 5 kVA. The following information is available:

1. The maximum efficiency of the transformer occurs when the output of the transformer is 3 kVA.
2. The transformer draws a current of 3 A and the power is 100 W when a 100-V supply is impressed on the low-voltage winding with the high-voltage winding open-circuit.

Find the rated efficiency of the transformer at 0.8 PF lagging.

Problem 3.B.19

A 50-kVA 400/2000-V single-phase transformer has the following equivalent circuit parameters referred to the primary:

$$Z_{eq} = 0.04 + j0.12 \ \Omega$$

Assume that the transformer supplies a load of 40 kVA at 2000 V and 0.8 PF lagging. Calculate the primary voltage of the transformer and its efficiency.

Problem 3.B.20

The load at the end of a feeder is 100 A at 0.85 PF and 132 V. The feeder's impedance is

$$Z_3 = 0.003 + j0.01 \ \Omega$$

Power is supplied to the feeder by a 10:1 transformer whose impedance referred to the primary

$$Z_2 = 4 + j8 \ \Omega$$

High-voltage power is provided to the transformer through a radial link with series impedance:

$$Z_3 = 1 + j4.8 \ \Omega$$

Find the voltage, current, and power factor at the sending end of the link. Calculate the efficiency of the system.

Problem 3.B.21

The no-load input power to a 50-kVA, 2300/230-V, single-phase transformer is 200 VA at 0.15 PF at rated voltage. The voltage drops due to resistance and leakage reactance are 0.012 and 0.018 times rated voltage when the transformer operates at rated load. Calculate the input power and power factor when the load is 30 kW at 0.8 PF lagging at rated voltage.

Problem 3.B.22

To identify the equivalent circuit parameters of a 100-kVA 4-kV/1-kV transformer, a short-circuit test is performed with the power input of 2.5 kW at

$$V_1 = 224 \ V \qquad \text{and} \qquad I_1 = 25 \ A$$

Determine the parameters R_{eq} and X_{eq} of the transformer referred to the primary.

Problem 3.B.23

A single-phase 2300/230-V 500-kVA distribution transformer is tested to find the equivalent-circuit parameters. The following date are obtained:

Open-circuit test: $V_2 = 230$ V, $I_2 = 94$ A, $P_2 = 2250$ W
Short-circuit test: $V_1 = 100$ V, $I_1 = 230$ A, $P_1 = 9.2$ kW

Find the equivalent-circuit parameters referred to the primary.

Problem 3.B.24

A single-phase 2300/230-V 500-kVA distribution transformer is tested to find the equivalent-circuit parameters. The following information is available:

Open-circuit test: $V_2 = 208$ V, $I_2 = 85$ A, $P_2 = 1800$ W
Short-circuit test: $V_1 = 95$ V, $I_1 = 218$ A, $P_1 = 8.2$ kW

Find the equivalent-circuit parameters referred to the primary.

Problem 3.B.25

For the transformer of Problem 3.B.24, rated load is delivered on the secondary side at unity power factor. Find the efficiency of the transformer for these conditions using the following models.

a. Series impedance and ideal transformer model neglecting the no-load losses and magnetizing effects.
b. Series impedance, shunt admittance (representing a no-load circuit), and ideal transformer.

Problem 3.B.26

A 100-kVA 60-Hz 12,000/240-V single-phase transformer is tested to obtain the equivalent circuit parameters using a short-circuit test. The voltage applied to the 12,000-V side is 600 V with a short-circuit current equal to the rated current value and short-circuit power of 1200 W.

a. Find the parameters R and X of the transformer.
b. Assume that the transformer supplies a load of 100 kVA at 240 V and 0.8 PF lagging on the secondary side. Find the voltage required at the primary side and the power factor.

Problem 3.B.27

Convert the parameters and operating conditions of the transformer of Problem 3.B.11 to the per unit system using the transformer rated kVA and the high-voltage side as base. Repeat the solution to the problem in the per unit system.

Problem 3.B.28

Repeat Problem 3.B.27 for the transformer of Problem 3.B.17.

Problem 3.B.29

Convert Problem 3.B.22 into a per unit system using the transformer rating and voltage as base, then find the parameters R_{eq} and X_{eq} in this system.

Problem 3.B.30

Two 230/115-V transformers are available. The rating of transformer A is 5 kVA, whereas that of transformer B is 4 kVA. Find the rating of the combination in the following configurations.

a. Primaries connected in series aiding and secondaries connected in series aiding.

b. Primaries connected in series aiding and secondaries connected in parallel.

Problem 3.B.31

A 5-kVA 230/115-V transformer is connected in a parallel–series configuration with a 3-kVA 230/23-V transformer. Find the rating of the combination.

Problem 3.B.32

Two 2300/230-V transformers are connected in parallel to supply a load of 500 kVA at 0.8 PF lagging. Find the power supplied by each transformer, given that

$$Z_A = 0.1 + j0.2$$

$$Z_B = 0.12 + j0.2$$

Impedances are given on the basis of the 2300-V side.

Problem 3.B.33

Two 2300/230-V transformers are connected in parallel. The following is the apparent power loading of each:

$$S_A = 300 \times 10^3 \underline{/40°}$$

$$S_B = 200 \times 10^3 \underline{/35°}$$

Assume that the series reactance of each transformer is 0.1 Ω. Find the series resistance of each transformer.

Problem 3.B.34

Repeat Example 3.8 for

$$I_2 = 40 \underline{/-30°}$$

$$I_3 = 40 \underline{/-35°}$$

Problem 3.B.35

Consider the three winding transformers of Example 3.8. Given that $V_1 = 400$, $S_1 = 32,000 \underline{/32.5°}$, and $V_2 = 394 \underline{/-0.7°}$, find the efficiency of the transformer.

Problem 3.B.36

A Y-connected balanced three-phase load consisting of three impedances of $10\ \underline{/30°}\ \Omega$ each is supplied with balanced line-to-neutral voltages:

$$V_{an} = 220\ \underline{/0}\ \text{V}$$

$$V_{bn} = 220\ \underline{/240°}\ \text{V}$$

$$V_{cn} = 220\ \underline{/120°}\ \text{V}$$

a. Calculate the phasor currents in each line.
b. Calculate the line-to-line phasor voltages.
c. Calculate the total active and reactive power supplied to the load.

Problem 3.B.37

Repeat Problem 3.B.36 as if the same three impedances were connected in a Δ connection.

Problem 3.B.38

Find the phase currents I_A, I_B, and I_C as well as the neutral current I_n for a three-phase network with an unbalanced load.

$$Z_A = 1 + j3$$

$$Z_B = 1 + j3$$

$$Z_C = 0.8 + j2.4\ \Omega$$

Assume that the applied three-phase voltage is balanced with magnitude of 100 V. Calculate the apparent power consumed by the load.

Problem 3.B.39

A 60-hp three-phase 400-V induction motor operates at 0.75 PF lagging. Find the active, reactive, and apparent power consumed per phase. Find the values of R and jX if the motor is modeled as a balanced three-phase impedance $Z = R + jX$ connected in a Y connection.

Problem 3.B.40

A three-phase bank of three single-phase transformers steps up the three-phase generator voltage of 13.8 kV (line to line) to a transmission voltage of 138 kV (line to line). The generator rating is 83 MVA. Specify the voltage, current, and kVA ratings of each transformer for the following connections:

a. Low-voltage windings, Δ, high-voltage windings Y.
b. Low-voltage windings Y, high-voltage windings Δ.
c. Low-voltage windings Y, high-voltage windings Y.
d. Low-voltage windings Δ, high-voltage windings Δ.

Problem 3.B.41

The equivalent impedance referred to the secondary of a 13.8/138-kV 83-MVA three-phase Δ/Y-connected transformer is

$$Z = 2 + j13.86 \ \Omega$$

Calculate the percentage voltage regulation when the transformer delivers rated capacity at 0.8 PF lagging at rated secondary voltage. Find the efficiency of the transformer at this condition given that core losses at rated voltages are 76.5 kW.

Problem 3.B.42

The load at the end of a high-voltage three-phase line is 30 MVA at 0.85 PF and 11 kV (line to line). Power is transmitted from the remote source through a step-up 11 kV/110 kV transformer T1 feeding the high-voltage line TL. At the end of the line is a step-down 110 kV/11 kV transformer T2. The equivalent circuit parameters of the individual components are

$$Z_{T1} = 1.45 + j26.6 \ \Omega$$

$$Z_{TL} = 10.2 + j34 \ \Omega$$

$$Z_{T2} = 6.8 + j106 \ \Omega$$

Find the sending end voltage and transmission efficiency of the system. Assume that impedance values are given on the basis of 110 kV.

Problem 3.B.43

Three identical 500-kVA 2300/230-V single-phase transformers are connected Δ–Y. Each transformer has the following equivalent impedance referred to the high-voltage side:

$$Z = 0.2 + j0.6 \ \Omega$$

Each transformer delivers its rated VA at 0.8 PF at its rated secondary voltage to a three-phase load. Find the primary line-to-line voltage.

Problem 3.B.44

A 50-kVA 2.4/0.6-kV transformer is connected as a step-up autotransformer from a 2.4-kV supply. Calculate the currents in each part of the transformer and the load rating. Neglect losses.

Problem 3.B.45

A step-up autotransformer is configured using a 2.4/0.6-kV single-phase transformer to supply a load of 180 kVA. Find the volt-ampere rating of the single-phase transformer, the primary winding current, and the input current.

Problem 3.B.46

Two identical single-phase transformers rated at 200 kVA, 13,200/2300 V are connected in an open-delta connection. Find the load apparent power that can be supplied. Find the apparent power loading of each transformer for rated load at 0.85 PF lagging.

Problem 3.B.47

The open-delta transformer of Problem 3.B.46 is operating such that transformer 1 does not supply any active power ($P_1 = 0$). The apparent power of transformer 2 is 150 kVA. Find the load power factor and its apparent power.

Problem 3.B.48

A Scott connection is used to transform a 230-V three-phase supply to a 230-V two-phase load of 100 kVA at unity power factor. Specify the voltages and currents in the main and teaser transformers.

Electromechanical Energy Conversion Principles

4.1 INTRODUCTION

In Chapter 3 we studied the transformer, which in essence is an energy conversion device with input and output being electric energy. The windings of a transformer are fixed in position relative to each other, in contrast to electromechanical energy conversion devices, which are the focus of our discussion in this and subsequent chapters.

An electromechanical energy conversion device is essentially a medium of transfer between an input side and an output side, as shown in Figure 4.1. In the case of an electric motor, the input is electrical energy drawn from the supply source and the output is mechanical energy supplied to the load, which may be a pump, fan, hoist, or any of a multitude of other mechanical loads. The electric generator is a device that converts mechanical energy supplied by a prime mover (such as a steam turbine or a hydro turbine) to electrical energy at the output side. Many electromechanical energy conversion devices can act as either generator or motor.

The present chapter is devoted to the development of principal relationships that govern the electromechanical energy conversion process. The relationships sought will be valuable in understanding the operation of electric machines studied in subsequent chapters.

4.2 ENERGY FLOW APPROACH

Let us consider an electromechanical energy conversion device operating as a motor from an energy flow point of view. Our intention is to develop a

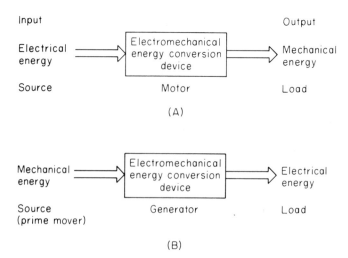

FIGURE 4.1 Functional block diagram of electromechanical energy conversion devices as   (a) motor;   (b) generator.

model of the process that is practical and easy to follow and therefore take a macroscopic approach based on the principle of energy conservation. The situation is best illustrated using the diagrams of Figure 4.2. We assume that an incremental change in electric energy supply dW_e has taken place. This energy flow into the device can be visualized as made up of three components, as shown in Figure 4.2(a). Part of the energy will be imparted to the magnetic field of the device and will result in an increase in the energy stored in the field, denoted by dW_f. A second component of energy will be expended as heat losses dW_{loss}. The third and most important component is that output energy made available to the load (dW_{mech}). The heat losses are due to ohmic (I^2R) losses in the stator (stationary member) and rotor (rotating member), iron or field losses through eddy current and hysteresis as discussed in Chapter 2, and mechanical losses in the form of friction and windage.

In part (b) of the figure, the energy flow is shown in a form that is closer to reality by visualizing Ampère's bonne homme making a trip through the machine. Starting in the stator, ohmic losses will be encountered, followed by field losses and a change in the energy stored in the magnetic field. Having crossed the air gap, our friend will witness ohmic losses in the rotor windings taking place, and in passing to the shaft, bearing frictional losses are also encountered. Finally, a mechanical energy output is available to the load. It should be emphasized here that the phenomena dealt with here are distributed in nature and what we are doing is simply developing an understanding in the form of mathematical expressions called

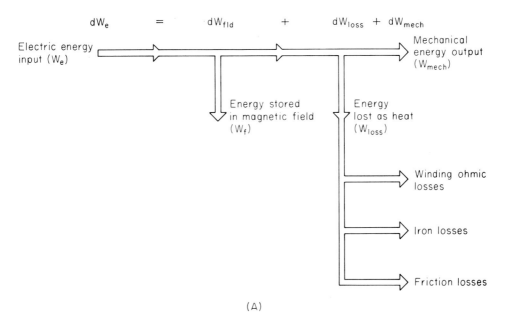

(A)

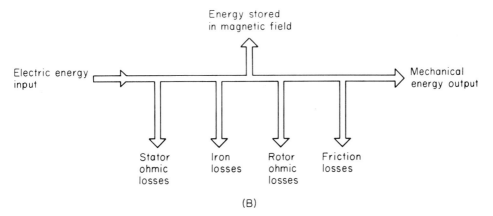

(B)

FIGURE 4.2 Energy flow in an electromechanical energy conversion device: (a) with losses segregated; (b) more realistic representation.

models. The trip by our Amperean friend can never take place in real life but is a helpful means in visualizing the process.

We are now ready to write our energy balance equation based on the foregoing arguments. Here we write

$$dW_e = dW_{fld} + dW_{loss} + dW_{mech}. \tag{4.1}$$

To simplify the treatment, let us assume that losses are negligible, as shown in Figure 4.3.

The electric power input $P_e(t)$ to the device is given in terms of the terminal voltage $e(t)$ and current $i(t)$ by

$$P_e(t) = e(t)i(t) \qquad (4.2)$$

The voltage $e(t)$ is related to the flux linkages λ by Faraday's law:

$$e(t) = \frac{d\lambda}{dt} \qquad (4.3)$$

As a result, we have

$$P_e(t)\, dt = i(t)\, d\lambda \qquad (4.4)$$

We recognize the left-hand side as being the increment in electric energy dW_e, and we therefore write

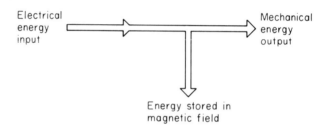

$$dW_e = dW_{fld} + dW_{mech}$$

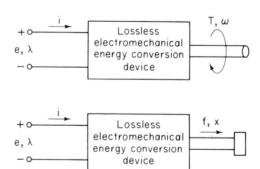

FIGURE 4.3 Schematic representations of lossless electromechanical energy conversion devices.

$$dW_e = i \, d\lambda \tag{4.5}$$

Assuming a lossless device, we can therefore write an energy balance equation which is a modification of Eq. (4.1):

$$dW_e = dW_{fld} + dW_{mech} \tag{4.6}$$

The increment in mechanical output energy can be expressed in the case of a translational (linear motion) increment dx and the associated force exerted by the field F_{fld} as

$$dW_{mech} = F_{fld} \, dx \tag{4.7}$$

In the case of rotary motion the force is replaced by torque T_{fld} and the linear increment dx is replaced by the angular increment $d\theta$:

$$dW_{mech} = T_{fld} \, d\theta \tag{4.8}$$

As a result, we have for the case of linear motion,

$$dW_{fld} = i \, d\lambda - F_{fld} \, dx \tag{4.9}$$

and for rotary motion

$$dW_{fld} = i \, d\lambda - T_{fld} \, d\theta \tag{4.10}$$

The foregoing results state that the net change in the field energy is obtained through knowledge of the incremental electric energy input ($i \, d\lambda$) and the mechanical increment of work done.

The field energy is a function of two states of the system. The first is the displacement variable x (or θ for rotary motion) and the second is either the flux linkages λ or the current i. This follows since knowledge of λ completely specifies i through the λ–i characteristic. Let us take first dependence of W_f on λ and x, and write

$$dW_f(\lambda, x) = \frac{\partial W_f}{\partial \lambda} \, d\lambda + \frac{\partial W_f}{\partial x} \, dx \tag{4.11}$$

The incremental increase in field energy W_f is made up of two components. The first is the product of $d\lambda$ and a (gain factor) coefficient equal to the partial derivative of W_f with respect to λ (x is held constant) and the second component is equal to the product of dx and the partial derivative of W_f

with respect to x (λ is held constant). This is a consequence of Taylor's series for a function of two variables. Comparing Eqs. (4.9) and (4.11), we conclude that

$$i = \frac{\partial W_f(\lambda, x)}{\partial \lambda} \tag{4.12}$$

$$F_{\text{fld}} = -\frac{\partial W_f(\lambda, x)}{\partial x} \tag{4.13}$$

This result states that if the energy stored in the field is known as a function of λ and x, then the electric force developed can be obtained by the partial differentiation shown in Eq. (4.13).

For rotary motion, we replace x by θ in the foregoing development to arrive at

$$T_{\text{fld}} = \frac{-\partial W_f(\lambda, \theta)}{\partial \theta} \tag{4.14}$$

Of course, W_f as a function of λ and θ must be available to obtain the developed torque. Our next task, therefore, is to determine the variations of the field energy with λ and x for linear motion and that with λ and θ for rotary motion.

4.3 FIELD ENERGY

To find the field force we need an expression for the field energy $W_f(\lambda_p, x_p)$ at a given state λ_p and x_p. This can be obtained by integrating the relation of Eq. (4.9), repeated here as

$$dW_f = i\, d\lambda - F_{\text{fld}}\, dx \tag{4.15}$$

from rest ($\lambda = 0$ and $x = 0$) to the given state λ_p and x_p. The path of integration does not have an effect on the final result, as we are dealing with a conservative (no-loss) system. We thus choose the path OAP shown in Figure 4.4. The segment OA is on the $\lambda = 0$ axis and thus $d\lambda = 0$ and $F_{\text{fld}} = 0$ since λ is zero. The segment AP is for a constant x value and hence $dx = 0$. As a result,

$$W_f(\lambda_p, x_p) = \int_0^A dW_f + \int_A^P dW_f$$

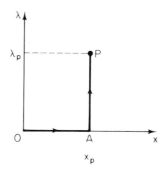

FIGURE 4.4   Integration of dW_f in the λ–x plane for 0 to P via A.

But we have $d\lambda$ and $F_{\text{fld}} = 0$ between 0 and A; thus

$$W_f(\lambda_p, x_p) = \int_A^P dW_f$$

$$= \int_A^P i\, d\lambda$$

As a result,

$$W_f(\lambda_p, x_p) = \int_0^{\lambda_p} i(\lambda, x_p)\, d\lambda \tag{4.16}$$

Equation (4.16) is the desired relationship.

If the λ–i characteristic is linear in i then

$$\lambda = Li \tag{4.17}$$

Thus Eq. (4.16) reduces to

$$W_f(\lambda_p, x_p) = \int_0^{\lambda_p} i(\lambda, x_p)\, d\lambda$$

$$= \frac{1}{L} \int_0^{\lambda_p} \lambda\, d\lambda$$

As a result, we conclude that

$$W_f(\lambda_p, x_p) = \frac{\lambda_p^2}{2L} \tag{4.18}$$

Note that L can be a function of x, as is evident from the study of Example 4.1.

Example 4.1   Consider the magnetic circuit shown in Figure 4.5. Find an expression for the force developed by the field. Suppose that the voltage applied to the coil is

$$e(t) = E_m \cos \omega t$$

Express the developed force as a function of time.

Solution

The reluctance of the magnetic structure is

$$\mathcal{R} = \mathcal{R}_c + \frac{2x}{\mu_0 A}$$

where $\mathcal{R}_c$ is the reluctance of the ferromagnetic structure. Thus the inductance is obtained using

$$L = \frac{N^2}{\mathcal{R}}$$

The energy stored in the field is thus given by Eq. (4.18) as

$$W_f(\lambda, x) = \frac{\lambda^2 \mathcal{R}}{2N^2} = \frac{\lambda^2}{2N^2} \left(\mathcal{R}_c + \frac{2x}{\mu_0 A} \right)$$

The force developed is therefore obtained using Eq. (4.13) as

$$F_{\text{fld}} = - \frac{\partial W_f}{\partial x}$$

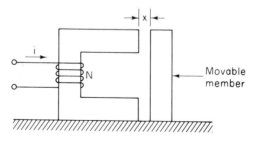

FIGURE 4.5   Magnetic structure for Example 4.1.

$$= -\frac{\lambda^2}{N^2\mu_0 A}$$

Given that the EMF (e) is sinusoidal, then using Faraday's law, we get

$$e(t) = \frac{d\lambda}{dt} = E_m \cos \omega t$$

Thus

$$\int_{\lambda(0)}^{\lambda} d\lambda = \int_0^t E_m \cos \omega t \, dt$$

$$\lambda(t) = \lambda(0) + \frac{E_m}{\omega} \sin \omega t$$

As a result,

$$F_{\text{fld}}(t) = \frac{-1}{N^2\mu_0 A} \left[\lambda(0) + \frac{E_m}{\omega} \sin \omega t \right]^2$$

The desired expression for the force developed by the magnetic field is thus obtained. If we assume that the initial flux linkages are zero,

$$\lambda(0) = 0$$

we have

$$F_{\text{fld}} = \frac{-E_m^2}{N^2\mu_0 A\omega^2} \sin^2 \omega t$$

Recall the identity

$$\sin^2\theta = \frac{1 - \cos 2\theta}{2}$$

As a result, we have

$$F_{\text{fld}} = \frac{-E_m^2}{2N^2\mu_0 A\omega^2} (1 - \cos 2\omega t)$$

This result tells us that the force has a sinusoidal component with a frequency that is twice the frequency of the supply.

The force developed by the field is such that it opposes an increase in x (increased reluctance) and can be written as

$$F_{\text{fld}} = -F_0(1 - \cos 2\omega t)$$

where

$$F_0 = \frac{E_m^2}{2N^2\mu_0 A\omega^2}$$

The next example is a natural extension of Example 4.1 and introduces dynamics in the development.

Example 4.2 The mass of the movable member of the structure of Example 4.1 is M. Find the current $i(t)$ assuming zero initial velocity.

Solution

We can write a dynamic equation for the mass as

$$M\frac{dv}{dt} = F_0(1 - \cos 2\omega t)$$

Thus

$$Mv(t) = F_0\left(t - \frac{1}{2\omega}\sin 2\omega t\right)$$

This assumes that $v(0) = 0$. The displacement $x(t)$ is now obtained as

$$x(t) = x(0) + \frac{F_0}{M}\int_0^t \left(t - \frac{\sin 2\omega t}{2\omega}\right) dt$$

$$= x(0) + \frac{F_0}{M}\left[\frac{t^2}{2} + \frac{1}{4\omega^2}(\cos 2\omega t - 1)\right]$$

Now we have

$$\lambda = Li$$

where L is the system's equivalent inductance, As a result, we obtain, substituting for λ and L,

$$i(t) = \frac{E_m \sin \omega t[\mathcal{R}_c + (2x/\mu_0 A)]}{\omega N^2}$$

Note that the expression of x in terms of F_0, M, ω, and t has to be used in conjunction with the last relation.

Coil Voltage

Using Faraday's law, we have

$$e(t) = \frac{d\lambda}{dt} = \frac{d}{dt}(Li)$$

Thus since L is time dependent, we have

$$e(t) = L\frac{di}{dt} + i\frac{dL}{dt}$$

But

$$\frac{dL}{dt} = \frac{dL}{dx}\left(\frac{dx}{dt}\right) = v\frac{dL}{dx}$$

As a result, we assert that the coil voltage is given by

$$e(t) = L\frac{di}{dt} + iv\frac{dL}{dx} \tag{4.19}$$

where $v = dx/dt$.

For the system of Example 4.1,

$$L = \frac{N^2}{\mathcal{R}_c + 2x/\mu_0 A}$$

Thus

$$\frac{dL}{dx} = \frac{-2N^2}{\mu_0 A(\mathcal{R}_c + 2x/\mu_0 A)^2}$$

$$= \frac{-2L^2}{\mu_0 A N^2}$$

As a result, for the system of Example 4.1 we have

$$e(t) = L\frac{di}{dt} - \frac{2L^2\,iv}{\mu_0 A N^2}$$

4.4 COENERGY

It is more convenient to choose the current i as the independent variable instead of the flux linkages λ. To achieve this we have to express λ as a function of i and x, and hence

$$d\lambda\,(i,\,x) = \frac{\partial\lambda}{\partial i}\,di + \frac{\partial\lambda}{\partial x}\,dx \tag{4.20}$$

Substituting in Eq. (4.15), we get

$$dW_{\text{fld}} = \left(\frac{\partial\lambda}{\partial i}\,di + \frac{\partial\lambda}{\partial x}\,dx\right)i - F_{\text{fld}}\,dx$$

As a result,

$$dW_{\text{fld}} = i\frac{\partial\lambda}{\partial i}\,di + \left(i\frac{\partial\lambda}{\partial x} - F_{\text{fld}}\right)dx \tag{4.21}$$

We also have

$$dW_{\text{fld}} = \frac{\partial W_f}{\partial i}\,di + \frac{\partial W_f}{\partial x}\,dx \tag{4.22}$$

By comparing the foregoing expressions, we conclude that

$$\frac{\partial W_f}{\partial i} = i\frac{\partial\lambda}{\partial i} \tag{4.23}$$

$$\frac{\partial W_f}{\partial x} = i\frac{\partial\lambda}{\partial x} - F_{\text{fld}} \tag{4.24}$$

The last expression can be rearranged into the form

$$F_{\text{fld}} = \frac{\partial}{\partial x}\,(i\lambda - W_f) \tag{4.25}$$

The quantity in parentheses is defined as the coenergy $W_c(i,\,x)$:

$$W_c(i,\,x) = i\lambda(i,\,x) - W_f(i,\,\lambda) \tag{4.26}$$

As a result, we have the alternative expression of the field force given by

$$F_{\text{fld}} = \frac{\partial W_c}{\partial x}(i, x) \tag{4.27}$$

Of course, for rotary motion, we have

$$T_{\text{fld}} = \frac{\partial W_c(i, \theta)}{\partial \theta} \tag{4.28}$$

The coenergy $W_c(i, x)$ is the area under the λ–i characteristic as shown in Figure 4.6. This follows from the defining relation. As a result, we have

$$W_c(i_p, x_p) = \int_0^{i_p} \lambda(i, x_p) \, di \tag{4.29}$$

For a linear λ–i characteristic we have

$$W_c(i_p, x_p) = \int_0^{i_p} Li \, di = \frac{1}{2} Li_p^2 \tag{4.30}$$

Note that W_f and W_c are numerically equal for a linear system.

It is important to emphasize that the field force (or torque) is the same whether it is calculated using the field energy or the coenergy. The signs in the equations are such that the field force acts in a direction to decrease the magnetic field energy stored at constant flux or to increase the coenergy at constant current.

Example 4.3   Find the force developed by the magnetic field of the magnetic structure of Example 4.1 in terms of current.

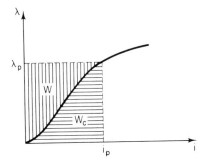

FIGURE 4.6 Defining the coenergy.

Solution

We have, by Eq. (4.30),

$$W_c = \tfrac{1}{2} L(x)i^2$$

The inductance was found to be

$$L = \frac{N^2}{\mathcal{R}} = \frac{N^2}{\mathcal{R}_c + 2x/\mu_0 A}$$

Thus

$$W_c = \frac{N^2 i^2}{2} \frac{1}{\mathcal{R}_c + 2x/\mu_0 A}$$

The force is thus obtained as

$$F_{\text{fld}} = \frac{\partial W_c}{\partial x} = -\frac{N^2 i^2}{\mu_0 A} \left(\frac{1}{\mathcal{R}_c + 2x/\mu_0 A} \right)^2$$

The direction of the force is opposite to that of increasing x.
 The force can be written as

$$F_{\text{fld}} = -\frac{N^2 i^2}{\mu_0 A \mathcal{R}^2}.$$

But

$$Ni = \phi\mathcal{R}$$

Thus

$$F_{\text{fld}} = -\frac{\phi^2}{\mu_0 A}$$

But

$$\lambda = N\phi$$

As a result,

$$F_{\text{fld}} = -\frac{\lambda^2}{N^2 \mu_0 A}$$

This is the same result as that obtained in Example 4.1.

    Another example should confirm the concepts discussed so far.

Example 4.4   The variation of inductance of the coil shown in Figure 4.7 with the distance x is found experimentally to fit the expression

$$L(x) = \frac{L_0}{1 + \alpha x^2}$$

where L_0 and α are constants. The position $x = 0$ corresponds to the core being in the center of the coil. Express the force developed as a function of the current i and the displacement x. If the terminal voltage $e(t)$ is given by

$$e(t) = E_m \cos \omega_s t$$

find the expression of the force in terms of x and time t.

Solution

The coenergy is given by

$$W_c = \tfrac{1}{2} L i^2$$

The force is thus

$$F_{\text{fld}} = \frac{\partial W_c}{\partial x}$$

$$= \tfrac{1}{2} i^2 \frac{\partial}{\partial x} \left(\frac{L_0}{1 + \alpha x^2} \right)$$

Thus we get

$$F_{\text{fld}} = -\frac{L_0}{2} i^2 \frac{2\alpha x}{(1 + \alpha x^2)^2}$$

$$= -\frac{L_0 i^2 \alpha x}{(1 + \alpha x^2)^2}$$

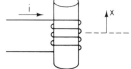

FIGURE 4.7 Coil and plunger for Example 4.4.

The terminal voltage $e(t)$ is given in terms of flux linkages as

$$e(t) = \frac{d\lambda}{dt}$$

As a result,

$$\lambda(t) = \lambda(0) + \int_0^t E_m \cos \omega_s t \, dt$$

$$= \lambda(0) + \frac{E_m}{\omega_s} \sin \omega_s t$$

The energy stored is given by

$$W = \frac{\lambda^2}{2L}$$

Thus the force is obtained as

$$F_{\text{fld}} = -\frac{\lambda^2}{2} \frac{\partial}{\partial x} \left(\frac{1 + \alpha x^2}{L_0} \right)$$

$$= -\frac{\lambda^2 \alpha x}{L_0}$$

As a result,

$$F_{\text{fld}} = -\frac{\alpha x [\lambda(0) + (E_m/\omega_s) \sin \omega_s t]^2}{L_0}$$

4.5   RELUCTANCE MOTORS

The preceding analysis of the electromechanical energy conversion process dealt with devices with single excitation coils. Our examples involved use of a linear motion device to illustrate the concepts discussed. In the preceding development we referred to the case of rotary motion and derived expressions for the torque developed by the electromagnetic field in terms of energy stored as well as in terms of coenergy. The reader may wish to consult Problems 4.A.1 and 4.A.2 for simple extensions of Examples 4.1 and 4.2 to rotational motion.

The present section deals with an important class of singly excited

$$L = a + b = L_{max}$$

At $\theta = \pi/2$,

$$L = a - b = L_{min}$$

As a result,

$$L = \tfrac{1}{2} [(L_{max} + L_{min}) + (L_{max} - L_{min}) \cos 2\theta] \qquad (4.35)$$

Figure 4.9 shows the variation of L with θ. It should be noted that this sinusoidal variation is an approximation of the actual variation. Design of the pole shapes is based on attempting to achieve this ideal situation.

The coenergy in this system is given by

$$W_c = \tfrac{1}{2} Li^2$$

The torque developed by field is therefore obtained using

$$T_{fld} = \frac{\partial W_c}{\partial \theta}$$

Thus we conclude that

$$T_{fld} = i^2(L_{min} - L_{max}) \sin 2\theta \qquad (4.36)$$

If we let

$$\Delta L = L_{max} - L_{min} \qquad (4.37)$$

the torque developed by the field is given by

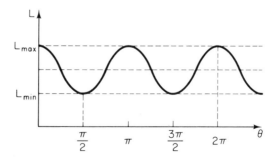

FIGURE 4.9   Variation of inductance with angular position θ for the electromechanical system of Figure 4.8.

$$T_{\text{fld}} = -i^2 \, \Delta L \sin 2\theta \qquad (4.38)$$

Note that if $\Delta L = 0$, no torque is developed. This takes place when the rotor is cylindrical and thus the air gap is uniform, resulting in $L_{\max} = L_{\min}$. For the unsymmetrical rotor case we refer to the torque developed as a reluctance torque.

Let us assume that the rotor is moving at an angular velocity ω_r starting at $t = 0$ with an initial angle θ_0. Therefore, we have

$$\theta = \omega_r t + \theta_0 \qquad (4.39)$$

Assume, moreover, that the current $i(t)$ is sinusoidal described by

$$i(t) = I_m \cos \omega_s t \qquad (4.40)$$

The torque developed by the field is therefore obtained as

$$T_{\text{fld}} = -\Delta L \, I_m^2 \cos^2 \omega_s t \sin [2(\omega_r t + \theta_0)]$$

Recall that

$$\cos 2\alpha = 2 \cos^2 \alpha - 1$$

Thus we have

$$T_{\text{fld}} = -\frac{\Delta L I_m^2}{2} (1 + \cos 2\omega_s t) \sin [2(\omega_r t + \theta_0)]$$

Recall that

$$\cos \alpha \sin \beta = \frac{\sin (\alpha + \beta) - \sin (\alpha - \beta)}{2}$$

Thus we have

$$T_{\text{fld}} = -\frac{\Delta L I_m^2}{2} \left(\sin [2(\omega_r t + \theta_0)] + \frac{1}{2} \sin \{2[(\omega_r + \omega_s)t + \theta_0]\} \right.$$
$$\left. - \frac{1}{2} \sin \{2[(\omega_s - \omega_r)t - \theta_0]\} \right) \qquad (4.41)$$

The three sinusoids in the expression above each have an average of zero except for the last two, which do not average to zero if $|\omega_s| = |\omega_r|$. Thus a necessary condition for nonzero average torque is

$$|\omega_r| = |\omega_s| \qquad (4.42)$$

The average torque is thus given by

$$T_{av} = -\frac{\Delta L I_m^2}{4} \sin 2\theta_0 \qquad (4.43)$$

This shows that an average torque will be produced for either direction of rotation if condition (4.42) is satisfied.

The principle of operation of the reluctance motor is used in an electromechanical energy conversion device commonly found in high-tech applications requiring precision movements required for robotic and disk positioning purposes. This is the subject of our next section.

4.6 VARIABLE-RELUCTANCE STEPPER MOTORS

A stepper motor is a device that produces a definite angular displacement of a driven shaft and is capable of holding its position against an applied torque. The stepper motor is controlled by digital command signals that apply current pulses to the motor's windings. In Figure 4.10, one method of constructing a variable-reluctance stepper motor is shown. The rotor has eight poles and the stator is composed of three segments a, b, and c each with eight poles, as shown. Segment b is identical with segment a but its poles are displaced through 15° in a counterclockwise direction. Segment c has its poles displaced a further 15° counterclockwise.

Let θ be the angular displacement of the rotor measured clockwise. The air-gap torque due to the current in segment a, denoted by i_a, is

$$T_a = \frac{1}{2} i_a^2 \frac{dL_a}{d\theta} \qquad (4.44)$$

The inductance L_a has a maximum (or direct-axis value) when the rotor poles are aligned with the stator poles. It will have a minimum when the rotor poles are midway between the stator poles. Assume that L_a varies sinusoidally to obtain

$$L_a = L_0 + L_m \cos 8\theta \qquad (4.45)$$

Thus the torque due to current i_a is

$$T_a = -4i_a^2 L_m \sin 8\theta \qquad (4.46)$$

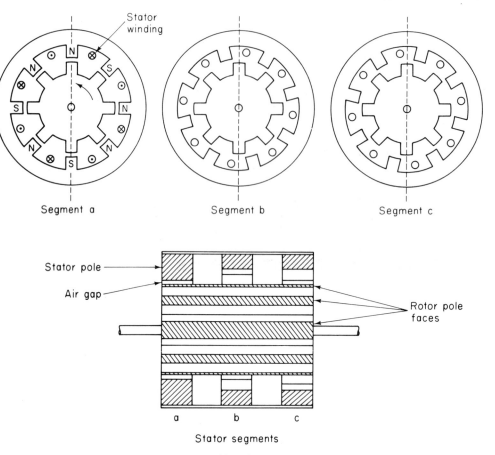

FIGURE 4.10 Variable reluctance stepper motor.

Note that this torque is zero for $\theta = 0$.

The inductance of segment b is

$$L_b = L_0 + L_m \cos 8(\theta - 15°) \qquad (4.47)$$

If a current i_b is flowing in segment b of the stator,

$$T_b = -4i_b^2 L_m \sin 8(\theta - 15°) \qquad (4.48)$$

It is clear that if segment a is energized and a torque causing a counterclockwise rotation θ is applied to the motor shaft, the inductance L_a is reduced and negative (clockwise) torque T_a will be developed by the motor. If current i_a is set to zero and current i_b is established, the motor will develop a torque rotating the shaft counterclockwise through 15°. If current i_b is set to zero and current i_c is set to zero and current i_c is established, a further 15° counterclockwise rotation results. Finally, if i_c is set to zero and current i_a is reestablished, then completion of a 45° rotation of the rotor results. Further current pulses in the sequence *a-b-c* produce further counterclockwise steps.

A more powerful system results if two segments are active simultaneously, such as $ab - bc - ca$ for positive rotation. The assumption of equal current, $i_a = i_b = i_c = i$, results in

$$T_{ab} = T_a + T_b = -4i^2 L_m[\sin 8\theta + \sin 8(\theta - 15)]$$

Recall that

$$\sin \alpha + \sin \beta = 2 \sin \frac{\alpha + \beta}{2} \cos \frac{\alpha - \beta}{2}$$

Thus

$$T_{ab} = -8i^2 L_m \sin 8(\theta - 7.5°) \cos (60°) \tag{4.49}$$

The developed torque is therefore given by

$$T_{ab} = -4 \sqrt{3} \, i^2 L_m \sin 8(\theta - 7.5°) \tag{4.50}$$

The torque is zero for $\theta = 7.5°$ and the peak value is $\sqrt{3}$ times that due to one segment acting alone.

The step angle of the rotor is determined by the number of poles. The typical step angles are 15°, 7.5°, 5°, 2.5°, and 2°. The stepper motor is used in digital control systems where the motor receives open-loop commands in the form of trains of pulses to turn the shaft specific angles.

4.7 MULTIPLY EXCITED SYSTEMS

Systems with more than one exciting winding are referred to as multiply excited systems. We encounter this situation in most rotating electromechanical energy conversion devices. The force (or torque) can be obtained by simple extension of the techniques discussed in the foregoing sections. We will consider a system with three windings as shown in Figure 4.11. Of course, our discussion can be simply extended to an arbitrary number of windings n.

The differential electric energy input is

$$dW_e = i_1 \, d\lambda_1 + i_2 \, d\lambda_2 + i_3 \, d\lambda_3 \tag{4.51}$$

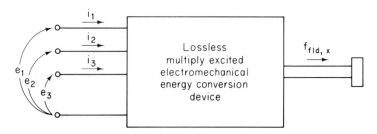

FIGURE 4.11 Multiply excited lossless electromechanical energy conversion device.

The mechanical energy increment is given by

$$dW_{mech} = F_{fld}\, dx$$

Thus the field energy increment is obtained as

$$dW_{fld} = dW_e - dW_{mech}$$
$$= i_1\, d\lambda_1 + i_2\, d\lambda_2 + i_3\, d\lambda_3 - F_{fld}\, dx \qquad (4.52)$$

If we express W_{fld} in terms of λ_1, λ_2, λ_3, and x, we have

$$dW_f = \frac{\partial W_f}{\partial \lambda_1}\, d\lambda_1 + \frac{\partial W_f}{\partial \lambda_2}\, d\lambda_2 + \frac{\partial W_f}{\partial \lambda_3}\, d\lambda_3 + \frac{\partial W_f}{\partial x}\, dx \qquad (4.53)$$

As a result,

$$F_{fld} = -\frac{\partial W_f(\lambda_1, \lambda_2, \lambda_3, x)}{\partial x} \qquad (4.54)$$

$$i_1 = \frac{\partial W_f(\lambda_1, \lambda_2, \lambda_3, x)}{\partial \lambda_1} \qquad (4.55)$$

$$i_2 = \frac{\partial W_f(\lambda_1, \lambda_2, \lambda_3, x)}{\partial \lambda_2} \qquad (4.56)$$

$$i_3 = \frac{\partial W_f(\lambda_1, \lambda_2, \lambda_3, x)}{\partial \lambda_3} \qquad (4.57)$$

The singly excited case can be seen to be included as a special case of our present result.

Similar expressions using the coenergy concept can be derived. The result is

$$F_{\text{fld}} = + \frac{\partial W_c(i_1, i_2, i_3, x)}{\partial x} \tag{4.58}$$

If we are dealing with a rotational system, we have

$$T_{\text{fld}} = - \frac{\partial W_f(\lambda_1, \lambda_2, \lambda_3, \theta)}{\partial \theta}$$

$$T_{\text{fld}} = + \frac{\partial W_c(i_1, i_2, i_3, \theta)}{\partial \theta} \tag{4.59}$$

We will need to find an expression for the field energy at a state corresponding to point P, where $\lambda_1 = \lambda_{1p}$, $\lambda_2 = \lambda_{2p}$, $\lambda_3 = \lambda_{3p}$, and $x = x_p$. Again, since we are dealing with a conservative system (no loss), we will be able to use a simple path of integration $OABCP$ as follows:

1. $OA:$ $\lambda_1 = \lambda_2 = \lambda_3 = 0$ and x varies from 0 to x_p. Thus $d\lambda_1 = d\lambda_2 = d\lambda_3 = 0$ and $F_{\text{fld}} = 0$, and we conclude that

$$\int_0^A dW_{\text{fld}} = \int_0^A (i_1 \, d\lambda_1 + i_2 \, d\lambda_2 + i_3 \, d\lambda_3 - F_{\text{fld}} \, dx)$$

$$= 0 \tag{4.60}$$

2. $AB:$ $\lambda_1 = 0$, $\lambda_2 = 0$, $x = x_p$, and λ_3 is allowed to vary from 0 to λ_{3p}. Thus $d\lambda_1 = d\lambda_2 = 0$, $dx = 0$, and we have

$$\int_A^B dW_{\text{fld}} = \int_A^B (i_1 \, d\lambda_1 + i_2 \, d\lambda_2 + i_3 \, d\lambda_3 - F_{\text{fld}} \, dx)$$

$$= \int_{\lambda_3 = 0}^{\lambda_{3p}} i_{3AB} \, d\lambda_3 \tag{4.61}$$

Here we have

$$i_{3AB} = i_3 \, (\lambda_1 = 0, \, \lambda_2 = 0, \, \lambda_3, \, x_p) \tag{4.62}$$

That is, λ_1 and λ_2 are held at zero while x is at x_p and the current is a function of λ_3 only.

3. $BC:$ $\lambda_1 = 0$, $\lambda_3 = \lambda_{3p}$, $x = x_p$, and λ_2 is varied from zero to λ_{2p}. Here we have $d\lambda_1 = d\lambda_3 = dx = 0$ and we obtain

$$\int_B^C dW_{\text{fld}} = \int_{\lambda_2 = 0}^{\lambda_{2p}} i_{2BC} \, d\lambda_2 \tag{4.63}$$

where

$$i_{2BC} = i_2 (\lambda_1 = 0, \lambda_2, \lambda_3 = \lambda_{3p}, x_p) \qquad (4.64)$$

4. CP: $\lambda_2 = \lambda_{2p}$, $\lambda_3 = \lambda_{3p}$, $x = x_p$, and λ_1 varies from zero to λ_{1p}. We have

$$\int_C^P dW_{\text{fld}} = \int_{\lambda_1=0}^{\lambda_{1p}} i_{1CP} d\lambda_1 \qquad (4.65)$$

with

$$i_{1CP} = i_1(\lambda_1, \lambda_2 = \lambda_{2p}, \lambda_3 = \lambda_{3p}, x = x_p) \qquad (4.66)$$

We can now conclude that

$$W_f(\lambda_{1p}, \lambda_{2p}, \lambda_{3p}, x_p) = \int_0^A dW_{\text{fld}} + \int_A^B dW_{\text{fld}} + \int_B^C dW_{\text{fld}} + \int_C^P dW_{\text{fld}}$$

This now can be written as

$$W_f(\lambda_{1p}, \lambda_{2p}, \lambda_{3p}, x_p)$$
$$= \int_0^{\lambda_{3p}} i_{3AB} \, d\lambda_3 + \int_0^{\lambda_{2p}} i_{2BC} \, d\lambda_2 + \int_0^{\lambda_{1p}} i_{1CP} \, d\lambda_1 \quad (4.67)$$

In an analogous manner we obtain the coenergy by replacing each λ by the corresponding i, and vice versa, to obtain

$$W_c(i_{1p}, i_{2p}, i_{3p}, x_p) = \int_0^{i_{3p}} \lambda_{3AB} \, di_3 + \int_0^{i_{2p}} \lambda_{2BC} \, di_2 + \int_0^{i_{1p}} \lambda_{1CP} \, di_1 \quad (4.68)$$

The expressions for λ in the integrands are

$$\lambda_{3AB} = \lambda_3 (i_1 = i_2 = 0, i_3, x_p) \qquad (4.69)$$

$$\lambda_{2BC} = \lambda_2(i_1 = 0, i_2, i_{3p}, x_p) \qquad (4.70)$$

$$\lambda_{1CP} = \lambda_3(i_1, i_{2p}, i_{3p}, x_p) \qquad (4.71)$$

It is clear that the three-winding case cannot be depicted easily in a multidimensional space. For a two-winding system we can show the path of integration by way of Figure 4.12.

To illustrate the algebraic details, we consider the case of a linear sys-

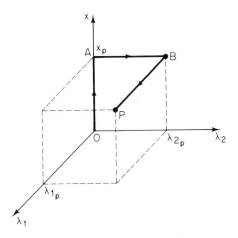

FIGURE 4.12 Path of integration for a doubly excited system.

tem where the relationships between the λ's and the i's are given in terms of inductances by

$$\lambda_1 = L_{11}i_1 + L_{12}i_2 + L_{13}i_3 \tag{4.72}$$
$$\lambda_2 = L_{21}i_1 + L_{22}i_2 + L_{23}i_3 \tag{4.73}$$
$$\lambda_3 = L_{31}i_1 + L_{32}i_2 + L_{33}i_3 \tag{4.74}$$

The inductances are functions of the displacement x or the angular position θ.

To calculate the coenergy, we have

$$\lambda_{3AB} = L_{33}i_3 \tag{4.75}$$

$$\lambda_{2BC} = L_{22}i_2 + L_{23}i_{3p} \tag{4.76}$$

$$\lambda_{1CP} = L_{11}i_1 + L_{12}i_{2p} + L_{13}i_{3p} \tag{4.77}$$

Now the integrals required for the coenergy are

$$\int_0^{i_{3p}} \lambda_{3AB} \, di_3 = \tfrac{1}{2} L_{33}i_{3p}^2$$

$$\int_0^{i_{2p}} \lambda_{2BC} \, di_2 = \tfrac{1}{2} L_{22}i_{2p}^2 + L_{23}i_{2p}i_{3p}$$

$$\int_0^{i_{1p}} \lambda_{1CP} \, di_1 = \tfrac{1}{2} L_{11}i_{1p}^2 + L_{12}i_{1p}i_{2p} + L_{13}i_{1p}i_{3p}$$

As a result, we have

$$W_c(i_{1p}, i_{2p}, i_{3p}, x_p) = \tfrac{1}{2}(L_{11}i_{1p}^2 + L_{22}i_{2p}^2 + L_{33}i_{3p}^2)$$
$$+ L_{12}i_{1p}i_{2p} + L_{23}i_{2p}i_{3p} + L_{13}i_{1p}i_{3p} \quad (4.78)$$

The result obtained for coenergy can be put in a compact form by defining the current vector $\mathbf{I}_p$ and the inductance matrix $\mathbf{L}$ by

$$\mathbf{I}_p = \begin{bmatrix} i_{1p} \\ i_{2p} \\ i_{3p} \end{bmatrix} \quad (4.79)$$

$$\mathbf{L} = \begin{bmatrix} L_{11} & L_{12} & L_{13} \\ L_{12} & L_{22} & L_{23} \\ L_{13} & L_{23} & L_{33} \end{bmatrix} \quad (4.80)$$

As a result, we have

$$W_c(\mathbf{I}_p, x_p) = \tfrac{1}{2}\mathbf{I}_p^T\mathbf{L}\mathbf{I}_p \quad (4.81)$$

where the superscript T indicates the transpose of a vector.

The energy stored in the field can be obtained using Eq. (4.67). The result is

$$W_f(\lambda_{1p}, \lambda_{2p}, \lambda_{3p}, x_p) = \tfrac{1}{2}(\Gamma_{11}\lambda_{1p}^2 + \Gamma_{22}\lambda_{2p}^2 + \Gamma_{33}\lambda_{3p}^2)$$
$$+ \Gamma_{12}\lambda_{1p}\lambda_{2p} + \Gamma_{23}\lambda_{2p}\lambda_{3p} + \Gamma_{13}\lambda_{1p}\lambda_{3p} \quad (4.82)$$

where

$$i_1 = \Gamma_{11}\lambda_1 + \Gamma_{12}\lambda_2 + \Gamma_{13}\lambda_3 \quad (4.83)$$
$$i_2 = \Gamma_{12}\lambda_1 + \Gamma_{22}\lambda_2 + \Gamma_{23}\lambda_3 \quad (4.84)$$
$$i_3 = \Gamma_{13}\lambda_1 + \Gamma_{23}\lambda_2 + \Gamma_{33}\lambda_3 \quad (4.85)$$

In compact form,

$$\mathbf{I} = \mathbf{\Gamma}\boldsymbol{\lambda} \quad (4.86)$$

where the flux linkage vector $\boldsymbol{\lambda}$ is

$$\boldsymbol{\lambda} = \begin{bmatrix} \lambda_1 \\ \lambda_2 \\ \lambda_3 \end{bmatrix} \quad (4.87)$$

The matrix $\mathbf{\Gamma}$ is the inverse of $\mathbf{L}$:

$$\mathbf{\Gamma} = \mathbf{L}^{-1} \tag{4.88}$$

With elements Γ_{11}, Γ_{12}, . . . ,

$$\mathbf{\Gamma} = \begin{bmatrix} \Gamma_{11} & \Gamma_{12} & \Gamma_{13} \\ \Gamma_{12} & \Gamma_{22} & \Gamma_{23} \\ \Gamma_{13} & \Gamma_{23} & \Gamma_{33} \end{bmatrix} \tag{4.89}$$

The stored energy in the field is therefore

$$W_f(\boldsymbol{\lambda}_p, x_p) = \tfrac{1}{2} \boldsymbol{\lambda}_p^T \mathbf{\Gamma} \boldsymbol{\lambda}_p \tag{4.90}$$

We can express the energy stored in the field in terms of currents and inductances by substituting:

$$\boldsymbol{\lambda} = \mathbf{L}\mathbf{I} \tag{4.91}$$

$$W_f(\boldsymbol{\lambda}_p, x_p) = \tfrac{1}{2} \mathbf{I}_p^T \mathbf{L}^T \mathbf{\Gamma} \mathbf{L} \mathbf{I}_p \tag{4.92}$$

Thus

$$W_f(\boldsymbol{\lambda}_p, x_p) = \tfrac{1}{2} \mathbf{I}_p^T \mathbf{L}^T \mathbf{I}_p \tag{4.93}$$

Note that the energy and coenergy are equal for this linear system.

4.8 DOUBLY EXCITED SYSTEMS

The theoretical discussion of the preceding section provides the basis for treating a multitude of rotating electric machines characterized by more than one exciting winding. As a typical example we take the system shown in Figure 4.13. Here we have a coil on the stator fed by an electric energy source 1 and a second coil mounted on the rotor and fed by source 2. This is an example that will prove useful as a prototype of many practical machines.

For this doubly excited system, we write the relation between flux linkages and currents as

$$\lambda_1 = L_{11}(\theta)i_1 + M(\theta)i_2 \tag{4.94}$$

$$\lambda_2 = M(\theta)i_1 + L_{22}(\theta)i_2 \tag{4.95}$$

The self-inductances L_{11} and L_{22} and the mutual inductance M can be obtained from

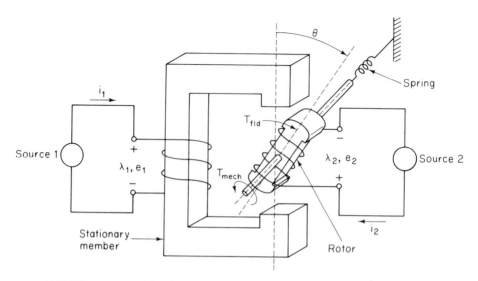

FIGURE 4.13   Doubly excited electromechanical energy conversion device.

$$L_{11}(\theta) = \left.\frac{\lambda_1}{i_1}\right|_{i_2=0} \tag{4.96}$$

$$M(\theta) = \left.\frac{\lambda_1}{i_2}\right|_{i_1=0} = \left.\frac{\lambda_2}{i_1}\right|_{i_2=0} \tag{4.97}$$

$$L_{22}(\theta) = \left.\frac{\lambda_2}{i_2}\right|_{i_1=0} \tag{4.98}$$

We note here that setting $i_2 = 0$ results in a structure excited by the winding on the stator. Therefore, the self-inductance L_{11} is expressed using Eq. (4.35) as

$$L_{11}(\theta) = L_1 + \Delta L_1 \cos 2\theta \tag{4.99}$$

where

$$L_1 = \tfrac{1}{2}(L_{max} + L_{min}) \tag{4.100}$$

$$\Delta L_1 = \tfrac{1}{2}(L_{max} - L_{min}) \tag{4.101}$$

The mutual inductance M is found by evaluating the flux linkages λ_2 of coil 2 due to the current i_1. For $\theta = 0$, the axis of coil 2 is aligned with the stator axis and as a result the mutual inductance is at a peak M_0. When $\theta = \pi/2$, the coil is perpendicular to the stator axis and zero flux linkages take place. For $\theta = \pi$, the rotor coil is now aligned with the stator flux axis but in the reverse direction; hence the mutual inductance is at a minimum of $-M_0$. The sinusoidal that fits this behavior is given by

$$M(\theta) = M_0 \cos \theta \qquad (4.102)$$

To evaluate the self-inductance of coil 2, we carry out the same reasoning process to arrive at

$$L_{22}(\theta) = L_2 + \Delta L_2 \cos 2\theta \qquad (4.103)$$

Note that in many practical applications ΔL_2 is considerably less than L_2 and we may conclude that L_{22} is independent of the rotor position.

The coenergy in this doubly excited system is given by

$$W_c = \tfrac{1}{2} L_{11} i_1^2 + M i_1 i_2 + \tfrac{1}{2} L_{22} i_2^2 \qquad (4.104)$$

The torque is therefore given by

$$T_{\text{fld}} = \frac{\partial W_c}{\partial \theta} \qquad (4.105)$$

$$= \tfrac{1}{2} i_1^2 \frac{\partial L_{11}}{\partial \theta} + i_1 i_2 \frac{\partial M}{\partial \theta} + \tfrac{1}{2} i_2^2 \frac{\partial L_{22}}{\partial \theta} \qquad (4.106)$$

Substituting from Eqs. (4.99), (4.102), and (4.103) into (4.106), we get

$$T_{\text{fld}} = - [(i_1^2 \, \Delta L_1 + i_2^2 \, \Delta L_2) \sin 2\theta + i_1 i_2 \, M_0 \sin \theta)] \qquad (4.107)$$

Let us define

$$T_R = i_1^2 \, \Delta L_1 + i_2^2 \, \Delta L_2 \qquad (4.108)$$

$$T_M = i_1 i_2 M_0 \qquad (4.109)$$

Thus the torque developed by the field is written as

$$T_{\text{fld}} = - (T_M \sin \theta + T_R \sin 2\theta) \qquad (4.110)$$

We note that for a round rotor, the reluctance of the air gap is constant and hence the self-inductances L_{11} and L_{22} are constant, with the result that $\Delta L_1 = \Delta L_2 = 0$. We thus see that for a round rotor $T_R = 0$, and in this case

$$T_{\text{fld}} = - T_M \sin \theta \qquad (4.111)$$

With this in mind we can see that for an unsymmetrical rotor, the torque is made up of a reluctance torque $T_R \sin 2\theta$ and the primary torque $T_M \sin \theta$.

4.9 SALIENT-POLE MACHINES

The majority of electromechanical energy conversion devices used in present-day applications are in the rotating electric machinery category with symmetrical stator structure. From a broad geometric configuration point of view, such machines can be classified as being either of the salient-pole or round-rotor (smooth air gap) classes. We now discuss the salient-pole type, as this class is a simple extension of the discussion of the preceding section.

In a salient-pole machine, one member (the rotor in our discussion) has protruding or salient poles and thus the air gap between stator and rotor is not uniform, as shown in Figure 4.14. It is clear that the results of Section 4.8 are applicable here and we simply modify these results to conform with common machine terminology, shown in Figure 4.14. Subscript 1 is replaced by s to represent stator quantities and subscript 2 is replaced by r to represent rotor quantities. Thus we rewrite Eq. (4.94) as

$$\lambda_s = (L_s + \Delta L_s \cos 2\theta)i_s + (M_0 \cos \theta)i_r \qquad (4.112)$$

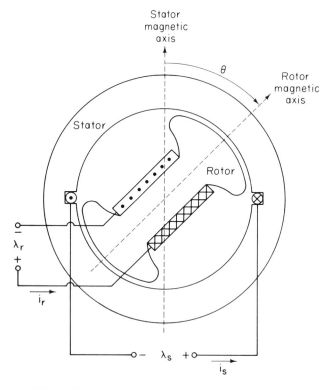

FIGURE 4.14 Two-pole single-phase salient-pole machine with saliency on the rotor.

Similarly, Eq. (4.95) is rewritten as

$$\lambda_r = (M_0 \cos \theta)i_s + L_r i_r \tag{4.113}$$

Note that we assume that L_{22} is independent of θ and is represented by L_r. Thus $\Delta L_2 = 0$ under this assumption. The developed torque given by Eq. (4.107) is thus written as

$$T_{\text{fld}} = -i_s i_r M_0 \sin \theta - i_s^2 \Delta L_s \sin 2\theta \tag{4.114}$$

We define the primary or main torque T_1 by

$$T_1 = -i_s i_r M_0 \sin \theta \tag{4.115}$$

We also define the reluctance torque T_2 by

$$T_2 = -i_s^2 \Delta L_s \sin 2\theta \tag{4.116}$$

Thus we have

$$T_{\text{fld}} = T_1 + T_2 \tag{4.117}$$

Let us assume that the source currents are sinusoidal, as indicated below:

$$i_s(t) = I_s \sin \omega_s t \tag{4.118}$$

$$i_r(t) = I_r \sin \omega_r t \tag{4.119}$$

Assume also that the rotor is rotating at an angular speed ω_m and hence

$$\theta(t) = \omega_m t + \theta_0 \tag{4.120}$$

Our intention is to examine the nature of the instantaneous torque developed under these conditions.

The primary or main torque T_1, expressed by Eq. (4.115), reduces to the following form under the assumptions of Eqs. (4.118) to (4.120):

$$T_1 = -\frac{I_s I_r M_0}{4} \{[(\omega_m + \omega_s - \omega_r)t + \theta_0] + \sin [(\omega_m - \omega_s + \omega_r)t + \theta_0]$$
$$- \sin [(\omega_m + \omega_s + \omega_r)t + \theta_0] - \sin [(\omega_m - \omega_s - \omega_r)t + \theta_0)]\} \tag{4.121}$$

An important characteristic of an electric machine is the average torque developed. Examining Eq. (4.121), we note that T_1 is made of four sinusoidal components each of zero average value if the coefficient of t is different from zero. It thus follows that as a condition for nonzero average of T_1, we must satisfy one of the following:

$$\omega_m = \pm \omega_s \pm \omega_r \qquad (4.122)$$

For example, when

$$\omega_m = -\omega_s + \omega_r$$

then

$$T_{1av} = -\frac{I_s I_r M_0}{4} \sin \theta_0$$

and when

$$\omega_m = \omega_s + \omega_r$$

$$T_{1av} = \frac{I_s I_r M_0}{4} \sin \theta_0$$

The reluctance torque T_2 of Eq. (4.116) can be written using Eqs. (4.118) to (4.120) as

$$T_2 = -\frac{I_s^2 \, \Delta L_s}{4} \{2 \sin (2\omega_m t + 2\theta_0) - \sin [2(\omega_m + \omega_s)t + 2\theta_0] - \sin [2(\omega_m - \omega_s)t + 2\theta_0]\} \qquad (4.123)$$

The reluctance torque will have an average value for

$$\omega_m = \pm \omega_s \qquad (4.124)$$

When either of the two conditions is satisfied,

$$T_{2av} = \frac{I_s^2 \, \Delta L_s}{4} \sin 2\theta_0$$

4.10 ROUND OR SMOOTH AIR-GAP MACHINES

The preceding section dealt with a machine with salient poles. Our discussion led to developing conditions for producing a torque with nonzero average value under sinusoidal excitation to the rotor and stator coils. A round-rotor machine is a special case of a salient-pole machine where the air gap between stator and rotor is (relatively) uniform. The term *smooth air gap* is an idealization of the situation illustrated in Figure 4.15. It is clear that for the case of a smooth air-gap machine the term ΔL_s is zero, as the reluctance does not vary with the angular displacement θ. Therefore, for the machine of Figure 4.15, we have

$$\lambda_s = L_s i_s + M_0 \cos \theta i_r \qquad (4.125)$$

$$\lambda_r = M_0 \cos \theta i_s + L_r i_r \qquad (4.126)$$

Under the assumptions of Eqs. (4.118) to (4.120), we obtain

$$T_{\text{fld}} = T_1 \qquad (4.127)$$

where T_1 is as defined in Eq. (4.121).

We have concluded that for an average value of T_1 to exist, one of the conditions of Eq. (4.122) must be satisfied:

$$\omega_m = \pm \omega_s \pm \omega_r \qquad (4.128)$$

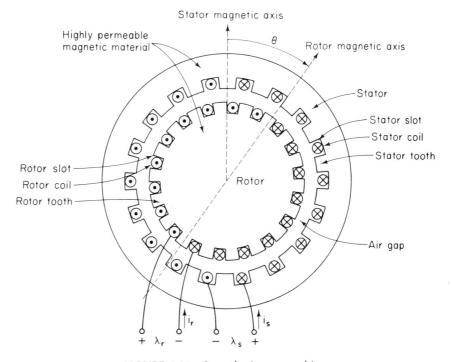

FIGURE 4.15   Smooth air-gap machine.

We have seen that for

$$\omega_m = -\omega_s + \omega_r \tag{4.129}$$

then

$$T_{av} = -\frac{I_s I_r M_0}{4} \sin \theta_0 \tag{4.130}$$

Now substituting Eq. (4.129) in (4.121), we get

$$T_{fld} = -\frac{I_s I_r M_0}{4} \{\sin \theta_0 + \sin [2(\omega_r - \omega_s)t + \theta_0]$$
$$- \sin (2\omega_r t + \theta_0) - \sin (-2\omega_s t + \theta_0)\} \tag{4.131}$$

We observe that the first term is a constant, whereas the other three terms are still sinusoidal time functions and each represents an alternating torque. Although these terms are of zero average value, they can cause speed pulsations and vibrations that may be harmful to the machine's operation and life. The alternating torques can be eliminated by adding additional windings to the stator and rotor, as discussed presently.

Two-Phase Machines

Consider the machine of Figure 4.16, where each of the distributed windings is represented by a single coil. It is clear that this is an extension of the machine of Figure 4.15 by adding one additional stator winding (bs) and one additional rotor winding (br) with the relative orientation shown in Figure 4.16(b). Our analysis of this machine requires first setting up the inductances required. This can be best done using the vector terminology of Section 4.7. We can write for this four-winding system:

$$\begin{bmatrix} \lambda_{as} \\ \lambda_{ar} \\ \lambda_{bs} \\ \lambda_{br} \end{bmatrix} = \begin{bmatrix} L_s & M_0 \cos \theta & 0 & -M_0 \sin \theta \\ M_0 \cos \theta & L_r & M_0 \sin \theta & 0 \\ 0 & M_0 \sin \theta & L_s & M_0 \cos \theta \\ -M_0 \sin \theta & 0 & M_0 \cos \theta & L_r \end{bmatrix} \begin{bmatrix} i_{as} \\ i_{ar} \\ i_{bs} \\ i_{br} \end{bmatrix} \tag{4.132}$$

or in the notation of Eq. (4.91),

$$\lambda = \mathbf{LI} \tag{4.133}$$

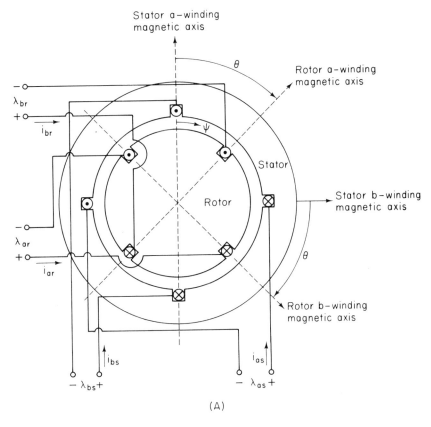

(A)

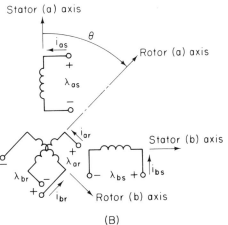

(B)

FIGURE 4.16   Two-phase smooth air-gap machine.

To explain Eq. (4.132), we note first that the upper left corner block of the matrix L represents the self-inductances and mutual inductances involved between the stator and rotor phase a coils as given in Eqs. (4.125) and (4.126). The lower right-hand corner block corresponds to phase b and we note that we assume that the phase a and b coils are similar in this discussion. What remains to be discussed is the upper right corner block (which is noted to be equal to the transponse of the lower left corner block), representing the mutual interaction between phase a and phase b. The mutual terms $L_{(as)(bs)}$ and $L_{(ar)(br)}$ are zero since phases a and b are at right angles. The term $L_{(as)(br)}$ is equal to $-M_0 \sin \theta$ by the use of projections of flux, and similarly, $L_{(ar)(bs)}$ is equal to $M_0 \sin \theta$.

The field energy and coenergy are the same as given by Eq. (4.81) or (4.93). The torque is obtained in the usual manner. Let us now assume that the terminal currents are given by the balanced, two-phase current sources

$$i_{as} = I_s \cos \omega_s t \qquad (4.134)$$

$$i_{bs} = I_s \sin \omega_s t \qquad (4.135)$$

$$i_{ar} = I_r \cos \omega_r t \qquad (4.136)$$

$$i_{br} = I_r \sin \omega_r t \qquad (4.137)$$

We also assume that

$$\theta(t) = \omega_m t + \theta_0 \qquad (4.138)$$

The torque is given by

$$T_{\text{fld}} = M_0[(i_{ar}i_{bs} - i_{br}i_{as}) \cos \theta - (i_{ar}i_{as} + i_{br}i_{bs}) \sin \theta] \qquad (4.139)$$

Substituting Eqs. (4.134) through (4.138) into (4.139), we obtain (after some manipulations)

$$T_{\text{fld}} = -M_0 I_s I_r \sin [(\omega_m - \omega_s + \omega_r)t + \theta_0] \qquad (4.140)$$

The condition for nonzero average torque is given by

$$\omega_m = \omega_s - \omega_r \qquad (4.141)$$

For this condition, we have

$$T_{\text{fld}} = -M_0 I_s I_r \sin \theta_0 \qquad (4.142)$$

The instantaneous torque in this case is constant in spite of the excitation being sinusoidal.

4.11 MACHINE-TYPE CLASSIFICATION

The results of the analysis leading to Eq. (4.128) for the machine configuration of Figure 4.15 and Eq. (4.141) for the machine configuration of Figure 4.16 provide a basis for defining conventional machine types.

Synchronous Machines

The two-phase machine of Figure 4.16 is excited with direct current applied to the rotor ($\omega_r = 0$) and balanced two-phase currents of frequency ω_s applied to the stator. Condition (4.141) with $\omega_r = 0$ yields

$$\omega_m = \omega_s \tag{4.143}$$

Thus, to produce a torque with nonzero average value, our machine's rotor should be running at the single value defined by the stator sources. This mode of operation yields a synchronous machine which is so named because it can convert average power only at one mechanical speed—the synchronous speed, ω_s. The synchronous machine is the main source of electric energy in modern power systems acting as a generator and is the subject of Chapter 6.

Induction Machines

Single-frequency alternating currents are fed into the stator circuits and the rotor circuits are all short circuited in a conventional induction machine. The machine of Figure 4.16 is used again for the analysis. Equations (4.134) and (4.135) still apply and are repeated here:

$$i_{a_s} = I_s \cos \omega_s t \tag{4.144}$$

$$i_{b_s} = I_s \sin \omega_s t \tag{4.145}$$

With the rotor circuits short-circuited,

$$v_{ar} = v_{br} = 0 \tag{4.146}$$

and the rotor is running according to Eq. (4.138):

$$\theta(t) = \omega_m t + \theta_0 \tag{4.147}$$

Conditions (4.146) are written as

$$v_{ar} = R_r i_{ar} + \frac{d\lambda_{ar}}{dt} = 0 \tag{4.148}$$

$$v_{br} = R_r i_{br} + \frac{d\lambda_{br}}{dt} = 0 \tag{4.149}$$

Here we assume that each rotor phase has a resistance of R_r Ω. We have, by Eq. (4.132),

$$\lambda_{ar} = M_0 \cos \theta i_{as} + L_r i_{ar} + M_0 \sin \theta i_{bs} \tag{4.150}$$

$$\lambda_{br} = -M_0 \sin \theta i_{as} + M_0 \cos \theta i_{bs} + L_r i_{br} \tag{4.151}$$

As a result, we have

$$0 = R_r i_{ar} + L_r \frac{di_{ar}}{dt} + M_0 I_s \frac{d}{dt} [\cos \omega_s t \cos (\omega_m t + \theta_0) + \sin \omega_s t \sin (\omega_m t + \theta_0)] \tag{4.152}$$

and

$$0 = R_r i_{br} + L_r \frac{di_{br}}{dt} + M_0 I_s \frac{d}{dt} [-\cos \omega_s t \sin (\omega_m t + \theta_0) + \sin \omega_s t \cos (\omega_m t + \theta_0)] \tag{4.153}$$

A few manipulations provide us with

$$M_0 I_s (\omega_s - \omega_m) \sin [(\omega_s - \omega_m)t - \theta_0] = L_r \frac{di_{ar}}{dt} + R_r i_{ar} \tag{4.154}$$

$$-M_0 I_s (\omega_s - \omega_m) \cos [(\omega_s - \omega_m)t - \theta_0] = L_r \frac{di_{br}}{dt} + R_r i_{br} \tag{4.155}$$

The right-hand sides are identical linear first-order differential operators. The left sides are sinusoidal voltages of equal magnitude but 90° apart in phase. The rotor currents will have a frequency of $(\omega_s - \omega_m)$, which satisfies condition (4.141), and thus an average power and an average torque will be produced by the induction machine.

The induction machine is the subject of detailed study in Chapter 5. For now let us emphasize the facts that the currents induced in the rotor have a frequency of $(\omega_s - \omega_m)$ and that average toruqe can be produced.

4.12   P-POLE MACHINES

The configuration of the magnetic field resulting from coil placement in the magnetic structure determines the number of poles in an electric machine. An important point to consider is the convention adopted for assigning polarities in schematic diagrams, which is discussed presently. Consider the bar magnet of Figure 4.17(a). The magnetic flux lines are shown as closed

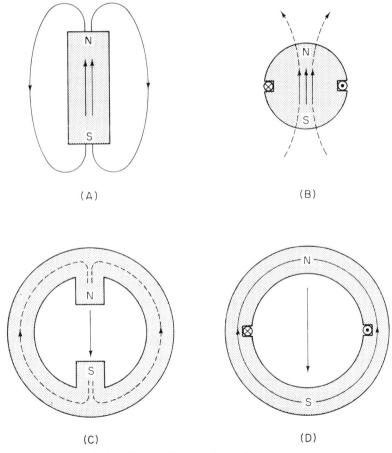

FIGURE 4.17   Two-pole configurations.

loops oriented from the south pole to the north pole within the magnetic material. Figure 4.17(b) shows a two-pole rotor with a single coil with current flowing in the direction indicated by the dot and cross convention. According to the right-hand rule, the flux lines are directed upward inside the rotor material, and as a result we assert that the south pole of the electromagnet is on the bottom part and that the north pole is at the top, as shown.

The situation with a two-pole stator is explained in Figure 4.17 (c) and (d). First consider part (c), showing a permanent magnet shaped as shown. According to our convention, the flux lines are oriented away from the south pole toward the north pole within the magnetic material (not in air gap). For part (d) we have an electromagnet resulting from the insertion of a single coil in slots on the periphery of the stator as shown. The flux lines are oriented in accordance with the right-hand rule and we conclude that the north and south pole orientations are as shown in the figure.

Consider now the situation illustrated in Figure 4.18, where two coils are connected in series and placed on the periphery of the stator in part (a) and on the rotor in part (b). An extension of the prior arguments concerning a two-pole machine results in assigning poles as shown in the figure. A four-pole machine results from the combination of the stator and rotor of Figure 4.18 and is shown in Figure 4.19 to illustrate the orientation of the magnetic axes of rotor and stator.

It is clear that any arbitrary even number of poles can be achieved by placing the coils of a given phase in symmetry around the periphery of

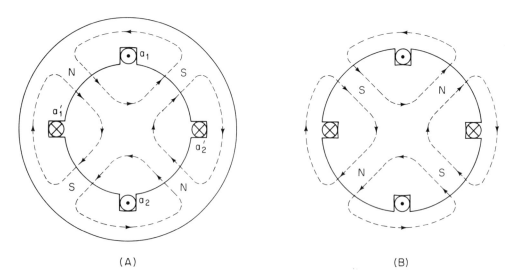

(A)   (B)

FIGURE 4.18 Four-pole configurations: (a) stator arrangement; (b) rotor arrangement.

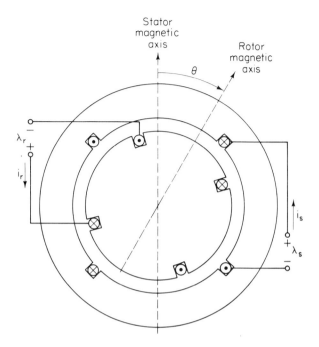

FIGURE 4.19   Four-pole single-phase machine.

stator and rotor of a given machine. The number of poles is simply the number encountered in one round trip around the periphery of the air gap. It is necessary for successful operation of the machine to have the same number of poles on the stator and rotor.

Consider the four-pole, single-phase machine of Figure 4.19. Because of the symmetries involved, the mutual inductance can be seen to be

$$M(\theta) = M_0 \cos 2\theta \qquad (4.156)$$

Compared with Eq. (4.102) for a two-pole machine, we can immediately assert that for a P-pole machine,

$$M(\theta) = M_0 \cos \frac{P\theta}{2} \qquad (4.157)$$

where P is the number of poles.

We note here that our treatment of the electric machines in Sections 4.9 through the present section was focused on two-pole configurations. It is clear that extending our analytic results to a P-pole machine can easily be done by replacing the mechanical angle θ in a relation developed for a two-pole machine by the angle $P\theta/2$ to arrive at the corresponding relation for a

P-pole machine. As an example, relations (4.125) and (4.126) for a P-pole machine are written as

$$\lambda_s = L_s i_s + \left(M_0 \cos \frac{P\theta}{2} \right) i_r \tag{4.158}$$

$$\lambda_r = \left(M_0 \cos \frac{P\theta}{2} \right) i_s + L_r i_r \tag{4.159}$$

Similarly, the torque expression (4.115) becomes

$$T_1 = -i_s i_r M_0 \sin \frac{P\theta}{2} \tag{4.160}$$

Note that θ in the expressions above is in mechanical degrees.

The torque T_1 under the sinusoidal excitation conditions (4.118) and (4.119) given by Eq. (4.121) is rewritten for a P-pole machine as

$$
\begin{aligned}
T_1 = -\frac{I_s I_r M_0}{4} \Bigg\{ &\sin\left[\left(\frac{P\omega_m}{2} + \omega_s - \omega_r \right) t + \frac{P\theta_0}{2} \right] \\
+ &\sin\left[\left(\frac{P\omega_m}{2} - \omega_s + \omega_r \right) t + \frac{P\theta_0}{2} \right] \\
- &\sin\left[\left(\frac{P\omega_m}{2} + \omega_s + \omega_r \right) t + \frac{P\theta_0}{2} \right] \\
- &\sin\left[\left(\frac{P\omega_m}{2} - \omega_s - \omega_r \right) t + \frac{P\theta_0}{2} \right] \Bigg\}
\end{aligned}
\tag{4.161}
$$

The conditions for average torque production of Eq. (4.122) are written for a P-pole machine as

$$\omega_m = \frac{2}{P} (\pm \omega_s \pm \omega_r) \tag{4.162}$$

Thus for given electrical frequencies the mechanical speed is reduced as the number of poles is increased.

A time saving and intuitively appealing concept in dealing with P-pole machines is that of electrical degrees. Let us define the angle θ_e corresponding to a mechanical angle θ for a P-pole machine as

$$\theta_e = \frac{P}{2} \theta \tag{4.163}$$

With this definition we see that all statements, including θ for a two-pole machine, apply to any P-pole machine with θ taken as an electrical angle.

Consider the first condition of Eq. (4.162) with $\omega_r = 0$ corresponding to synchronous machine operation:

$$\omega_m = \frac{2}{P} \omega_s \qquad (4.164)$$

The stator angular speed ω_s is related to frequency f_s in hertz by

$$\omega_s = 2\pi f_s \qquad (4.165)$$

The mechanical angular speed ω_m is related to the mechanical speed n in revolutions per minute by

$$\omega_m = \frac{2\pi n}{60} \qquad (4.166)$$

Combining Eqs. (4.164) to (4.166), we obtain

$$f_s = \frac{Pn}{120} \qquad (4.167)$$

This is an important relation in the analysis of rotating electric machines.

SOME SOLVED PROBLEMS

Problerm 4.A.1

Consider the electromechanical energy conversion device shown in Figure 4.20. Assume that the reluctance of the structure varies with the angle θ as

$$\mathcal{R} = \mathcal{R}_o + \mathcal{R}_m\theta$$

Find expressions for the torque developed by the field using first the current as an independent variable and then using the flux linkages as independent variable. Show that the two expressions yield the same result.

Solution

In terms of current, we have

$$W_c = \frac{1}{2} Li^2 = \frac{N^2 i^2}{2\mathcal{R}}$$

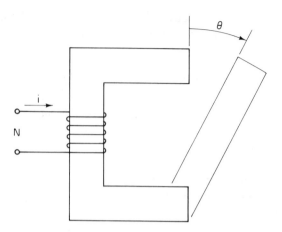

FIGURE 4.20 Magnetic structure for Problem 4.A.1.

The torque is thus given by

$$T_{\text{fld}} = \frac{\partial W_c}{\partial \theta} = -\frac{N^2 i^2}{2\mathscr{R}^2} \frac{\partial \mathscr{R}}{\partial \theta}$$

$$= -\frac{N^2 i^2}{2\mathscr{R}^2} \mathscr{R}_m$$

We conclude that in terms of current and angular displacement,

$$T_{\text{fld}} = \frac{-N^2 i^2 \mathscr{R}_m}{2(\mathscr{R}_0 + \mathscr{R}_m \theta)^2}$$

In terms of flux linkages, we have

$$W = \frac{\lambda^2}{2L} = \frac{\lambda^2 \mathscr{R}}{2N^2}$$

Thus

$$T_{\text{fld}} = -\frac{\partial W}{\partial \theta} = -\frac{\lambda^2 \mathscr{R}_m}{2N^2}$$

Thus the desired relationship is obtained.

The last expression for the torque can be rewritten using $\lambda = N\phi$ as

$$T_{\text{fld}} = -\frac{\phi^2 \mathscr{R}_m}{2}$$

But

$$\phi = \frac{Ni}{\mathcal{R}}$$

As a result,

$$T_{\text{fld}} = -\frac{N^2 i^2}{2\mathcal{R}^2} \mathcal{R}_m$$

which is the same result as that obtained using coenergy.

Problem 4.A.2

Assume that the excitation current for the device of Problem 4.A.1 is a direct current $i(t) = I_o$ and that the angular displacement is

$$\theta(t) = a + bt + \frac{1}{2} ct^2$$

where a, b, and c are constant. Find the voltage $e(t)$, neglecting the coil resistance.

Solution

The voltage $e(t)$ is obtained from

$$e(t) = \frac{d\lambda}{dt} = \frac{d}{dt} Li$$

$$= L \frac{di}{dt} + i \frac{dL}{dt}$$

But i is constant and thus

$$e(t) = I \frac{dL}{dt}$$

We invoke the chain rule to obtain

$$e(t) = I \frac{dL}{d\mathcal{R}} \left(\frac{d\mathcal{R}}{d\theta} \right) \frac{d\theta}{dt}$$

Now

$$L = \frac{N^2}{\mathscr{R}}$$

Thus

$$\frac{dL}{d\mathscr{R}} = -\frac{N^2}{\mathscr{R}^2}$$

Also,

$$\mathscr{R} = \mathscr{R}_o + \mathscr{R}_m \theta$$

Thus

$$\frac{d\mathscr{R}}{d\theta} = \mathscr{R}_m$$

We also have

$$\frac{d\theta}{dt} = b + ct$$

As a result, we obtain

$$e(t) = -\frac{N^2 I \mathscr{R}_m (b + ct)}{\mathscr{R}^2}$$

Substituting for $\mathscr{R}$, we get

$$e(t) = -\frac{N^2 I \mathscr{R}_m (b + ct)}{[\mathscr{R}_o + \mathscr{R}_m(a + bt + \tfrac{1}{2} ct^2)]^2}$$

This is the desired expression for the voltage impressed on the coil terminals.

Problem 4.A.3

Assume that the movable part of the structure of Examples 4.1 to 4.3 is attached to a spring with stiffness k and is subject to friction with coefficient B. Describe the system dynamics first in terms of flux linkages λ, displacement x, and velocity v, and then in terms of current, displacement x, and velocity v. Assume that the coil resistance is r.

Solution

At any time t, the following force equation applies:

$$M \frac{dv}{dt} - F_{\text{fld}} + Bv + K(x - x_{\text{o}}) = 0$$

where

$$\frac{dx}{dt} = v$$

The voltage equation of the coil is

$$V(t) = ri(t) + e(t)$$

where

$$e(t) = \frac{d\lambda}{dt}$$

$$\lambda = Li$$

The voltage impressed on the coil is $V(t)$.
  In terms of λ, we have

$$F_{\text{fld}} = - \frac{\lambda^2}{N^2 \mu_{\text{o}} A}$$

Therefore, the required set of equations is

$$\frac{dv}{dt} = \frac{K}{M}(x_{\text{o}} - x) - \frac{B}{M} v - \frac{\lambda^2}{N^2 \mu_{\text{o}} A}$$

$$\frac{dx}{dt} = v$$

$$\frac{d\lambda}{dt} = V(t) - \frac{r\lambda}{L}$$

or

$$\frac{d\lambda}{dt} = V(t) - \frac{r\lambda[\mathcal{R}_c + (2x/\mu_{\text{o}} A)]}{N^2}$$

In terms of i, we have

$$F_{\text{fld}} = -\frac{N^2 i^2}{\mu_o A \mathcal{R}^2}$$

$$= -\frac{N^2 i^2}{\mu_o A}\left(\mathcal{R}_c + \frac{2x}{\mu_o A}\right)^2$$

Thus our required equations are

$$\frac{dv}{dt} = \frac{K}{M}(x_o - x) - \frac{B}{M}v - \frac{N^2 i^2}{\mu_o A}\left(\mathcal{R}_c + \frac{2x}{\mu_o A}\right)^2$$

$$\frac{dx}{dt} = v$$

$$\frac{di}{dt} = -\frac{r}{L}i + \frac{2Liv}{\mu_o A N^2} + \frac{V(t)}{L}$$

In the above,

$$L = \frac{N^2}{\mathcal{R}} = \frac{N^2}{\mathcal{R}_c + 2x/\mu_o A}$$

We note that the three first-order equations are nonlinear.

Problem 4.A.4

A rotating electromechanical energy conversion device has two coils with constant self-inductances. The mutual inductance between the stator and rotor coils is given by

$$L_{12} = M_1 \cos \theta + M_3 \cos 3\theta$$

where θ is the angular displacement between the stator and rotor coil axes. Find the torque developed for

$$i_1(t) = I_1 \sin \omega_1 t$$

$$i_2(t) = I_2 \sin \omega_2 t$$

Assume that

$$\theta(t) = \omega_m t + \alpha$$

Solution

We have

$$\frac{\partial L_{12}}{\partial \theta} = - M_1 \sin \theta - 3M_3 \sin 3\theta$$

As a result,

$$T = i_1 i_2 \frac{\partial L_{12}}{\partial \theta}$$

$$= - I_1 I_2 \sin \omega_1 t \sin \omega_2 t \, (M_1 \sin \theta + 3M_3 \sin 3\theta)$$

The result above can be modified to

$$T = \frac{I_1 I_2}{2} [\cos (\omega_+ t) - \cos (\omega_- t)] \, (M_1 \sin \theta + 3M_3 \sin 3\theta)$$

where

$$\omega_+ = \omega_1 + \omega_2$$

$$\omega_- = \omega_1 - \omega_2$$

We can write T as the sum

$$T = T_1 + T_3$$

where

$$T_1 = \frac{I_1 I_2 M_1}{2} (\cos \omega_+ t - \cos \omega_- t) \sin \theta$$

$$T_3 = \frac{3 I_1 I_2 M_3}{2} (\cos \omega_+ t - \cos \omega_- t) \sin 3\theta$$

The torque T_1 can further be written as

$$T_1 = \frac{I_1 I_2 M_1}{4} [\sin (\omega_+ t + \theta)$$
$$+ \sin (\theta - \omega_+ t)$$
$$- \sin (\omega_- t + \theta)$$
$$- \sin (\theta - \omega_- t)]$$

or

$$T_1 = \frac{I_1 I_2 M_1}{4} \{\sin [(\omega_1 + \omega_2 + \omega_m)t + \alpha]$$
$$+ \sin [(\omega_m - \omega_1 - \omega_2)t + \alpha]$$
$$- \sin [(\omega_1 - \omega_2 + \omega_m)t + \alpha]$$
$$- \sin [(\omega_m - \omega_1 + \omega_2)t + \alpha]\}$$

The torque T_3 is treated similarly, with the result

$$T_3 = \frac{3I_1 I_2 M_3}{4} \{\sin [(\omega_1 + \omega_2 + 3\omega_m)t + 3\alpha]$$
$$+ \sin [(3\omega_m - \omega_1 - \omega_2)t + 3\alpha]$$
$$- \sin [(\omega_1 - \omega_2 + 3\omega_m)t + 3\alpha]$$
$$- \sin [(3\omega_m - \omega_1 + \omega_2)t + 3\alpha]$$

PROBLEMS

Problem 4.B.1

The relationship between current, displacement, and flux linkages in a conversative electromechanical device is given by

$$i = \lambda[0.8\lambda + 2.2(\lambda - 1)^7]$$

Find expressions for the stored energy and the magnetic field force in terms of λ and x. Find the force for $x = 0.75$.

Problem 4.B.2

Repeat Problem 4.B.1 for the relationship

$$i = \lambda^3 + \lambda(0.1\lambda + x)$$

Problem 4.B.3

Find an expression for the force exerted by the magnetic field in an electromechanical device with the following nonlinear characteristic:

$$i = I_o \frac{a\lambda + b\lambda^3}{1 + cx}$$

where I_o, a, b and c are given constants.

Problem 4.B.4

A plunger-type solenoid is characterized by the relation

$$\lambda = \frac{10i}{1 + 10^4 x/2.54}$$

Find the force exerted by the field for $x = 2.54 \times 10^{-3}$ and $i = 11$ A.

Problem 4.B.5

Find expressions for coenergy and force exerted by the field for a conservative system with the following characteristic:

$$\lambda = \frac{i^{1/3} + i^{1/2}}{0.5(x + 1)}$$

Problem 4.B.6

An electromagnet used to lift steel slabs is shown in Figure 4.21. Show that the minimum current required to lift a slab of mass M is given by

$$i = \frac{\mathcal{R}_c + (2x/\mu_o A)}{N} \sqrt{\mu_o M A g}$$

In the above, $\mathcal{R}_c$ is the reluctance of the magnetic core, A is the cross-sectional area of one air gap, and g is the acceleration of gravity.

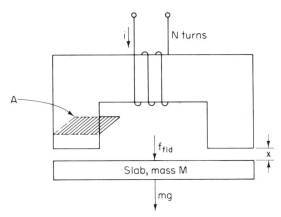

FIGURE 4.21 Electromagnet for Problem 4.B.6.

Problem 4.B.7

Find the minimum value of the current i required to lift a slab with $M = 100$ kg using the electromagnet of Problem 4.B.6, given that

$$A = 16 \times 10^{-4} \text{ m}^2 \qquad x = 0.15 \times 10^{-2} \text{ m}$$
$$N = 400 \text{ turns}$$

Neglect the reluctance of the magnetic core.

Problem 4.B.8

Consider the magnetic structure shown in Figure 4.22. Show that the dynamics of the system are described by

$$\frac{d\lambda(t)}{dt} = \frac{-(g_1 l_1 + x l_2)R}{\mu_o N^2 h l_1 l_2} \lambda(t) + RI(t)$$

$$\frac{d^2 x}{dt^2} = \frac{-\lambda^2}{2\mu_o M N^2 h l_1} + g$$

where g is the acceleration of gravity. Neglect core reluctance.

Problem 4.B.9

Assume that the voltage applied to the coil of the magnetic structure shown in Figure 4.23 is given by

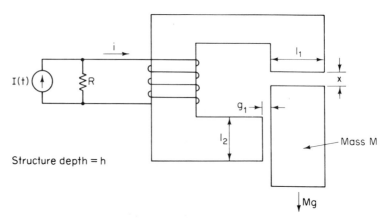

FIGURE 4.22   Structure for Problem 4.B.8.

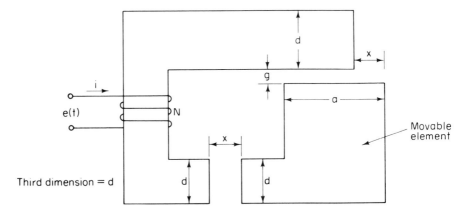

FIGURE 4.23 Magnetic structure for Problem 4.B.9.

$$e(t) = E_m \cos \omega t$$

Assume that $\lambda(0) = 0$, and find the force exerted on the movable part in terms of the structure geometry, E_m, ω, and time.

Problem 4.B.10

For the structure of Problem 4.B.9, assume that $E_m = 10$ V and that the supply frequency is 60 Hz. The coil has 500 turns. The gap length $g = 0.1$ cm, $d = 10$ cm, and $a = 15$ cm. Find the force in terms of x and time. Calculate the average force in terms of x.

Problem 4.B.11

The inductance of a coil used with a plunger-type electromechanical device is given by

$$L = \frac{1.6 \times 10^{-5}}{x}$$

where x is the plunger displacement. Assume that the current in the coil is given by

$$i(t) = 5 \sin \omega t$$

where $\omega = 2\pi(60)$. Find the force exerted by the field for $x = 10^{-2}$ m. Assume that x is fixed and find the necessary voltage applied to the coil terminals given that its resistance is 1 Ω.

Problem 4.B.12

The reluctance of the magnetic structure for a plunger-type relay is given by

$$\mathcal{R} = \frac{400}{x(0.02 - x)}$$

Assume that the exciting coil has 1000 turns and that the exciting current is

$$i = 4 \cos t$$

Find the magnetic field force as a function of time and its average value if the displacement is given by

$$x(t) = 0.01 \cos t$$

Problem 4.B.13

The inductance of the exciting coil of a certain reluctance motor is given by

$$L = \frac{L_0}{1 + a \cos 4\theta}$$

Find an expression for the energy stored in the magnetic field and hence find an expression for the torque developed in terms of flux linkages and the angular displacement θ and the system parameters. Find the torque developed for $\theta = 15°$ and $i = 10$ A. Assume that $L_0 = 0.05$ H and $a = -0.3$.

Problem 4.B.14

Repeat Problem 4.B.13 using the coenergy concept.

Problem 4.B.15

Repeat Problem 4.B.13 for the inductance expressed as

$$L = \frac{L_0}{1 + a(\cos 4\theta + \cos 8\theta)}$$

The value of a in this case is -0.25.

Problem 4.B.16

Repeat Problem 4.B.14 for the conditions stated in Problem 4.B.15.

Problem 4.B.17

The reluctance of the magnetic circuit of a reluctance motor is given by

$$\mathcal{R} = \mathcal{R}_o(1 + a \cos 4\,\theta)$$

Let $\mathcal{R}_o = 2000$ and assume that the number of turns on the exciting coil is 10 turns. Find the developed torque for $i = 10$ A, $\theta = 15°$ assuming that $a = -0.3$.

Problem 4.B.18

The developed torque for the motor of Problem 4.B.13 is given by

$$T = -12\lambda^2 \sin 4\,\theta$$

Assume that

$$\lambda(t) = 0.5 \cos \omega t$$

and that

$$\theta(t) = \omega_r t + \alpha$$

Find the instantaneous torque developed and the condition for nonzero average torque.

Problem 4.B.19

A rotating electromechanical conversion device has a stator and rotor, each with a single coil. The inductances of the device are

$$L_{11} = 0.4 \text{ H} \qquad\qquad L_{22} = 2.4 \text{ H}$$
$$L_{12} = 1.25 \cos \theta \text{ H}$$

where the subscript 1 refers to stator and the subscript 2 refers to rotor. The angle θ is the rotor angular displacement from the stator coil axis. Express the torque as a function of currents i_1, i_2, and θ and compute the torque for $i_1 = 3$ A and $i_2 = 1$ A.

Problem 4.B.20

Assume for the device of Problem 4.B.19 that

$$L_{11} = 0.4 + 0.1 \cos 2\theta$$

All other parameters are unchanged. Find the torque in terms of θ for $i_1 = 3$ A and (a) $i_2 = 0$; (b) $i_2 = 1$ A.

Problem 4.B.21

Assume for the device of Problem 4.B.19 that the stator and rotor coils are connected in series, with the current being

$$i(t) = I_m \sin \omega t$$

Find the instantaneous torque and its average value over one cycle of the supply current in terms of I_m and ω.

Problem 4.B.22

A rotating electromechanical energy conversion device has the following inductances in terms of θ in radians (angle between rotor and stator axes):

$$L_{11} = 0.764\theta$$

$$L_{22} = -0.25 + 1.719\theta$$

$$L_{12} = -0.75 + 1.337\theta$$

Find the torque developed for the following excitations.

(a) $i_1 = 15$ A, $i_2 = 0$.

(b) $i_1 = 0$ A, $i_2 = 15$ A.

(c) $i_1 = 15$ A, $i_2 = 15$ A.

(d) $i_1 = 15$ A, $i_2 = -15$ A.

Problem 4.B.23

For the machine of Problem 4.B.19, assume that the rotor coil terminals are shorted ($e_2 = 0$) and that the stator current is given by

$$i_1(t) = I \sin \omega t$$

Find the torque developed as a function of I, θ, and time.

Problem 4.B.24

For the device of Problem 4.B.23, the rotor coil terminals are connected to a 10-Ω resistor. Find the rotor current in the steady state and the torque developed.

Induction Motors

5.1 INTRODUCTION

An induction machine (Figure 5.1) is one in which alternating currents are supplied directly to stator windings and by transformer action (induction) to the rotor. The flow of power from stator to rotor is associated with a change of frequency, as shown in Chapter 4, with an output being mechanical power transmitted to the load connected to the motor shaft. The induction motor is the most widely used motor in industrial and commercial utilization of electric energy. Reasons for the popularity of induction motors include simplicity, reliability, and low cost, combined with reasonable overload capacity, minimal service requirements, and good efficiency. The rotor of an induction motor may be one of two types. In the wound-rotor motor, distributed windings are employed with terminals connected to insulated slip rings mounted on the motor shaft. The second type is called the squirrel-cage rotor, where the windings are simply conducting bars embedded in the rotor and short-circuited at each end by conducting end rings. The rotor terminals are thus inaccessible in a squirrel-cage construction, whereas the rotor terminals are made available through carbon brushes bearing on the slip rings for the wound-rotor construction.

The stator of a three-phase induction motor carries three sets of windings that are displaced by 120° in space to constitute a three-phase winding set. The application of a three-phase voltage to the stator winding results in the appearance of a rotating magnetic field, as we will find out in the next section. It is clear that induced currents in the rotor (due to the rotating field) interact with the field to produce a torque that rotates the rotor in the direction of the rotating field.

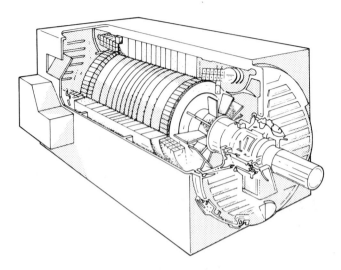

FIGURE 5.1   Cutaway of a three-phase cage-type induction motor.

5.2 MMF WAVES AND THE ROTATING MAGNETIC FIELD

Consider an N-turn coil which spans 180 electrical degrees shown in Figure 5.2. This coil is referred to as a full-pitch coil (as opposed to fractional-pitch coils spanning less than 180 electrical degrees) and is embedded in the stator slots as shown. We will assume that the permeability of the stator and rotor iron paths is much greater than that of the air and therefore that all the reluctance of the magnetic circuit is in the two air gaps. The magnetic field intensity H in the air gap at angle θ is the same in magnitude as that at $\theta + \pi$ due to symmetry, but the fields are in the opposite direction. The MMF around any closed path in the vicinity of the coil side is Ni. The MMF is uniformly distributed around the sides and hence we can conclude that the MMF is constant at $Ni/2$ between the sides ($\theta = -\pi/2$ to $+ \pi/2$) and abruptly changes to $-Ni/2$, as shown in Figure 5.2. In part (a) of the figure the physical arrangement is shown, and in part (b) the rotor and stator are laid flat so that the corresponding waveforms can be shown in part (c) in relation to the physical position on the periphery.

The MMF waveform of the concentrated full-pitch coil is rectangular, as shown in Figure 5.2(c), and one needs to resolve it into a sum of sinusoidals. There are two reasons for this. The first is that electrical engineers prefer to use sinusoids due to the wealth of available techniques enabling convenient analysis of such waveforms. The second is that in the design of ac machines, the windings are distributed such that a close approximation

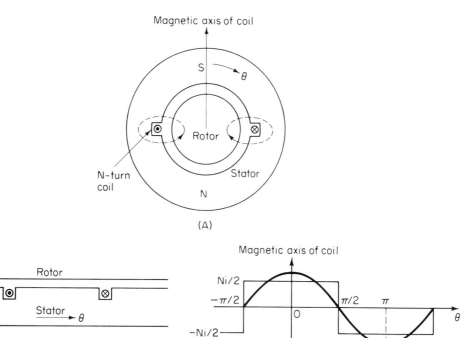

FIGURE 5.2   MMF of a concentrated full-pitch coil: (a) physical arrangement; (b) rotor and stator laid flat;   (c) MMF waves.

to a sinusoid is obtained. In resolving a periodic waveform into a sum of sinusoids, one uses the celebrated Fourier series, which for our present purposes states that the rectangular waveform can be written

$$F_\theta = \frac{4Ni}{2\pi} \left[\cos\theta + \tfrac{1}{3}Ni \cos 3\theta + \cdots \right]$$

$$F_1 = \frac{4}{\pi} \frac{Ni}{2} \cos\theta \tag{5.1}$$

If the remaining harmonics are negligible, we see that the MMF of the coil is approximated by a sinusoidal space waveform.

Distributed Windings

Windings that are spread over a number of slots around the air gap are referred to as distributed windings and are shown in Figure 5.3. Phase a is shown in detail in the figure and we assume a two-pole three-phase ac con-

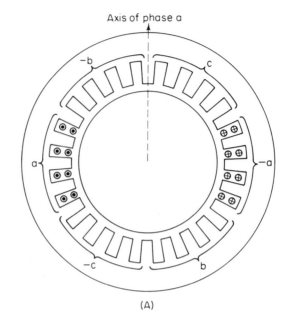

(A)

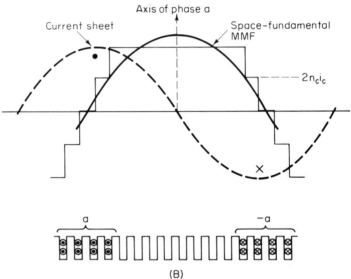

(B)

FIGURE 5.3   MMF wave for one phase (full-pitch coil).

figuration. The empty slots are occupied by phases b and c. We will deal with phase a only, whose winding is arranged in two layers such that each coil of n_c turns has one side in the top of a slot and the other coil side in the bottom of a slot pole pitch (180° electrical) away. The coil current is i_c.

The MMF wave is a series of steps each of height $2n_ci_c$ equal to the ampere conductors in the slot. The fundamental component of the Fourier series expansion of the MMF wave is shown by the sinusoid in the figure. Note that this arrangement provides a closer approximation to a sinusoidal distribution. The fundamental component is given by

$$F_{a_1} = K i_a \cos \theta \qquad (5.2)$$

where

$$K = \frac{4}{\pi} k_w \frac{N_{\text{ph}}}{P} \qquad (5.3)$$

The factor k_w accounts for the distribution of the winding, N_{ph} is the number of turns per phase, and P is the number of poles. It is assumed that the coils are series connected and thus $i_c = i_a$.

The MMF is a standing wave with a sinusoidal spatial distribution around the periphery. Assume that the current i_a is sinusoidal such that

$$i_a(t) = I_m \cos \omega t \qquad (5.4)$$

Thus we have

$$F_{a_1}(t) = K I_m \cos \theta \cos \omega t \qquad (5.5)$$

The relation above can be written as

$$F_{a_1}(t) = A_{a(p)} \cos \theta \qquad (5.6)$$

where $A_{a(p)}$ is the amplitude of the MMF component wave at time t,

$$A_{a(p)} = F_{\text{max}} \cos \omega t \qquad (5.7)$$

with

$$F_{\text{max}} = K I_m \qquad (5.8)$$

Rotating Magnetic Fields

An understanding of the nature of the magnetic field produced by a polyphase winding is necessary for the analysis of polyphase ac machines. We will consider a two-pole three-phase machine. The windings of the individual phases are displaced by 120 electrical degrees in space. This is shown in Figure 5.4. The magnetomotive forces developed in the air gap due to currents in the windings will also be displaced 120 electrical degrees in space.

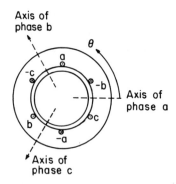

FIGURE 5.4   Simplified two-pole three-phase stator winding.

Assuming sinusoidal, balanced three-phase operation, the phase currents are displaced by 120 electrical degrees in time. The instantaneous values of the currents are

$$i_a = I_m \cos \omega t \qquad (5.9)$$

$$i_b = I_m \cos (\omega t - 120°) \qquad (5.10)$$

$$i_c = I_m \cos (\omega t - 240°) \qquad (5.11)$$

where I_m is the maximum value of the current and the time origin is arbitrarily taken as the instant when the phase a current is a positive maximum. The phase sequence is assumed to be abc. The instantaneous currents are shown in Figure 5.5. As a result, we have

$$A_{a(p)} = F_{\max} \cos \omega t \qquad (5.12)$$

$$A_{b(p)} = F_{\max} \cos (\omega t - 120°) \qquad (5.13)$$

$$A_{c(p)} = F_{\max} \cos (\omega t - 240°) \qquad (5.14)$$

where $A_{a(p)}$ is the amplitude of the MMF component wave at time t.

At time t, all three phases contribute to the air-gap MMF at a point P (whose spatial angle is θ). We thus have the resultant MMF given by

$$A_p = A_{a(p)} \cos \theta + A_{b(p)} \cos (\theta - 120°) + A_{c(p)} \cos (\theta - 240°) \qquad (5.15)$$

Using Eqs. (5.12) to (5.14), we have

$$A_p = F_{\max}[\cos \theta \cos \omega t + \cos (\theta - 120°) \cos (\omega t - 120°) \\ + \cos (\theta - 240°) \cos (\omega t - 240°)] \qquad (5.16)$$

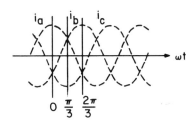

FIGURE 5.5   Instantaneous three-phase currents.

Equation (5.16) can be simplified using the following trigonometric identity:

$$\cos \alpha \cos \beta = \tfrac{1}{2} [\cos (\alpha - \beta) + \cos (\alpha + \beta)]$$

As a result, we have

$$
\begin{aligned}
A_p = \tfrac{1}{2} [& \cos (\theta - \omega t) + \cos (\theta + \omega t) \\
& + \cos (\theta - \omega t) + \cos (\theta + \omega t - 240°) \\
& + \cos (\theta - \omega t) + \cos (\theta + \omega t - 480°)] F_{max}
\end{aligned}
$$

The three cosine terms involving $(\theta + \omega t)$, $(\theta + \omega t - 240°)$, and $(\theta + \omega t - 480°)$ are three equal sinusoidal waves displaced in phase by 120° with a zero sum. Therefore,

$$A_p = \tfrac{3}{2} [F_{max} \cos (\theta - \omega t)] \qquad (5.17)$$

The wave of Eq. (5.17) depends on the spatial position θ as well as time. The angle ωt provides rotation of the entire wave around the air gap at the constant angular velocity ω. At time t_1 the wave is sinusoid with its positive peak displaced ωt_1 from the point P (at θ); at a later instant (t_2), the wave has its positive peak displaced ωt_2 from the same point. We thus see that a polyphase winding excited by balanced polyphase currents produces the same effect as a permanent magnet rotating within the stator.

5.3   SLIP

When the stator is supplied by a balanced three-phase source, it will produce a magnetic field that rotates at synchronous speed as determined by the number of poles and applied frequency f_s. Thus according to Eq. (4.167),

$$n_s = \frac{120 f_s}{P} \quad \text{r/min} \tag{5.18}$$

The rotor runs at a steady speed n_r r/min in the same direction as the rotating stator field. The speed n_r is very close to n_s when the motor is running light and is lower as the mechanical load is increased. The difference $(n_s - n_r)$ is termed the slip and is commonly defined as a per unit value s. Thus

$$s = \frac{n_s - n_r}{n_s} \tag{5.19}$$

As a result of the relative motion between stator and rotor, induced voltages will appear in the rotor with a frequency f_r called the slip frequency. Thus

$$f_r = s f_s \tag{5.20}$$

From the above we can conclude that the induction motor is simply a transformer with a secondary frequency f_r.

Example 5.1   Determine the number of poles, the slip, and the frequency of the rotor currents at rated load for three-phase 5-hp induction motors rated at:

   a. 220 V, 50 Hz, 1440 r/min.
   b. 120 V, 400 Hz, 3800 r/min.

Solution

We use Eq. (5.18) to obtain P, using n_r, the rotor speed given.
   (a). We have

$$P = \frac{120 \times 50}{1440} = 4.17$$

But P should be an even number. Therefore, take $P = 4$. Hence the synchronous speed is

$$n_s = \frac{120 f}{P} = \frac{120 \times 50}{4} = 1500 \text{ r/min}$$

The slip is thus given by

$$s = \frac{n_s - n_r}{n_s} = \frac{1500 - 1440}{1500} = 0.04$$

The rotor frequency is calculated as

$$f_r = sf_s = 0.04 \times 50 = 2 \text{ Hz}$$

(b) We have

$$P = \frac{120 \times 400}{3800} = 12.63$$

Take $P = 12$, to obtain

$$n_s = \frac{120 \times 400}{12} = 4000 \text{ r/min}$$

$$s = \frac{4000 - 3800}{4000} = 0.05$$

$$f_r = 0.05 \times 400 = 20 \text{ Hz}$$

The analysis of the performance of a three-phase induction motor can be greatly facilitated through the use of equivalent circuits, which is the subject of the next section.

5.4 EQUIVALENT CIRCUITS

An equivalent circuit of the three-phase induction motor can be developed on the basis of the foregoing considerations and transformer models treated in Chapter 3. Looking into the stator terminals, we find that the applied voltage V_s will supply the resistive drop $I_s R_1$ as well as the inductive voltage $jI_s X_1$ and the counter EMF E_1, where I_s is the stator current and R_1 and X_1 are the stator effective resistance and inductive reactance, respectively. In a manner similar to that employed for the analysis of the transformer, we model the magnetizing circuit by the shunt conductance G_c and inductive susceptance $-jB_m$.

The induced voltage in the rotor circuit E_2 is referred to the stator side using the ratio of the number of stator conductors N_s to the number of rotor conductors to obtain

$$E_{2s} = \frac{N_s}{N_r} E_2 \tag{5.21}$$

The rotor's induced voltage referred to the stator E_{2s} is related to the stator EMF by

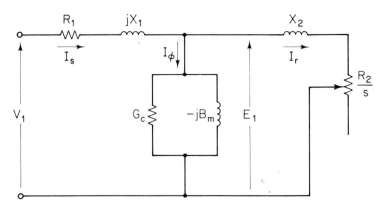

FIGURE 5.6   Equivalent circuit for a three-phase induction motor.

$$E_{2s} = sE_1 \tag{5.22}$$

The inclusion of s in the equation above is necessary to account for the relative motion between stator and rotor. The rotor current referred to the stator side is denoted by I_r. The induced EMF E_{2s} supplies the resistive voltage component I_rR_2 and inductive component $jI_r(sX_2)$. R_2 is the rotor resistance, and X_2 is the rotor inductive reactance on the basis of the stator frequency and voltage. Thus.

$$E_{2s} = I_rR_2 + jIL_r(sX_2)$$

or   (5.23)

$$sE_1 = I_rR_2 + jI_r(sX_2)$$

From the above we conclude that

$$\frac{E_1}{I_r} = \frac{R_2}{s} + jX_2 \tag{5.24}$$

The complete equivalent circuit of the induction motor is shown in Figure 5.6. The rotor circuit impedance referred to the stator side is a series combination of a resistance (R_2/s) and the rotor reactance. Both R_2 and X_2 are referred to the stator side.

Power Considerations

If we consider the active power flow into the induction machine, we find that the input power P_s supplies the stator I^2R losses as well as the core losses. The remaining power denoted by the air-gap power is expended as

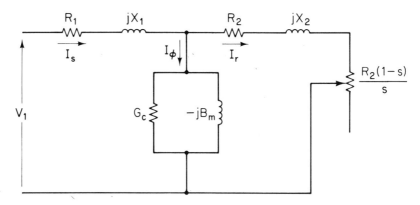

FIGURE 5.7 Modified equivalent circuit of the induction motor.

rotor I^2R losses, with the remainder being the mechanical power delivered to the motor shaft. We can express the air-gap power as

$$P_g = 3I_r^2 \frac{R_2}{s} \qquad (5.25)$$

The rotor I^2R losses are given by

$$P_{1r} = 3I_r^2 R_2 \qquad (5.26)$$

As a result, the mechanical power output (neglecting mechanical losses) is

$$P_r = P_g - P_{1r}$$

$$= 3I_r^2 \frac{1-s}{s} R_2 \qquad (5.27)$$

Formula (5.27) suggests a splitting of R_2/s into the sum of R_2 representing the rotor resistance and a resistance R_e:

$$R_e = \frac{1-s}{s} R_2 \qquad (5.28)$$

which is the equivalent resistance of the mechanical load. As a result, it is customary to modify the equivalent circuit to the form shown in Figure 5.7.

Efficiency

The input power to the motor P_s is given by

$$P_s = 3V_1I_s \cos \phi_s \qquad (5.29)$$

where V_1 is the phase voltage and $\cos \phi_s$ is the motor power factor. The efficiency of the motor is given by

$$\eta = \frac{P_o}{P_s} \tag{5.30}$$

The output power is equal to rotor developed mechanical power minus the mechanical losses due to friction P_f.

$$P_o = P_r - P_f \tag{5.31}$$

Torque

The torque T developed by the motor is related to P_r by

$$T = \frac{P_r}{\omega_r} \tag{5.32}$$

with ω_r being the angular speed of the rotor. Thus

$$\omega_r = \omega_s(1 - s) \tag{5.33}$$

The angular synchronous speed ω_s is given by

$$\omega_s = \frac{2\pi n_s}{60} \tag{5.34}$$

As a result, the torque is given by

$$T = \frac{3I_r^2 R_2}{s\omega_s} \tag{5.35}$$

The torque is slip dependent. An example is appropriate now.

Example 5.2   The power input to a three-phase 220-V 60-Hz three-phase induction motor at full load is 20,800 W. The corresponding line current is 64 A, and the speed is 830 r/min at that load. The motor is rated at 25 hp and has a no-load speed of 895 r/min. Calculate the slip, power factor, torque, and efficiency at full load.

Solution

To find the number of poles, we use

$$n_s \simeq 895 = \frac{120f}{P} = \frac{120(60)}{P}$$

Thus $P \simeq 8.044$. Only integers are allowed; thus $P = 8$ and

$$n_s = \frac{120(60)}{8} = 900 \text{ r/min}$$

Thus at full load

$$s = \frac{900 - 830}{900} = 0.0778$$

The power factor is obtained from

$$\cos \phi = \frac{P_s}{\sqrt{3} \, V_1 \, I_L} = \frac{20{,}800}{220 \sqrt{3} \, (64)} = 0.853$$

The output power at full load is

$$P_o = 25 \times 746 \text{ W}$$

Thus the torque is calculated as

$$T = \frac{25 \times 746}{(2\pi/60)(830)} = 214.57 \text{ N} \cdot \text{m}$$

The efficiency is

$$\eta = \frac{25 \times 746}{20{,}800} = 0.8966$$

5.5 SIMPLIFIED EQUIVALENT CIRCUITS

It is customary to utilize a simplified equivalent circuit for the induction motor in which the shunt branch is moved to the voltage source side. This situation is shown in Figure 5.8. The stator resistance and shunt branch can be neglected in many instances. A total reactance X_T is defined as the sum of X_1 and X_2 as shown.

On the basis of the approximate equivalent circuit, we can find the rotor current as

$$I_r = \frac{V_1}{R_1 + R_2/s + jX_T} \qquad (5.36)$$

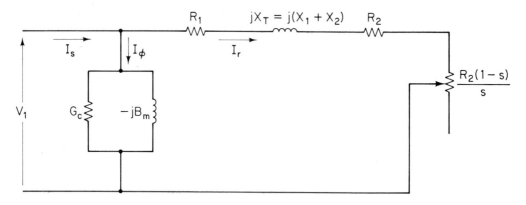

FIGURE 5.8 Approximate equivalent circuit of the induction motor.

At starting, we have $\omega_r = 0$; thus $s = 1$. The rotor starting current is hence given by

$$I_{r_{st}} = \frac{V_1}{(R_1 + R_2) + jX_T} \tag{5.37}$$

It is clear that the motor starting current is much higher than the normal (or full-load) current. Depending on the motor type, the starting current can be as high as six to seven times the normal current. We now consider two examples.

Example 5.3 A 50-hp 440-V three-phase 60-Hz six-pole Y-connected induction motor has the following parameters per phase:

$$R_2 = 0.12 \ \Omega$$

$$R_1 = 0.1 \ \Omega$$

$$G_c = 6.2 \times 10^{-3} \ S$$

$$X_T = 0.75 \ \Omega$$

$$B_m = 0.07 \ S$$

The rotational losses are equal to the stator hysteresis and eddy-current losses. For a slip of 3%, find the following

    a. The line current and power factor.
    b. The horsepower output.
    c. The starting torque.

Solution

(a) Using the equivalent circuit of Figure 5.8, we obtain the current I_r using Eq. (5.36) as

$$I_r = \frac{440/\sqrt{3}}{(0.1 + 0.12/0.03) + j0.75}$$

$$= 60.948 \underline{/-10.366°}$$

The no-load current is

$$I_\phi = \frac{400}{\sqrt{3}} (6.2 \times 10^{-3} - j0.07)$$

$$= 1.575 - j17.78$$

Thus the stator current is

$$I_s = I_r + I_\phi$$

$$= 67.9137 \underline{/-25.0446°}$$

Thus

$$\phi_s = 25.0°$$

$$\cos \phi_s = 0.906 \text{ lagging}$$

(b) The air-gap power is given by

$$P_g = 3I_r^2 \frac{R_2}{s}$$

$$= 3(60.948)^2 \left(\frac{0.12}{0.03}\right) = 44.576 \times 10^3 \text{ W}$$

The mechanical shaft power is

$$P_m = (1 - s)P_g = 43.239 \times 10^3 \text{ W}$$

The core losses are

$$P_c = 3\left(\frac{400}{\sqrt{3}}\right)^2 (6.2 \times 10^{-3}) = 1200 \text{ W}$$

Thus

$$P_{r\ell} = 1200 \text{ W}$$

$$P_o = P_m - P_{r\ell}$$

$$= 42.038 \times 10^3 \text{ W}$$

$$= 56.35 \text{ hp}$$

(c) At starting $s = 1$,

$$I_r = \frac{440/\sqrt{3}}{(0.1 + 0.12) + j0.75} = 325.02 \underline{/-73.652°}$$

$$P_g = 3(325.02)^2(0.12) = 38.029 \text{ kW}$$

$$\omega_s = \frac{2\pi(60)}{3} = 40\pi$$

As a result, we conclude that

$$T = \frac{P_g}{\omega_s} = 302.63 \text{ N} \cdot \text{m}$$

Example 5.4   The rotor resistance and reactance at standstill of a three-phase four-pole 110-V induction motor are 0.18 Ω and 0.75 Ω, respectively. Find the rotor current at starting as well as when the speed is 1720 r/min. Assume that the frequency is 60 Hz. Calculate the equivalent resistance of the load.

Solution

$$R_2 = 0.18 \ \Omega \qquad X_2 = 0.75 \ \Omega$$

$$n_s = \frac{120f}{P} = \frac{120 \times 60}{4} = 1800 \text{ r/min}$$

$$s = \frac{1800 - 1720}{1800} = 0.0444$$

At standstill, $s = 1$,

$$I_r = \frac{V}{\sqrt{R_2^2 + X_2^2}} = \frac{110/\sqrt{3}}{\sqrt{(0.18)^2 + (0.75)^2}} = 82.34 \text{ A}$$

For speed of 1720 r/min, we get

$$I_r = \frac{V}{\sqrt{(R_2/s)^2 + X_2^2}} = 15.419 \text{ A}$$

The equivalent resistance of the load is

$$R_L = R_2 \frac{1 - s}{s}$$

$$= 3.87 \text{ }\Omega$$

5.6   TORQUE CHARACTERISTICS

The torque developed by the motor can be derived in terms of the motor parameters and slip using the expressions given before.

$$T = \frac{3|V_1|^2}{\omega_s} \frac{R_2/s}{(R_1 + R_2/s)^2 + X_T^2} \tag{5.38}$$

The maximum torque occurs for

$$\frac{\partial T}{\partial (R_2/s)} = 0 \tag{5.39}$$

The result is

$$R_1^2 + X_T^2 = \left(\frac{R_2}{s}\right)^2 \tag{5.40}$$

Thus we obtain the slip at which maximum torque occurs as

$$s_{\max T} = \frac{R_2}{\sqrt{R_1^2 + X_T^2}} \tag{5.41}$$

The value of maximum torque is

$$T_{\max} = \frac{3|V_1|^2}{2\omega_s \left(R_1 + \sqrt{R_1^2 + X_T^2}\right)} \tag{5.42}$$

If we neglect the resistance of the stator,

$$s_{maxT} = \frac{R_2}{X_T} \tag{5.43}$$

and

$$T_{max} = \frac{3|V_1|^2}{2\omega_s X_T} \tag{5.44}$$

The maximum torque does not depend on the rotor resistance. The torque–slip variations are shown in Figure 5.9. We now take an example.

Example 5.5   The rotor resistance and reactance of a squirrel-cage induction-motor at standstill are 0.12 Ω per phase and 0.7 Ω per phase, respectively. Assuming a transformer ratio of unity, from the eight-pole stator having a phase voltage of 254 at 60 Hz to the rotor secondary, calculate the following.

    a. Rotor starting current per phase.
    b. The value of slip producing maximum torque.

Solution

(a) At starting $s = 1$; as a result,

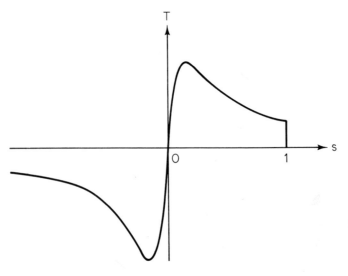

FIGURE 5.9   Torque–slip characteristics for induction motor.

$$I_r = \frac{254}{0.12 + j0.7} = 357.64 \, \underline{/-80.27°} \text{ A}$$

(b)

$$s_{\max T} = \frac{R_2}{X_T} = \frac{0.12}{0.7} = 0.17143$$

Example 5.6 The maximum torque of a 220-V three-phase four-pole 60-Hz induction motor is 180 N · m, at a slip of 0.15. If the motor is operated on a 50-Hz 220-V supply, find the value of the maximum torque, slip, and speed at which this occurs. Neglect stator impedance.

Solution

We have

$$T_{\max} = \frac{3|V|^2}{2\omega_s X_T}$$

In terms of frequency, we have

$$T_{\max} = \frac{3|V|^2}{(2\pi/60)(120f/P)(2\pi f L_T)}$$

Thus

$$T_{\max} = \frac{3|V|^2 P}{8\pi^2 f^2 L_T}$$

The number of poles does not change and the inductance L_T is assumed unchanged. As a result,

$$\frac{T_{\max 2}}{T_{\max 1}} = \frac{|V_2|^2 f_1^2}{|V_1|^2 f_2^2}$$

We thus obtain

$$T_{\max 2} = 180 \left[\frac{200(60)}{220(50)} \right]^2$$

$$= 214.21 \text{ N} \cdot \text{m}$$

The slip at maximum torque is

$$S_{\max T} = \frac{R_2}{X_T} = \frac{R_2}{2\pi f L_T}$$

Thus

$$\frac{s_2}{s_1} = \frac{f_1}{f_2}$$

$$s_2 = 0.15\left(\frac{60}{50}\right) = 0.18$$

The slip at maximum torque for 50 Hz is 18%. The synchronous speed in this case is

$$n_s = \frac{120(50)}{4} = 1500 \text{ r/min}$$

Thus the speed at maximum torque is

$$n_r = 1500(1 - 0.18) = 1230 \text{ r/min}$$

This is lower than that for maximum torque at 60 Hz.

$$n_r = 1800(1 - 0.15) = 1530 \text{ r/min}$$

5.7 SOME USEFUL RELATIONS

In many electric machine problems, knowledge of variables such as current and torque under a certain operating condition can be used to obtain the corresponding values under another operating condition. To start, we consider the torque equation

$$T = \frac{3I_r^2 R_2}{s\omega_s}$$

For operating condition 1, we have

$$T_1 = \frac{3I_1^2 R_2}{s_1\omega_s}$$

For a different operating condition, we have

$$T_2 = \frac{3I_2^2 R_2}{s_2\omega_s}$$

As a result

$$\frac{T_1}{T_2} = \frac{s_2}{s_1} \left(\frac{I_1}{I_2}\right)^2 \tag{5.45}$$

If for example, condition 1 is starting the motor ($s = 1$) and condition 2 is full-load operation s_f, then

$$\frac{T_{st}}{T_f} = s_f \left(\frac{I_{st}}{I_f}\right)^2 \tag{5.46}$$

Consider now the rotor current equation written for condition 1 as

$$I_1^2 = \frac{V^2}{(R_2/s_1)^2 + X_T^2}$$

For condition 2, we have

$$I_2^2 = \frac{V^2}{(R_2/s_2)^2 + X_T^2}$$

Thus

$$\left(\frac{I_1}{I_2}\right)^2 = \frac{(R_2/s_2)^2 + X_T^2}{(R_2/s_1)^2 + X_T^2}$$

Recall that

$$\frac{R_2}{X_T} = s_{maxT}$$

As a result,

$$\left(\frac{I_1}{I_2}\right)^2 = \frac{1 + (s_{maxT}/s_2)^2}{1 + (s_{maxT}/s_1)^2} \tag{5.47}$$

Let condition 1 be starting the motor and condition 2 be that for full load; then

$$\left(\frac{I_{st}}{I_f}\right)^2 = \frac{1 + (s_{maxT}/s_f)^2}{1 + s_{maxT}^2}$$

Similarly,

$$\left(\frac{I_{\max T}}{I_f}\right)^2 = \frac{1 + (s_{\max T}/s_f)^2}{2}$$

We can combine Eqs. (5.45) and (5.47) to get

$$\frac{T_1}{T_2} = \frac{s_2}{s_1} \frac{1 + (s_{\max T}/s_2)^2}{1 + (s_{\max T}/s_1)^2} \tag{5.48}$$

From the relation above, we can conclude that first

$$\frac{T}{T_{\max}} = \frac{2s\,s_{\max T}}{s_{\max T}^2 + s^2} \tag{5.49}$$

This provides a relation between torque at slip s and the maximum torque in terms of slip. We can further write

$$\frac{T_{st}}{T_{\max}} = \frac{2s_{\max T}}{1 + s_{\max T}^2} \tag{5.50}$$

Finally,

$$\frac{T_{st}}{T_f} = s_f \frac{1 + (s_{\max T}/s_f)^2}{1 + s_{\max T}^2} \tag{5.51}$$

The following example illustrates the concept of operating conditions comparison.

Example 5.7 The full-load slip of a squirrel-cage induction motor is 0.05, and the starting current is four times the full-load current. Neglecting the stator core and copper losses as well as the rotational losses, obtain:

   a. The ratio of starting torque to the full-load torque.
   b. The ratio of maximum to full-load torque and the corresponding slip.

Solution

(a) We have

$$s_{fld} = 0.05 \qquad I_{st} = 4I_{fld}$$

We have, neglecting stator resistance, using Eq. (5.46):

$$\frac{T_{st}}{T_{fld}} = \left(\frac{I_{st}}{I_{fld}}\right)^2 s_{fld}$$

Thus

$$\frac{T_{st}}{T_{fld}} = (4)^2(0.05) = 0.8$$

(b) We have

$$I_{fld}^2 = \frac{V_1^2}{(R_2/s)^2 + X_T^2}$$

$$I_{st}^2 = \frac{V_1^2}{R_2^2 + X_T^2}$$

Thus we get

$$\left(\frac{I_{st}}{I_{fld}}\right)^2 = \frac{(R_2/0.05)^2 + X_T^2}{R_2^2 + X_T^2} = (4)^2$$

Thus we have

$$400R_2^2 + X_T^2 = 16R_2^2 + 16X_T^2$$

$$384R_2^2 = 15X_T^2$$

As a result,

$$s_{maxT} = \frac{R_2}{X_T} = \sqrt{\frac{15}{384}} = 0.19764$$

We now have

$$T_{fld} = \frac{3|V_1|^2(R_2/s)}{(R_2/s)^2 + X_T^2}$$

Substituting

$$R_2 = X_T s_{maxT}$$

we get

$$T_{fld} = \frac{3|V_1|^2(s_{maxT}/s)}{X_T[1 + (s_{maxT}/s)^2]}$$

As a result, we use Eq. (5.44) to get

$$\frac{T_{max}}{T_{fld}} = \frac{s_{fld}}{s_{maxT}} \frac{(s_{max}/s_{fld})^2 + 1}{2}$$

$$= \frac{0.05}{0.19764} \frac{(0.19764/0.05)^2 + 1}{2}$$

$$= 2.1$$

5.8 INTERNAL MECHANICAL POWER

The internal mechanical power developed by an induction motor can be analyzed in a manner similar to the torque. The internal mechanical power is given by

$$P_m = 3I_r^2 R_2 \frac{1 - s}{s} \tag{5.52}$$

Substituting for the rotor current, we obtain

$$P_m = 3V_1^2 \frac{R_2(1 - s)/s}{(R_1 + R_2/s)^2 + X_T^2} \tag{5.53}$$

Let

$$a = \frac{R_2}{s} \tag{5.54}$$

Thus

$$P_m = 3V_1^2 \frac{a - R_2}{(R_1 + a)^2 + X_T^2} \tag{5.55}$$

The maximum internal mechanical power is obtained by setting $\partial P_m/\partial a = 0$, to obtain

$$(R_1 + a)^2 + X_T^2 + 2(R_1 + a)(R_2 - a) = 0$$

This reduces to

$$a^2 - 2aR_2 - [X_T^2 + R_1(R_1 + 2R_2)] = 0 \tag{5.56}$$

Define a new reactance:

$$\widetilde{X}_T^2 = X_T^2 + R_1(R_1 + 2R_2) \tag{5.57}$$

Clearly, if stator resistance is negligible, then $\widetilde{X}_T$ and X_T are equal. As a result, the condition for maximum power is

$$a^2 - 2aR_2 - \widetilde{X}_T^2 = 0$$

Thus the required value of a is

$$a = R_2 + \sqrt{R_2^2 + \widetilde{X}_T^2} \tag{5.58}$$

As a result, the slip at maximum power is obtained as

$$s_{\max P} = \frac{R_2}{R_2 + Z_B} \tag{5.59}$$

where Z_B is the blocked rotor impedance defined by

$$Z_B = \sqrt{R_2^2 + \widetilde{X}_T^2} \tag{5.60}$$

The maximum internal power is now found as

$$P_{m\max} = \frac{3V_1^2}{2(R_1 + a)}$$

or
$$\tag{5.61}$$

$$P_{m\max} = \frac{3V_1^2}{2(R_1 + R_2 + Z_B)}$$

where

$$Z_B = \sqrt{R_1^2 + X_T^2 + R_2(R_2 + 2R_1)} \tag{5.62}$$

The slip at which maximum internal power is developed is less than the slip for maximum torque. This follows since

$$s_{\mathrm{max}T} = \frac{R_2}{\sqrt{R_1^2 + X_T^2}} \tag{5.63}$$

As a result,

$$\frac{s_{\mathrm{max}T}}{s_{\mathrm{max}P}} = \frac{R_2 + Z_B}{\sqrt{R_1^2 + X_T^2}} \tag{5.64}$$

This ratio is greater than 1 since Z_B is greater than $\sqrt{R_1^2 + X_T^2}$.

In most applications it is convenient to neglect the stator resistance R_1. This yields simpler expressions. In this case we have

$$Z_B = \sqrt{R_2^2 + X_T^2} \tag{5.65}$$

$$s_{\mathrm{max}P} = \frac{R_2}{R_2 + Z_B} \tag{5.66}$$

$$P_{m\mathrm{max}} = \frac{3V_1^2}{2(R_2 + Z_B)} \tag{5.67}$$

The relation of slip at maximum torque to that at maximum power reduces to

$$\frac{s_{\mathrm{max}T}}{s_{\mathrm{max}P}} = \frac{R_2 + \sqrt{R_2^2 + X_T^2}}{X_T} \tag{5.68}$$

Thus we have the good-looking result

$$\frac{1}{s_{\mathrm{max}P}} = \sqrt{\left(\frac{1}{s_{\mathrm{max}T}}\right)^2 + 1} + 1 \tag{5.69}$$

Example 5.8   An eight-pole 60-Hz induction motor has rotor resistance and reactance at standstill of 0.1 Ω and 0.8 Ω per phase, respectively. Assume that the voltage per phase is 120 V and neglect stator and core effects. Find the slip at maximum power, the maximum power developed, and the torque at maximum power. Compare the maximum torque to that obtained for maximum power.

Solution

We have $R_2 = 0.1$ and $X_T = X_2 = 0.8$. Thus

$$Z_B = \sqrt{(0.1)^2 + (0.8)^2} = 0.8062$$

The slip for maximum power is

$$s_{\text{max}P} = \frac{0.1}{0.1 + 0.8062} = 0.1103$$

The maximum power is now obtained as

$$P_{m\text{max}} = \frac{3(120)^2}{2(0.1 + 0.8062)} = 23{,}835.12 \text{ W}$$

The corresponding torque is obtained using

$$T = \frac{P}{\omega_s(1 - s)}$$

Now

$$n_s = \frac{120(60)}{8} = 900 \text{ r/min}$$

Thus

$$T = \frac{23{,}835.12}{(2\,\pi/60)(900)(1 - 0.1103)}$$

$$= 284.2667 \text{ N} \cdot \text{m}$$

The slip for maximum torque is obtained as

$$s_{\text{max}T} = \frac{R_2}{X_T} = \frac{0.1}{0.8} = 0.125$$

The corresponding rotor current is obtained from

$$I_r = \frac{120}{\sqrt{(0.1/0.125)^2 + (0.8)^2}} = 106.066 \text{ A}$$

The mechanical power is obtained for maximum torque as

$$P_m = 3I_r^2 R_2 \frac{1 - s}{s}$$

$$= 3(106.066)^2 \left(\frac{0.1}{0.125}\right) (1 - 0.125)$$

$$= 23,625 \text{ W}$$

This is clearly lower than the maximum power. The maximum torque is now obtained as

$$T_{max} = \frac{P_m}{\omega_s (1 - s_{maxT})}$$

$$= \frac{23,625}{(2\pi/60)(900)(1 - 0.125)} = 286.479 \text{ N} \cdot \text{m}$$

This is higher than the torque at maximum power.

5.9   EFFECTS OF ROTOR IMPEDANCE

Speed control of induction motors of the wound-rotor type can be achieved by inserting additional rotor resistance (R_a). In addition to this, torque control at a given speed can be achieved using this method. Let

$$K = \frac{3V^2}{\omega_s} \tag{5.70}$$

$$a_1 = \frac{R_2}{s_1} \tag{5.71}$$

$$a_2 = \frac{R_2 + R_a}{s_2} \tag{5.72}$$

$$\alpha = \frac{T_2}{T_1}$$

where the subscript 1 refers to operating conditions without additional rotor resistance and the subscript 2 refers to conditions with the additional rotor resistance. Neglect stator resistance in the present analysis and we have

$$T = \frac{3V^2(R_2/s)}{\omega_s[(R_2/s)^2 + X_T^2]} \tag{5.73}$$

or

$$T = K\frac{a}{a^2 + X_T^2} \tag{5.74}$$

Introducing a resistance in the rotor winding will change a. Let the original torque be T_1. Thus

$$T_1 = K\frac{a_1}{a_1^2 + X_T^2} \tag{5.75}$$

Let the new torque with additional resistance be T_2. Thus

$$T_2 = K\frac{a_2}{a_2^2 + X_T^2} \tag{5.76}$$

By the definition of the torque ratio α, we get

$$\alpha = \frac{a_1^2 + X_T^2}{a_2^2 + X_T^2}\frac{a_2}{a_1} \tag{5.77}$$

Equation (5.77) provides us with means to derive the value of additional resistance R_a required to obtain the same torque ($T_1 = T_2$) at two different values of slip (or rotor speed). To do this we set $\alpha = 1$, to obtain

$$a_2(a_1^2 + X_T^2) = a_1(a_2^2 + X_T^2)$$

or

$$(a_2 - a_1)(X_T^2 - a_1a_2) = 0 \tag{5.78}$$

Thus for $T_1 = T_2$, we get $a_1 = a_2$ or

$$\frac{R_2}{s_1} = \frac{R_2 + R_a}{s_2}$$

or

$$R_a = R_2\left(\frac{s_2}{s_1} - 1\right) \tag{5.79}$$

Let us note here that for equal torque, the rotor circuit currents are the same, since

$$I_r = \frac{V}{\sqrt{(R_2/s)^2 + X_T^2}} = \frac{V}{\sqrt{a^2 + X_T^2}} \qquad (5.80)$$

$$\left(\frac{I_{r1}}{I_{r2}}\right)^2 = \frac{a_2^2 + X_T^2}{a_1^2 + X_T^2} = 1 \qquad \text{for} \qquad T_1 = T_2 \qquad (5.81)$$

Example 5.9   The rotor resistance and reactance of a wound-rotor induction motor at standstill are 0.1 Ω per phase and 0.8 Ω per phase, respectively. Assuming a transformer ratio of unity, from the eight-pole stator having phase voltage of 120 V at 60 Hz to the rotor secondary, find the additional rotor resistance required to produce maximum torque at:

    a. Starting $s = 1$.
    b. A speed of 450 r/min.

Solution

Given that

$$R_2 = 0.1 \ \Omega$$

$$X_2 = 0.8 \ \Omega$$

we get for maximum torque operation (condition 1),

$$s_1 = s_{\text{max}T} = \frac{R_2}{X_2} = 0.125$$

(a) The second condition of operation is that at starting,

$$s_2 = s_{\text{st}} = 1$$

Thus we use

$$R_a = R_2 \left(\frac{s_2}{s_1} - 1\right)$$

to get

$$R_a = 0.1 \ [(0.125)^{-1} - 1]$$

$$= 0.7 \ \Omega$$

(b) The synchronous speed is

$$n_s = \frac{120f}{P}$$

$$= \frac{120\,(60)}{8} = 900 \text{ r/min}$$

For $n_r = 450$ r/min, we have

$$s_2 = \frac{900 - 450}{900} = 0.5$$

As a result, we calculate

$$R_a = 0.1 \left(\frac{0.5}{0.125} - 1 \right)$$

$$= 0.3 \ \Omega$$

The slip at maximum torque is proportional to the rotor resistance, whereas the maximum torque is independent of the rotor resistance. This can be verified by reference to Eqs. (5.43) and (5.44). As a result, we assert that an additional resistance inserted in the rotor circuit will increase the slip at maximum torque (lower rotor speed); the maximum torque however is unchanged. The torque–slip characteristics of a three-phase induction motor with varying rotor resistance are shown in Figure 5.10.

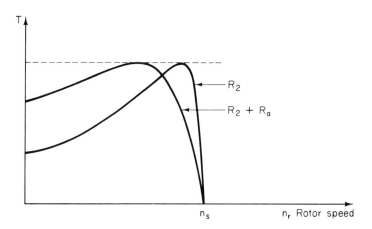

FIGURE 5.10   Effect of rotor resistance on torque–speed characteristics of a three-phase induction motor.

Effect of Capacitor Bank

Inserting a capacitor bank in the rotor circuit of a wound-rotor induction motor leads to the equivalent circuit representation shown in Figure 5.11. Evidently, all expressions for the performance of the motor can be obtained by replacing X_T by

$$X_T - \frac{X_c}{s^2}$$

Thus

$$I_r = \frac{V}{(R_1 + R_2/s) + j(X_T - X_c/s^2)} \tag{5.82}$$

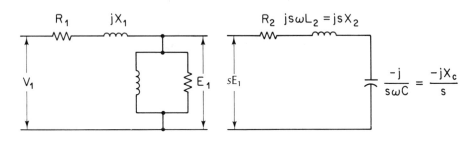

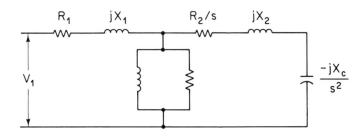

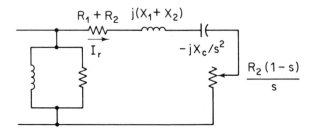

FIGURE 5.11 Development of equivalent circuit for an induction motor with a capacitor bank in the rotor circuit.

The mechanical power is thus

$$P_m = \frac{(3|I_r|^2)R_2(1 - s)}{s} \tag{5.83}$$

and the torque is

$$T = \frac{3V^2R_2}{\omega_s} \frac{1}{s[(R_1 + R_2/s)^2 + (X_T - X_c/s^2)^2]} \tag{5.84}$$

The maximum power factor (neglecting the magnetizing circuit) occurs when

$$X_T = \frac{X_c}{s^2} \tag{5.85}$$

In this case,

$$I_{r_{maxPF}} = \frac{V}{R_1 + R_2/s}$$

$$= \frac{V}{R_1 + R_2\sqrt{X_T/X_c}} \tag{5.86}$$

Obviously, for a given speed, there is an X_c that yields the maximum power factor.

Example 5.10 A 400-V four-pole three-phase 50-Hz wound-rotor induction motor has the following parameters:

$$R_1 = 10 \ \Omega/\text{phase} \qquad X_1 = 24 \ \Omega/\text{phase}$$
$$R_2 = 10 \ \Omega/\text{phase} \qquad X_2 = 24 \ \Omega/\text{phase}$$

A three-phase Y-connected capacitor bank with 20 Ω capacitive reactance per phase is inserted in the rotor circuit. It is required to:

a. Find the starting current and torque.
b. Find the current and torque at a slip of 0.05.

Neglect magnetizing and core effects.

Solution

We use

$$f = \frac{Pn_s}{120}$$

$$\frac{50 \times 120}{4} = n_s$$

As a result,

$$\omega_s = \frac{2\pi\,(50)\,(2)}{4}$$

$$= 50\pi$$

From the problem specifications

$$X_T = 48\ \Omega$$

The rotor impedance is thus obtained as

$$Z_2(s) = \left(10 + \frac{10}{s}\right) + j\left(48 - \frac{20}{s^2}\right)$$

The rotor current is therefore given by

$$I'_r = \frac{V}{Z_2(s)} = \frac{400/\sqrt{3}}{Z_2(s)}$$

(a) At starting $s = 1$, and thus

$$Z_2(1) = 20 + j28 = 34.41\ \underline{/54.46^\circ}$$

The rotor current at starting is

$$I'_r = \frac{400/\sqrt{3}}{34.41\ \underline{/54.66^\circ}} = 6.71\ \underline{/-54.46^\circ}\ \text{A}$$

The starting torque is now obtained as

$$T_{st} = \frac{(400)^2\,(10)}{50\pi}\frac{1}{(34.41)^2} = 8.6\ \text{N}\cdot\text{m}$$

(b) For $s = 0.05$,

$$Z_2(0.05) = \left(10 + \frac{10}{0.05}\right) + j\left[48 - \frac{20}{(0.05)^2}\right]$$

$$= 210 - j7952 = 7954.77 \underline{/-88.49°}$$

As a result,

$$I'_r = \frac{400/\sqrt{3}}{7954.77 \underline{/-88.49°}} = 0.03 \underline{/88.49° \text{ A}}$$

The corresponding torque is obtained as

$$T = \frac{3(400/\sqrt{3})^2(10)}{50\pi} \frac{1}{0.05} \frac{1}{(7954.77)^2}$$

$$= 3.21937 \times 10^{-3} \text{ N} \cdot \text{m}$$

5.10   CLASSIFICATION OF INDUCTION MOTORS

Integral-horsepower three-phase squirrel-cage motors are available from manufacturers' stock in a range of standard ratings up to 200 hp at standard frequencies, voltages, and speeds. (Larger motors are regarded as special-purpose.) Several standard designs are available to meet various starting and running requirements. Representative torque–speed characteristics of four designs are shown in Figure 5.12. These curves are typical of 1800-r/min (synchronous-speed) motors in ratings from 7.5 to 200 hp. Induction motors are classified as follows:

Class A: Normal starting torque, normal starting current, low slip. This design has a low-resistance single-cage rotor. It provides good running performance at the expense of starting. The full-load slip is low and the full-load efficiency is high. The maximum torque usually is over 200% of full-load torque and occurs at a small slip (less than 20%). The starting torque at full voltage varies from about 200% of full-load torque in small motors to about 100% in large motors. The high starting current (500 to 800% of full-load current when started at rated voltage) is the disadvantage of this design.

Class B: Normal starting torque, low starting current, low slip. This design has approximately the same starting torque as the class A, with

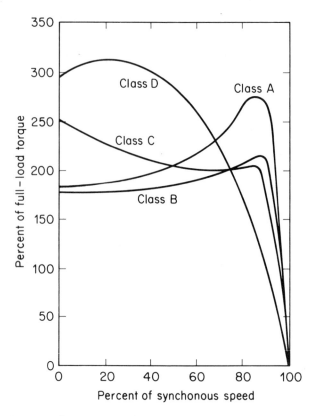

FIGURE 5.12 Typical torque–speed curves for 1800-r/min general-purpose induction motors.

only 75% of the starting current. The full-load slip and efficiency are good (about the same for the class A). However, it has a slightly decreased power factor and a lower maximum torque (usually only slightly over 200% of full-load torque being obtainable). This is the commonest design in the size range 7.5 to 200 hp used for constant-speed drives where starting-torque requirements are not severe.

Class C: High starting torque, low starting current. This design has a higher starting torque with low starting current but somewhat lower running efficiency and higher slip than the class A and class B designs.

Class D: High starting torque, high slip. This design produces very high starting torque at low starting current and a high maximum torque at 50 to 100% slip, but runs at a high slip at full load (7 to 11%) and consequently has low running efficiency.

5.11 SPEED CONTROL OF INDUCTION MOTORS

The speed of a three-phase induction motor driving a given load is obtained by matching the torque–speed characteristics of the motor and load as shown in Figure 5.13. It is clear that the operating point is the intersection of the two characteristics and is fixed for a given set of motor variables. If it is desired to change the motor speed while carrying the same load, it is clear that a change in the torque–speed characteristic of the motor is required. There are two major means for doing this: (1) changing the synchronous speed n_s, and (2) changing the slip at which the operating point takes place. There are many methods of achieving the foregoing requirements.

Changing Synchronous Speed

It is clear from Eq. (5.38) that changing ω_s results in a change in the torque–speed characteristics. Recall that

$$n_s = \frac{120f}{P}$$

We can conclude that a change in n_s can be obtained either through a change in the number of stator poles or by a change in supply frequency. We thus have

1. *Pole-changing method:* This method enables us to change n_s in discrete steps. A two-pole stator set of windings is changed to four-pole by a simple switching operation, as shown in Figure 5.14. Of course, this

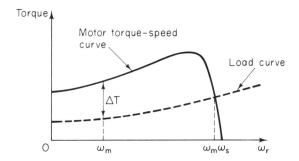

FIGURE 5.13 Induction-motor torque–speed curve and curve of load torque.

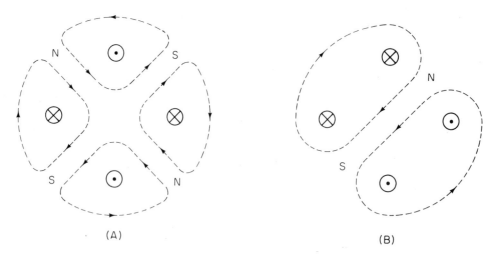

FIGURE 5.14 Illustrating four pole-to-two pole switching: (a) four-pole mode; (b) two-pole mode.

can be done for any nymber of poles for a ratio of $2:1$. This mechanism is simple to implement in squirrel-cage rotor motors but for the wound-rotor type, we must also rearrange the rotor windings.

2. *Line frequency control:* By varying the line frequency, the synchronous speed is varied. To maintain approximately constant flux density, the line voltage should also be varied directly with the frequency. A wound-rotor induction motor can be used as a frequency changer to supply the operating motor. Developments in solid-state power converters provide efficient means for supplying variable-frequency variable-voltage power sources, as discussed later.

We now turn our attention to the second major category of speed-control techniques.

Changing Slip

Changing the slip is a major category that involves changing the torque–speed characteristics. Means for achieving this include:

1. *Line voltage control:* The torque developed varies with the square of the applied voltage. As shown in Figure 5.15, the speed can be varied simply by changing the voltage applied to the stator. This method is common for squirrel-cage motors.

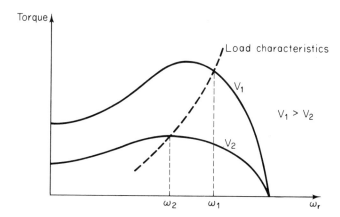

FIGURE 5.15   Effect of line voltage on speed of induction motor.

2. *Rotor resistance control:* This is suitable for the wound-rotor type and was discussed in detail in Section 5.9. The main drawback of this method is that efficiency is reduced at lower speeds. In practice, this method is used for applications with short periods of lower-speed operation.

3. *Rotor slip energy recovery:* The input power to the rotor is given by

$$P_{\text{in}_r} = \frac{3I_r^2 R_2}{s}$$

The rotor power output is given by

$$P_{\text{out}_r} = 3I_r^2 R_2 \frac{1 - s}{s}$$

As a result, the theoretical rotor efficiency is

$$\eta_r = 1 - s$$

It is thus clear that speed control by increasing s results in an increase in rotor I^2R losses and hence a decreased efficiency. There exist many methods for recovering this energy from the rotor (at rotor frequency $f_r = sf_s$), converting it to supply frequency, and subsequently returning it to the source. The devices used to achieve this are solid-state power devices, which will be discussed later.

5.12 STARTING INDUCTION MOTORS

In Section 5.7 we have seen that the ratio of current at starting to that at full load can be obtained from

$$\left(\frac{I_{st}}{I_f}\right)^2 = \frac{1 + (s_{maxT}/s_f)^2}{1 + s_{maxT}^2}$$

It is clear that the rotor current at starting is much higher than that at full load. Although induction motor designs provide for a rugged construction that can withstand large starting currents, it is undesirable to start induction motors with full applied voltage. One reason for this is simply the current overloading of the source circuit, which may not be able to withstand the current inrush. A second reason is that the large drop in voltage (due to feeder impedence) may result in a reduced voltage supply at the stator terminals that does not result in sufficient starting torque. The motor may not start at all.

Methods for reducing the starting current in an induction motor can be broadly classified into two categories: methods that apply reduced voltages to the stator at starting, and those that increase the resistance of the rotor circuit at starting and decrease it as the motor speed is increased.

Reduced-Voltage Starting

For low-rotor-resistance squirrel-cage motors a three-phase autotransformer is employed. At starting a low value of voltage is applied to the stator, which is then restored to full voltage when the rotor speed is within 25% of full-load value. In Figure 5.16 contacts a and c are closed at starting and are subsequently open while contacts b are closed.

An alternative method is the Y–Δ switching technique, which requires that both ends of each phase of the stator windings be accessible. At standstill the stator is connected in Y and as full speed is approached, a switch is activated to connect the stator in a Δ connection. The effect of Y-Δ switching is equivalent to an autotransformer starter with a ratio of $1/\sqrt{3}$.

Variable-Resistance Starting

The basic principle involved in variable-resistance starting is to cause the rotor resistance at standstill to be several times as much as its value when the motor is running at full speed. There are two basic methods to achieve this:

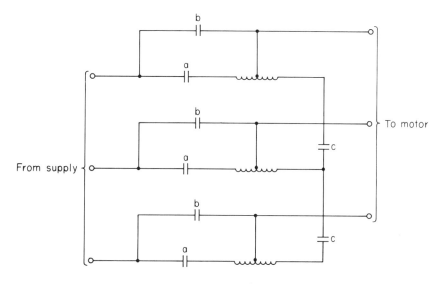

FIGURE 5.16   Autotransformer starting.

1. *External-resistance starting:* This method is used for the wound-rotor
   type of induction motor, where a three-phase bank of resistors is connected to the rotor slip rings at starting. As a result, a reduced starting
   current is obtained simultaneously with an increased starting torque.
   It is clear that high starting torques can be achieved simply by inserting additional rotor resistance of value corresponding to maximum
   torque production at starting. One advantage of this method is that
   energy lost as heat is dissipated through the environment, not within
   the motor itself.

2. *Deep-bar and double-squirrel-cage rotors:* The main idea in this technique is that the frequency of induced voltages in the rotor is equal to
   the product of slip and synchronous frequency. At standstill, the slip
   is unity and 60-Hz frequency currents flow in the rotor circuit, while
   at full speed, a rotor frequency of 2 to 3 Hz is common. It is therefore
   clear that if the rotor circuit is arranged such that a higher effective
   resistance is offered by the rotor at starting, whereas a lower-value of
   effective resistance is obtained at running, the desired effect is obtained.

   For the latter, two methods are available: (1) deep-bar rotors and
   (2) double-cage rotors. In the deep-bar-rotor design, the rotor conductors are formed of deep, narrow bars which can be considered as a
   large number of layers of differential depth. The leakage inductance
   of a bottom layer is greater than that for a top layer. The current in
   the low-reactance upper layers will be greater than that in the high-

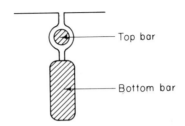

Top bar

Bottom bar

FIGURE 5.17   Double-cage arrangement.

reactance bottom layers when alternating currents are considered. As a result, the effective resistance of the rotor at stator frequency is much higher than its value for lower frequencies. As the speed increases, the rotor frequency decreases and therefore the rotor effective resistance decreases.

In the double-cage arrangement the squirrel cage winding is made of layers of bars short-circuited by end rings. The lower bars have a larger cross section than the upper bars, as shown in Figure 5.17 and consequently have lower resistance. The inductance of the lower bars is greater than that of the upper ones. At standstill, the current in the rotor is concentrated mainly in the upper bars because the lower bars have a high reactance. Thus at starting the rotor offers high impedance corresponding to the high-resistance upper layer. At low rotor frequencies, with low slip, the impedance is negligible and the rotor resistance approaches that of the two layers in parallel.

Example 5.11   A 230-V three-phase four-pole 60-Hz double-cage motor has the following parameters:

$$Upper\ cage:\ R_u = 2.4\ \Omega,\ X_u = 1\ \Omega$$
$$Lower\ cage:\ R_\ell = 0.24\ \Omega,\ X_\ell = 4.20\ \Omega$$

Assume that mutual reactance between the cages is negligible. It is required to find:

a. The currents in the upper and lower cages at starting and at a full-load slip of 0.04.

b. The torques developed by the upper and lower cages at starting and at a full-load slip of 0.04.

Solution

(a) The approximate equivalent circuit of the motor, neglecting stator circuit and no-load variables, is shown in Figure 5.18. The impedance of the upper cage is

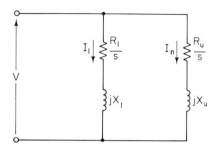

FIGURE 5.18 Equivalent circuit of a double-cage motor.

$$Z_u = \frac{2.4}{s} + j1$$

The impedance of the lower cage is

$$Z_\ell = \frac{0.24}{s} + j4.2$$

At starting, $s = 1$,

$$Z_u = 2.4 + j1 = 2.6 \,\underline{/22.62°}\; \Omega$$

$$Z_\ell = 0.24 + j4.2 = 4.2069 \,\underline{/86.73°}\; \Omega$$

Thus at starting

$$I_u = \frac{230 \,\underline{/-22.62°}}{\sqrt{3}\,(2.6)} = 51.073 \,\underline{/-22.62°}\; A$$

$$I_\ell = \frac{230 \,\underline{/-86.73°}}{\sqrt{3}\,(4.2069)} = 31.565 \,\underline{/-86.73°}\; A$$

Note that at starting the current in the upper cage is larger than that in the lower cage. The motor starting current is

$$I_{st} = I_u + I_\ell = 70.8 \,\underline{/-46.26°}$$

(b) The torques are obtained using

$$T = \frac{3I^2R}{s\omega_s}$$

The motor has four poles; thus

$$\omega_s = \frac{2\pi}{60} \left(\frac{120 \times 60}{4} \right) = 60\pi$$

At starting,

$$T_u = \frac{3(51.073)^2(2.4)}{60\pi} = 99.637 \text{ N} \cdot \text{m}$$

$$T_\ell = \frac{3(31.565)^2(0.24)}{60\pi} = 3.8058 \text{ N} \cdot \text{m}$$

The total starting torque is

$$T_{st} = T_u + T_\ell = 103.44 \text{ N} \cdot \text{m}$$

The ratio of upper to lower cage torques at starting is 26.18.
At full load, $s = 0.04$,

$$Z_u = 60 + j1 = 60.008 \underline{/0.955°} \ \Omega$$

$$Z_\ell = 6 + j4.2 = 7.3239 \underline{/34.992°} \ \Omega$$

Thus the cage currents are

$$I_u = \frac{230}{\sqrt{3} \ Z_u} = 2.2129 \underline{/-0.9555°}$$

$$I_\ell = \frac{230}{\sqrt{3} \ Z_\ell} = 18.731 \underline{/-34.992°}$$

The current in the upper cage is much less than that in the lower cage at full load. The torques are obtained as

$$T_u = \frac{3(2.213)^2(2.4)}{0.04 \ (60\pi)} = 4.6762 \text{ N} \cdot \text{m}$$

$$T_\ell = \frac{3(18.131)^2(0.24)}{0.04 \ (60\pi)} = 31.392 \text{ N} \cdot \text{m}$$

The motor full-load torque is therefore

$$T_{fld} = 36.068 \text{ N} \cdot \text{m}$$

The major portion of the torque is due to the lower cage on full load.

SOME SOLVED PROBLEMS

Problem 5.A.1

The full-load speed of a 50-Hz induction motor is 460 r/min. Find the number of poles and the slip at full load.

Solution

We use Eq. (5.18), with

$$n_s \simeq n_r = 460$$

Thus

$$P = 13.043$$

We cannot have an odd number of poles. Thus take

$$P = 12$$

Thus

$$n_s = \frac{120(50)}{12} = 500 \text{ r/min}$$

The slip at full load is

$$s = \frac{500 - 460}{500} = 0.08$$

Problem 5.A.2

A 230-V six-pole 60-Hz three-phase induction motor has a Δ-connected stator and a Y-connected rotor. The number of rotor conductors is 75% of that on the stator. Find the voltage and frequency between the rotor slip rings under the following conditions.

a. Rotor at rest.

b. Slip of 4%.

c. Rotor is driven by another motor at a speed of 800 r/min in the opposite direction of the rotating field.

Solution

(a) The rotor-induced voltage per phase at rest is given by

$$E_2 = E_1' = 0.75(230)$$

Here we use the stator *EMF* referred to the rotor side. Between the slip rings of the Y-connected rotor, we get

$$E_{2L} = \sqrt{3}\, E_2 = \sqrt{3}\, (0.75)(230) = 298.78 \text{ V}$$

The frequency of induced voltage in rotor is 60 Hz.
   (b) For a slip of 0.04, we get

$$E_{2Ls} = sE_{2L} = (0.04)(298.78) = 11.95 \text{ V}$$

The rotor frequency is

$$f_r = sf_s = 0.04(60) = 2.4 \text{ Hz}$$

   (c) The synchronous speed is

$$n_s = \frac{120f}{P} = 1200 \text{ r/min}$$

$$s = \frac{n_s - n_r}{n_s} = \frac{1200 + 800}{1200} = 1.67$$

Thus

$$E_{2L_s} = 1.67(298.78) = 498.96 \text{ V}$$
$$f_r = sf_s = 1.67(60) = 100 \text{ Hz}$$

Problem 5.A.3

A three-phase 440-V 60-Hz induction motor is operated at a frequency of 25 Hz. Find the required supply voltage to maintain the air-gap flux density at its rated value.

Solution

We use the induced EMF equation

$$E = 4.44 f N \phi$$

Thus

$$\frac{E_A}{E_B} = \frac{f_A}{f_B}$$

$$E_B = 440 \left(\frac{25}{60} \right) = 183.33 \text{ V}$$

Problem 5.A.4

A three-phase 220-V 60-Hz six-pole induction motor has the following parameters:

$$R_1 = 0.2 \ \Omega \qquad X_1 = 0.5 \ \Omega$$
$$R_2 = 0.1 \ \Omega \qquad X_2 = 0.2 \ \Omega$$
$$G_c = 0 \qquad B_m = 0.05 \text{ S}$$

Assume that fixed losses are 350 W. Find the stator current, the output power and torque, and the efficiency of the motor for a slip of 2.5%. Find the current and torque at starting with rated voltage applied.

Solution

For a slip of 0.025, the input impedance of the stator circuit is given by

$$Z_i = 0.2 + j0.5 + \frac{j20[0.1/0.025 + j0.2]}{(0.1/0.025) + j(20 + 0.2)}$$
$$= 4.228 \ \underline{/19.98°} \ \Omega$$

As a result, the stator current is

$$I_s = \frac{V}{Z_i} = \frac{220/\sqrt{3}}{4.228 \ \underline{/19.988°}} = 30.043 \ \underline{/-19.988°} \text{ A}$$

The input power is obtained as

$$P_s = \sqrt{3}\, V_1 I_s \cos \phi$$
$$= \sqrt{3}\,(220)(30.043) \cos 19.988$$
$$= 10.758 \times 10^3 \text{ W}$$

The power to the rotor circuit is now obtained as

$$P_r = P_s - 3I_s^2 R_1$$
$$= 10.758 \times 10^3 - 3(30.043)^2(0.2) = 10.216 \times 10^3 \text{ W}$$

The mechanical power is obtained from

$$P_m = (1 - s)P_r = 0.975(10.216 \times 10^3) = 9{,}961.25 \text{ W}$$

The net output is

$$P_o = P_m - P_f = 9{,}961.25 - 350 = 9611.25 \text{ W}$$

As a result, the efficiency is found to be

$$\eta = \frac{P_o}{P_s} = \frac{9611.25}{10{,}758} = 0.8934$$

The synchronous speed is

$$n_s = \frac{120f}{P} = \frac{120(60)}{6} = 1200 \text{ rpm}$$

As a result, the output torque is obtained as

$$T_o = \frac{P_o}{\omega_r} = \frac{9611.25}{(2\pi/60)(1200)(1 - 0.025)} = 78.445 \text{ N} \cdot \text{m}$$

At starting, $s = 1$, the stator input impedance is now obtained as

$$Z_i = 0.2 + j0.5 + \frac{j20(0.1 + j0.2)}{0.1 + j20.2}$$
$$= 0.7594 \,\underline{/66.8940°}\ \Omega$$

Thus

$$I_{st} = \frac{220/\sqrt{3}}{0.7594} = 167.254 \text{ A}$$

The starting torque is obtained using

$$T_{st} = \frac{3|V_1|^2 R_2}{\omega_s[(R_1 + R_2)^2 + (X_1 + X_2)^2]}$$
$$= \frac{(220)^2(0.1)}{(2\pi/60)(1200)\ [(0.3)^2 + (0.7)^2]}$$
$$= 66.406 \text{ N} \cdot \text{m}$$

Problem 5.A.5

A three-phase 220-V four-pole 5-hp 60-Hz induction motor has an effective stator resistance per phase of 0.3 Ω. The following test results are available:

> *No-load test:* $V_1 = 220$ V, $P_{nld} = 310$ W, $I_{nld} = 6.2$ A.
> *Blocked rotor test:* $V_1 = 48$ V, $P_{BR} = 610$ W, $I_{BR} = 18$ A.
> *Load test:* $V_1 = 220$ V, $P_s = 3650$ W, $I = 11.3$ A,
> $n_r = 1710$ r/min.

a. Find the fixed losses of the motor and hence determine from the load test data, the stator power factor and ohmic losses under load, rotor power input, rotor ohmic loss under load, rotor power output, horsepower, torque, and efficiency.
b. From the blocked rotor test data, find the equivalent resistance and reactance of stator and rotor as well as rotor resistance and reactance.
c. Find the motor starting torque for 110-V and 220-V stator voltages.

Solution

(a) The no-load power input supplies the stator losses and the fixed losses. Thus we have

$$P_f = 310 - 3(6.2)^2(0.3) = 275.4 \text{ W}$$

Under load, the power factor is obtained from

$$\cos \phi_s = \frac{3650}{\sqrt{3}\ (220)(11.3)} = 0.8477$$

The stator ohmic losses under load are obtained from

$$3I_s^2 R_1 = 3(11.3)^2(0.3) = 114.921 \text{ W}$$

The rotor power input is

$$P_{rg} = P_s - P_f - 3I_s^2 R_1$$
$$= 3650 - 275.4 - 114.921$$
$$= 3259.675 \text{ W}$$

The synchronous speed is

$$n_s = \frac{120 \times 60}{4} = 1800 \text{ r/min}$$

Thus, under load, the slip is

$$s = \frac{1800 - 1710}{1800} = 0.05$$

The rotor ohmic losses are given by

$$3I_r^2 R_2 = sP_{rg}$$
$$= 0.05(3259.675) = 162.984 \text{ W}$$

Thus the rotor output is

$$P_r = (1 - s)P_{rg} = 3096.69 \text{ W}$$

The horsepower output is now obtained as

$$\text{hp} = \frac{3096.69}{746} = 4.1511 \text{ hp}$$

The torque is

$$T = \frac{P_r}{(2\pi/60)(1710)} = 17.2931 \text{ N} \cdot \text{m}$$

The efficiency is

$$\eta = \frac{P_r}{P_s} = \frac{3096.69}{3650} = 0.8484$$

(b) The equivalent resistance R_T is obtained from

$$R_T = \frac{P_{BR}}{3I^2} = \frac{610}{3(18)^2} = 0.6276 \ \Omega$$

Thus

$$R_2 = R_T - R_1 = 0.6276 - 0.3 = 0.3276 \ \Omega$$

The equivalent impedance Z_T is

$$Z_T = \frac{48/\sqrt{3}}{18} = 1.5396 \ \Omega$$

Thus

$$X_T = \sqrt{Z_T^2 - R_T^2} = 1.4059 \ \Omega$$

We take the rotor reactance as half of X_T:

$$X_2 = \frac{1.4059}{2} = 0.70295 \ \Omega$$

(c) At starting,

$$T_{st} = \frac{3V^2 R_2}{\omega_s(R_T^2 + X_T^2)}$$

For $V = 220/\sqrt{3}$, we get

$$T_{st} = \frac{(220)^2(0.3276)}{(2\pi/60)(1800)[(0.6276)^2 + (1.4059)^2]}$$

$$= 35.4862 \ \text{N} \cdot \text{m}$$

For $V = 110/\sqrt{3}$, we get

$$T_{st} = 8.871 \ \text{N} \cdot \text{m}$$

Problem 5.A.6

A 50-hp 440-V three-phase, induction motor is used to determine the power rating of a load. With the load connected to the motor shaft, the input power is 30.5 kW and the stator current is 43.5 A at rated voltage. The blocked rotor test on the motor yields a current of 44.3 A at 52.5 V and a power input of 2.1 kW. When running with no load at rated voltage, the motor takes 16.3 A and a power of 1.16 kW at rated voltage. Find the horsepower rating of the load.

Solution

With the load on, we can get the stator power factor using

$$P = \sqrt{3}VI_s \cos \phi_s$$

$$30.5 \times 10^3 = \sqrt{3}\,(440)(43.5) \cos \phi_s$$

Thus

$$\phi_s = -23.071°$$

$$I_s = 43.5 \,\underline{/-23.071°}\, \text{A}$$

The no-load power factor is also obtained from

$$1.16 \times 10^3 = \sqrt{3}(440)(16.3) \cos \phi_0$$

Thus

$$\phi_0 = -84.642°$$

$$I_o = 16.3 \,\underline{/-84.642°}\, \text{A}$$

The rotor current with the load on is thus obtained as

$$I_r = I_s - I_o$$

$$= 38.507 \,\underline{/-1.216°}\, \text{A}$$

From the blocked rotor test,

$$R_T = \frac{2.1 \times 10^3}{(44.3)^2} = 1.0701 \; \Omega$$

As a result, the output power of the motor with the load on is given by

$$P_o = P_s - P_0 - 3I_r^2 R_T$$

$$= 30.5 \times 10^3 - 1.16 \times 10^3 - 3(38.507)^2(1.0701)$$

$$= 24.58 \times 10^3 \; \text{W}$$

Thus the horsepower rating of the load is

$$\text{hp} = \frac{24.58 \times 10^3}{746} = 32.949 \; \text{hp}$$

Problem 5.A.7

A 7.5-hp 440-V 1730-r/min induction motor develops a starting torque of 2.5 times rated full-load torque when rated voltage is applied. Find the starting torque when 230 V is applied for starting.

Solution

The torque is proportional to the square of the voltage.

$$T_2 = T_1 \left(\frac{V_2}{V_1}\right)^2$$

$$= 2.5 T_{fld} \left(\frac{230}{440}\right)^2$$

$$= 0.6831 T_{fld}$$

$$T_{fld} = \frac{7.5 \times 746}{(2\pi/60)(1730)} = 30.8834 \text{ N} \cdot \text{m}$$

Thus

$$T_2 = 21.0965 \text{ N} \cdot \text{m}$$

PROBLEMS

Problem 5.B.1

Determine the number of poles, the slip, and the frequency of the rotor currents at rated load for three-phase induction motors rated at:

a. 2200 V, 60 Hz, 588 r/min.
b. 120 V, 600 Hz, 873 r/min.

Problem 5.B.2

The full-load speed of a 25-Hz induction motor is 720 r/min, whereas its no-load speed is 745 r/min. Find the slip at full load.

Problem 5.B.3

The full-load slip of a 60-Hz 10-pole induction motor is 0.075. Find the speed of the rotor relative to the rotating field and that of the rotating field relative to the stator.

Problem 5.B.4

The full-load slip of a four-pole 60-Hz induction motor is 0.05. Find the speed of the rotating field relative to the stator structure and the frequency of the rotor current.

Problem 5.B.5

The rotor-induced EMF per phase at standstill for a three-phase induction motor is 100 V rms. The rotor circuit resistance and leakage reactance per phase are 0.3 Ω and 1 Ω, respectively.

    a. Find the rotor current and power factor at standstill.
    b. Find the rotor current and power factor at a slip of 0.06. Compute the rotor developed power.

Problem 5.B.6

A 15-hp 220-V three-phase 60-Hz six-pole Y-connected induction motor has the following parameters per phase:

$$R_1 = 0.128 \ \Omega \qquad G_c = 5.4645 \times 10^{-3}$$
$$R_2 = 0.0935 \ \Omega \qquad B_m = 0.125 \ \text{S}$$
$$X_T = 0.496 \ \Omega$$

The rotational losses are equal to the stator hysteresis and eddy-current losses. For a slip of 3%, find the following.

    a. The line current and power factor.
    b. The horsepower output.
    c. The starting torque.

Problem 5.B.7

A three-phase 100-hp 440-V four-pole, 60-Hz induction motor has a no-load input power of 3420 W and a no-load current of 45 A. The stator resistance is 0.06 Ω and that of the rotor is 0.08 Ω. The total reactance is 0.6 Ω. Find the line current, output torque, and horsepower for a motor speed of 1764 rpm.

Problem 5.B.8

A 60-Hz four-pole 208-V three-phase induction motor has the following parameters:

$$R_1 = 0.3 \ \Omega \qquad\qquad X_1 = 1 \ \Omega$$
$$R_2 = 0.2 \ \Omega \qquad\qquad X_2 = 1 \ \Omega$$
$$G_c = 20 \times 10^{-3} \ \text{S} \qquad B_m = 60 \times 10^{-3} \ \text{S}$$

Find the speed and stator current for a slip of 2%. For a load slip of 5%, find the motor power factor and output power. Use the approximate equivalent circuit.

Problem 5.B.9

The horsepower output of a three-phase induction motor is 100 hp at a speed of 1732 r/min. The motor has four poles and operates on a 60-Hz supply. The friction loss is 900 W and the iron loss is 4200 W. Find the power input to the motor assuming stator ohmic loss to be 2700 W.

Problem 5.B.10

The efficiency of a three-phase induction motor at a full load of 15 hp is 0.89. The corresponding power factor is 0.9 for operation on 440-V supply. Neglect stator impedance and the exciting circuit. Find the slip at full load and the starting current.

Problem 5.B.11

The rotor resistance and reactance of a four-pole 230-V 60-Hz three-phase induction motor are 0.15 Ω and 0.5 Ω, respectively. Find the output power and torque for a slip of 5%.

Problem 5.B.12

The no-load input to a three-phase induction motor is 10,100 W and the current is 15.3 A for a stator voltage of 2000 V. Friction loss is estimated at 2000 W and the stator resistance is 0.22 Ω. From a blocked rotor test with 440 V, the input power is 36.4 kW and the current is 170 A. Find the parameters of the equivalent circuit.

Problem 5.B.13

The no-load power input to a three-phase induction motor is 2300 W and the current is 21.3 A at rated voltage of 440 V. On full load, the motor takes 35 kW and 51 A at a speed of 1152 r/min. Assume that stator resistance is 0.125 Ω and that the motor has six poles and operates on a 60-Hz supply. Assume that stator core losses are equal to rotational losses. Find (a) the power factor and (b) the motor efficiency at full load.

Problem 5.B.14

The rotor resistance and reactance of a squirrel-cage induction motor at standstill are 0.1 Ω per phase and 0.8 Ω per phase respectively. Assuming a transformer ratio of unity, from the eight-pole stator having a phase voltage of 120 V at 60 Hz to the rotor secondary, calculate the following.

  a. Rotor starting current per phase.
  b. The value of slip producing maximum torque.

Problem 5.B.15

A three-phase 50-Hz four-pole induction motor has a blocked rotor reactance that is four times the rotor resistance per phase. Find the speed at maximum torque.

Problem 5.B.16

A 15-hp 440-V three-phase induction motor has a starting current of 132 A using rated voltage. Find the required starting voltage if the starting current is limited to 60 A.

Problem 5.B.17

An induction motor develops a starting torque of 80 N · m when 50% of rated voltage is applied. Determine the starting torque when 65% of rated voltage is applied.

Problem 5.B.18

A six-pole 150-hp 4000-V three-phase induction motor was tested to determine its equivalent-circuit parameters. The following information is available:

No-load test: $V = 4000$ V, $I_{nld} = 7$ A, $P_{nld} = 7$ kW.
Blocked rotor test: $V = 1340$ V, $I_{BR} = 32$ A, $P_{BR} = 18.2$ kW.

Stator resistance is 0.7 Ω. Neglect rotational loss.

  a. Find the motor parameters and slip at maximum torque using the approximate equivalent circuit.
  b. Find the eficiency and motor power factor at rated load.
  c. Find the output power when the speed is 1152 r/min.

Problem 5.B.19

The full-load slip of a squirrel-cage induction motor is 0.05, and the starting current is five times the full-load current. Neglecting the stator core and copper losses as well as the rotational losses, obtain:

a. The ratio of starting torque (st) to the full-load torque (fld).
b. The ratio of maximum (max) to full-load torque and the corresponding slip.

Problem 5.B.20

A three-phase six-pole 60-Hz induction motor develops a torque of 144 N · m at a speed of 1000 r/min. Assume that the rotor resistance is 0.2 Ω. Neglect stator impedance and the exciting circuit. Given that the maximum torque is 180 N · m, find the slip at maximum torque and the applied voltage.

Problem 5.B.21

The following data pertains to a 500-hp three-phase 2200-V 25-Hz 12-pole wound-rotor induction motor:

$$R_1 = 0.225 \ \Omega \qquad\qquad R_2 = 0.235 \ \Omega$$
$$X_T = 1.43 \ \Omega \qquad\qquad G_c = 1.2821 \times 10^{-3}$$
$$B_m = 31.447 \times 10^{-3} \ \text{S}$$

a. Find the no-load line current and input power with rated applied voltage. Assume that rotational losses are equal to the iron losses.
b. Find the applied voltage and input power for a blocked rotor condition such that the line current is 228 A.
c. Find the maximum torque and the corresponding slip, line current, and power factor.
d. Find the required additional rotor resistance so that maximum torque is developed at starting and hence find the value of the maximum torque.

Problem 5.B.22

The slip at maximum torque for an induction motor is 0.18. Find the power factor of the motor under these conditions, neglecting stator impedance. Suppose that it is desired to obtain maximum torque at a slip of 0.26. Find the ratio of the required additional rotor resistance to the original rotor resistance.

Problem 5.B.23

A three-phase induction motor has the following parameters:

$$R_1 = 0.09 \ \Omega \qquad X_1 = 0.12 \ \Omega$$
$$R_2 = 0.033 \ \Omega \qquad X_2 = 0.11 \ \Omega$$

Find the slip at maximum torque. If it is desired to have maximum torque developed at standstill, find the required rotor resistance.

Problem 5.B.24

The rotor resistance is a 4 Ω and the total reactance is 10 Ω for a four-pole 60-Hz induction motor. Find the applied stator voltage if the starting torque is 95 N · m. Find the starting torque if the rotor resistance is increased to 6 Ω.

Problem 5.B.25

The ratio of starting torque to full-load torque of a three-phase induction motor is 0.3 whereas the ratio of starting current to full-load current is 2.5. Find the slip at full load neglecting stator impedance and exciting circuit.

Problem 5.B.26

The rotor resistance and reactance of a wound-rotor induction motor at standstill are 0.12 Ω per phase and 0.7 Ω per phase, respectively. Assuming a transformer ratio of unity, from the eight-pole stator having a phase voltage of 254 V at 60 Hz to the rotor secondary, find the additional rotor resistance required to produce maximum torque at:

   a. Starting $s = 1$.
   b. A speed of 450 r/min.

Problem 5.B.27

Find the required additional rotor resistance to limit starting current to 45 A for a three-phase 600-V induction motor. Assume that

$$R_T = 1.66 \ \Omega \qquad X_T = 4.1 \ \Omega$$

Neglect the no-load circuit.

Problem 5.B.28

The rotor ohmic losses at maximum torque are 6.25 times that at full load with 0.032 slip for a three-phase induction motor. Find the slip at maximum torque and the ratio of maximum to full-load torque.

Problem 5.B.29

The rotor ohmic losses at starting are 6.25 times that at full load with slip of 0.035 for a three-phase induction motor. Find the slip at maximum torque and the ratio of starting to full-load torques.

Problem 5.B.30

For the motor of Problem 5.B.7, find the slip at maximum mechanical power and the value of the maximum horsepower. Compare the rated horsepower to the theoretical maximum.

Problem 5.B.31

Repeat Problem 5.B.30, neglecting stator resistance, and compare the results.

Problem 5.B.32

Find the maximum power and the slip at which it occurs for the motor of Problem 5.B.11.

Problem 5.B.33

The slip at maximum torque of an induction motor is 0.125. Find the slip for maximum power.

Problem 5.B.34

The slip for maximum power of an induction motor is 0.11, and the rotor impedance is 0.9 Ω. Find the rotor resistance and reactance. Neglect stator impedance.

Problem 5.B.35

A three-phase 50-Hz four-pole induction motor has a blocked rotor reactance that is four times the rotor resistance per phase. Find the speed at maximum power. What would be the speed at maximum power if the rotor resistance were increased by 50%?

Problem 5.B.36

The ratio of starting to full-load current for an induction motor is 5. The full-load slip is 0.05. Find the slip at maximum power.

Problem 5.B.37

A four-pole 60-Hz three-phase induction motor has a maximum power of 220 hp and a maximum torque of 1140 N · m. Find the rotor parameters, neglecting stator impedance and exciting circuit.

Problem 5.B.38

The maximum mechanical power output of a three-phase induction motor is 33 kW when operating on a 220-V 60-Hz supply. With 200 V and 50 Hz, the maximum mechanical power output is 31.54 kW. Find the rotor resistance and the reactance at 50 Hz and 60 Hz. Neglect stator impedance.

Problem 5.B.39

A 230-V three-phase four-pole 60-Hz double-cage motor has the following parameters:

       Upper cage: $R_u = 2.2\ \Omega,\ X_u = 0.8\ \Omega$
       Lower cage: $R_\ell = 0.24\ \Omega,\ X_\ell = 4.2\ \Omega$

Find the currents and torques at starting and a full-load slip at 0.04.

Problem 5.B.40

The lower cage resistance is 0.22 Ω and the upper cage reactance at 60 Hz is 1 Ω for a double-cage motor. The ratio of current in the upper cage to that in the lower cage at starting is 2 and the corresponding torque ratio is 25. Find the resistance of the upper cage and reactance of the lower cage.

Problem 5.B.41

The ratio of torque developed by the upper cage to that developed by the lower cage at a slip of 0.04 is 0.24. The corresponding current ratio is 0.2. Assume that the lower cage impedance is given by

$$Z_\ell = 0.24 + j4$$

Find the impedance of the upper cage.

Problem 5.B.42

The resistance of the upper and lower cages of a double-cage motor are 1.44 Ω and 0.24 Ω, respectively. The ratio of upper cage torque to lower cage torque is 0.24 at a slip of 0.04, whereas this ratio is 0.21 at a slip of 0.03. Find the reactances of the upper and lower cages.

Problem 5.B.43

Find the slip at which the upper cage torque is 0.22 times that of the lower cage for the motor of Problem 5.B.42.

Synchronous Machines

6.1 INTRODUCTION

Almost every daily function of today's civilization depends on electric power produced by electric utility systems. The backbone of a utility system consists of a number of generating stations that are interconnected in a grid and operate in parallel. The largest single-unit electric machine for electric energy production is the synchronous machine. Generators with power ratings of several hundred to over a thousand megavoltamperes (MVA) are fairly common in many utility systems. A synchronous machine provides a reliable and efficient means for energy conversion.

The operation of a synchronous generator is (like all other electromechanical energy conversion devices) based on Faraday's law of electromagnetic induction. The term *synchronous* refers to the fact that this type of machine operates at speed proportional to the system frequency under normal conditions. Synchronous machines are capable of operating as motors, in which case the electric energy supplied at the armature terminals of the unit is converted into mechanical form, as discussed in Chapter 4. Another important function of this versatile machine is as a synchronous condenser where the unit is operated as a motor running without mechanical load and supplying or absorbing reactive power.

The term *armature* in rotating machinery refers to the machine part in which an alternating voltage is generated as a result of relative motion with respect to a magnetic flux field. In a synchronous machine, the armature winding is on the stator and the field winding is on the rotor, as shown in Figure 6.1. The field is excited by direct current that is conducted through carbon brushes bearing on slip (or collector) rings. The dc source is called

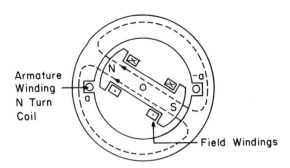

FIGURE 6.1   Simplified sketch of a synchronous machine.

the exciter and is often mounted on the same shaft as the synchronous machine. Various excitation systems with ac exciters and solid-state rectifiers are used with large turbine generators. The main advantages of these systems include the elimination of cooling and maintenance problems associated with slip rings, commutators, and brushes. The pole faces are shaped such that the radial distribution of the air-gap flux density B is approximately sinusoidal.

The armature winding includes many coils. One coil is shown in Figure 6.1 and has two coil sides (a and $-a$) placed in diametrically opposite slots on the inner periphery of the stator with conductors parallel to the shaft of the machine. The rotor is turned at a constant speed by a mechanical power source connected to its shaft. As a result, the flux waveform sweeps by the coil sides a and $-a$. The induced voltage in the coil is a sinusoidal time function. It is evident that for each revolution of the two poles, the coil voltage passes through a complete cycle of values. The frequency of the voltage in cycles per second (hertz) is the same as the rotor speed in revolutions per second. Thus a two-pole synchronous machine must revolve at 3600 r/min to produce a 60-Hz voltage, common in North America. In systems requiring 50-Hz voltage, the two-pole machine runs at 3000 r/min.

P-Pole Machines

Many synchronous machines have more than two poles. A P-pole machine satisfies the following relation derived in Chapter 4 and used repeatedly in Chapter 5.

$$f = \frac{P_n}{120}$$

(6.1)

The frequency f is proportional to the speed in revolutions per minute. Note that P is the number of poles of the machine.

In treating P-pole synchronous machines, it is more convenient to express angles in electrical degrees rather than in the more familiar mechanical units. Here, we conceptually concentrate on a single pair of poles and recognize that the conditions associated with any other pair are simply repetitions of those of the pair under consideration. As discussed in Chapter 4, we have

$$\theta_e = \frac{P}{2}\theta_m \tag{6.2}$$

where θ_e and θ_m denote angles in electrical and mechanical degrees, respectively.

Cylindrical versus Salient-Pole Construction

Machines like the ones illustrated in Figure 6.1 have rotors with salient poles. There is another type of rotor, which is shown in Figure 6.2. The machine with such a rotor is called a cylindrical rotor or nonsalient-pole machine. The choice between the two designs (salient or nonsalient) for a specific application depends on the proposed prime mover. For hydroelectric generation, a salient-pole construction is employed. This is because hydraulic turbines run at relatively low speeds, and in this case, a large number of poles are required to produce the desired frequency, as indicated by Eq. (6.1). On the other hand, steam and gas turbines perform better at relatively high speeds, and two-or four-pole cylindrical rotor turboalternators are used in this case. This will avoid the use of protruding parts on the rotor, which at high speeds will give rise to dangerous mechanical stresses.

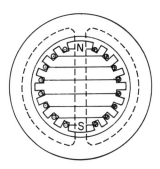

FIGURE 6.2 Cylindrical rotor two-pole machine.

Three-Phase Machines

With very few exceptions, synchronous generators (commonly referred to as alternators) are three-phase machines. For the production of a set of three voltages phase-displaced by 120 electrical degrees in time, it follows that a minimum of three coils phase-displaced 120 electrical degrees in space must be used. An elementary three-phase two-pole machine with one coil per phase is shown in Figure 6.3.

For clarity of presentation, we represent each coil as a separate generator. An immediate extension of single-phase circuits would be to carry the power from the three generators along six wires. However, for the sake of economy, instead of having a return wire from each load to each generator, a single wire is used for the return of all three. The current in the return wire will be $I_a + I_b + I_c$; and for a balanced load, these will cancel out, as may be seen by inspecting the phasor diagram in Figure 6.4. If the load is unbalanced, the return current will still be small compared to either I_a, I_b or I_c. Thus the return wire could be made smaller than the other three. This connection is known as a four-wire three-phase system. It is desirable for safety and system protection to have a connection from the electrical system to ground. A logical point for grounding is the generator neutral point, the junction of the Y.

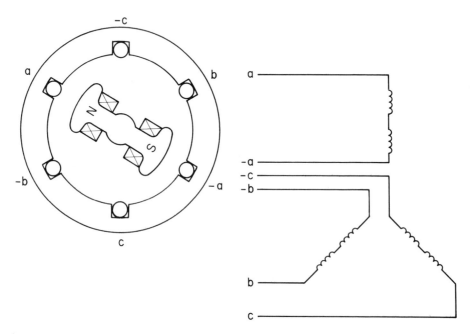

FIGURE 6.3 Elementary three-phase two-pole machine.

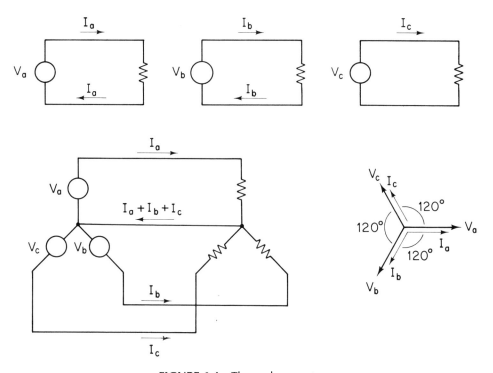

FIGURE 6.4   Three-phase system.

6.2   ROUND-ROTOR MACHINES: EQUIVALENT CIRCUIT

Consider a three-phase synchronous machine with a round rotor operating in the steady state. The flux linkages λ_a for phase a windings are written as

$$\lambda_a = L_{aa}i_a + L_{ab}i_b + L_{ac}i_c + L_{af}I_f \qquad (6.3)$$

The subscript f refers to the field windings which are placed on the rotor and carry a direct current I_f. For the present modeling purposes, the inductances required for our equation are given by

$$L_{aa} = L_0 + L_1 \qquad (6.4)$$

$$L_{ab} = L_{ac} = \frac{-L_0}{2} \qquad (6.5)$$

$$L_{af} = M \cos \theta \qquad (6.6)$$

The angle θ is that between the stator and rotor magnetic axes. The self-inductance L_{aa} is the sum of a component L_0 due to main air-gap flux and L_1 due to the armature leakage flux. The mutual inductances L_{ab} and L_{ac} are assumed to be due only to the main air-gap flux. The factor $-\frac{1}{2}$ is due to the 120° spatial displacement between phases ($\cos \pm 120° = -\frac{1}{2}$). The mutual inductance between phase a and the field varies as $\cos \theta$, as discussed in Chapter 4.

Let us assume balanced three-phase operation and hence

$$i_a + i_b + i_c = 0 \tag{6.7}$$

Combining Eq. (6.3) to (6.7), we get

$$\lambda_a = (L_1 + L_\phi)i_a + MI_f \cos \theta \tag{6.8}$$

Where L_ϕ is given by

$$L_\phi = \tfrac{3}{2} L_0 \tag{6.9}$$

The rotor is assumed to be running at constant synchronous speed and thus the angular displacement is

$$\theta(t) = \omega_s t + \theta_0 \tag{6.10}$$

As a result, the rate of change of the phase a flux linkages is given by

$$\frac{d\lambda_a}{dt} = (L_1 + L_\phi) \frac{di_a}{dt} + \omega_s MI_f \cos \left(\omega_s t + \theta_0 + \frac{\pi}{2} \right) \tag{6.11}$$

The terminal voltage of phase a is therefore given by

$$v_t = R_a i_a + \frac{d\lambda_a}{dt} \tag{6.12}$$

The resistance of phase a is denoted by R_a. Equation (6.12) relates to the machine operating as a motor. For the machine operating as a generator, we have

$$\frac{d\lambda_a}{dt} = v_t + R_a i_a \tag{6.13}$$

In this case, we have

$$\lambda_a = -(L_1 + L_\phi)i_a + MI_f \cos \theta \tag{6.14}$$

Let us define the field excitation voltage magnitude in effective (rms) value by·

$$E_f = \frac{\omega_s M I_f}{\sqrt{2}} \tag{6.15}$$

We also define the synchronous inductance L_s by

$$L_s = L_1 + L_\phi \tag{6.16}$$

As a result for the motor case, Eqs. (6.11) and (6.12) reduce to

$$v_t = R_a i_a + L_s \frac{di_a}{dt} + \sqrt{2} E_f \cos\left(\omega_s t + \theta_0 + \frac{\pi}{2}\right) \tag{6.17}$$

For the generator case, we find that Eqs. (6.13) and (6.14) reduce to

$$v_t = \sqrt{2} E_f \cos\left(\omega_s t + \theta_0 + \frac{\pi}{2}\right) - \left(R_a i_a + L_s \frac{di_a}{dt}\right) \tag{6.18}$$

Equations (6.17) and (6.18) provide the basis for equivalent circuits and phasor diagrams for the round-rotor synchronous machine operating in the steady state as either motor or generator.

Let the phase a current be a sinusoidal described by the steady-state expression

$$i_a(t) = \sqrt{2} I_a \cos \omega_s t \tag{6.19}$$

As a result, Eq. (6.17) for the motor case yields

$$v_t = \sqrt{2} I_a \left[R_a \cos \omega_s t + \omega_s L_s \cos\left(\omega_s t + \frac{\pi}{2}\right) \right]$$
$$+ \sqrt{2} E_f \cos\left(\omega_s t + \theta_0 + \frac{\pi}{2}\right) \tag{6.20}$$

It is evident that v_t is sinusoidal as well. We can therefore write the following complex equivalent of Eq. (6.20) with I_a as reference:

$$\mathbf{V}_t = (R_a + jX_s)\mathbf{I}_a + \mathbf{E}_f \tag{6.20a}$$

The phasor $\mathbf{E}_f$ is given by the polar form

$$\mathbf{E}_f = E_f \underline{/\theta_0 + \pi/2^\circ} \tag{6.21}$$

The reference current is

$$\mathbf{I}_a = I_a \underline{/0°} \tag{6.22}$$

In Eq. (6.20), X_s denotes the *synchronous reactance* given by

$$X_s = \omega_s L_s \tag{6.23}$$

Recall Eq. (6.16), it is then clear that X_s can be expressed as the sum of the leakage reactance X_1 and the reactance X_ϕ, commonly known as the armature reaction reactance:

$$X_s = X_1 + X_\phi \tag{6.24}$$

with

$$X_1 = \omega_s L_1 \tag{6.25}$$

$$X_\phi = \omega_s L_\phi \tag{6.26}$$

At this point, we also define the synchronous impedance Z_s by

$$Z_s = R_a + jX_s \tag{6.27}$$

We are now ready to detail equivalent circuits and phasor diagrams of the round-rotor machine.

The excitation voltage E_f is represented as an independent sinusoidal source as shown in Figure 6.5(a). The synchronous impedance Z_s is in series with the source. It is clear that Eq. (6.20) is satisfied by this equivalent circuit. The phasor diagram of Figure 6.5(b) is drawn with I_a as the reference phasor. We assume θ_0 to be negative in this construction. E_f leads I_a

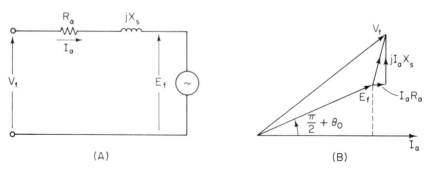

FIGURE 6.5 Round-rotor synchronous machine as a motor: (a) equivalent circuit; (b) phasor diagram.

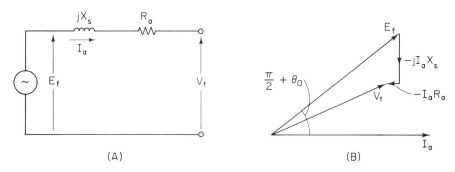

(A)   (B)

FIGURE 6.6 Round-rotor synchronous machine as a generator: (a) equivalent circuit;   (b) phasor diagram.

by the angle $(\theta_0 + \pi/2)$. The terminal voltage V_t is obtained by adding the voltage drop $I_a Z_s$ to E_f as shown in the phasor diagram.

The equivalent circuit for the case of the machine operating as a generator is shown in Figure 6.6(a) and can be seen to satisfy Eq. (6.18) written in the phasor form

$$E_f = V_t + I_a Z_s \tag{6.28}$$

The phasor diagram shown in Figure 6.6(b) is similar in construction to that of Figure 6.5(b) except that $I_a Z_s$ is subtracted from E_f to obtain V_t.

The choice of I_a as reference was dictated by simplicity of the resulting trignometric manipulations in the foregoing analysis. It is more practical, however, to take the terminal voltage V_t as reference and proceed given the terminal power factor (cosine of angle between V_t and I_a) to find E_f as shown in Figure 6.7(a) for the motor case and Figure 6.7(b) for the generator case. The angle between E_f and V_t is commonly known as the torque or power angle δ. An example is taken up at this point.

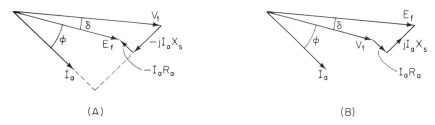

(A)   (B)

FIGURE 6.7 Round-rotor machine phasor diagrams with V_t as reference: (a) motor case;   (b) generator case.

Example 6.1 A 1250-kVA three-phase Y-connected 4160-V 10-pole 60-Hz synchronous generator has an armature resistance of 0.126 Ω per phase and a synchronous reactance of 3 Ω per phase. Find the full-load generated voltage per phase at 0.8 PF lagging.

Solution

We have

$$I_a = \frac{1250 \times 10^3}{\sqrt{3}\,(4160)} = 173.48 \text{ A}$$

Thus in polar form

$$I_a = 173.48 \underline{/-36.87°} \text{ A}$$

Given that

$$Z_s = 0.126 + j3$$

$$= 3.00264 \underline{/87.595°} \ \Omega$$

we apply Eq. (6.28) to obtain

$$E_f = \frac{4160}{\sqrt{3}} + (173.48 \underline{/-36.87°})(3.00264 \underline{/87.595°}) = 2761.137 \underline{/8.397°} \text{ V}$$

We will take up another example at this point but we deal with a motor.

Example 6.2 The rotational losses for a 1000-hp synchronous motor are found to be 18 kW. The motor operates from a 2300-V three-phase supply at 0.8 PF lagging. Find the excitation voltage at full load assuming that the machine's synchronous reactance is 1.9 Ω and that the armature resistance is negligible.

Solution

The total input power is the sum of the output power and losses. As a result,

$$P_{in} = 746 \times 10^3 + 18 \times 10^3 = 764 \times 10^3 \text{ W}$$

The current is thus obtained as

$$I_a = \frac{764 \times 10^3}{\sqrt{3} \, (2300)(0.8)} = 239.726 \; \underline{/-36.87°} \; \text{A}$$

We now have

$$E_f = V_t - jI_aX_s$$

$$= \frac{2300}{\sqrt{3}} - j(239.726 \; \underline{/-36.87°})(1.9)$$

$$= 1115.793 \; \underline{/-19.061°} \; \text{V}$$

This is the required excitation voltage per phase.

6.3 ARMATURE REACTION

We have seen in Chapter 5 that a rotating magnetic field results from a three-phase balanced current in a set of three-phase windings that are placed 120° apart spatially. It is therefore clear that the armature currents in a three-phase synchronous machine create a rotating MMF. This MMF wave is commonly called armature reaction MMF. This MMF wave rotates at synchronous speed and is directly opposite to phase a at the instant when phase a has its maximum current ($t = 0$). The dc field winding produces a sinusoid F with an axis 90° ahead of the axis of phase a in accordance with Faraday's law.

The resultant magnetic field in the machine is the sum of the two contributions from the field and armature reaction. Figure 6.8 shows a sketch of the armature and field windings of a cylindrical rotor generator.

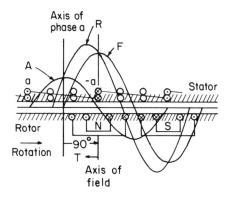

FIGURE 6.8 Spatial MMF waves in a cylindrical rotor synchronous generator.

The space MMF produced by the field winding is shown by the sinusoid F. This is shown for the specific instant when the electromotive force (EMF) of phase a due to excitation has its maximum value. The time rate of change of flux linkages with phase a is a maximum under these conditions, and thus the axis of the field is 90° ahead of phase a. The armature-reaction wave is shown as the sinusoid A in the figure. This is drawn opposite phase a because at this instant both I_a and the EMF of the field E_f (also called excitation voltage) have their maximum value. The resultant magnetic field in the machine is denoted by R and is obtained by graphically adding the F and A waves.

Sinusoids can conveniently be handled using phasor methods. We can thus perform the addition of the A and F waves using phasor notation. Figure 6.9 shows a space phasor diagram where the fluxes ϕ_f (due to the field, ϕ_{ar} (due to armature reaction), and ϕ_r (the resultant flux) are represented. It is clear that under the assumption of a uniform air gap and no saturation, these are proportional to the MMF waves F, A, and R, respectively. The figure is drawn for the case when the armature current is in phase with the excitation voltage. The situation for the case when the armature current lags the excitation voltage E_f is shown in Figure 6.10.

The effect of the armature-reaction flux is represented by an inductive reactance. The basis for this is shown in Figure 6.11, where the phasor diagram of component fluxes and corresponding voltages is given. The field flux ϕ_f is added to the armature-reaction flux ϕ_{ar} to yield the resultant airgap flux ϕ_r. The armature-reaction flux ϕ_{ar} is in phase with the armature current I_a. The excitation voltage E_f is generated by the field flux, and E_f lags ϕ_f by 90°. Similarly, E_{ar} and E_r are generated by ϕ_{ar} and ϕ_r, respectively, with each of the voltages lagging the flux causing it by 90°.

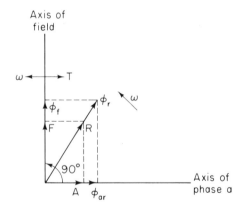

FIGURE 6.9 Space phasor diagram for armature current in phase with excitation voltage.

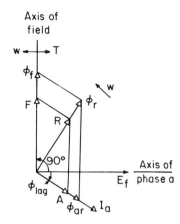

FIGURE 6.10 Space phasor diagram for armature current lagging the excitation voltage.

If we introduce the constant of proportionality x_ϕ that relates the rms values of E_{ar} and I_a, we can write

$$E_{ar} = -jx_\phi I_a \tag{6.29}$$

where the $-j$ emphasizes the 90° lagging effect. We therefore have for the generator case

$$E_r = E_f - jx_\phi I_a \tag{6.30}$$

The equivalent circuit of Figure 6.6(a) is modified to show the armature reaction effects as shown in Figure 6.12. The inductive reactance x_ϕ, known as the magnetizing reactance of the machine, accounts for the armature-reaction effects and is the same as the reactance x_ϕ defined in Eq. (6.26).

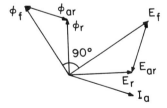

FIGURE 6.11 Phasor diagram for fluxes and resulting voltages in a synchronous machine.

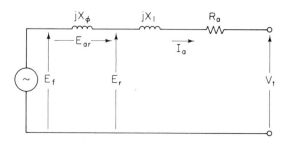

FIGURE 6.12 Equivalent circuit of a round-rotor synchronous generator showing armature-reaction effects.

Example 6.3   A 10-MVA 13.8-kV 60-Hz two-pole Y-connected three-phase alternator has an armature winding resistance of 0.07 Ω per phase and a leakage reactance of 1.9 Ω per phase. The armature-reaction EMF for the machine is related to the armature current by

$$E_{ar} = -j19.91I_a$$

Assume that the generated EMF is related to the field current by

$$E_f = 60I_f$$

a. Compute the field current required to establish rated voltage across the terminals of a load when rated armature current is delivered at 0.8 PF lagging.

b. Compute the field current needed to provide rated terminal voltage to a load that draws 100% of rated current at 0.85 PF lagging.

Solution

The terminal voltage (line to neutral) is given by

$$V_t = \frac{13{,}800}{\sqrt{3}} = 7967.43 \text{ V}$$

(a) The rated armature current is obtained as

$$I_a = \frac{10 \times 10^6}{\sqrt{3}(13{,}800)} = 418.37 \text{ A}$$

From the equivalent circuit of Figure 6.12,

$$E_r = V_t + I_a(R_a + jX_1)$$
$$= 7967.43 + (418.37\underline{/-\cos^{-1}0.8})(0.07 + j1.9)$$

Thus

$$E_f = 8490.35 \; \underline{/4.18°} \; V$$

From the specifications,

$$E_{ar} = -j(19.91)(418.37 \; \underline{/-\cos^{-1}0.8°})$$

Thus

$$E_{ar} = -8329.75 \; \underline{/53.13°} \; V$$

Now, the field exciation is

$$E_f = E_r - E_{ar}$$

Thus we obtain

$$E_f = 8490.35 \; \underline{/4.18°} + 8329.75 \; \underline{/53.13°}$$
$$= 15{,}308.61 \; \underline{/28.40°} \; V$$

From the given relation, we obtain the required field current:

$$|I_f| = \frac{15{,}308.61}{60} = 255.14 \; A$$

(b) We have in this case

$$I_a = (418.37)(1.0) \; \underline{/-\cos^{-1}0.85°}$$
$$= 418.37 \; \underline{/-31.79°} \; A$$

As a result,

$$E_r = 7967.43 + (418.37 \; \underline{/-31.79°})(0.07 + j1.9)$$
$$= 8436.94 \; \underline{/4.49°} \; V$$

We now calculate

$$E_{ar} = -j(19.91)(418.37 \underline{/-31.79°})$$

$$= -8329.74 \underline{/58.21°} \text{ V}$$

As a result,

$$E_f = E_r - E_{ar}$$

$$= 8436.94 \underline{/4.48°} + 8329.74 \underline{/58.21°}$$

$$= 14{,}957.72 \underline{/31.16°} \text{ V}$$

The field current is thus obtained as

$$|I_f| = \frac{E_f}{60} = 249.30 \text{ A}$$

6.4 PRINCIPAL STEADY-STATE CHARACTERISTICS

The model of a round-rotor synchronous machine obtained in the form of the equivalent circuit of Section 6.2 is now used to obtain performance characteristics of the machine operating in the steady state as a generator. These describe the interrelations among terminal voltage, field current, armature current, and power factor.

We begin by considering a synchronous generator delivering power to a constant-power-factor load at a constant frequency. A compounding curve shows the variation of the field current required to maintain rated terminal voltage with the load, as represented by armature current. Since excitation voltage E_f is assumed to be proportional to the field current I_f, it is then only necessary to examine the variation of $|E_f|$ with I_a to obtain the compounding curve for a given power factor. This can be done on the basis of Eq. (6.28) with V_t fixed at its rated value and varying I_a and computing $|E_f|$.

The nature of the compounding curves can be determined from the following equation derived from Eq. (6.28):

$$|E_f|^2 = V_t^2 + 2|Z_s| \, |I_a|V_t \cos (\psi + \phi) + |Z_s|^2|I_a|^2 \qquad (6.28a)$$

where it is assumed that ψ is the angle of Z_s and that ϕ is the phase angle between I_a and V_t.

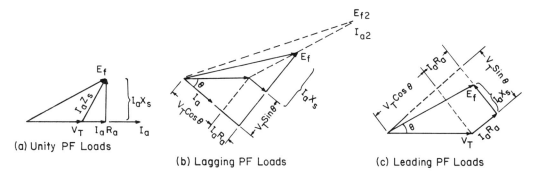

(a) Unity PF Loads

(b) Lagging PF Loads

(c) Leading PF Loads

FIGURE 6.13 Phasor diagrams for a synchronous machine operating at different power factors are: (a) unity PF loads; (b) lagging PF loads; (c) leading PF loads.

For no load, $I_a = 0$, the excitation voltage $|E_f|$ is equal to terminal rated voltage. As the load (I_a) is increased, we see that $|E_f|$ increases provided that ϕ is negative (lagging power factor) or zero, as $\cos(\psi + \phi)$ will be positive in this case. For leading power factors $\phi > 0$, and since ψ is close to 90°, we find that $\cos(\psi + \phi)$ can be negative. In this case, an increase in I_a can result in a decrease in the excitation voltage. The phasor diagrams shown in Figure 6.13 help illustrate the points made here. Typical compounding curves for a round-rotor generator are shown in Figure 6.14. An example is given now to illustrate the various points on a compounding curve.

Example 6.4 A 10-MVA three-phase Y-connected 13,800-V two-pole 60-Hz turbogenerator has an armature resistance of 0.06 Ω per phase and a synchronous reactance of 1.8 Ω per phase. Find the excitation voltage at no load, half load, and full load for power factors of (a) 0.8 leading, (b) unity, and (c) 0.8 lagging with rated terminal voltage.

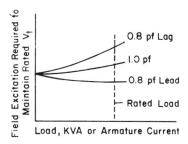

FIGURE 6.14 Synchronous-machine compounding curves.

Solution

We have

$$Z_s = 0.06 + j1.8 = 1.801 \underline{/88.09°}$$

The full-load current is obtained as

$$I_{af} = \frac{10 \times 10^6}{\sqrt{3}\,(13,800)} = 418.37 \text{ A}$$

The terminal voltage per phase is

$$V_t = \frac{13,800}{\sqrt{3}} = 7967.434 \text{ V}$$

We thus have

$$E_f = 7967.434 + \sigma(418.37)(1.801) \underline{/88.09° + \phi}$$

where σ is the fraction of full load considered and ϕ is the phase angle.
    (a) For a power factor of 0.8 leading,

$$\phi = 36.87°$$

At no load, $\sigma = 0$,

$$E_f = 7967.434 \text{ V}$$

At half load, $\sigma = 0.5$,

$$E_f = 7967.434 + 0.5(418.37)(1.801) \underline{/124.961°}$$

$$= 7757.702 \underline{/2.281°} \text{ V}$$

At full load, $\sigma = 1$, and we calculate

$$E_f = 7560.935 \underline{/4.685°} \text{ V}$$

Note that $|E_f|$ is decreased as loading is increased.
    (b) For a unity power factor load, $\phi = 0$: At no load, $\sigma = 0$, $E_f = 7967.434$ V. At half load, $\sigma = 0.5$,

$$E_f = 7967.434 + 0.5(418.37)(1.801) \underline{/88.09°}$$

$$= 7988.863 \underline{/2.701°} \text{ V}$$

At full load, we get

$$E_f = 8027.935 \underline{/5.383°} \text{ V}$$

Note that $|E_f|$ is increased as loading is increased.

(c) For a power factor of 0.8 lagging, $\phi = -36.87°$. At half load, we have

$$E_f = 7967.434 + 0.5(418.37)(1.801) \underline{/51.221°}$$

$$= 8208.65 \underline{/2.05°} \text{ V}$$

At full load, we have

$$E_f = 8459.772 \underline{/3.981°} \text{ V}$$

Again, we note the increase in $|E_f|$ as loading is increased.

The second characteristic of a round-rotor synchronous machine considered here is the variation of the machine terminal voltage as the armature current is varied for fixed field current corresponding to producing full-load armature current at rated terminal voltage. Each characteristic curve is obtained for a given power factor. The nature of the characteristic can be examined by rewriting Eq. (6.28a) as

$$V_t^2 + [2|Z_s| \, |I_a| \cos (\psi + \phi)] \, V_t + |Z_s|^2|I_a|^2 - |E_f|^2 = 0 \quad (6.28b)$$

Clearly, at no load, the value of V_t is equal to $|E_f|$. Note, however, that the value of $|E_f|$ used in establishing this characteristic depends on the power factor chosen and is picked from the corresponding compounding curve. The general trend from Eq. (6.28b) is that of a parabolic variation as shown in Figure 6.15. Note that for a leading power factor for currents above rated

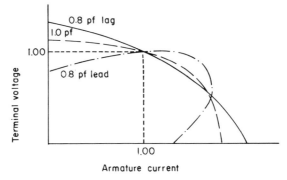

FIGURE 6.15   Constant-field-current volt-ampere characteristic of synchronous machine.

value, the curve yields two values for the resulting terminal voltage. An example will illustrate how this characteristic is constructed.

Example 6.5   Consider the machine of Example 6.4 operating at 0.8 PF leading. Construct the V_t versus I_a curve by computing selected points on the curve.

Solution

From Example 6.4 at 0.8 PF leading, we found that for rated terminal voltage, we need an excitation voltage of

$$E_f = 7560.935 \text{ V}$$

Substituting in Eq. (6.28b), the characteristic is described by

$$V_t^2 - 2.064V_tI_a + (3.244 \ I_a^2 - 5.7168 \times 10^7) = 0$$

The first point to find is at no load, $I_a = 0$, where we find that

$$V_t = E_f = 7560.935 \text{ V}$$

The full-load current is 418.37 A and we now compute V_t for half load as

$$V_{t1/2} = 7770.5 \text{ V}$$

The other root of the equation is negative. At full load, we get

$$V_{tf} = 7967.434 \text{ V}$$

For two times full load, we get

$$V_{t2} = 8322.9 \text{ V}$$

Note that the terminal voltage increases with an increase in load.
    The voltage V_t will be single-valued when the quadratic has two equal roots and this takes place at

$$(2.064I_a)^2 - 4(3.244I_a^2 - 5.7168 \times 10^7) = 0$$

As a result, the current at the knee (k) of the curve is

$$I_{ak} = 5122.3 \text{ A}$$

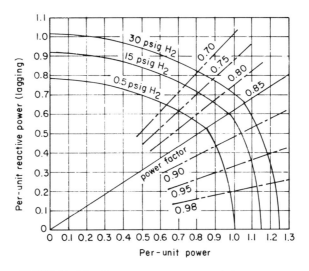

FIGURE 6.16   Generator reactive-capability curves.

This is almost 12 times full load. The corresponding value of V_t is obtained as

$$V_{t_k} = 5286.5 \text{ V}$$

The final point to obtain is that for which $V_t = 0$, and is found to be at

$$I_a = 4198.2 \text{ A}$$

That is about 10 times full-load current.

Important characteristics of the synchronous machine are given by the reactive-capability curves. These give the maximum reactive power loadings corresponding to various active power loadings for rated voltage operation. Armature heating constraints govern the machine for power factors from rated to unity. Field heating represents the constraints for lower power factors. Figure 6.16 shows a typical set of curves for a large turbine generator.

6.5 POWER ANGLE CHARACTERISTICS AND THE INFINITE BUS CONCEPT

Consider the simple circuit shown in Figure 6.17. The impedance Z connects the sending end, whose voltage is E, and the receiving end, with voltage V. Let us assume that in polar form, we have

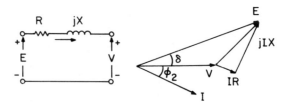

FIGURE 6.17 Equivalent circuit and phasor diagram for a simple link.

$$E = E \: \underline{/\delta}$$

$$V = V \: \underline{/0}$$

$$Z = Z \: \underline{/\psi}$$

We therefore conclude that the current I is given by

$$I = \frac{E - V}{Z}$$

This reduces to

$$I = \frac{E}{Z} \: \underline{/\delta - \psi} - \frac{V}{Z} \: \underline{/-\psi} \tag{6.31}$$

The complex power S_1 at the receiving end is

$$S_1^\star = E^\star I$$

Similarly, the complex power S_2 at the receiving end is

$$S_2^\star = V^\star I$$

Using Eq. (6.31), we thus have

$$S_1^\star = \frac{E^2}{Z} \: \underline{/-\psi} - \frac{EV}{Z} \: \underline{/-\psi - \delta} \tag{6.32}$$

$$S_2^\star = \frac{EV}{Z} \: \underline{/\delta - \psi} - \frac{V^2}{Z} \: \underline{/-\psi} \tag{6.33}$$

Recall that

$$S^\star = P - jQ$$

We thus have the power equations:

$$P_1 = \frac{E^2}{Z} \cos(-\psi) - \frac{EV}{Z} \cos(\psi + \delta) \tag{6.34}$$

$$Q_1 = \frac{E^2}{Z} \sin \psi - \frac{EV}{Z} \sin(\psi + \delta) \tag{6.35}$$

$$P_2 = \frac{EV}{Z} \cos(\delta - \psi) - \frac{V^2}{Z} \cos \psi \tag{6.36}$$

$$Q_2 = \frac{EV}{Z} \sin(\psi - \delta) - \frac{V^2}{Z} \sin \psi \tag{6.37}$$

An important case is when the resistance is negligible; then

$$\psi = 90° \tag{6.38}$$

$$Z = X \tag{6.39}$$

Here we have Eqs. (6.34) to (6.37) reducing to

$$P_1 = P_2 = \frac{EV}{X} \sin \delta \tag{6.40}$$

$$Q_1 = \frac{E^2 - EV \cos \delta}{X} \tag{6.41}$$

$$Q_2 = \frac{EV \cos \delta - V^2}{X} \tag{6.42}$$

In large-scale power systems, a three-phase synchronous machine is paralleled through an equivalent system reactance (X_e) to the network, which has a high generating capacity relative to any single unit. We often refer to the network or system as an infinite bus when a change in input mechanical power or in field excitation to the unit does not cause an appreciable change in system frequency or terminal voltage. Figure 6.18 shows such a situation, where V is the infinite bus voltage.

The foregoing analysis shows that in the present case, we have for power transfer,

$$P = P_{max} \sin \delta \tag{6.43}$$

with

$$P_{max} = \frac{EV}{X_t} \tag{6.44}$$

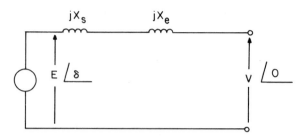

FIGURE 6.18   Synchronous machine connected to an infinite bus.

and

$$X_t = X_s + X_e \qquad (6.45)$$

It is clear that if an attempt were made to advance δ further than 90° (corresponding to maximum power transfer) by increasing the mechanical power input, the electrical power output would decrease from the P_{max} point. Therefore, the angle δ increases further as the machine accelerates. This drives the machine and system apart electrically. The value P_{max} is called the steady-state stability limit or pull-out power. We will consider the following example, which illustrates the utility of the above.

Example 6.6   The synchronous reactance of a cylindrical rotor machine is 1.2 per unit (p.u.). The machine is connected to an infinite bus whose voltage is 1 p.u. through an equivalent reactance of 0.3 p.u. For a power output of 0.7 p.u., the power angle is found to be 30°.

(a). Find the excitation voltage E_f and the pull-out power.

(b). For the same power output, the power angle is to be reduced to 25°. Find the value of the reduced equivalent reactance connecting the machine to the bus to achieve this. What would be the new pull-out power?

Solution

(a).

$$X_t = 1.5 \text{ p.u.}$$

$$P = P_{max} \sin \delta$$

$$0.7 = P_{max} \sin 30°$$

Hence

$$P_{max} = 1.4 \text{ p.u.}$$

Thus

$$1.4 = \frac{E_f(1)}{1.5}$$

$$E_f = 2.1$$

(b).

$$0.7 = P_{max} \sin 25°$$

$$P_{max} = 1.66$$

$$1.66 = \frac{E_f V}{X_{new}} = \frac{2.1(1)}{X_{new}}$$

$$X_{new} = \frac{2.1}{1.66} = 1.27 \text{ p.u.}$$

$$x_e = 1.27 - 1.2 = 0.07 \text{ p.u.}$$

Reactive Power Generation

Inspection of Eq. (6.42) reveals that the generator produces reactive power $(Q_2 > 0)$ if

$$E \cos \delta > V$$

In this case, the generator appears to the network as a capacitor. This condition applies for high magnitude E, and the machine is said to be overexcited. On the other hand, the machine is underexcited if it consumes reactive power $(Q_2 < 0)$. Here we have

$$E \cos \delta < V$$

Figure 6.19 shows phasor diagrams for both cases. The overexcited synchronous machine is normally employed to provide synchronous condenser action, where usually no real load is carried by the machine $(\delta = 0)$. In this case we have

$$Q_2 = \frac{V(E - V)}{X}$$

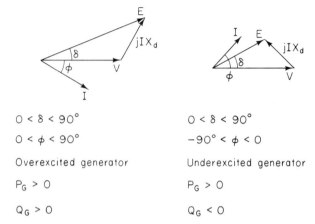

$$0 < \delta < 90°$$

$$0 < \phi < 90°$$

Overexcited generator

$$P_G > 0$$

$$Q_G > 0$$

$$0 < \delta < 90°$$

$$-90° < \phi < 0$$

Underexcited generator

$$P_G > 0$$

$$Q_G < 0$$

FIGURE 6.19   Phasor diagrams for overexcited and underexcited synchronous machines.

Control of reactive power generation is carried out by simply changing E, by varying the dc excitation. An example will help demonstrate the use of these concepts.

Example 6.7   A cylindrical rotor machine is supplying a load of 0.8 PF lagging at an infinite bus. The ratio of the excitation voltage to the infinite bus voltage is found to be 1.25. Compute the power angle δ.

Solution

For a power factor of 0.8, we have

$$\frac{Q}{P} = \tan \phi = 0.75$$

Using the active and reactive power formulas, (6.40) and (6.42), we have

$$\frac{Q}{P} = \frac{\cos \delta - \dfrac{V}{E}}{\sin \delta}$$

$$0.75 = \frac{\cos \delta - 0.8}{\sin \delta}$$

Cross-multiplying, we have

$$0.8 + 0.75 \sin \delta = \cos \delta$$

Using

$$\cos^2\delta = 1 - \sin^2\delta$$

We get

$$[(0.75)^2 + 1] \sin^2\delta + 2(0.8)(0.75) \sin\delta + [(0.8)^2 - 1] = 0$$

Consequently,

$$\sin\delta = 0.23$$

$$\delta = 13.34°$$

Static Stability Limit Curves

Let us consider a machine with synchronous reactance X_s connected to an infinite bus of voltage V through the reactance X_e, as shown in Figure 6.20. We wish to determine the relationship between P, Q, V_t, and δ as V and E are obtained on the basis of Eqs. (6.40) and (6.42).

$$P = \frac{EV_t \sin(\delta - \theta)}{X_s} \tag{6.46}$$

$$Q = \frac{EV_t \cos(\delta - \theta) - V_t^2}{X_s} \tag{6.47}$$

To eliminate E from the relationships above, we note that two additional expressions for P can be obtained since active power losses are not present. The first is based on the transfer from node 1 to node 3 across the

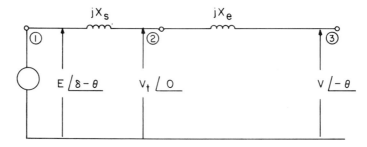

FIGURE 6.20 Equivalent circuit for a synchronous machine connected through an external network to an infinite bus.

total reactance X_t. The second is based on the transfer between nodes 2 and 3 across the reactance X_e. We therefore have

$$P = \frac{EV \sin \delta}{X_t} = \frac{V_t V \sin \theta}{X_e} \qquad (6.48)$$

where X_t is given by Eq. (6.45). Thus we conclude that

$$E = V_t \frac{X_t \sin \theta}{X_e \sin \delta} \qquad (6.49)$$

Using Eq. (6.49) in Eqs. (6.46) and (6.47), we get

$$P = \frac{V_t^2 X_t}{X_s X_e \sin \delta} [\sin \theta \sin (\delta - \theta)] \qquad (6.50)$$

$$Q + \frac{V_t^2}{X_s} = \frac{V_t^2 X_t}{X_s X_e \sin \delta} [\sin \theta \cos (\delta - \theta)] \qquad (6.51)$$

To eliminate θ we will recall the following two trigonometric identities:

$$\sin \alpha \sin \beta = \tfrac{1}{2} [\cos (\alpha - \beta) - \cos (\alpha + \beta)] \qquad (6.52)$$

$$\sin \alpha \cos \beta = \tfrac{1}{2} [\sin (\alpha + \beta) + \sin (\alpha - \beta)] \qquad (6.53)$$

Applying Eq. (6.52) to Eq. (6.50), we obtain

$$P = \frac{V_t^2 X_t}{2 X_s X_e \sin \delta} [\cos (\delta - 2\theta) - \cos \delta]$$

Moreover, Eq. (6.53) applied to Eq. (6.51) gives

$$Q + \frac{V_t^2}{X_s} = \frac{V_t^2 X_t}{2 X_s X_e \sin \delta} [\sin \delta - \sin (\delta - 2\theta)]$$

Rearranging, we get

$$P + \frac{V_t^2 X_t}{2 X_s X_e \tan \delta} = \frac{V_t^2 X_t}{2 X_s X_e \sin \delta} \cos (\delta - 2\theta) \qquad (6.54)$$

$$Q - \frac{V_t^2 (X_s - X_e)}{2 X_s X_e} = \frac{-V_t^2 X_t}{2 X_s X_e \sin \delta} \sin (\delta - 2\theta) \qquad (6.55)$$

Squaring both sides of Eqs. (6.54) and (6.55) and adding, we obtain the desired result:

$$\left(P + \frac{V_t^2 X_t}{2X_s X_e \tan \delta}\right)^2 + \left[Q - \frac{V_t^2(X_s - X_e)}{2X_s X_e}\right]^2 = \left[\frac{V_t^2 X_t}{2X_s X_e \sin \delta}\right]^2 \quad (6.56)$$

Equation (6.56) indicates that the locus of P and Q delivered by the machine is a circle with center at (P_0, Q_0) and radius R_0, where

$$P_0 = \frac{-V_t^2 X_t}{2X_s X_e \tan \delta} \quad (6.57)$$

$$Q_0 = \frac{V_t^2}{2}\left(\frac{1}{X_e} - \frac{1}{X_s}\right) \quad (6.58)$$

$$R_0 = \frac{V_t^2}{2 \sin \delta}\left(\frac{1}{X_s} + \frac{1}{X_e}\right) \quad (6.59)$$

The static stability limit curve for the machine is obtained from Eq. (6.56) by setting δ to 90°. Hence we have

$$P^2 + \left[Q - \frac{V_t^2}{2}\left(\frac{1}{X_e} - \frac{1}{X_s}\right)\right]^2 = \left[\frac{V_t^2}{2}\left(\frac{1}{X_s} + \frac{1}{X_e}\right)\right]^2 \quad (6.60)$$

The static stability limit curve is commonly referred to as the pull-out curve of P and Q and will determine the minimum permissible output var for output watts and terminal voltage specifications. Figure 6.21 shows such a curve.

Example 6.8   Given a generator and a system with reactances of $X_s = 1.2$ and $X_e = 0.2$, both on a 100-MVA base. Assume a generator terminal voltage of 0.95 p.u. The infinite bus voltage is unknown. Find the minimum permissible output var for the p.u. output watts varying from zero to 1 in steps of 0.25.

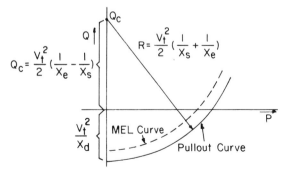

FIGURE 6.21   Static stability limit curve for a synchronous machine.

Solution

The given parameters inserted in the static stability limit curve equation (6–60) yield

$$P^2 + (Q - 1.88)^2 = 6.93$$

For $P = 0$, we have by simple substitution:

$$Q = -0.752$$

Similarly, we get

For $P = 0.25$: $Q = -0.740$.
For $P = 0.5$: $Q = -0.704$.
For $P = 0.75$: $Q = -0.643$.
For $P = 1.00$: $Q = -0.555$.

Power Angle Characteristics

The discussion of the present section shows that the active power of a round-rotor synchronous machine with negligible armature resistance is given by Eq. (6.43) written as

$$P = \frac{EV}{X_s} \sin \delta \qquad (6.61)$$

It is clear, therefore, that for fixed field excitation and terminal voltage, the active power variation with the power angle δ is as shown in Figure 6.22. For positive δ, the excitation voltage leads the terminal voltage and

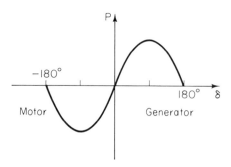

FIGURE 6.22 Synchronous machine power angle characteristic for round-rotor type.

the machine acts as a generator. For negative δ, the excitation voltage lags the terminal voltage and the machine acts as a motor, as discussed in the next section.

6.6 SYNCHRONOUS MOTOR OPERATION

A synchronous machine can be operated as a motor by supplying three-phase voltage to the armature terminals and direct current to the rotor (field windings). It should be noted, however, that unlike an induction machine, a synchronous machine is not self-starting. Arrangements for starting a stationary synchronous motor must be made as discussed later in this section. Once the machine is running at synchronous speed, it will continue to do so. We discuss presently an important performance characteristic of the synchronous motor, V curves.

The V curves of a synchronous motor for a given terminal voltage relate the armature current I_a to the field excitation voltage E_f. Each curve is constructed for a constant active power output. Assuming that armature resistance is negligible, we write

$$V_t = E_f + jI_aX_s \tag{6.62}$$

Therefore,

$$|E_f| \underline{/\delta} = V_t - j|I_a| X_s \underline{/\phi}$$

As a result, separating real and imaginary parts yields

$$E_f \cos \delta = V_t + I_aX_s \sin \phi \tag{6.63}$$

$$E_f \sin \delta = -I_aX_s \cos \phi \tag{6.64}$$

Note that δ is negative for motor operation. The phasor diagrams for leading and lagging power factors are shown in Figure 6.23. The drop I_aX_s is at right angles to I_a.

The condition of constant power operation can be seen on the basis of Eq. (6.61) or (6.64) to require that

$$E_f \sin \delta = \text{constant}$$

As a result, we conclude that for constant power, the tip of the phasor E_f is on the horizontal line *abcd* in the phasor diagrams of Figure 6.24.

Point *a* on the phasor diagram corresponds to E_f lagging V_t by 90°,

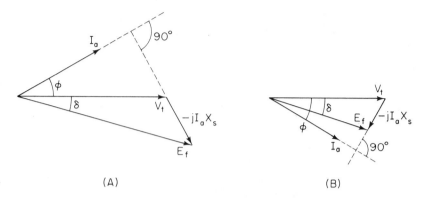

FIGURE 6.23 Phasor diagrams of round-rotor synchronous machine as motor: (a) leading power factor; (b) lagging power factor.

the maximum permissible power angle. For this condition, $\delta = -90°$ and the value of $|I_a|$ is maximum, as can be seen from Figure 6.25. Point b corresponds to a lagging-power-factor mode of operation, where it is clear that $|I_a|$ is decreased while $|E_f|$ is increased.

Point c corresponds to unity-power-factor operation (V_t and I_a are in phase) and $|I_a|$ is seen to have a minimum value.

Point d corresponds to a leading power factor. A representative set of V curves is shown in Figure 6.25.

An analytical procedure for establishing a V curve can be followed from a study of the following example.

Example 6.9   Consider the motor of Example 6.2. Establish the following points ($|I_a|$ and $|E_f|$) on the V curve corresponding to full-load power output:

    a. Stability limit, $\delta = -90°$.
    b. Power factor of 0.8 lagging.
    c. Unity power factor.
    d. Power factor of 0.8 leading.

Solution

We recall the following specifications from Example (6.2) on a per phase basis.

$$V_t = \frac{2300}{\sqrt{3}} = 1327.91 \text{ V}$$

$$P = \frac{764 \times 10^3}{3} \text{ W}$$

$$X_s = 1.9 \ \Omega$$

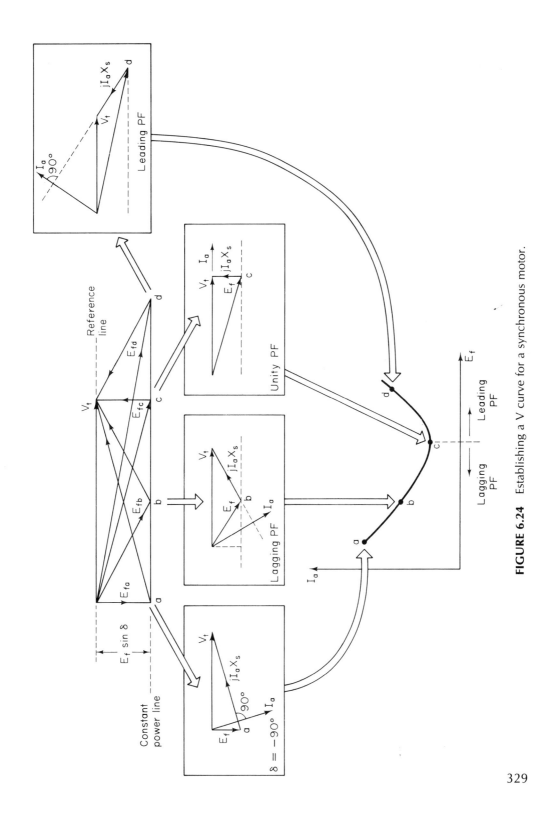

FIGURE 6.24 Establishing a V curve for a synchronous motor.

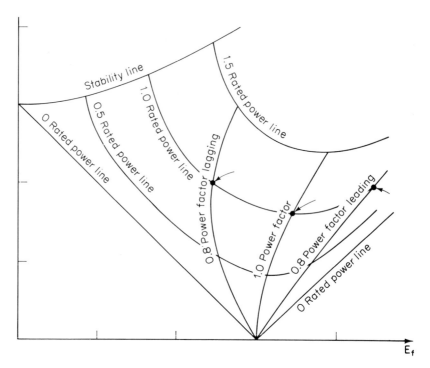

FIGURE 6.25 V curve of a round-rotor synchronous motor.

As a result, we have

$$\frac{764 \times 10^3}{3} = \frac{-1327.91}{1.9} |E_f| \sin \delta$$

Thus

$$|E_f| \sin \delta = -364.38 \text{ V}$$

(a) For $\delta = -90°$, we get

$$|E_f| = 364.38 \text{ V}$$

We find I_a by using

$$I_a = \frac{V_t - E_f}{jX_s} = \frac{1327.91 - 364.38 \underline{/-90°}}{j1.9}$$

$$= 724.73 \underline{/-74.66°} \text{ A}$$

(b) For $\phi = -\cos^{-1} 0.8 = -36.87°$,

$$I_a = \frac{764 \times 10^3}{\sqrt{3}(2300)(0.8)} = 239.73 \underline{/-36.87°} \text{ A}$$

We find E_f from

$$E_f = V_t - jI_aX_s$$
$$= 1327.91 - (239.73 \underline{/-36.87°})(1.9 \underline{/90°})$$
$$= 1115.793 \underline{/-19.061°} \text{ V}$$

(c) For unity power factor

$$I_a = \frac{764 \times 10^3}{2300 \sqrt{3}} = 191.78 \underline{/0} \text{ A}$$

We can get the same result from

$$-|I_a| X_s = |E_f| \sin \delta = -364.38 \text{ V}$$

The value of E_f is obtained from

$$E_f = 1327.91 - 191.78(1.9)\underline{/90°}$$
$$= 1377 \underline{/-15.34°} \text{ V}$$

(d) For a power factor of 0.8 leading, we have

$$I_a = 239.73 \underline{/36.87°} \text{ A}$$
$$E_f = 1327.91 - (239.73 \underline{/36.87°})(1.9 \underline{/90°})$$
$$= 1642.14 \underline{/-12.82°} \text{ V}$$

This concludes this example. We now discuss methods for starting a synchronous motor.

Starting

A synchronous motor must be brought up to a speed close to synchronous speed so that the rotor locks into synchronism with the rotating field. There is more than one method to carry this out. The most common method is as an induction motor by using amortisseur or damper windings on the rotor of the machine. The dc field winding is not energized at starting and the damper windings act as a squirrel cage to provide starting torque. The rotor is brought to a speed close to synchronous speed (small slip value) due to induction motor action. Subsequent to this, the dc field winding is energized and the magnetic fluxes lock into synchronism. It is clear that an induction-motor starting mechanism such as that discussed in Chapter 5 must be employed.

For synchronous machines without damper winding a dc motor coupled to the synchronous motor shaft may be used. It is often practical to use the field exciter generator as a dc motor for starting. An alternative method of starting involves use of a small induction motor with at least one pair of poles less than the synchronous motor's poles to compensate for the loss in induction motor speed due to slip.

Synchronous Capacitors

An important application of a synchronous motor is its use for power factor improvement as a synchronous capacitor (or condenser). The motor is manufactured without a shaft extension, as it is operated without a mechanical load. The motor is overexcited without load and the armature current leads the terminal voltage by close to 90° and hence appears to the network as a capacitor. The synchronous capacitor is connected in parallel with loads characterized by a low power factor such as induction motor. The result is an increase in the power factor at the network terminals. The following example illustrates a number of relevant points.

Example 6.10   An 800-hp three-phase induction motor operates at 0.7 PF lagging from a 2300-V supply.

    a. Find the line current.

    b. The synchronous motor of Example 6.2 is connected in parallel with the induction motor and is operated as a capacitor (zero power factor). Find the necessary excitation voltage to improve the line power factor to 0.9 lagging.

    c. Repeat part (b) for an improved power factor of 0.95 lagging.

Solution

(a) The line current with the induction motor operating alone is

$$I_{L_o} = \frac{800 \times 746}{\sqrt{3}(2300)(0.7)}$$

$$= 214.01 \,\underline{/-45.57°} \text{ A}$$

(b) To improve the line power factor to 0.9 PF lagging, we have the phasor diagram of Figure 6.26. The armature current of the synchronous motor is denoted by I_c and the line current is I_{L_1}. Thus

$$I_{L_1} = I_{L_0} + I_c$$

As a result,

$$|I_{L_1}|\underline{/-\cos^{-1}0.9} = 214.01\,\underline{/-45.57°} + j|I_c|$$

The real part of the above yields

$$0.9|I_{L_1}| = 149.81$$

Thus

$$|I_{L_1}| = 166.46 \text{ A}$$

The imaginary part yields

$$0.44|I_{L_1}| = 152.84 - |I_c|$$

Thus

$$|I_c| = 80.28 \text{ A}$$

To obtain the required excitation voltage, we use

$$E_f = V_t - jIX_s$$

$$= \frac{2300}{\sqrt{3}} - j(80.28\,\underline{/90°})(1.9)$$

$$= 1480.44 \text{ V}$$

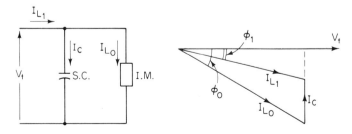

FIGURE 6.26   Equivalent circuit and phasor diagram for Example 6.10.

(c) To improve the power factor to 0.95 lagging, we repeat the above to obtain

$$|I_{L_2}| = 157.69 \text{ A}$$

$$|I_c| = 103.6 \text{ A}$$

and

$$E_f = 1524.74 \text{ V}$$

We note that the magnitude of the line current is reduced by inserting the synchronous capacitor. A higher improvement in power factor requires a larger field excitation.

6.7 EQUIVALENT-CIRCUIT PARAMETER ESTIMATION

The determination of the equivalent-circuit parameters of a round-rotor synchronous machine involves two tests: the open-circuit and short-circuit tests. These two tests are of importance to appropriately account for saturation effects as well as for the determination of machine constants. The open-circuit characteristic is a curve of the armature terminal voltage on an open circuit as a function of the field excitation with the machine running at synchronous speed. An experimental setup for the test is shown in Figure 6.27(a).

As the test name implies, in a short-circuit test the armature terminals of the synchronous machine are short-circuited through ammeters, and the field current is gradually increased until a maximum safe armature current value is reached. This is shown in Figure 6.27(b).

Determination of Synchronous Reactance

The synchronous reactance X_s can be determined on the basis of Eq. (6.28) provided that terminal voltage V_t, generated voltage E_f, and the corresponding current phasor are available.

A conveniently fast alternative method makes use of the short-circuit and open-circuit characteristics of the machine. To understand this, recall that for a short-circuit condition,

$$V_t = 0$$

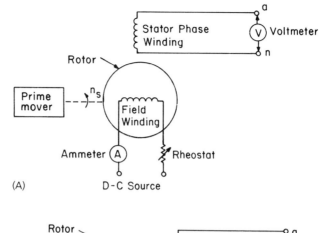

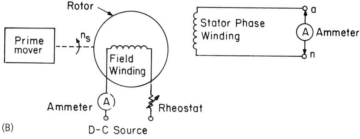

FIGURE 6.27 Experimental setup for the characteristics of a synchronous machine: (a) open-circuit; (b) short-circuit.

Consequently,

$$Z_s = \frac{E_f}{I_{asc}}$$

If we assume, as is usual, that the armature resistance is negligible, we assert that

$$X_s = \frac{E_f}{I_{asc}}$$

An open-circuit characteristic and a short-circuit characteristic are shown in Figure 6.28. The short-circuit armature current is directly proportional to the field current up to almost 150% of rated armature current. The unsaturated value of the synchronous reactance is found using E_f from the air-gap line in Figure 6.29 corresponding to I_{asc}. For operation near rated terminal voltage, we assume that the machine is equivalent to an unsatu-

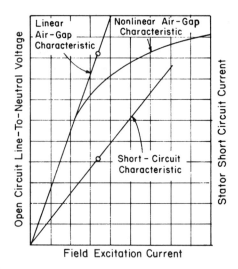

FIGURE 6.28 Typical open-circuit and short-circuit characteristics of a synchronous machine.

rated one with a straight-line magnetization curve through the origin and the rated voltage point on the open-circuit characteristic as shown by the dashed line in Figure 6.29. Accordingly,

$$X_s = \frac{V_t}{I'_{asc}}$$

Example 6.11 The short-circuit characteristic of a synchronous generator is such that rated armature current is obtained by 0.6 per unit (p.u.) excitation. Find the value of the unsaturated synchronous reactance.

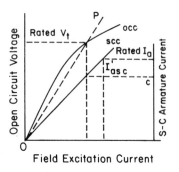

FIGURE 6.29 Defining the synchronous reactance.

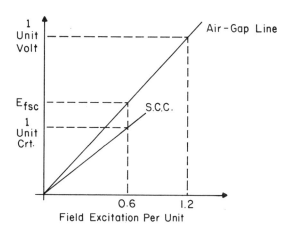

FIGURE 6.30   Geometry of characteristics for Example 6.11.

Solution

The geometry of the problem is shown in Figure 6.30. From similarity of triangles, we deduce that

$$\frac{E_{f\text{sc}}}{0.6} = \frac{1}{1.2}$$

Consequently,

$$E_{f\text{sc}} = 0.5 \text{ p.u.}$$

from which we calculate

$$X_s = \frac{E_{f\text{sc}}}{I_a} = \frac{0.5}{1} = 0.5 \text{ p.u.}$$

6.8   SALIENT-POLE MACHINES

The magnetic reluctance of the air gap in a machine with salient poles is not uniform due to the presence of protruding field poles. The reluctance along the polar axis is appreciably less than that along the interpolar axis. We often refer to the polar axis as the direct axis and the interpolar as the quadrature axis. This effect can be taken into account by resolving the armature current I_a into two components, one in time phase and the other in time quadrature with the excitation voltage as shown in Figure 6.31. The com-

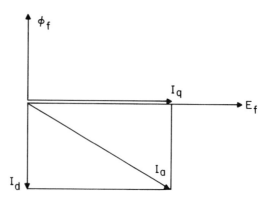

FIGURE 6.31   Resolution of armature current in two components.

ponent I_d of the armature current is along the direct axis (the axis of the field poles), and the component I_q is along the quadrature axis.

Let us consider the effect of the direct-axis component alone. With I_d lagging the excitation EMF E_f by 90°, the resulting armature-reaction flux ϕ_{ad} is directly opposite the field poles, as shown in Figure 6.32. The effect of the quadrature-axis component is to produce an armature-reaction flux ϕ_{aq}, which is in the quadrature-axis direction, as shown in Figure 6.32. The phasor diagram with both components present is shown in Figure 6.33.

We recall that in the cylindrical-rotor machine case, we employed the synchronous reactance X_s to account for the armature-reaction EMF in an equivalent circuit. The same argument can be extended to the salient-pole case. With each of the components currents I_d and I_q, we associate component synchronous-reactance voltage drops, jI_dx_d and jI_qx_q, respectively. The

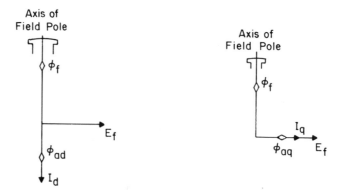

FIGURE 6.32   Direct-axis and quadrature-axis air-gap fluxes in a salient-pole synchronous machine.

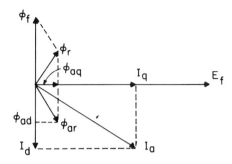

FIGURE 6.33 Phasor diagram for a salient-pole synchronous machine.

direct-axis synchronous reactance x_d and the quadrature-axis synchronous reactance x_q are given by

$$x_d = x_1 + x_{\phi d}$$

$$x_q = x_1 + x_{\phi q}$$

where x_1 is the armature leakage reactance and is assumed to be the same for direct-axis and quadrature-axis currents. The direct-axis and quadrature-axis magnetizing reactances $x_{\phi d}$ and $x_{\phi q}$ account for the inductive effects of the respective armature-reaction flux. Figure 6.34 shows a phasor diagram implementing the result.

$$E_f = V_t + I_a r_a + jI_d x_d + jI_q x_q$$

In many instances, the power factor angle Φ at the machine terminals is explicitly known rather than the internal power factor angle $(\phi + \delta)$, which is required for the resolution of I_a into its direct-axis and quadrature-axis components. We can circumvent this difficulty by recalling that in phasor notation,

$$I_a = I_q + I_d \tag{6.66}$$

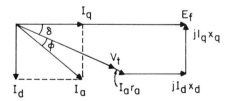

FIGURE 6.34 Phasor diagram for a synchronous machine.

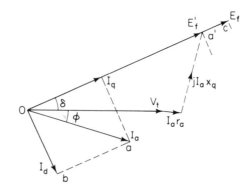

FIGURE 6.35   Modified phasor diagram for a salient-pole synchronous machine.

Substitution of Eq. (6.66) into Eq. (6.65), for I_q and rearranging, we obtain

$$E_f = V_t + I_a(r_a + jx_q) + jI_d(x_d - x_q) \qquad (6.67)$$

Let us define

$$E_f' = V_t + I_a(r_a + jx_q) \qquad (6.68)$$

E_f' as defined is in the same direction as E_f since jI_d is also along the same direction. Our procedure, then, is to obtain E_f' as given by Eq. (6.68) and then obtain the component I_d based on the phase angle of E_f'. Finally, we find E_f as a result of

$$E_f = E_f' + jI_d(x_d - x_q) \qquad (6.69)$$

This is shown in Figure 6.35.

Table 6.1 gives typical ranges for values of x_d and x_q for synchronous machines. Note that x_q is less than x_d because of the greater reluctance of the air gap in the quadrature axis. An example is taken up at this point.

TABLE 6.1

Typical Per-Unit Values of Machine Reactances

	Two-Pole Turbine Generators	Four-Pole Turbine Generators	Salient-Pole Generators	Condensers (Air-Cooled)
x_d	0.85–1.45	1.00–1.45	0.6–1.5	1.25–2.20
x_q	0.92–1.42	0.92–1.42	0.4–0.8	0.95–1.3

Note: Machine kVA rating as base.

Example 6.12   The reactances x_d and x_q of a salient-pole synchronous generator are 0.95 and 0.7 p.u., respectively. The armature resistance is negligible. The generator delivers rated kVA at unity PF and rated terminal voltage. Calculate the excitation voltage.

Solution

We apply

$$E_f' = V_t + jI_a x_q$$

As a result,

$$E_f' = 1 + j(1)(0.7) = 1.22 \underline{/34.99°}$$

We now have

$$I_d = I_a \sin 34.99°$$

$$= 0.57 \text{ p.u.}$$

We can thus compute

$$|E_f| = |E_f'| + |I_d(x_d - x_q)|$$

$$= 1.22 + 0.57(0.25)$$

$$= 1.36$$

$$\delta = 34.99°$$

Figure 6.36 pertains to this example.

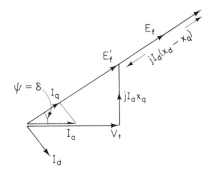

FIGURE 6.36   Phasor diagram for Example 6.12.

Power Angle Characteristics

The power angle characteristics for a salient-pole machine connected to an infinite bus of voltage V through a series reactance of x_e can be arrived at by considering the phasor diagram shown in Figure 6.37. The active power delivered to the bus per phase is

$$P = (I_d \sin \delta + I_q \cos \delta) \, V \qquad (6.70)$$

Similarly, the delivered reactive power Q is

$$Q = (I_d \cos \delta - I_q \sin \delta) V \qquad (6.71)$$

To eliminate I_d and I_q, we need the following identities obtained from inspection of the phasor diagram:

$$I_d = \frac{E_f - V \cos \delta}{X_d} \qquad (6.72)$$

$$I_q = \frac{V \sin \delta}{X_q} \qquad (6.73)$$

where

$$X_d = x_d + x_e \qquad (6.74)$$

$$X_q = x_q + x_e \qquad (6.75)$$

Substitution of Eqs. (6.72) and (6.73) into (6.70) and (6.71) yields

$$P = \frac{VE_f}{X_d} \sin \delta + \frac{V^2}{2}\left(\frac{1}{X_q} - \frac{1}{X_d}\right) \sin 2\delta \qquad (6.76)$$

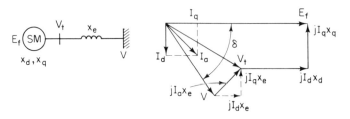

FIGURE 6.37 Salient-pole machine connected to an infinite bus through an external impedance.

$$Q = \frac{VE_f}{X_d} \cos \delta - V^2 \left(\frac{\cos^2\delta}{X_d} + \frac{\sin^2\delta}{X_q} \right) \tag{6.77}$$

Equations (6.76) and (6.77) contain six quantities—the two variables P and δ and the four parameters E_f, V, X_d, and X_q—and can be written in many different ways. The following form illustrates the effect of saliency. Define P_d and Q_d as

$$P_d = \frac{VE_f}{X_d} \sin \delta \tag{6.78}$$

and

$$Q_d = \frac{VE_f}{X_d} \cos \delta - \frac{V^2}{X_d} \tag{6.79}$$

The equations above give the active and reactive power generated by a round-rotor machine with synchronous reactance X_d. We thus have

$$P = P_d + \frac{V^2}{2} \left(\frac{1}{X_q} - \frac{1}{X_d} \right) \sin 2\delta \tag{6.80}$$

$$Q = Q_d - V^2 \left(\frac{1}{X_q} - \frac{1}{X_d} \right) \sin^2\delta \tag{6.81}$$

The second term in the two equations above introduces the effect of salient poles, and in the power equation, the term corresponds to the reluctance torque. Note that if $X_d = X_q$, as in a uniform-air-gap machine, the second terms in both equations are zero. Figure 6.38 shows the power angle characteristics of a typical salient-pole machine.

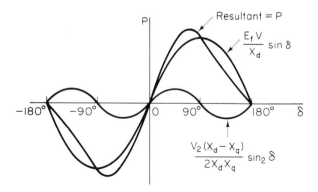

FIGURE 6.38 Power angle characteristics of a salient-pole synchronous machine.

The pull-out power and power angle for the salient-pole machine can be obtained by solving Eq. (6.82), requiring the partial derivative of P with respect to δ be equal to zero.

$$\frac{\partial P}{\partial \delta} = 0 \tag{6.82}$$

The actual value of pull-out power can be shown to be higher than that obtained assuming nonsaliency.

Example 6.13 The machine of Example 6.12 is connected to an infinite bus through a link with reactance of 0.2 p.u. The excitation voltage is 1.3 p.u. and the infinite bus voltage is maintained at 1 p.u. For a power angle of 25°, compute the active and reactive power supplied to the bus.

Solution

We have

$$X_d = 0.95 + 0.2 = 1.5 \text{ p.u.}$$

$$X_q = 0.7 + 0.2 = 0.9 \text{ p.u.}$$

$$E_f = 1.3 \qquad \delta = 25° \qquad V_t = 1$$

As a result, we use Eq. (6.76) to obtain

$$P = \frac{E_f V}{X_d} \sin \delta + \frac{V^2}{2} \left(\frac{1}{X_q} - \frac{1}{X_d} \right) \sin 2\delta$$

$$= \frac{1.3(1)}{1.15} \sin 25° + \frac{1}{2} \left(\frac{1}{0.9} - \frac{1}{1.15} \right) \sin 50°$$

$$= 0.57 \text{ p.u.}$$

Equation (6.77) yields

$$Q = \frac{E_f V}{X_d} \cos \delta - V^2 \left(\frac{\cos^2 \delta}{X_d} + \frac{\sin^2 \delta}{X_q} \right)$$

$$= \frac{1.3(1)}{1.15} \cos 25° - \left(\frac{\cos^2 25}{1.15} + \frac{\sin^2 25}{0.9} \right)$$

$$= 0.11$$

which are the required answers.

Example 6.14   A synchronous machine is supplied from a constant-voltage source. At no load, the motor armature current is found to be negligible when the excitation is 1.0 p.u. The per-unit motor constants are $x_d = 1.0$ and $x_q = 0.6$.

      a. If the machine loses synchronism when the angle between the quadrature axis and the terminal voltage phasor direction is 60 electrical degrees, what is the per-unit excitation at pull-out?

      b. What is the load on the machine at pull-out? Assume the same excitations as in part (a).

Solution

(a)

$$P = \frac{VE_f}{X_d} \sin \delta + \frac{V^2}{2X_dX_q}(X_d - X_q) \sin 2\delta$$

$$= \frac{E_f(1)}{1} \sin \delta + \frac{1}{2 \times 0.6}(1 - 0.6) \sin 2\delta$$

$$= E_f \sin \delta + \tfrac{1}{3} \sin 2\delta$$

For pull-out power, we have

$$\frac{\partial P}{\partial \delta} = E_f \cos \delta + \tfrac{2}{3} \cos 2\delta = 0$$

The pull-out angle is

$$\delta_m = 60°$$

Hence we obtain

$$E_f = \tfrac{2}{3} \text{ p.u.}$$

(b) Consequently, the pull-out load is found to be

$$P = \tfrac{2}{3} \sin 60° + \tfrac{1}{3} \sin 120° = 0.866 \text{ p.u.}$$

SOME SOLVED PROBLEMS

Problem 6.A.1

The following table lists information available for four conditions of operation of a round-rotor synchronous machine connected to an infinite bus. The bus voltage is kept constant at the unknown value V. The reactive power delivered to the bus is denoted by Q. Neglect armature resistance and assume that the total reactance between the excitation voltage E and the bus is X. Complete the table and find V and X.

Condition	P	Q	E	$\delta°$
A	0.772	0.696	?	25
B	?	?	2.2	30
C	0.801	0.58	1.9	?
D	?	-0.16	?	37°

Solution

The relevant equations are

$$P = \frac{EV}{X} \sin \delta \qquad (1)$$

$$Q = \frac{(E \cos \delta - V)V}{X} \qquad (2)$$

We have two equations for each condition and therefore can solve for eight unknowns.

Condition A

$$0.772 = \frac{E_A V}{X} \sin 25° \qquad (A.1)$$

$$0.696 = \frac{(E_A \cos 25° - V)V}{X} \qquad (A.2)$$

Condition B

$$P_B = \frac{2.2V}{X} \sin 30° \qquad (B.1)$$

$$Q_B = \frac{(2.2 \cos 30° - V)V}{X} \qquad (B.2)$$

Condition C

$$0.801 = \frac{1.9V}{X} \sin \delta_C \qquad (C.1)$$

$$0.58 = \frac{(1.9 \cos \delta_C - V)V}{X} \qquad (C.2)$$

Condition D

$$P_D = \frac{E_D V}{X} \sin 37°$$ (D.1)

$$-0.16 = \frac{(E_D \cos 37° - V)V}{X}$$ (D.2)

The eight unknowns are E_A, V, X, P_B, Q_B, δ_C, P_D, and E_D. Substituting Eq. (A.1) in (A.2), we get

$$0.696 = \frac{0.772}{\tan 25°} - \frac{V^2}{X}$$

Thus

$$\frac{X}{V} = 1.0421 \, V$$ (A.3)

Thus we write Eq. (C.1) as

$$\frac{0.801}{1.9} (1.0421V) = \sin \delta_C$$

or (C.3)

$$0.43935V = \sin \delta_C$$

We also write Eq. (C.2) as

$$0.58(1.0421V) = 1.9 \cos \delta_C - V$$

Thus

$$0.844V = \cos \delta_C$$ (C.4)

Squaring and adding Eqs. (C.3) and (C.4), we get

$$[(0.8444)^2 + (0.43935)^2]V^2 = 1$$

Thus

$$V = 1.0505 \approx 1.05$$

Using Eq. (A.3), we conclude that

$$X = 1.1501 \approx 1.15$$

Having obtained V and X, we systematically compute the remaining unknowns. From Eq. (A.1), we get

$$E_A = \frac{0.772(1.1501)}{1.0505 \sin 25}$$

$$= 1.9999 \simeq 2.00$$

Equation (A.2) serves for verification purposes:

$$\text{right-hand side} = \frac{(2 \cos 25° - 1.05)1.05}{1.15} = 0.6963 \simeq 0.696$$

From Eq. (B.1), we get

$$P_B = \frac{2.2(1.05)}{1.15} \sin 30° = 1.0043$$

$$Q_B = \left(\frac{2.2 \cos 30° - 1.05}{1.15}\right)1.05 = 0.78089$$

From Eq. (C.3),

$$\delta_C = \sin^{-1}(0.43935)(1.05) = 27.472°$$

From Eq. (D.2),

$$-0.16(1.15) = (E_D \cos 37° - 1.05)1.05$$

Thus

$$E_D = 1.0953 \simeq 1.1$$

From Eq. (D.1), we get

$$P_D = \frac{1.1(1.05)}{1.15} \sin 37° = 0.60443$$

The completed table is as follows:

Condition	P	Q	E	δ
A	0.772	0.696	2.000	25°
B	1.0043	0.781	2.200	30°
C	0.801	0.58	1.9	27.472°
D	0.60443	−0.16	1.1	37°

with

$$V = 1.05 \qquad X = 1.15$$

Problem 6.A.2

A salient-pole synchronous machine is delivering power to an infinite bus whose voltage is fixed at 1 per unit. The excitation voltage is fixed at 1.9 per unit. The direct-axis and quadrature-axis reactances are to be determined with the help of the following table. Find x_d and x_q and complete the table.

Condition	P	Q	δ
1	0.96	?	32°
2	?	0.424	35°

Solution

The relevant equations are

$$P = \frac{VE_f}{X_d} \sin \delta + \frac{V^2}{2} \left(\frac{1}{X_q} - \frac{1}{X_d} \right) \sin 2\delta$$

$$Q = \frac{VE_f}{X_d} \cos \delta - \frac{V^2}{X_d} - V^2 \left(\frac{1}{X_q} - \frac{1}{X_d} \right) \sin^2 \delta$$

For condition 1, we have

$$0.96 = \frac{1.9 \sin 32°}{X_d} + \frac{1}{2} \left(\frac{1}{X_q} - \frac{1}{X_d} \right) \sin 64°$$

For condition 2, we have

$$0.424 = \frac{1.9 \cos 35°}{X_d} - \frac{1}{X_d} - \left(\frac{1}{X_q} - \frac{1}{X_d} \right) \sin^2 35°$$

The equations above reduce to

$$0.96 = \frac{0.557}{X_d} + \frac{0.449}{X_q}$$

$$0.424 = \frac{0.885}{X_d} - \frac{0.329}{X_q}$$

Solving for the values of X_d and X_q, we get

$$X_d = 1.148$$

$$X_q = 0.947$$

The value of Q for condition 1 is

$$Q = \frac{1.9 \cos 32}{1.148} - \frac{1}{1.148} - \left(\frac{1}{0.949} - \frac{1}{1.148}\right) \sin^2 32°$$

$$= 0.481$$

The value of P for condition 2 is

$$P_2 = \frac{1.9 \sin 35°}{1.148} + \frac{1}{2}\left(\frac{1}{0.949} - \frac{1}{1.148}\right) \sin 35°(2)$$

$$= 1.036$$

Our completed table is thus:

Condition	P	Q	δ
1	0.96	0.481	32°
2	1.036	0.424	35°

Problem 6.A.3

The condition for pull-out for a salient-pole machine is given by Eq. (6.82), with P as given in Eq. (6.80). Show that the angle δ at which pull-out occurs satisfies

$$\cos^2\delta + A \cos \delta - 0.5 = 0$$

with

$$A = \frac{E_f X_q}{2V(X_d - X_q)}$$

Show also that if $2/A^2 < 1$, then approximately, we have

$$\cos \delta_m \simeq \frac{V(X_d - X_q)}{E_f X_q}$$

and that

$$P_{max} = \frac{VE_f}{X_d} + \frac{V^3}{E_f}\frac{1}{X_d}\left(\frac{X_d}{X_q} - 1\right)^2$$

Solution

Write Eq. (6.80) as

$$P = P_F \sin \delta + P_R \sin 2\delta$$

where

$$P_F = \frac{VE_f}{X_d}$$

$$P_R = \frac{V^2}{2}\left(\frac{1}{X_q} - \frac{1}{X_d}\right)$$

The condition for pull-out is

$$\frac{\partial P}{\partial \delta} = 0$$

Thus, differentiating, we get

$$P_F \cos \delta + 2P_R \cos 2\delta = 0$$

Recalling that

$$\cos 2\delta = 2 \cos^2\delta - 1$$

we thus have

$$4P_R \cos^2\delta + P_F \cos \delta - 2P_R = 0$$

or

$$\cos^2\delta + \frac{P_F}{4P_R} \cos \delta - 0.5 = 0$$

$$\cos^2\delta + \frac{E_fX_q}{2V(X_d - X_q)} \cos \delta - 0.5 = 0$$

Let us put

$$A = \frac{E_f X_q}{2V(X_d - X_q)}$$

Thus

$$\cos^2\delta + A \cos \delta - 0.5 = 0$$

The solution of the quadratic is

$$\cos \delta_m = \frac{-A \pm \sqrt{A^2 \pm 2}}{2}$$

$$\cos \delta_m = \frac{A}{2}\left(-1 \pm \sqrt{1 - \frac{2}{A^2}}\right)$$

For

$$\frac{2}{A^2} < 1$$

$$\cos \delta_m = \frac{A}{2}\left[-1 \pm \left(1 + \frac{1}{A^2}\right)\right]$$

Take the positive sign,

$$\cos \delta_m = \frac{A}{2}\frac{1}{A^2}$$

$$= \frac{1}{2A}$$

$$\cos \delta_m = \frac{V(X_d - X_q)}{E_f X_q}$$

Note that since δ_m is close to 90°, we may assume that

$$\sin \delta_m = 1$$

Now

$$P_{max} = \frac{VE_f}{X_d}\sin \delta_m + \frac{V^2}{2}\left(\frac{1}{X_q} - \frac{1}{X_d}\right)\sin 2\delta_m$$

Thus

$$P_{max} = \frac{VE_f}{X_d} + V^2 \left(\frac{1}{X_q} - \frac{1}{X_d} \right) \cos \delta_m$$

$$= \frac{VE_f}{X_d} + V^2 \left(\frac{X_d - X_q}{X_q X_d} \right) \frac{V(X_d - X_q)}{E_f X_q}$$

$$= \frac{VE_f}{X_d} + \frac{V^3}{E_f} \frac{(X_d - X_q)^2}{X_d X_q^2}$$

$$= \frac{VE_f}{X_d} + \frac{V^3}{E_f} \frac{1}{X_d} \left(\frac{X_d}{X_q} - 1 \right)^2$$

Problem 6.A.4

A salient-pole machine has the following parameters:

$$X_d = 1.1 \qquad X_q = 0.9$$
$$E_f = 1.2 \qquad V = 1$$

Calculate the exact value of the pull-out angle and compare with the approximate result using the expression of Problem 6.A.3.

Solution

We have

$$X_d = 1.1 \qquad X_q = 0.9$$
$$E_f = 1.2 \qquad V = 1$$

$$A = \frac{1.2(0.9)}{2(0.2)} = 2.7$$

$$\frac{2}{A^2} = 0.27 < 1$$

Exact calculation:

$$\cos^2\delta + 2.7 \cos \delta - 0.5 = 0$$

$$\cos \delta = \frac{-2.7 \pm \sqrt{(2.7)^2 + 2}}{2}$$

$$= 0.174$$

$$\delta_m = 79.98°$$

Approximately,

$$\cos \delta_m = \frac{1(1.1 - 0.9)}{1.2(0.9)} = 0.1852$$

$$\delta_m = 79.3281°$$

PROBLEMS

Problem 6.B.1

A 180-kVA three-phase Y-connected 440-V 60-Hz synchronous generator has a synchronous reactance of 1.6 Ω and a negligible armature resistance. Find the full-load generated voltage per phase at 0.8 PF lagging.

Problem 6.B.2

A 9375-kVA 13,800-kV 60-Hz two-pole Y-connected synchronous generator is delivering rated current at rated voltage and unity PF. Find the armature resistance and synchronous reactance given that the field excitation voltage is 11,935.44 V and leads the terminal voltage by an angle 47.96°.

Problem 6.B.3

The magnitude of the field excitation voltage for the generator of Problem 6.B.2 is maintained constant at the value specified above. Find the terminal voltage when the generator is delivering rated current at 0.8 PF lagging.

Problem 6.B.4

The armature resistance and synchronous reactance of a three-phase synchronous motor are 0.8 and 6 Ω, respectively. The terminal voltage is 400 V (line to line). Find the excitation voltage when the motor draws a line current of 18 A at 0.8 PF leading.

Problem 6.B.5

The excitation voltage of a synchronous motor is 6370.5 V per phase and lags the terminal voltage by 4.5°. Assume that the terminal voltage is 11,000 V (line to line). The synchronous reactance is 10 Ω and armature resistance is negligible. Find the armature current and power factor for this condition of operation.

Problem 6.B.6

The power input in a three-phase synchronous motor is 400 kW. The excitation voltage is 5200 $\underline{/-9°}$ (line to line) with terminal voltage taken as reference. Assume that armature resistance is negligible and that the synchronous reactance is 8 Ω. Find the terminal voltage and the power factor of the motor.

Problem 6.B.7

A 9375-kVA three-phase Y-connected 13,800-V (line to line) two-pole 60-Hz turbine generator has an armature resistance of 0.064 Ω per phase and a synchronous reactance of 1.79 Ω per phase. Find the full-load generated voltage per phase at:

  a. Unity power factor.
  b. A power factor of 0.8 lagging.
  c. A power factor of 0.8 leading.

Problem 6.B.8

A 5-kVA 220-V 60-Hz six-pole Y-connected synchronous generator has a leakage reactance per phase of 0.78 Ω and negligible armature resistance. The armature-reaction EMF for this machine is related to the armature current by

$$E_{ar} = -j16.88(I_a)$$

Assume that the generated EMF is related to field current by

$$E_f = 25I_f$$

  a. Compute the field current required to establish rated voltage across the terminals of a unity power factor load that draws rated generator armature current.
  b. Determine the field current needed to provide rated terminal voltage to a load that draws 125% of rated current at 0.8 PF lagging.

Problem 6.B.9

For the machine of Example 6.3, the field current is adjusted to 250 A and the armature current is 400 A. Find the power factor at the machine's terminals.

Problem 6.B.10

For the machine of Example 6.3, the field current is adjusted to 250 A and the machine's terminal power factor is 0.85 lagging. Find the armature current.

Problem 6.B.11

Repeat Example 6.5 for a power factor of 0.8 lagging.

Problem 6.B.12

The synchronous reactance of a cylindrical rotor synchronous generator is 0.9 p.u. If the machine is delivering active power of 1 p.u. to an infinite bus whose voltage is 1 p.u. at unity PF, calculate the excitation voltage and the power angle.

Problem 6.B.13

A synchronous generator with a synchronous reactance of 1.3 p.u. is connected to an infinite bus whose voltage is 1 p.u. through an equivalent reactance of 0.2 p.u. The maximum permissible output is 1.2 p.u.

   a. Compute the excitation voltage E.
   b. The power output is gradually reduced to 0.7 p.u. with fixed field excitation. Find the new current and power angle δ.

Problem 6.B.14

Compute the reactive power generated by the machine of Problem 6.B.13 under the conditions in part (b). If the machine is required to generate a reactive power of 0.4 p.u. while supplying the same active power by changing the field excitation, find the new excitation voltage and power angle δ.

Problem 6.B.15

The apparent power delivered by a cylindrical rotor synchronous machine to an infinite bus is 1.2 p.u. The excitation voltage is 1.3 p.u. and the power angle is 20°. Compute the synchronous reactance of the machine given that the infinite bus voltage is 1 p.u.

Problem 6.B.16

The synchronous reactance of a cylindrical rotor generator is 1 p.u. and its terminal voltage is 1 p.u. when connected to an infinite bus through a reactance of 0.4 p.u. Find the minimum permissible output vars for zero-output active power and unity-output active power.

Problem 6.B.17

A cylindrical rotor machine is delivering active power of 0.8 p.u. and reactive power of 0.6 p.u. when the excitation voltage is 1.2 p.u. and the power angle is 25°. Find the terminal voltage and synchronous reactance of the machine.

Problem 6.B.18

A cylindrical rotor machine is delivering active power of 0.8 and reactive power of 0.6 p.u. at a terminal voltage of 1 p.u. If the power angle is 20°, compute the excitation voltage and the machine's synchronous reactance.

Problem 6.B.19

The synchronous reactance of a cylindrical rotor machine is 0.8 p.u. The machine is connected to an infinite bus through two identical parallel transmission links with a reactance of 0.4 p.u. each. The excitation voltage is 1.4 p.u and the machine is supplying a load of 0.8 p.u.

a. Compute the power angle δ for the outlined conditions.
b. If one link is opened with the excitation voltage maintained at 1.4 p.u., find the new power angle to supply the same load as in part (a).

Problem 6.B.20

The synchronous reactance of a cylindrical rotor machine is 0.9 p.u. The machine is connected to an infinite bus through two identical parallel transmission links with a reactance of 0.6 p.u. each. The excitation voltage is 1.5 p.u., and the machine is supplying a load of 0.8 p.u.

a. Compute the power angle δ for the given conditions.
b. If one link is opened with the excitation voltage maintained at 1.5 p.u., find the new power angle to supply the same load as in part (a).

Problem 6.B.21

A round-rotor synchronous machine has a synchronous reactance of 1 p.u. The machine is connected to an infinite bus whose voltage is maintained at a voltage V. The following table lists information available about two conditions of operation. Complete the table and find V.

Condition	P	Q	E	δ
A	0.63	−0.1	?	35°
B	?	−0.08	1.2	?

Problem 6.B.22

A round-rotor synchronous machine is connected to an infinite bus whose voltage is maintained at 1.05 p.u. Find the synchronous reactance of the machine and complete the following table.

Condition	P	Q	E	δ
A	1.07	−0.31	1.15	?
B	0.95	?	?	40°

Problem 6.B.23

A round-rotor synchronous machine has a synchronous reactance of 0.9 p.u. The machine is connected to an infinite bus whose voltage is maintained at a value of 1.1 p.u. Complete the following table for two conditions of operation.

Condition	P	Q	E	δ
A	?	0.04	?	25°
B	?	?	1.2	30°

Problem 6.B.24

Repeat Example 6.9 for 80% of full-load power.

Problem 6.B.25

The synchronous reactance of a round-rotor synchronous motor is 2 Ω. The motor draws 240 A at a lagging power factor for an excitation voltage of 1120 $\underline{/-20°}$, phase value. The terminal voltage is taken as reference. Find the power factor of the motor and the terminal voltage.

Problem 6.B.26

A round-rotor synchronous machine has a synchronous reactance of 1.8 Ω. The machine is operated as a synchronous condenser on a 2000-V bus and draws a current of 250 A. Find the excitation voltage and reactive power of the motor.

Problem 6.B.27

A 200-hp induction motor operates at 0.75 PF lagging. Find the reactive power required to improve the power factor to 0.9 lagging.

Problem 6.B.28

An induction motor delivers 150 hp to a mechanical load. The power factor is 0.7 lagging. A synchronous condenser is connected in parallel with the motor and supplies a leading reactive power of 800 kvar. Find the power factor of the combination. Neglect losses in the induction motor.

Problem 6.B.29

An induction motor delivers 100 hp to a load. Losses are neglected in the motor. A synchronous condenser in parallel with the motor delivers 45 kvar at a leading power factor of 0.05. Find the power factor of the combination.

Problem 6.B.30

For the synchronous condenser of Problem 6.B.27, assume that the line voltage is 2300 V. Find the required field excitation assuming that the synchronous reactance is 2 Ω.

Problem 6.B.31

A 5-kVA 220-V 60-Hz six-pole Y-connected synchronous generator has a leakage reactance per phase of 0.8 Ω and negligible armature resistance. The air-gap line is described by $E = 18I_f$ and the short-circuit characteristic is described by $I_{sc} = 6I_f$.

    a. Find the value of the unsaturated synchronous reactance.
    b. Find the value of armature reaction reactance.
    c. Find the value of the field current needed to yield rated terminal voltage at rated current for a unity-PF load.

Problem 6.B.32

For the generator of Problem 6.B.31, find the maximum power output if the terminal voltage and field excitation voltage are kept constant at the values defined in Problem 6.B.31. If the power output is gradually reduced to 3 kW, with fixed field excitation, find the new current and power angle δ.

Problem 6.B.33

Repeat Problem 6.B.32 for the generator of Problem 6.B.31 with power output reduced to 150 kW.

Problem 6.B.34

    a. For the generator of Problem 6.B.32, compute the reactive power generated by the machine.

b. If the machine is required to generate a reactive power of 600 var while supplying the same active power by changing the field excitation, find the new excitation voltage and power angle.

Problem 6.B.35

Repeat Problem 6.B.34 for the generator operating under the conditions of Problem 6.B.33. Assume the new reactive power to be 90 kvar.

Problem 6.B.36

A 180-kVA 400-V 300-r/min 60-Hz three-phase Y-connected cylindrical rotor synchronous generator has the following particulars: $r_a = 0$ and $x_1 = 0.296 \ \Omega$. The air-gap line is described by $E = 17I_f$, expressed per phase. The short-circuit characteristic is described by $I_{sc} = 10.75I_f$.

a. Find the value of the unsaturated synchronous reactance.
b. Find the value of the armature reaction reactance.
c. Find the value of the field current needed to yield rated terminal voltage at rated current for a 0.8 lagging PF load.

Problem 6.B.37

For the machine of Problem 6.B.36, it is required to compute the field current needed to provide rated voltage when rated current is delivered to the load at:

a. A power factor of 0.8 leading.
b. A power factor of 0.8 lagging.

Assume that E_{ar} at rated current corresponds to 305 equivalent field volts.

Problem 6.B.38

A 40,000-kVA 14,000-V Y-connected alternator has negligible armature resistance. Some pertinent data are as follows:

Short-circuit characteristic, $I_a = 7I_f$
Air-gap line, volts per phase; $E = 33I_f$
Open-circuit characteristic volts per phase:

$$E = \frac{21,300I_f}{430 + I_f}$$

Find both the unsaturated and saturated synchronous reactances for this machine.

Problem 6.B.39

A 5-kVA 220-V Y-connected three-phase salient-pole synchronous generator is used to supply power to a unity-PF load. The direct-axis synchronous reactance is 12 Ω and the quadrature-axis synchronous reactance is 7 Ω. Assume that rated current is delivered to the load at rated voltage and that armature resistance is negligible. Compute the excitation voltage and the power angle.

Problem 6.B.40

A three-phase synchronous machine has the following parameters:

$$r_a = 0 \qquad X_d = 9.13 \ \Omega$$

$$X_q = 7.03 \ \Omega$$

The machine delivers 5 kW at unity power factor at 220 V. Find the excitation voltage and power angle δ.

Problem 6.B.41

The excitation voltage (line to line) in a 5-kVA 220-V Y-connected three-phase salient-pole synchronous generator is found to be 300 V when delivering rated current at unity power factor. The power angle δ is found to be 36°. Find the direct-axis and quadrature-axis synchronous reactances. Neglect armature resistance.

Problem 6.B.42

A salient-pole synchronous machine is connected to an infinite bus through a link with reactance of 0.2 p.u. The direct-axis and quadrature-axis reactances of the machine are 0.9 and 0.65 p.u., respectively. The excitation voltage is 1.3 p.u., and the voltage of the infinite bus is maintained at 1 p.u. For a power angle of 30°, compute the active and reactive power supplied to the bus.

Problem 6.B.43

A salient-pole synchronous machine is connected to an infinite bus through a link with reactance of 0.2 p.u. The direct-axis reactance of the machine is 0.95 p.u.; the quadrature-axis reactance is yet to be determined. The excitation voltage is 1.35 and the voltage of the infinite bus is maintained at 1 p.u. For a power angle of 35° and active power of 0.79 p.u., find the quadrature-axis reactance of the machine and the reactive power delivered.

Problem 6.B.44

A salient-pole machine supplies a load of 1.2 p.u. at unity power factor to an infinite bus whose voltage is maintained at 1.05 p.u. The machine excitation voltage is computed to be 1.4 p.u. when the power angle is 25°. Evaluate the direct-axis and quadrature-axis synchronous reactances.

Problem 6.B.45

A salient-pole machine is connected to an infinite bus whose voltage is kept constant at 1 p.u. The direct-axis and quadrature-axis reactances of the machine are $X_d = 1.2$ and $X_q = 0.8$. The following table details two operating conditions. Complete the table, neglecting armature resitance.

Condition	P	Q	E_f	δ
1	1.0	?	?	36°
2	?	0.3	1.7	?

Problem 6.B.46

A salient-pole synchronous machine supplies a load at unity power factor to an infinite bus whose voltage is maintained at 1.05 p.u. Excitation voltage is given by 1.4 for a certain power angle. The direct axis reactance is 0.93 p.u. and that of the quadrature axis is 0.43 p.u. Find the active power delivered P and the power angle δ for the specified operating conditions.

Problem 6.B.47

A 5-kVA 220-V Y-connected three-phase salient-pole synchronous generator supplies rated current at rated voltage at 0.8 PF lagging to a local load. The direct-axis synchronous reactance is 12 Ω and the quadrature-axis synchronous reactance of the machine is 7 Ω. Compute the excitation voltage and the power angle, neglecting armature resistance.

Problem 6.B.48

The reactances x_d and x_q of a salient-pole synchronous generator are 1.00 and 0.6 p.u., respectively. The excitation voltage is 1.77 p.u. and the infinite bus voltage is maintained at 1 p.u. For a power angle of 19.4°, compute the active and reactive power supplied to the bus.

Problem 6.B.49

For the machine of Problem 6.B.48, assume that the active power supplied to the bus is 0.8 p.u. Compute the power angle and the reactive power supplied to the bus. (*Hint:* Assume that cos δ ≃ 1 for an approximation.)

Problem 6.B.50

For the machine of Problem 6.B.48, assume that the reactive power supplied to the bus is 0.6 p.u. Compute the power angle and the active power supplied to the bus.

Problem 6.B.51

A salient-pole machine supplies a load of 1.2 p.u. at unity PF to an infinite bus. The direct-axis and quadrature-axis synchronous reactances are

$$x_d = 0.9283 \qquad x_q = 0.4284$$

The power angle δ is 25°. Evaluate the excitation and terminal voltages.

Direct-Current Machines

7.1 INTRODUCTION

From the historical point of view, the direct-current (dc) machine was the earliest electromechanical energy conversion device and heralded the dawn of the electrical age. The famous copper disk experiments of Faraday led to further inventions that evolved the dc machine as the first electric energy source used for illumination purposes.

The most distinctive feature of the dc machine is its versatility. The machine is reversible, so it can operate either as generator converting mechanical energy into electrical energy in the form of direct current and voltage, or as a motor converting electrical energy into useful mechanical work. Direct-current generators are not in widespread use at present because alternating-current generation, transmission, and distribution are more advantageous from an economic point of view. Direct-current generators can be used in situations where electric energy consumption takes place very close to the generation site and where output voltages are required to follow closely prespecified patterns. The latter application area can also be met by solid-state-controlled rectification devices.

The operating characteristics of dc motors offer distinct advantages that makes them attractive for many industrial applications. A wide variety of speed–torque characteristics can be obtained as a result of the various combinations of separately excited, series, shunt, and compound field winding connections. Wide ranges of speed and precise control can be obtained easily using systems of dc machines.

This chapter treats dc machines from an application point of view, and therefore emphasis is given more to operating characteristics rather than design details and procedures.

7.2 CONSTRUCTION FEATURES

A dc machine is a rotating electromechanical energy conversion device that
has a stator with salient poles that are excited by one or more field windings.
The armature winding of a dc machine is on the rotor, with current conducted from or to it by means of carbon brushes in contact with copper
commutator segments. A cutaway view of a dc motor is shown in Figure
7.1, and a schematic representation of a two-pole dc machine is given in
Figure 7.2.

The armature windings consist of many coil sides placed on the rotor
with the conductors parallel to the shaft. The field windings are fed with
direct current, and as a result the air-gap flux is almost constant under each
of the salient poles. The rotor is turned at a constant speed and as a result,
induced voltages appear in the armature coils. The induced voltages are
alternating and must be rectified.

Rectification is traditionally carried out using a commutator, which is
a cylinder mounted on the rotor and formed of copper segments insulated
from each other and from the rotor shaft. Commutator segments are in contact with stationary carbon brushes. For the configuration of Figure 7.2, the
commutator connects the coil side which is under the south pole to the
positive brush and that which is under the north pole to the negative brush
at all times. The commutator is essentially a mechanical full-wave rectifier.

The magnetic field due to the direct current in the field winding is
stationary with respect to the stator. The armature currents create a stationary magnetic flux distribution whose axis is at right angles to that of the
field flux. The interaction of the two flux distributions creates the torque in
the dc machine.

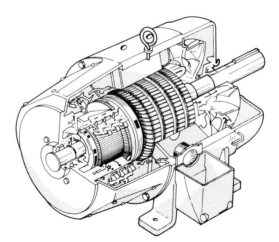

FIGURE 7.1   Direct-current motor. (Courtesy of General Electric Company.)

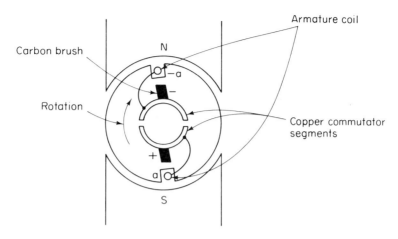

FIGURE 7.2   Schematic representation of a dc machine.

7.3   DIRECT-CURRENT GENERATORS

The armature conductors are rotating at a constant speed n r/min relative to the field flux. The flux linkages λ can be written

$$\lambda(t) = -N\phi \cos \omega t$$

where N is the number of coil turns, ϕ the flux per pole, and ω the angular velocity in electrical radians per second. The induced voltage $e(t)$ is given by Faraday's law as

$$e(t) = \frac{d\lambda}{dt} = N\omega\phi \sin \omega t$$

Due to commutator action, the induced voltage is rectified with an average value given by

$$E_{av} = \frac{1}{\pi} \int_0^\pi e(t)\, d\omega t$$

As a result,

$$E_{av} = \frac{2N\phi\omega}{\pi}$$

The average voltage equation can be rearranged to yield a practical formula by using the speed in r/min instead of ω and expressing N in terms of the number of armature conductors. First recall that

$$\omega = \frac{\omega_m P}{2} = \frac{2\pi n}{60}\frac{P}{2}$$

Also, if Z is the number of armature conductors and a is the number of parallel paths, we have the number of coils in a parallel path given by

$$N = \frac{Z}{2a}$$

We therefore have

$$E_g = \frac{ZnP\phi}{60a} \tag{7.1}$$

Equation (7.1) is a practical formula that relates the generated voltage E_g to the speed of rotation, n in r/min, flux per pole ϕ, number of armature conductors Z, number of poles P, and number of armature parallel paths a. From an analysis point of view for a given generator, Z, P, and a are constants and we therefore write

$$E_g = K_g\phi\omega \tag{7.2}$$

It is clear that the armature-generated voltage depends on the flux and rotational speed. The flux, in turn, varies with the field current I_f according to the magnetization curve pertinent to the machine. In general, the curve is not linear, especially in the saturation region, as discussed earlier. Analysis of dc machines including saturation effects can be done effectively by using approximate straight-line segments, as is done in Problem 7.A.1. For most practical application purposes, a simple straight-line approximation of the magnetization curve provides reasonable procedures and results. We will therefore rewrite Eq. (7.2) as

$$E_g = K\omega I_f \tag{7.3}$$

where I_f is the field current.

Armature Circuit Model

The armature of a dc generator is modeled by a simple circuit consisting of an ideal voltage source E_g in series with a resistance representing the armature resistance R_a for steady-state analysis purposes. For transient analysis studies, appropriate representations of the winding inductances are necessary. The armature circuit model described above is shown in Figure 7.3 and is simply the familiar Thévenin equivalent of the machine looking into the armature terminals.

The terminal voltage of the armature is denoted by V_a, as shown in Figure 7.3, and the armature current is I_a. We thus write for the generator case

$$V_a = E_g - I_a R_a \tag{7.4}$$

From an electric power point of view, we can rewrite Eq. (7.4) as

$$V_a I_a = E_g I_a - I_a^2 R_a$$

or

$$P_a = P_{\text{in}} - I_a^2 R_a \tag{7.5}$$

Thus the output power of the armature $P_a = V_a I_a$ is the net of the input power $P_{\text{in}} = E_g I_a$ and the armature circuit ohmic losses $I_a^2 R_a$. Note that this representation does not account for losses, such as rotational, eddy-current, and hysteresis losses.

Field Circuit Model

The field windings are represented simply by a resistance R_f as shown in Figure 7.3. Note that the source of the field excitation voltage is left unspecified, for there are various ways of obtaining this voltage, as discussed next.

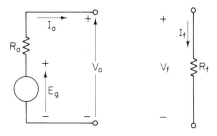

FIGURE 7.3 Armature circuit and field circuit models for a dc generator.

7.4 DC-GENERATOR CONNECTIONS

Direct-current generators can be classified according to the source of field excitation voltage as either separately excited or self-excited. In a separately excited generator, the field voltage is supplied by an external dc source such as another dc generator, batteries, or a solid-state dc power supply. A self-excited generator provides the necessary field voltage through one of various means of connections of the field windings to the armature circuit.

Separately Excited DC Generators

The schematic diagram of Figure 7.4 shows a separately excited dc generator. It is clear that the terminal voltage V_T of the machine is the same as the armature developed voltage, and thus we have

$$V_T = E_g - I_a R_a \qquad (7.6)$$

The armature developed voltage E_g is given by

$$E_g = K\omega I_f \qquad (7.7)$$

It is clear that for a given load, the voltage V_T can be controlled by either a change in the speed of the prime mover ω or the field excitation I_f.

The main operating characteristic of a dc generator is that of the load curve relating the terminal voltage V_T to the load current. For a separately excited dc generator the load current I_L is the same as the armature current I_a. The load curve for a given developed voltage E_g is evidently a straight line, as indicated by Eq. (7.6). As a result we have the characteristic shown in Figure 7.5. The experimentally obtained characteristic does not follow precisely the straight-line approximation pattern shown, due to the nonlin-

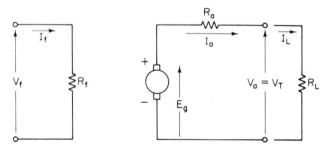

FIGURE 7.4 Schematic representation of a separately excited dc generator.

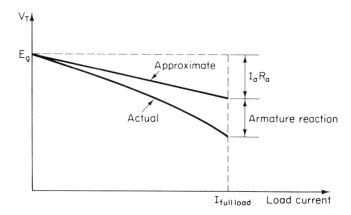

FIGURE 7.5   Load curve of a separately excited dc generator.

earity of the magnetization characteristic and to the effect of the demagne-
tization due to the armature reaction MMF. However, Eq. (7.6) provides
us with reasonably accurate results for practical purposes. Another effect
that should be mentioned is the fact that since the carbon brushes and commutator segments are not perfect conductors, a brush voltage drop is encountered. This is of a relatively small magnitude (2 to 5 V) and is subtracted from the actual value of the armature developed voltage E_g in our
calculations. It is appropriate to take some examples now.

Example 7.1   The generated armature voltage of a separately excited dc generator is 151
V at a speed of 1450 r/min when the field current is 2.8 A.

    a. Find the generated armature voltage for a field current of 2.4 A at
       1450 r/min.

    b. Find the generated armature voltage for a field current of 2.1 A at a
       speed of 1600 r/min.

Solution

Let

$$E_{g1} = 151 \text{ V} \qquad I_{f_1} = 2.8 \text{ A} \qquad n_1 = 1450 \text{ r/min}$$

(a) For $I_{f_2} = 2.4$ A and $n_2 = 1450$ r/min, we get

$$E_{g2} = E_{g1}\frac{n_2}{n_1}\frac{I_{f2}}{I_{f1}}$$

$$= 151\left(\frac{2.4}{2.8}\right) = 129.43 \text{ V}$$

(b) For $I_{f3} = 2.1$ A and $n_3 = 1600$ r/min, we get

$$E_{g3} = E_{g1}\frac{n_3}{n_1}\frac{I_{f3}}{I_{f1}}$$

$$= 151\left(\frac{1600}{1450}\right)\left(\frac{2.1}{2.8}\right) = 124.97 \text{ V}$$

Example 7.2 The armature resistance of a 50-kW 250-V separately excited dc generator is 0.025 Ω. The generator delivers rated load at rated terminal voltage. Find the armature current and the generated armature voltage at rated load. If the terminal voltage is maintained at 250 V but the output is reduced to 40 kW, find the corresponding generated armature voltage.

Solution

The rated armature current is

$$I_a = \frac{50 \times 10^3}{250} = 200 \text{ A}$$

Thus we have

$$E_g = V_T + I_a R_a$$

$$= 250 + 200(0.025) = 255 \text{ V}$$

For the second condition of operation, we have

$$I_a = \frac{40 \times 10^3}{250} = 160 \text{ A}$$

Thus we have

$$E_g = 250 + 160(0.025) = 254 \text{ V}$$

Example 7.2 is straightforward to solve. The next example explores some variants on the same theme.

Example 7.3   The generator of Example 7.2 delivers 40 kW with the generated armature voltage being 255 V. Find the terminal voltage and the corresponding armature current. If the terminal voltage is maintained at 253 V and the generated armature voltage is 257 V, what would be the load power?

Solution

We have

$$I_a = \frac{40 \times 10^3}{V_T}$$

Thus

$$V_T = 255 - \frac{40,000}{V_T}(0.025)$$

As a result,

$$V_T^2 - 255V_T + 1000 = 0$$

The solution is

$$V_T = 251.02 \text{ V}$$

Thus

$$I_a = \frac{40,000}{251.02} = 159.35 \text{ A}$$

Under the second set of conditions, we get

$$I_a = \frac{E_g - V_T}{R_a} = \frac{257 - 253}{0.025} = 160 \text{ A}$$

As a result,

$$P_T = V_T I_a = 253(160) = 40.48 \times 10^3 \text{ W}$$

Self-Excited DC Generators

In a self-excited dc machine there is no need for an independent source for the field excitation, as the field windings are supplied by energy derived from the armature circuit. Self-excitation can be done in many ways, as shown in Figure 7.6.

1. *Series generator:* The field windings are connected in series with the armature circuit and the field current is the same as the armature current, as shown in Figure 7.6(a).

2. *Shunt generator:* The field windings are connected across (shunt) the armature terminals. The field voltage is the same as the armature voltage, as shown in Figure 7.6(b).

3. *Compound generator:* In this case two field windings are used. A series field winding made of wire with relatively large cross-sectional area is connected in series with the armature circuit. A shunt field winding (finer wire) is connected across the combination of series field and armature. This is called a *long shunt compound* connection and is shown in Figure 7.6(c). If the shunt field is connected directly across the armature terminals and the series field is connected in series with the combination, we have a *short shunt compound* connection, as shown in Figure 7.6.(d).

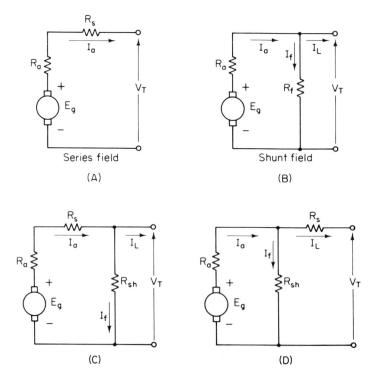

FIGURE 7.6 Self-excited generator connections: (a) series generator; (b) shunt generator; (c) long shunt compound; (d) short shunt compound.

When the two field windings are connected such that the fluxes aid each other, the machine is referred to as a *commulative compound* machine. If the two windings oppose each other, we have a *differential compound* machine. The relative number of turns of the series field winding affects the relation between voltage at no load and that at full load. If the number of series field turns is high enough so that the full-load voltage is higher than the no-load voltage, we have an *over-cummulative compound* generator. Clearly, a *flat-compound* generator has equal voltages at no full and no load. An *under-compound* machine has a lower full-load voltage than the no-load voltage.

7.5   SELF-EXCITED-GENERATOR PERFORMANCE

The present section treats the performance characteristics of the categories of self-excited generators discussed in the preceding section. ·

Series Generator

Based on the equivalent circuit of Figure 7.6(a), we have

$$V_T = E_g - I_a(R_a + R_s) \tag{7.8}$$

We also have by Eq. (7.7), since $I_f = I_a$,

$$E_g = K\omega I_a \tag{7.9}$$

As a result,

$$V_T = [K\omega - (R_a + R_s)]I_a \tag{7.10}$$

The idealized load curve of a series generator showing V_T versus I_a is a straight line passing through the origin as shown in Figure 7.7(a). Accounting for the actual magnetization characteristic leads to the characteristic shown in Figure 7.7(b). A series generator operating at virtually no load produces an armature generated voltage due to the residual magnetism available. As the load current is increased, E_g is increased until the maximum value determined by saturation. The generator acts as a constant-current source beyond that region.

The following example illustrates some aspects of the operation of a series generator.

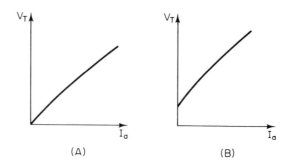

FIGURE 7.7 Series generator load curve: (a) ideal characteristic; (b) actual characteristic.

Example 7.4   A 10-kW 125-V dc series generator has a series field resistance of 0.05 Ω.

     a. Find the armature resistance given that at rated armature current and terminal voltage, the generated armature voltage is 137 V.

     b. Find the generated armature voltage for 75% of rated load and rated terminal voltage.

     c. Assume that the generated armature voltage is 136 V and the generator is delivering 8 kW. Find the terminal voltage.

Solution

(a)
$$I_a = \frac{10,000}{125} = 80 \text{ A}$$

$$R_s + R_a = \frac{E_g - V_T}{I_a} = \frac{137 - 125}{80} = 0.15 \text{ Ω}$$

Thus

$$R_a = 0.15 - R_s = 0.1 \text{ Ω}.$$

(b)
$$I_a = 0.75 \times 80 = 60 \text{ A}$$

$$E_g = 125 + 60(0.15) = 134 \text{ V}$$

(c)
$$E_g = V_T + I_a(R_a + R_s)$$

$$136 = V_T + \frac{0.15(8000)}{V_T}$$

Thus

$$V_T^2 - 136V_T + 1200 = 0$$

As a result,

$$V_T = 126.51 \text{ V}$$

Shunt Generator

Based on the equivalent circuit of Figure 7.6(b), we have

$$V_a = V_T = E_g - I_a R_a \qquad (7.11)$$

$$I_f = \frac{V_T}{R_f} \qquad (7.12)$$

Thus, using Eq. (7.7), we get, assuming a linear magnetization curve,

$$E_g = K\omega \frac{V_T}{R_f} \qquad (7.13)$$

We also have

$$I_L = I_a - I_f \qquad (7.14)$$

The equations above are the basis for a linear analysis of the steady-state performance of a shunt generator. The following example illustrates this.

Example 7.5   A 30-kW, 250-V dc shunt generator has an armature resistance of 0.12 Ω and a shunt field resistance of 40 Ω. The generated armature voltage is 267 V and the terminal voltage is 250 V when the generator is supplying a load. Find the output power and efficiency of the generator.

Solution

We have

$$I_a = \frac{E_g - V_T}{R_a} = \frac{267 - 250}{0.12} = 141.667 \text{ A}$$

A line current is

$$I_L = I_a - \frac{V_T}{R_f}$$

$$= 141.667 - \frac{250}{40} = 135.417 \text{ A}$$

Thus the output power is

$$P_0 = V_T I_L = 33,854.167 \text{ W}$$

The input power is

$$P_i = E_g I_a = 37,825.0 \text{ W}$$

As a result the efficiency is found as

$$\eta = \frac{P_o}{P_i} = 0.895$$

Buildup of Self-Excited Shunt Generators

The field current is the ratio of the terminal or armature voltage to the field resistance, as indicated in Eq. (7.12). A field rheostat is normally connected in series with the shunt field winding. As a result, the field current can be adjusted by adjusting the value of the field rheostat. Graphically, Eq. (7.12) is represented as shown in Figure 7.8. The slope of the field resistance line is the total resistance of the field circuit.

The process of buildup of generated voltage for a dc shunt generator can be explained in terms of the representation shown in Figure 7.9. The dc magnetization curve is drawn together with the field resistance line on the same set of axes ($V_a - I_f$). The following are the steps involved.

Step 1:   The prime mover rotates at rated speed and the generated armature voltage is E_o due to the residual magnetism. In response to E_o, the field current rises from zero to a value I_o such that $I_o = E_o/R_f$.

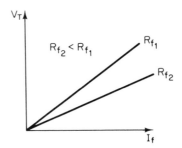

FIGURE 7.8   Field resistance lines for a dc shunt generator.

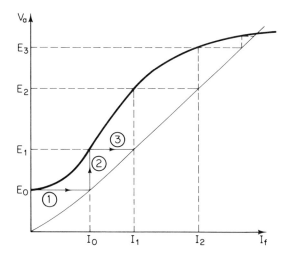

FIGURE 7.9 Build-up of voltage in a dc shunt generator.

Step 2: The generated voltage rises from E_0 to E_1 in accordance with the magnetization curve requirements. This is due to the increase in MMF from zero to $N_f I_0$.

Step 3: The field current increases from I_0 to I_1 since $I_1 = E_1/R_f$.

Step 4: As a result of I_1, the generated voltage increases to E_2, and so on.

The effect of field resistance on the steady value of the generated voltage is shown in Figure 7.10. Note that a smaller field resistance results in a higher generated voltage. An increase in the field resistance to a value higher than R_c, as shown, will result in failure to build up voltage.

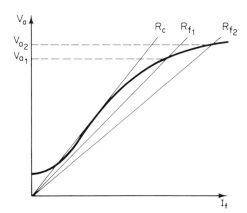

FIGURE 7.10 Effect of field resistance on generated voltage.

A high field resistance (including an open circuit in the field winding) is one reason for a dc shunt generator failing to build up voltage. Other reasons include lack (or low) of residual magnetism, field circuit reversed with respect to the armature circuit, and an open connection in the armature circuit.

The load curve of a dc shunt generator is shown in Figure 7.11. The voltage drops off somewhat with load but not appreciably for practical purposes.

Compound Generators

Let us consider the cummulative compound case. The equivalent circuit of the long shunt machine given in Figure 7.6(c) leads to the following relations:

$$V_T = E_g - I_a(R_a + R_s) \tag{7.15}$$

$$I_f = \frac{V_T}{R_{sh}} \tag{7.16}$$

Now we have, by Eq. (7.2),

$$E_g = K\omega\phi \tag{7.17}$$

Note that ϕ is the result of the MMF of the series and shunt fields. Thus, assuming linearity, we have for the cummulative case

$$\phi = K_s I_a + K_{sh} I_f \tag{7.18}$$

K_s and K_{sh} are constants that depend on the winding number of turns and other magnetic circuit parameters.

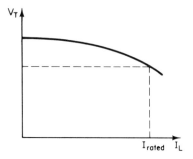

FIGURE 7.11   Load curve of dc shunt generator.

For the short shunt case we have, from Figure 7.6(d),

$$V_a = E_g - I_a R_a \qquad (7.19)$$

$$I_f = \frac{V_a}{R_{sh}} \qquad (7.20)$$

$$I_L = I_a - I_f \qquad (7.21)$$

$$\phi = K_s I_L + K_{sh} I_f \qquad (7.22)$$

The equations above are enough for a steady-state analysis of the performance of a compound generator.

Depending on the relative additional aiding that the series field provides to the dominant shunt field MMF, the three types of compounding emerge as discussed earlier. The load curves for the three cases are compared with that of the shunt generator and a differential compound generator in Figure 7.12. Note that for a differential compound generator, the value of $K_s I_a$ in Eq. (7.18) is negative.

Example 7.6   A 600-V 100-kW dc long shunt compound generator has a series field resistance of 0.02 Ω and a shunt field resistance of 200 Ω. When the generator is delivering rated output, the power input is 103.5 kW. Find the armature resistance.

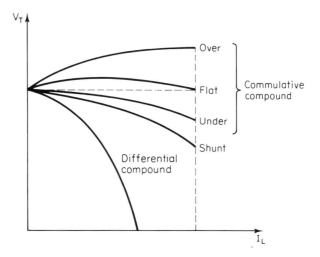

FIGURE 7.12   Load characteristics of compound generators.

Solution

We have

$$I_L = \frac{100 \times 10^3}{600} = 166.67 \text{ A}$$

$$I_f = \frac{600}{200} = 3\text{A}$$

Thus

$$I_a = I_L + I_f = 169.67 \text{ A}$$

Now

$$E_g = \frac{P_i}{I_a} = \frac{103.5 \times 10^3}{169.67} = 610.008 \text{ V}$$

As a result,

$$R_a + R_s = \frac{E_g - V_T}{I_a}$$

$$= \frac{610.008 - 600}{169.67} = 0.059$$

We can now conclude that the armature resistance is given by

$$R_a = 0.059 - 0.02 = 0.039 \text{ } \Omega$$

7.6 DC MOTORS

A dc machine is reversible in the sense that it can be operated either as a generator or as a motor. From an analytical point of view, the models adopted for the generator case are equally applicable to the motor case. We note that electric power is supplied to the motor and the output is mechanical power represented by a torque (T) in the armature and a shaft rotational speed. It is, therefore, clear that the power flow in a dc motor is the reverse of that in a dc generator. The difference, then, from a model point of view, is the form of the basic armature voltage relationship. The armature developed voltage denoted by E_g for the generator case is referred to as the back electromotive force (EMF) E_c.

Armature Circuit Model

The armature circuit of a direct-current motor is represented by the model of Figure 7.13. The voltage applied to the armature is denoted by V_a and the counter (back) electromotive force (EMF) is denoted by E_c. The resistance of the armature windings is denoted by R_a. The armature current I_a can be obtained from the basic voltage relationship

$$V_a = E_c + R_a I_a \tag{7.23}$$

The back EMF E_c varies with the field flux ϕ_f and armature speed ω according to

$$E_c = K_1 \phi_f \omega \tag{7.24}$$

The constant of proportionality K_1 is a machine parameter that depends on the number of conductors in the armature, the number of poles, and the number of winding connections.

It is clear that at standstill, the motor speed ω is zero and thus at starting, the motor's back EMF E_c is zero. It is then evident that the armature current at starting will be of large value unless a reduced value of V_a is applied to the armature at that time. Special attention should be paid to this aspect, which is discussed later in the chapter.

The power input to the armature circuit is given by

$$P_{a_i} = V_a I_a \tag{7.25}$$

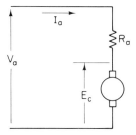

FIGURE 7.13   Armature circuit model for a dc motor.

This power supplies the armature power losses and the armature power P_a. Thus

$$P_{a_i} = I_a^2 R_a + P_a \qquad (7.26)$$

The developed armature power P_a is given by

$$P_a = E_c I_a \qquad (7.27)$$

The net output power P_o is obtained by subtracting the motor's rotational losses P_{rot} from P_a:

$$P_o = P_a - P_{\text{rot}} \qquad (7.28)$$

The net output power P_o is the available shaft power.

The output torque in newton meters is obtained from the relation

$$T_o = \frac{P_o}{\omega} \qquad (7.29)$$

where ω is the armature speed in radians per seconds. An example is in order at this time.

Example 7.7   A voltage of 230 V applied to the armature of a dc motor results in a full-load armature current of 205 A. Assume that the armature resistance is 0.2 Ω. Find the back EMF E_c, the net output power, and the torque assuming that the rotational losses are 1445 W at a full-load speed of 1750 r/min.

Solution

The armature voltage and current are specified as

$$V_a = 230 \text{ V}$$

$$I_a = 205 \text{ A}$$

The back EMF is obtained as

$$E_c = V_a - I_a R_a$$
$$= 230 - 205(0.2)$$
$$= 189 \text{ V}$$

The power developed by the armature is thus

$$P_a = E_c I_a$$
$$= 189(205) = 38{,}745 \text{ W}$$

The net output power is thus obtained by subtracting the rotational losses from the armature developed power:

$$P_o = P_a - P_{\text{rot}}$$
$$= 38{,}745 - 1445$$
$$= 37{,}300 \text{ W}$$

The net output torque is now calculated as

$$T_o = \frac{P_o}{\omega}$$
$$= \frac{37{,}300}{\dfrac{2\pi}{60}(1750)}$$
$$= 203.536 \text{ N} \cdot \text{m}$$

Field Circuit Model

A full analysis of the performance of a dc motor requires knowledge of the field circuit connection. We note here that the back EMF depends on the field flux ϕ_f as indicated in Eq. (7.24). The field flux varies with the field current in a manner similar to the saturation characteristic of the magnetic material of the machine. It is common practice to assume that the machine is operated such that the field flux is proportional to the field current. In this case we have

$$\phi_f = K_2 I_f \tag{7.30}$$

The field current is denoted by I_f. As a result, we have

$$E_c = K_1 K_2 I_f \omega \tag{7.31}$$

The field circuit windings are modeled by a resistance R_f.

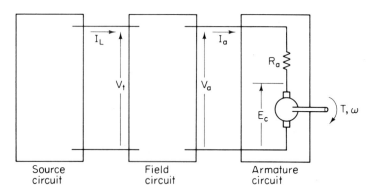

FIGURE 7.14 Functional block diagram of a dc motor.

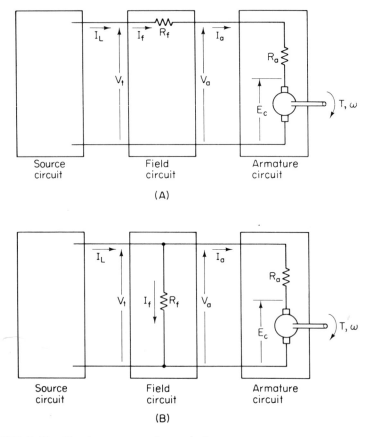

FIGURE 7.15 Circuit representations of direct current: (a) series motor; (b) shunt motor.

A complete circuit model of a dc motor is shown in Figure 7.14. The dc source voltage is denoted by V_t and its current is denoted by I_L. The field circuit is placed between the source and the armature circuit. In a series motor, the field circuit is connected in series with the armature, as shown in Figure 7.15(a). In a shunt motor, the field circuit is connected in parallel (shunt) with the armature circuit, as shown in Figure 7.15(b). Combinations of the two connections result in compound motors.

7.7 DC SERIES MOTORS

As mentioned earlier, in a dc series motor the field windings are connected in series with the armature windings. The resistance of the field is denoted by R_s, as shown in the schematic diagram of Figure 7.16. It is clear that the armature current, field current, and line current are the same. The terminal voltage V_t is equal to the sum of the armature voltage and the voltage drop across the series field resistance

$$I_L = I_f = I_a \tag{7.32}$$

$$V_t = V_a + I_a R_s \tag{7.33}$$

As a result, we can write

$$V_t = E_c + I_a(R_a + R_s) \tag{7.34}$$

Equation (7.31) can be written as

$$E_c = K_1 K_2 I_a \omega \tag{7.35}$$

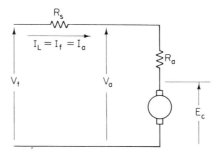

FIGURE 7.16 Schematic diagram of a dc series motor.

We are now in a position to derive performance analysis equations for the dc series motor.

Basic performance characteristics of a dc motor involve the variation of output variables such as armature developed power P_a, torque T, and speed ω with the armature current I_a. In practice, it is also desired to obtain performance characteristics displaying the variation of armature current I_a, torque T, and speed ω versus output power P_o. These practical characteristics are first obtained as variations with armature developed power P_a and are then corrected for rotational losses to obtain the required characteristics in terms of P_o. Important economic performance characteristics of a motor are naturally the variation of power losses efficiency with loading. These characteristics can be obtained using our basic relations discussed above.

P_a–I_a Characteristic

The armature developed power (internal power) is given by Eq. (7.27) as

$$P_a = E_c I_a \tag{7.36}$$

Substituting for E_c from Eq. (7.12), we obtain

$$P_a = [V_t - I_a(R_a + R_s)]I_a \tag{7.37}$$

It is clear that for a specified terminal voltage and machine parameters, the power developed by the armature follows a parabolic (second-order) variation with the armature current. In fact, for a given armature power, P_a, there are two solutions of Eq. (7.37) for I_a, obtained from the following modified form of Eq. (7.37):

$$(R_a + R_s)I_a^2 - V_t I_a + P_a = 0$$

The solutions for I_a are given by

$$I_a = \frac{V_t \pm \sqrt{V_t^2 - 4 P_a(R_a + R_s)}}{2(R_a + R_s)} \tag{7.38}$$

From Eq. (7.38) it is clear that there is a value of P_a for which there is only one solution for I_a. This condition requires that the quantity under the square-root sign be zero, which results in

$$P_{a_m} = \frac{V_t^2}{4(R_a + R_s)} \tag{7.39}$$

with the corresponding armature current given by

$$I_{a_m} = \frac{V_t}{2(R_a + R_s)} \qquad (7.40)$$

Note that the subscript m refers to the fact that this power value is the maximum attainable by the motor. To verify this statement, we differentiate Eq. (7.37) with respect to I_a and set the result to zero, to obtain

$$\frac{\partial P_a}{\partial I_a} = V_t - 2(R_a + R_s)I_a = 0 \qquad (7.41)$$

Clearly, this leads to Eq. (7.40). A typical sketch of the variation of P_a with I_a for a dc series motor is shown in Figure 7.17.

The operational range of the motor is well below that corresponding to maximum developed armature power. In this range we take the negative sign in Eq. (7.38), with the result that

$$I_a = \frac{V_t - \sqrt{V^2_t - 4P_a(R_a + R_s)}}{2(R_a + R_s)} \qquad (7.42)$$

An approximation for the value of I_a is obtained by neglecting the armature and series field resistances, to yield

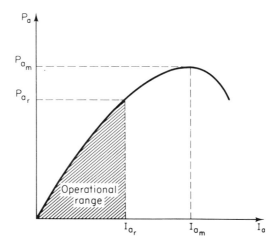

FIGURE 7.17 Variation of armature developed power with armature current in a dc series motor.

$$I_a \simeq \frac{P_a}{V_t} \tag{7.43}$$

The approximate expression yields a lower value for I_a than the more detailed expression of Eq. (7.42). The error, however, is small for lower values of P_a. An example is appropriate at this time.

Example 7.8   The combined armature and series field resistance of a 10-hp 240-V dc series motor is 0.6 Ω. Neglecting rotational losses, it is required to calculate:

  a. Armature current at full rated load.
  b. The value of maximum developed armature power and the corresponding armature current.
  c. The armature current at half-rated load using Eqs. (7.42) and (7.43).

Solution

(a) At full-rated load, neglecting rotational losses, the armature developed power is

$$P_a = 10 \times 746 = 7460 \text{ W}$$

The armature current is obtained as

$$I_a = \frac{240 - \sqrt{(240)^2 - 4(7460)(0.6)}}{2(0.6)}$$

$$= 33.968 \text{ A}$$

Note that the second solution for I_a is 366.032 A if we take the positive sign in Eq. (7.38). The approximate solution of Eq. (7.43) is 31.083 A with an error of 2.885 A.

   (b) For maximum armature power we have

$$I_{am} = \frac{240}{2(0.6)} = 200 \text{ A}$$

The value of P_a at this current is obtained using Eq. (7.39) as

$$P_{am} = \frac{(240)^2}{4(0.6)} = 24{,}000 \text{ W}$$

   (c) At half-rated load we have

$$P_a = 5 \times 746 = 3730 \text{ W}$$

The corresponding armature current is

$$I_a = \frac{240 - \sqrt{(240)^2 - 4(3730)(0.6)}}{2(0.6)}$$

$$= 16.198 \text{ A}$$

The approximate equation (7.43) yields

$$I_a \simeq \frac{3730}{240} = 15.542 \text{ A}$$

The error in this case is 0.656 A.

   The output (or shaft) power of the motor is obtained by subtracting the rotational power losses from the armature developed power. The resulting characteristic is shown in Figure 7.18.

Torque and Speed Characteristics

The armature developed torque (internal torque) is obtained using the basic relation

$$T_a = \frac{P_a}{\omega} \tag{7.44}$$

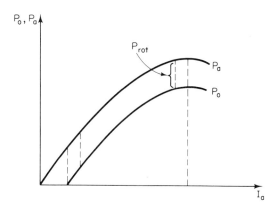

FIGURE 7.18 Obtaining the variation of output power with armature current for a dc series motor.

Using Eqs. (7.35) and (7.36), we conclude that

$$T_a = K_1 K_2 I_a^2 \qquad (7.45)$$

Thus, ideally, the torque varies as the square of armature current as shown in Figure 7.19.

The motor speed in radians per second ω is obtained in terms of armature current by combining Eqs. (7.34) and (7.35) to eliminate E_c to obtain

$$V_t = K_1 K_2 I_a \omega + I_a (R_a + R_s)$$

As a result,

$$\omega = \frac{V_t}{K_1 K_2 I_a} - \frac{R_a + R_s}{K_1 K_2} \qquad (7.46)$$

Equation (7.46) tells us that for small values of armature current, the motor speed is high. Thus at no load on the motor a runaway condition (extremely high speed) exists. A series motor is used only in applications where a load is always connected to the shaft. Examples include hoists and cranes as well as railway motors. A centrifugal switch must be used to protect a series motor against runaway. A typical characteristic of speed versus armature current for a dc series motor is shown in Figure 7.20.

The armature current in terms of armature speed is obtained from Eq. (7.46) as

$$I_a = \frac{V_t}{K_1 K_2 \omega + (R_a + R_s)} \qquad (7.47)$$

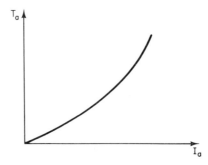

FIGURE 7.19 Variation of armature developed torque with armature current for a dc series motor.

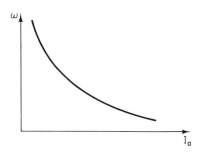

FIGURE 7.20   Variation of speed with armature current for a dc series motor.

Substituting Eq. (7.47) in (7.45), we obtain

$$T_a = \frac{K_1 K_2 V_t^2}{[K_1 K_2 \omega + (R_a + R_s)]^2} \tag{7.48}$$

Equation (7.48) provides us with the torque–speed characteristic of the dc series motor shown in Figure 7.21.

Example 7.9   A 600-V 150-hp dc series motor operates at its full-rated load at 600 r/min. The armature resistance is 0.12 Ω and the series field resistance is 0.04 Ω. The motor draws 200 A at full load.

   a. Find the armature back emf at full load.
   b. Find the armature developed power and internal developed torque,
   c. Assume that a change in load results in the line current dropping to 150 A. Find the new speed in r/min and the new developed torque.

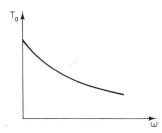

FIGURE 7.21   Torque–speed characteristic of a dc series motor.

Solution

(a) The armature back EMF is obtained from

$$E_{c_1} = V_t - I_{a_1}(R_a + R_s)$$

$$= 600 - 200(0.12 + 0.04)$$

$$= 568 \text{ V}$$

(b) The power developed by the armature is

$$P_{a_1} = E_{c_1}I_{a_1}$$

$$= 568(200) = 113.6 \times 10^3 \text{ W}$$

As a result, we calculate the internal developed torque as

$$T_{a_1} = \frac{P_{a_1}}{\omega_1} = \frac{113.6 \times 10^3}{(2\pi/60)(600)} = 1808 \text{ N} \cdot \text{m}$$

(c) For the new line current, we can calculate a new value for the back EMF:

$$E_{c_2} = 600 - 150(0.16) = 576 \text{ V}$$

From Eq. (7.35) we have

$$E_{c_1} = K_1 K_2 I_{a_1}\omega_1$$

$$E_{c_2} = K_1 K_2 I_{a_2}\omega_2$$

As a result,

$$\frac{\omega_2}{\omega_1} = \frac{I_{a_1}}{I_{a_2}} \frac{E_{c_2}}{E_{c_1}}$$

Thus

$$\omega_2 = \frac{200}{150}\left(\frac{576}{568}\right)\omega_1$$

or

$$n_2 = \frac{200}{150}\left(\frac{576}{568}\right)(600)$$

$$= 811.268 \text{ r/min}$$

In a similar manner, from Eq. (7.45), we can write

$$\frac{T_2}{T_1} = \left(\frac{I_2}{I_1}\right)^2$$

Thus

$$T_2 = 1808\left(\frac{150}{200}\right)^2 = 1017 \text{ N} \cdot \text{m}$$

Starting

It is clear that at standstill, the motor speed and hence the armature back EMF E_c are zero. The application of voltage to the armature circuit will result in a high value of current. This can be seen from Eq. (7.34) with $E_c = 0$, giving

$$I_{a_{\text{st}}} = \frac{V_t}{R_a + R_s} \tag{7.49}$$

To limit the value of starting current one of two means must be used, the most logical being applying a reduced value of voltage through a variable dc supply, the other being the conventional method of inserting a starter resistor R_{st} which limits the value of I_a at starting to an acceptable level. This resistance is commonly a variable resistance that is gradually reduced to zero as the motor picks up speed. If the starting resistor is left in the armature circuit as the motor picks up speed, an inferior performance will result. It should be realized that dc series motors are characterized by a high value of starting torque, as can be seen from Eq. (7.48). An example is in order here.

Example 7.10 The series motor of Example 7.9 is required to be started using a starting resistor R_{st} such that the starting current is limited to 150% of rated value:

    a. Find the required value of R_{st} and the starting torque,
    b. If the starting resistor is left in the armature circuit and the motor line current drops to its rated value of 200 A, find the armature back EMF and the speed of the motor.

Solution

(a) At starting $E_c = 0$,

$$V_t = I_a(R_a + R_s + R_{st})$$

$$600 = 300(0.16 + R_{st})$$

Thus

$$R_{st} = 1.84 \ \Omega$$

We know from Example 7.9 that for an armature current of 200 A, the corresponding torque is 1808 N · m. Thus for a starting current of 300 A, we have

$$T_{st} = T_{fld}\left(\frac{I_{st}}{I_{fld}}\right)^2$$

$$= 1808\left(\frac{300}{200}\right)^2 = 4068 \ \text{N} \cdot \text{m}$$

(b) With starting resistor left in the armature circuit, we have

$$E_c = V_t - I_a(R_a + R_s + R_{st})$$

$$= 600 - 200(2)$$

Thus

$$E_c = 200 \ \text{V}$$

We now have

$$\frac{E_{c2}}{E_{c1}} = \frac{n_2}{n_1}$$

or

$$\frac{200}{568} = \frac{n_2}{600}$$

As a result,

$$n_2 = 211.268 \ \text{rpm}$$

Power Loss and Efficiency

The input power to a dc series motor is given by

$$P_{\text{in}} = V_t I_a \tag{7.50}$$

The power losses in the motor include losses in the armature and series field resistances as well as the rotational losses, which are considered constant P_{rot}. Thus

$$P_\ell = I_a^2(R_a + R_s) + P_{\text{rot}} \tag{7.51}$$

The output power is, therefore, given by

$$P_{\text{o}} = P_{\text{in}} - P_\ell \tag{7.52}$$

The efficiency of the motor is given by

$$\eta = \frac{P_{\text{o}}}{P_{\text{in}}} \tag{7.53}$$

It is clear that the efficiency varies with the armature current and hence with the motor load.

The efficiency increases as the armature current is increased up to the point of maximum efficiency and then drops as the armature current is increased further. To obtain the value of armature current corresponding to maximum efficiency, we set to zero the derivative of the efficiency with respect to armature current:

$$\frac{\partial \eta}{\partial I_a} = 0$$

As a result, we have

$$P_{\text{in}} \frac{\partial P_{\text{o}}}{\partial I_a} - P_{\text{o}} \frac{\partial P_{\text{in}}}{\partial I_a} = 0 \tag{7.54}$$

Using Eq. (7.52), we get

$$P_{\text{in}} \left(\frac{\partial P_{\text{in}}}{\partial I_a} - \frac{\partial P_\ell}{\partial I_a} \right) - (P_{\text{in}} - P_\ell) \frac{\partial P_{\text{in}}}{\partial I_a} = 0$$

This reduces to

$$P_{\text{in}} \frac{\partial P_\ell}{\partial I_a} = P_\ell \frac{\partial P_{\text{in}}}{\partial I_a} \tag{7.55}$$

The required derivatives are

$$\frac{\partial P_\ell}{\partial I_a} = 2\, I_a(R_a + R_s)$$

$$\frac{\partial P_{in}}{\partial I_a} = V_t$$

Thus the condition for maximum efficiency is

$$2V_t I_a^2(R_a + R_s) = V_t[I_a^2(R_a + R_s) + P_{rot}]$$

Simplifying, we get

$$P_{rot} = I_a^2(R_a + R_s) \tag{7.56}$$

The efficiency is a maximum when the fixed losses represented by the rotational losses in this case is equal to the I^2R losses. Thus

$$I_{a_{max}} = \sqrt{\frac{P_{rot}}{R_a + R_s}} \tag{7.57}$$

The maximum value of the efficiency is given by

$$\eta_{max} = \frac{P_{in_m} - 2P_{rot}}{P_{in_m}} \tag{7.58}$$

The formula for maximum efficiency can also be written

$$\eta_{max} = 1 - \frac{2[P_{rot}(R_a + R_s)]^{1/2}}{V_t} \tag{7.59}$$

From the development above, we can conclude that the maximum efficiency is a function of rotational losses, armature and series field resistances, and terminal voltage. As can reasonably be anticipated, the maximum efficiency of the motor is increased by an increase in the terminal voltage and by reducing the armature and series field resistances.

Example 7.11 Consider the dc series motor of Example 7.9. The full-load output is 150 hp with full-load current given by 200 A.

    a. Find the rotational losses.

b. Find the armature current and power output at maximum efficiency as well as the value of maximum efficiency.

c. Find the full-load efficiency.

Assume that rotational losses are fixed.

Solution

(a) The output power at full load is

$$P_o = 150 \times 746 = 111.9 \times 10^3 \text{ W}$$

From Example 7.9, the armature developed power at full load

$$P_a = 113.6 \times 10^3 \text{ W}$$

As a result,

$$P_{rot} = P_a - P_o$$
$$= 1.7 \times 10^3 \text{ W}$$

(b) For maximum efficiency, the armature current is obtained as

$$I_a = \sqrt{\frac{1.7 \times 10^3}{0.16}} = 103.1 \text{ A}$$

The back EMF for this current is

$$E_c = 600 - 103.1(0.16)$$
$$= 583.5 \text{ V}$$

The power developed by the armature is thus

$$P_a = 583.5 \times 103.1 = 60.16 \times 10^3 \text{ W}$$

As a result,

$$P_o = 60.16 \times 10^3 - 1.7 \times 10^3$$
$$= 58.46 \times 10^3 \text{ W}$$

The corresponding input power is

$$P_{\text{in}} = 600 \times 103.1$$

$$= 61.86 \times 10^3 \text{ W}$$

As a result, the maximum efficiency is

$$\eta_{\text{max}} = \frac{58.46 \times 10^3}{61.86 \times 10^3} = 0.945$$

(c) The full-load input power is

$$P_{\text{in}} = 600 \times 200 = 120 \times 10^3 \text{ W}$$

The output power at full load is

$$P_{\text{o}} = 150 \times 746 = 111.9 \times 10^3 \text{ W}$$

Thus the full-load efficiency is

$$\eta = \frac{111.9 \times 10^3}{120 \times 10^3} = 0.9325$$

Note that at full load, the motor is operating beyond the point of maximum efficiency.

7.8 DC SHUNT MOTORS

In a dc shunt motor, the field winding is connected in parallel with the armature circuit. The resistance of the field is denoted by R_f, as shown in the schematic diagram of Figure 7.22. The terminal voltage V_t is identical with both field voltage V_f and armature voltage V_a:

$$V_t = V_f = V_a \tag{7.60}$$

The line current I_L is the sum of armature and field currents:

$$I_L = I_a + I_f \tag{7.61}$$

The field current I_f is given by

$$I_f = \frac{V_t}{R_f} \tag{7.62}$$

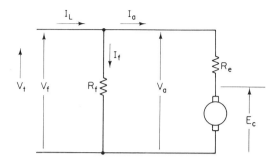

FIGURE 7.22   Schematic diagram of a dc shunt motor.

As a result, we can write

$$V_t = E_c + I_a R_a \tag{7.63}$$

Equation (7.31) can be written as

$$E_c = K_1 K_2 \frac{V_t}{R_f} \omega \tag{7.64}$$

Let us define

$$K_{sh} = K_1 K_2 \frac{V_t}{R_f} \tag{7.65}$$

Thus we have

$$E_c = K_{sh}\omega \tag{7.66}$$

For a fixed terminal voltage and field resistance, the back EMF is proportional to the armature speed ω. We will now consider performance analysis equations along the same lines adopted in the analysis of a dc series motor.

P_a–I_a Characteristics

The armature developed power is

$$P_a = E_c I_a$$

Using Eq. (7.63), we get

$$P_a = (V_t - I_a R_a)I_a \tag{7.67}$$

Note that Eq. (7.67) is of the same form as that of Eq. (7.37) with $R_s = 0$. We can thus conclude that for a given armature power there are two values of I_a:

$$I_a = \frac{V_t \pm \sqrt{V_t^2 - 4P_aR_a}}{2R_a} \qquad (7.68)$$

The maximum value of armature developed power is

$$P_{am} = \frac{V_t^2}{4R_a} \qquad (7.69)$$

with the corresponding armature current

$$I_{am} = \frac{V_t}{2R_a} \qquad (7.70)$$

A typical sketch of the variation of P_a with I_a for a dc shunt motor is shown in Figure 7.23.

Again, in the range of operation, we take the negative sign in applying Eq. (7.68), to obtain

$$I_a = \frac{V_t - \sqrt{V_t^2 - 4P_aR_a}}{2R_a} \qquad (7.71)$$

An approximation to I_a is obtained by neglecting I_aR_a in Eq. (7.67), to obtain

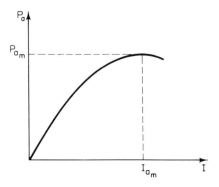

FIGURE 7.23 Variation of armature developed power with armature current for a dc shunt motor.

$$I_a \simeq \frac{P_a}{V_t} \tag{7.72}$$

Thus we can conclude that the armature current varies approximately in proportion to the armature developed power.

Torque and Speed Characteristics

The armature developed torque is obtained using

$$T_a = \frac{P_a}{\omega}$$

From Eq. (7.66), we conclude that

$$T_a = K_{sh}I_a \tag{7.73}$$

It is clear that the internal developed torque is proportional to the armature current in a dc shunt motor as shown in Figure 7.24.

The motor speed can be obtained in terms of armature current by substituting Eq. (7.66) in (7.63), to obtain

$$\omega = \frac{V_t - I_aR_a}{K_{sh}} \tag{7.74}$$

At no load, the armature current is zero and thus

$$\omega_0 = \frac{V_t}{K_{sh}} \tag{7.75}$$

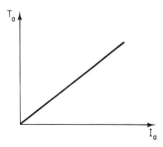

FIGURE 7.24   Variation of internal torque with armature current in a dc shunt motor.

Thus

$$\omega = \omega_0 - \frac{R_a}{K_{sh}} I_a \qquad (7.76)$$

The coefficient R_a/K_{sh} is usually too small and thus a dc shunt motor can be assumed to be a constant-speed motor for all practical purposes. The variation of ω with I_a is shown in Figure 7.25.

The torque–speed variation is obtained from Eqs. (7.73) and (7.74) as

$$T_a = \frac{K_{sh}}{R_a} (V_t - K_{sh}\omega)$$

Using Eq. (7.75), we get

$$T_a = \frac{K_{sh}^2}{R_a} (\omega_0 - \omega) \qquad (7.77)$$

where ω_0 is the speed at no load. The inverse relation specifying the speed in terms of torque is given by

$$\omega = \frac{V_t}{K_{sh}} - \frac{T_a R_a}{K_{sh}^2}$$

or   (7.78)

$$\omega = \omega_0 - \frac{T_a R_a}{K_{sh}^2}$$

The speed–torque characteristic of a dc shunt motor is shown in Figure 7.26 and is simply the result of a scaling of Figure 7.25, since torque is proportional to armature current in this case.

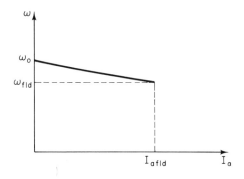

FIGURE 7.25   Variation of speed with armature current for a dc shunt motor.

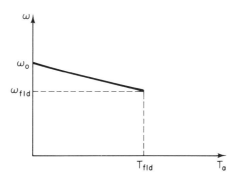

FIGURE 7.26   Speed–torque characteristic for a dc shunt motor.

Example 7.12   The armature resistance of a 10-hp 230-V dc shunt motor is 0.3 Ω. The field resistance is 160 Ω. The motor draws a line current of 3.938 A on no load at a speed of 1200 r/min. At full load, the armature current is 40 A.

  a. Find the armature current at no load.
  b. Find the power developed by the armature on no load.
  c. Find the full-load efficiency of the motor.
  d. Find the full-load speed of the motor.

Solution

(a) The field current is obtained as

$$I_f = \frac{230}{160} = 1.438 \text{ A}$$

Thus the armature current is obtained as

$$I_a = I_L - I_f = 3.938 - 1.438 = 2.5 \text{ A}$$

(b) The back EMF on no load is calculated as

$$E_c = V_t - I_a R_a$$
$$= 230 - 2.5(0.3) = 229.25 \text{ V}$$

Thus we obtain the no-load armature power as

$$P_{\text{no load}} = E_c I_a$$
$$= 229.25(2.5) = 573.125 \text{ W}$$

(c) At full load, we have

$$E_c = 230 - 40(0.3) = 218 \text{ V}$$

The armature power is thus calculated as

$$P_a = E_c I_a = 218(40) = 8720 \text{ W}$$

The net power output is obtained by subtracting the no-load power (rotational losses) from part (b) from the armature power:

$$P_o = 8720 - 573.125 = 8146.875 \text{ W}$$

The power input is found as

$$P_{\text{in}} = V_t I_L$$
$$= 230(41.438) = 9530.625 \text{ W}$$

As a result, we calculate the efficiency:

$$\eta = \frac{8146.875}{9530.625}$$
$$= 0.855$$

(d) From part (c), the full-load back EMF is given by

$$E_{c2} = 218 \text{ V}$$

The no-load back EMF from part (b) is

$$E_{c1} = 229.25 \text{ V}$$

Thus we can find the full-load speed as

$$n_{\text{fld}} = n_{\text{no load}} \frac{E_{c2}}{E_{c1}}$$

$$= 1200\left(\frac{218}{229.25}\right) = 1141.112 \text{ rpm}$$

Power Loss and Efficiency

The power input to the motor is given by

$$P_{\text{in}} = V_t I_L$$

Recall that the line current is the sum of field current I_f and armature current I_a:

$$I_L = I_a + I_f$$

Thus

$$P_{\text{in}} = V_t I_a + P_f \tag{7.79}$$

where P_f is the power loss in the shunt field circuit resistance defined by

$$P_f = \frac{V_t^2}{R_f} = V_t I_f \tag{7.80}$$

The power input to the armature is $V_t I_a$ and supplies the armature power loss and the rotational losses with a net mechanical power output given by

$$P_\text{o} = V_t I_a - I_a^2 R_a - P_{\text{rot}} \tag{7.81}$$

The motor's efficiency is thus obtained as

$$\eta = \frac{P_\text{o}}{P_{\text{in}}}$$

Clearly, the motor's efficiency depends on the armature current.
    The point at which maximum efficiency occurs is obtained by setting the derivative of η with respect to current to zero. As a result, we obtain

$$P_{\text{in}} \frac{\partial P_\text{o}}{\partial I_a} - P_\text{o} \frac{\partial P_{\text{in}}}{\partial I_a} = 0$$

Performing the required operations, we obtain

$$\frac{\partial P_\text{o}}{\partial I_a} = V_t - 2 I_a R_a$$

$$\frac{\partial P_{\text{in}}}{\partial I_a} = V_t$$

We can thus obtain the condition for maximum efficiency:

$$(V_t I_a + P_f)(V_t - 2I_a R_a) - (V_t I_a - I_a^2 R_a - P_{rot})V_t = 0$$

Simplifying, we get

$$R_a I_a^2 + 2I_a I_f R_a - (P_{rot} + P_f) = 0 \qquad (7.82)$$

The solution to the foregoing quadratic equation in I_a yields the desired value at which maximum efficiency occurs.

A practical approximation to the formula above results if we consider that the product $I_f R_a$ is usually small and as a result

$$R_a I_{a_{max}}^2 \simeq P_{rot} + P_f \qquad (7.83)$$

This result is similar to that obtained for the dc series motor, where it was concluded that for maximum efficiency, the fixed losses are equal to the armature losses.

Example 7.13 A 230-V 25-hp dc shunt motor draws an armature current of 90 A at full rated load. Assume that the armature resistance is 0.2 Ω and that the shunt field resistance is 216 Ω. Find the rotational losses at full load and the motor efficiency in this case.

Solution

At full rated armature current of 90 A, the power developed by the armature is given by

$$P_a = (V_t - I_a R_a)I_a$$

$$= [230 - 90(0.2)](90) = 19,080 \text{ W}$$

The output power at full load is

$$P_o = 25 \times 746 = 18,650 \text{ W}$$

As a result, the rotational losses are obtained from

$$P_{rot} = P_a - P_o$$

$$= 19,080 - 18,650 = 430 \text{ W}$$

The line current is given by

$$I_L = I_a + \frac{V_t}{R_f}$$

$$= 90 + \frac{230}{216} = 91.065$$

The input power is thus

$$P_{\text{in}} = V_t I_L = 230(91.065)$$

$$= 20{,}944.91 \text{ W}$$

The efficiency can thus be computed as

$$\eta = \frac{P_o}{P_{\text{in}}} = \frac{18{,}650}{20{,}944.91} = 0.89$$

Example 7.14   The rotational losses for a 230-V 25-hp dc shunt motor are found to be 430 W. The armature resistance is 0.2 Ω and the shunt field resistance is 216 Ω. Obtain the armature current corresponding to maximum efficiency using both the exact and approximate expressions developed in the text. Calculate the output power in both cases and the corresponding maximum efficiency.

Solution

The exact formula is given by Eq. (7.82) as

$$R_a I_a^2 + 2I_a I_f R_a - (P_{\text{rot}} + P_f) = 0$$

We have

$$R_a = 0.2 \ \Omega$$

$$I_f = \frac{230}{216} = 1.065 \text{ A}$$

$$P_f = \frac{(230)^2}{216} = 244.907 \text{ W}$$

$$P_{\text{rot}} = 430 \text{ W}$$

Thus we have

$$0.2I_a^2 + 2I_a(1.065)(0.2) - (430 + 244.907) = 0$$

Alternatively,

$$I_a^2 + 2.13I_a - 3374.537 = 0$$

The solution for I_a is obtained as

$$I_a = \frac{-2.13 \pm 116.201}{2}$$

We take

$$I_{a_{\max \eta 1}} = 57.036 \text{ A}$$

The output power is given by Eq. (7.81) as

$$
\begin{aligned}
P_o &= V_t I_a - I_a^2 R_a - P_{\text{rot}} \\
&= 230(57.036) - (57.036)^2(0.2) - 430 \\
&= 12{,}037.66 \text{ W}
\end{aligned}
$$

The input power is given by Eq. (7.79) as

$$
\begin{aligned}
P_{\text{in}} &= V_t I_a + P_f \\
&= 230(57.036) + 244.907 = 13{,}363.187 \text{ W}
\end{aligned}
$$

As a result, the maximum efficiency on the basis of the exact calculation is

$$\eta_{\max 1} = \frac{12{,}037.66}{13{,}363.187} = 0.901$$

The approximate formula (7.83) provides the value of armature current for maximum efficiency as

$$
\begin{aligned}
I_{a_{\max \eta 2}} &= \sqrt{\frac{P_{\text{rot}} + P_f}{R_a}} \\
&= 58.091 \text{ A}
\end{aligned}
$$

The output power is calculated as

$$P_o = V_t I_a - I_a^2 R_a - P_{\text{rot}}$$

$$= 230(58.091) - (58.091)^2(0.2) - 430$$

$$= 12,256.017 \text{ W}$$

The input power is calculated as

$$P_{\text{in}} = V_t I_a + P_f$$

$$= 230(58.09) + 244.907$$

$$= 13,605.837 \text{ W}$$

The corresponding maximum efficiency is

$$\eta_{\text{max}2} = \frac{12,256.017}{13,605.837} = 0.901$$

The final answers are the same to the assumed accuracy.

7.9 COMPOUND MOTORS

In a dc compound motor both shunt and series field windings are employed. The resulting field is a combination of the contributions of the two windings. The shunt field contribution predominates that of the series field. When the series field aids the shunt field, the compounding is *cumulative*. When the series field opposes the shunt field, the compounding is *differential*. Note also that the series field may be connected in series with armature in the long shunt case, or in series with the line in the short shunt case. Our analysis will assume that a long shunt connection is employed as shown in Figure 7.27.

The field flux ϕ_f in a *cumulative* compound machine is given by

$$\phi_f = K_3 I_{\text{sh}} + K_4 I_a \qquad (7.84)$$

The constants K_3 and K_4 pertain to the shunt and series field parameters assuming that the motor operates in the linear region of its magnetization characteristic. For a *differential* compound motor, we have

$$\phi_f = K_3 I_{\text{sh}} - K_4 I_a. \qquad (7.85)$$

We observe that the first term in both Eqs. (7.84) and (7.85) is essentially constant while the second term varies linearly with the armature current and hence with the load on the motor.

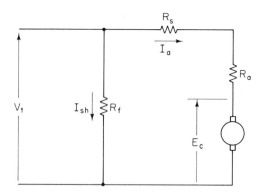

FIGURE 7.27   Schematic of a dc compound motor.

From the circuit model of Figure 7.27, we conclude that

$$V_t = E_c + I_a(R_s + R_a) \tag{7.86}$$

$$I_{\text{sh}} = \frac{V_t}{R_f} \tag{7.87}$$

The power developed by the armature is given by

$$P_a = E_c I_a \tag{7.88}$$

The developed torque is given by

$$T_a = \frac{P_a}{\omega} \tag{7.89}$$

Recalling that the back EMF is given by

$$E_c = K_1 \phi_f \omega \tag{7.90}$$

we conclude from Eqs. (7.84) and (7.85) that

$$E_c = K_1 (K_3 I_{\text{sh}} \pm K_4 I_a)\omega \tag{7.91}$$

The plus sign pertains to a cumulative compound motor and the minus sign pertains to a differential compound motor. We now have the ingredients necessary for performance analysis of a compound motor.

The variation of P_a with I_a is easily obtained from Eqs. (7.86) and (7.88) as

$$P_a = [V_t - I_a(R_s + R_a)]I_a \qquad (7.92)$$

This is of the same form as for the series and shunt ($R_s = 0$) motor cases, with identical conclusions as to maximum value of armature developed power and linear variation of P_a with I_a for small-resistance voltage drops.

Using Eqs. (7.88) to (7.90), we have

$$T_a = K_1\phi_f I_a \qquad (7.93)$$

Substituting for ϕ_f from Eq. (7.84) or (7.85), we conclude that the armature developed (internal) torque is given by

$$T_a = K_1(K_3 I_{\text{sh}} \pm K_4 I_a)I_a \qquad (7.94)$$

Equation (7.94) shows us that the torque developed by a compound motor exhibits a characteristic that combines both a shunt motor and a series motor characteristic, as shown in Figure 7.28. For a cumulative compound motor the torque will be higher than that from a shunt motor with the same armature current. For a differential compound motor, the developed torque is lower than that for a shunt motor with the same armature current.

The motor speed ω can be expressed in terms of armature current using Eqs. (7.91) and (7.86) as

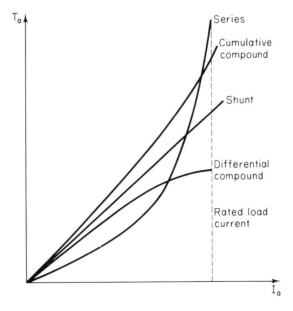

FIGURE 7.28 Comparison of developed torque versus armature current for dc motor connections.

$$\omega = \frac{V_t - I_a(R_s + R_a)}{K_1(K_3 I_{sh} \pm K_4 I_a)} \qquad (7.95)$$

For a cumulative compound motor (with plus sign), as the motor load is increased, I_a increases and the numerator of Eq. (7.95) decreases while the denominator increases. Thus the drop in speed of a cumulative compound motor is at a faster rate than the speed of the shunt motor, as shown in Figure 7.29.

A differential compound connection is rarely used in practice, as this connection is *inherently unstable*, as is shown presently. Let us rewrite Eq. (7.95) for a differential connection:

$$\omega = \frac{V_t - I_a(R_s + R_a)}{K_1(K_3 I_{sh} - K_4 I_a)} \qquad (7.96)$$

The denominator will decrease as I_a (motor load) is increased. The speed will increase and a runaway condition can exist if $(K_3 I_{sh} - K_4 I_a)$ approaches zero.

Some useful relationship can be derived from the torque and speed equations (7.94) and (7.95). Let us rewrite Eq. (7.94) for a cumulative motor as

$$T_a = K_1 K_3 I_{sh}(1 + \beta I_a)I_a \qquad (7.97)$$

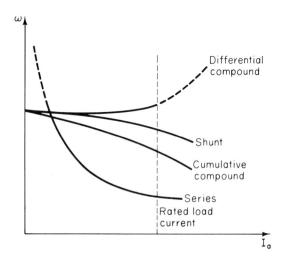

FIGURE 7.29 Comparison of speed versus armature current characteristics for dc motor connections.

where

$$\beta = \frac{K_4}{K_3 I_{\text{sh}}} \tag{7.98}$$

If the motor develops a torque T_{a1} for an armature current I_{a1} and a torque T_{a2} for an armature current I_{a2}, Eq. (7.97) leads to

$$\frac{T_{a1}}{T_{a2}} = \frac{(1 + \beta I_{a1}) I_{a1}}{(1 + \beta I_{a2}) I_{a2}} \tag{7.99}$$

Similarly, we write Eq. (7.95) as

$$\omega = \frac{V_t - I_a(R_s + R_a)}{K_1 K_3 I_{\text{sh}}(1 + \beta I_a)} \tag{7.100}$$

From Eq. (7.100) we can conclude that

$$\frac{\omega_1}{\omega_2} = \frac{V_t - I_{a1}(R_s + R_a)}{V_t - I_{a2}(R_s + R_a)} \frac{1 + \beta I_{a2}}{1 + \beta I_{a1}} \tag{7.101}$$

In Eq. (7.101), ω_1 is the speed corresponding to armature current I_{a1} and ω_2 is the speed corresponding to armature current I_{a2}.

The relations above are based on the assumption that β of Eq. (7.98) is available. The value of β can be obtained using the following procedure. Assume that the motor is run as a shunt motor with a torque T_{sh} corresponding to an armature current I_a. In this case we have

$$T_{\text{sh}} = K_1 K_3 I_{\text{sh}} I_a \tag{7.102}$$

The motor is then connected as a cumulative compound connection and the armature current is adjusted to the same value I_a obtained with shunt operation. The corresponding torque is T_c and is given by

$$T_c = K_1 K_3 I_{\text{sh}}(1 + \beta I_a) I_a \tag{7.103}$$

Equations (7.102) and (7.103) yield

$$\frac{T_c}{T_{\text{sh}}} = 1 + \beta I_a \tag{7.104}$$

Thus β is found as

$$\beta = \frac{1}{I_a}\left(\frac{T_c}{T_{sh}} - 1\right) \qquad (7.105)$$

An example is appropriate at this junction.

Example 7.15   A cumulative compound motor is operated as a shunt motor and develops a torque of 2000 N · m when the armature current is 140 A. When reconnected as a cumulative compound motor at the same current it develops a torque of 2400 N · m. Find the torque when the compound motor load is increased such that the armature current is increased by 10%.

Solution

We use Eq. (7.105) to obtain

$$\beta I_{a1} = \frac{2400}{2000} - 1 = 0.2$$

We now have

$$I_{a2} = I_{a1}(1.1)$$

As a result, we use Eq. (7.99) to obtain

$$\frac{T_{a1}}{T_{a2}} = \frac{(1 + 0.2)140}{[1 + 1.1(0.2)](1.1 \times 140)}$$

$$= \frac{1.2}{1.1(1.22)} = 0.8942$$

But

$$T_{a1} = 2400 \text{ N} \cdot \text{m}$$

Thus

$$T_{a2} = \frac{2400}{0.8942} = 2684 \text{ N} \cdot \text{m}$$

Clearly, the developed torque increases with an increase in armature current. The following example deals with the speed variations for the motor considered in this example.

Example 7.16 Assume that the combined armature and series field resistance of the motor of Example 7.15 is 0.16 Ω. Assume that the terminal voltage is 600 V. Let the motor speed be 1200 rpm when operating as a cumulative compound motor with an armature current of 140 A. Find the speed corresponding to an armature current of 110% of 140 A.

Solution

A direct application of Eq. (7.101) is all that we need.

$$\frac{\omega_1}{\omega_2} = \frac{600 - 0.16(140)}{600 - 0.16(140 \times 1.1)} \left[\frac{1 + 0.2(1.1)}{1 + 0.2} \right]$$

$$= 1.004$$

With

$$n_1 = 1200 \text{ rpm}$$

then

$$n_2 = \frac{n_1}{1.004} = 1195.346 \text{ rpm}$$

Clearly, the motor speed drops with an increase in armature current.

Torque–Speed Characteristic

The torque–speed characteristic for a compound motor is slightly more involved than that of the series and shunt cases. The torque developed can be written on the basis of Eq. (7.94) as

$$T_a = T_{sh} + T_s$$

where

$$T_{sh} = K_1 K_3 I_{sh} I_a$$

$$T_s = K_1 K_4 I_a^2$$

The armature current can be obtained from Eq. (7.95) as a function of ω as

$$I_a = \frac{V_t - K_1 K_3 I_{sh}\omega}{R_a + R_s + K_1 K_4\omega} \tag{7.106}$$

As a result, we conclude that

$$T_a = K_1 \left[K_3 I_{sh} \frac{V_t - K_1 K_3 I_{sh}\omega}{R_a + R_s + K_1 K_4 \omega} + K_4 \frac{(V_t - K_1 K_3 I_{sh}\omega)^2}{(R_a + R_s + K_1 K_4 \omega)^2} \right] \quad (7.107)$$

The torque is clearly composed of two components T_{sh} and T_s related to shunt and series field contributions. It must be observed, however, that T_{sh} contains a component due to the series field effect and similarly for T_s.

7.10 MOTOR AND LOAD MATCHING

The torque–speed characteristic of a dc motor depends on the motor connection. For a series motor we concluded in Eq. (7.48) that

$$T_a = \frac{K_1 K_2 V_t^2}{[K_1 K_2 \omega + (R_a + R_s)]^2} \quad (7.108)$$

The constant $K_1 K_2$ is obtained from the defining relation for back EMF given by

$$E_c = K_1 K_2 I_a \omega \quad (7.109)$$

For a shunt motor we have, by Eq. (7.77),

$$T_a = \frac{K_{sh}^2}{R_a} (\omega_0 - \omega) \quad (7.110)$$

where

$$\omega_0 = \frac{V_t}{K_{sh}} \quad (7.111)$$

The constant K_{sh} can be obtained on the basis of

$$E_c = K_{sh}\omega \quad (7.112)$$

The cumulative compound motor torque–speed characteristic is given by

$$T_a = K_1 \left[K_3 I_{sh} \frac{V_t - K_1 K_3 I_{sh}\omega}{R_a + R_s + K_1 K_4 \omega} + K_4 \frac{(V_t - K_1 K_3 I_{sh}\omega)^2}{(R_a + R_s + K_1 K_4 \omega)^2} \right] \quad (7.113)$$

where the constants are defined by the back EMF relation

$$E_c = K_1(K_3 I_{sh} + K_4 I_a)\omega \tag{7.114}$$

Mechanical loads driven by motors have their own torque–speed characteristics depending on the application. The load torque–speed characteristic is determined experimentally and can be classified as follows:

1. Constant torque load as in extruder drives
2. Torque that increases linearly with speed
3. Constant power load as in reeling drives
4. Torque varying as speed squared as in fan drives

In general, the load torque–speed characteristic is written as

$$T_\ell = f_\ell(\omega) \tag{7.115}$$

The intersection of the two torque–speed characteristics for motor and load provide us with the values of torque and the corresponding speed for operation as shown in Figure 7.30. Examples should clarify these points.

Example 7.17   The combined armature and series field resistance of a dc series motor is 0.15 Ω. The motor operates from a 250-V supply, drawing an armature current of 85 A at a speed of 62.83 rad/s. Establish the torque–speed characteristic of the motor. Find the torque and speed when driving a constant power load with the characteristic

$$T_\ell = \frac{21.334 \times 10^3}{\omega}$$

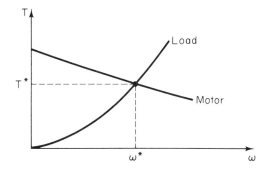

FIGURE 7.30   Matching motor and load torque–speed characteristics.

Solution

We note at the outset that we can solve the problem by assuming that armature developed power is 21.334×10^3 W and apply techniques already known to us. We opt, however, for the following solution procedure.

We have

$$V_t = 250 \text{ V} \qquad I_a = 85 \text{ A at } 62.83 \text{ rad/s}$$

Thus

$$E_c = 250 - (85)(0.15) = 237.25 \text{ V}$$

$$K_1 K_2 = \frac{237.25}{85(62.83)} = 44.424 \times 10^{-3}$$

As a result,

$$T_a = \frac{2.7765 \times 10^3}{(44.424 \times 10^{-3}\omega + 0.15)^2}$$

To match the load, we require that

$$\frac{2.7765 \times 10^3}{(44.424 \times 10^{-3}\omega + 0.15)^2} = \frac{21.334 \times 10^3}{\omega}$$

Thus

$$1.9735 \times 10^{-3}\omega^2 + 13.326 \times 10^{-3}\,\omega + 22.5 \times 10^{-3}$$
$$= 130.14 \times 10^{-3}\omega$$

or

$$1.9735 \times 10^{-3}\omega^2 - 116.82 \times 10^{-3}\omega + 22.5 \times 10^{-3} = 0$$

The solution for ω is

$$\omega^\star = 59.00 \text{ rad/s}$$

The corresponding torque is

$$T^\star = 361.59 \text{ N} \cdot \text{m}$$

Example 7.18 A 120-V dc shunt motor has an armature resistance of 0.1 Ω and develops a torque of 111.16 N · m when running at 1100 r/min. Neglecting rotational losses, it is required to:

    a. Establish the torque–speed characteristic of this motor.
    b. Find the developed torque at a speed of 1150 rpm.
    c. Find the motor speed when the developed torque is 30 N · m.
    d. Find the motor speed when it drives a load with the following torque–speed characteristic:

$$T_\ell = 6.2778 \sqrt{\omega}$$

    e. Calculate the torque, armature developed power, and armature current for the conditions of part (d).
    f. Assuming that the field resistance is 120 Ω, find the line current and the power input for the load conditions of part (d).

Solution

(a) We have

$$V_t = 120 \text{ V} \qquad T = 111.16 \text{ N} \cdot \text{m at } 1100 \text{ r/min}$$

$$R_a = 0.1 \ \Omega$$

The power developed by the armature can be calculated as

$$P_a = 111.16 \left(\frac{2\pi}{60} 1100 \right) = 12.805 \times 10^3 \text{ W}$$

But

$$P_a = (V_t - I_a R_a) I_a$$

Thus

$$12.805 \times 10^3 = (120 - 0.1 I_a) I_a$$

As a result,

$$0.1 I_a^2 - 120 I_a + 12.805 \times 10^3 = 0$$

$$I_a = 118.39 \text{ A}$$

$$E_c = 120 - 0.1(118.39) = 108.16 \text{ V}$$

$$K_{sh} = \frac{E_c}{\omega} = \frac{108.16}{(2\pi/60)(1100)} = 938.97 \times 10^{-3}$$

$$\omega_0 = \frac{V_t}{K_{sh}} = 127.8 \text{ rad/s}$$

$$\frac{K_{sh}^2}{R_a} = 8.8166$$

We can conclude that

$$T_a = 8.8166(127.8 - \omega)$$

(b) For $n = 1150$ r/min, we get

$$T_a = 8.8166\left[127.8 - \frac{2\pi}{60}(1150)\right] = 64.998 \text{ N} \cdot \text{m}$$

(c) For $T_a = 30 \text{ N} \cdot \text{m}$,

$$30 = 8.8166(127.8 - \omega)$$

As a result,

$$\omega = 124.4 \text{ rad/s}$$

(d) For a load with the torque–speed characteristic

$$T_\ell = 6.2778 \sqrt{\omega}$$

to match, we have

$$6.2778 \sqrt{\omega} = 8.8166(127.8 - \omega)$$

Thus

$$0.507 \times 10^{-3}\omega = (127.8 - \omega)^2$$

or

$$\omega^2 - 256.11\omega + 16.333 \times 10^3 = 0$$

The solution is

$$\omega = 119.98 \text{ rad/s}$$

(e) The torque is obtained from

$$T_\ell = 6.2778 \sqrt{119.98} = 68.764 \text{ N} \cdot \text{m}$$

The armature developed power is

$$P_a = T_\ell \omega = 8.2503 \times 10^3 \text{ W}$$

Thus we use

$$P_a = (V_t - I_a R_a) I_a$$

to obtain I_a:

$$8.2503 \times 10^3 = (120 - 0.1 I_a) I_a$$

or

$$0.1 I_a^2 - 120 I_a + 8.250 \times 10^3 = 0$$

As a result,

$$I_a = 73.22 \text{ A}$$

(f) The field current is calculated as

$$I_f = \frac{120}{120} = 1 \text{ A}$$

Thus

$$I_L = I_a + I_f$$
$$= 74.22 \text{ A}$$

The power input is

$$P_{\text{in}} = V_t I_L$$
$$= 120(74.22)$$
$$= 8.9064 \times 10^3 \text{ W}$$

7.11 SPEED CONTROL OF DC MOTORS

There are many applications where the speed of a dc motor is required to vary within a prescribed range for a given load. A number of methods are available for controlling the speed of a dc motor. The general idea stems from the basic relations

$$E_c = V_a - I_a R_a \qquad (7.116)$$

$$E_c = K_1 \phi_f \omega \qquad (7.117)$$

As a result, we can see that the angular speed is given by

$$\omega = \frac{V_a - I_a R_a}{K_1 \phi_f} \qquad (7.118)$$

It is clear that a change in the applied armature voltage will result in a change in speed. Similarly, a change in field flux will result in a change in motor speed. Two major means are available to achieve these changes. The first involves use of an adjustable voltage source, while the next employs variable resistors to affect the required changes. Next, we discuss some of the options, starting with the dc series motor.

Series Motor Speed Control

The methods used to control the speed of a dc series motor include: (1) series resistance voltage control, (2) armature shunt resistance current control, (3) series and shunt armature resistance control, (4) shunt diverter in parallel with the series field, and (5) line voltage control. We discuss each of these methods in turn.

Series Resitance Voltage Control

The simplest means for adjusting the speed of a dc series motor is to insert a variable resistance R_b in series with the armature and series field circuit as shown in Figure 7.31. The effect of the additional resistance can be seen from the torque–speed relation

$$T_a = \frac{K_1 K_2 V_t^2}{[K_1 K_2 \omega_A + (R_a + R_s)]^2} \qquad (7.119)$$

With resistance R_b, we get

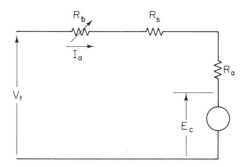

FIGURE 7.31   Dc series motor speed control by series resistance.

$$T_a = \frac{K_1 K_2 V_t^2}{[K_1 K_2 \omega_B + (R_a + R_s + R_b)]^2} \tag{7.120}$$

It is clear that the torque–speed characteristic is shifted to a lower position by increasing R_b. Stated differently, for the same torque, we have

$$K_1 K_2 \omega_A + (R_a + R_s) = K_1 K_2 \omega_B + (R_a + R_s + R_b)$$

Thus

$$\omega_B = \omega_A - \frac{R_b}{K_1 K_2} \tag{7.121}$$

The angular speed ω_B is less than ω_A corresponding to no additional resistances, as shown in Figure 7.32.

Example 7.19   A dc series motor is rated at 250 V and has a combined armature and series field resistance of 0.15 Ω. The constant $K_1 K_2$ is found to be

$$K_1 K_2 = 44.424 \times 10^{-3} \text{ V/A} \cdot \text{rad/s}$$

The motor drives a load with characteristic

$$T_\ell = \frac{21.334 \times 10^3}{\omega}$$

It is desired to drive the load at 55 rad/s. Find the required value of a series resistance R_b to achieve this requirement.

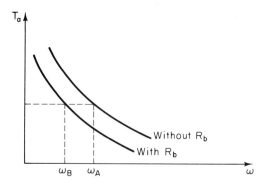

FIGURE 7.32 Effect of series resistance on torque–speed characteristic of series motor.

Solution

The motor torque with R_b inserted in series with the field is

$$T_a = \frac{44.424 \times 10^{-3}(250)^2}{(0.15 + R_b + 44.424 \times 10^{-3}\omega)^2}$$

With $\omega = 55$ rad/s, the load torque is

$$T_\ell = \frac{21.334 \times 10^3}{55} = 387.89 \text{ N} \cdot \text{m}$$

To match the load, we have

$$387.89 = \frac{2.7765 \times 10^3}{(R_b + 2.5933)^2}$$

Thus we get

$$R_b = 0.0821 \ \Omega$$

Armature Shunt Resistance Current Control

In this method a shunt resistance R_c is placed in parallel with the armature circuit, as shown in Figure 7.33. The line current passes through the series field winding. The armature current I_a is less than the series field current due to the presence of R_c.

   The armature's developed torque is given by

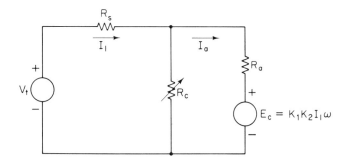

FIGURE 7.33 Armature shunt resistance current control equivalent circuit.

$$T_a = \frac{E_c I_a}{\omega} \tag{7.122}$$

The back EMF is given by

$$E_c = K_1 \phi_f \omega = K_1 K_2 I_\ell \omega \tag{7.123}$$

Note that the field flux is proportional to the current I_ℓ passing through the field series resistance. As a result, the torque is given by

$$T_a = K_1 K_2 I_\ell I_a \tag{7.124}$$

We can obtain I_ℓ and I_a using the following two loop equations obtained from the circuit of Figure 7.33:

$$V_t = I_\ell R_s + I_a R_a + K_1 K_2 \omega I_\ell \tag{7.125}$$

$$V_t = I_\ell R_s + (I_\ell - I_a) R_c \tag{7.126}$$

Solving the two equations, we get

$$I_a = \frac{V_t(R_c - K_1 K_2 \omega)}{\Delta} \tag{7.127}$$

$$I_\ell = \frac{V_t(R_a + R_c)}{\Delta} \tag{7.128}$$

where

$$\Delta = R_a(R_s + R_c) + R_c(R_s + K_1 K_2 \omega) \tag{7.129}$$

As a result, the torque is given by the compact form

$$T_a = \frac{K_1 K_2 V_t^2}{\Delta^2} (R_a + R_c)(R_c - K_1 K_2 \omega) \qquad (7.130)$$

The torque can be written as

$$T_a = \frac{K_1 K_2 V_t^2 (R_c - K_1 K_2 \omega)}{(R_a + R_c)\{R_s + [R_c/(R_a + R_c)] (R_a + K_1 K_2 \omega)\}^2} \qquad (7.131)$$

This form allows us to explore some cases of interest. First, without R_c, we should expect the familiar series motor torque–speed formula. With R_c open circuit, we have R_c approaching infinity. The resulting expression is

$$\lim_{R_c \to \infty} T_a = \lim_{R_c \to \infty} \frac{K_1 K_2 V_t^2 (1 - K_1 K_2 \omega / R_c)}{(1 + R_a/R_c)\{R_s + [1/(1 + R_a/R_c)] (R_a + K_1 K_2 \omega)\}^2}$$

$$= \frac{K_1 K_2 V_t^2}{(R_s + R_a + K_1 K_2 \omega)^2} \qquad (7.132)$$

This is exactly the result for a dc series motor. The second case of interest is for $R_c = 0$. The result is

$$\lim_{R_c \to 0} T_a = \frac{-(K_1 K_2)^2 V_t^2 \, \omega}{R_a R_s^2}$$

This result leads us to conclude that the direction of rotation may be reversed for small values of R_c. This is not surprising since for $R_c = 0$, as can be seen from the equivalent circuit,

$$I_\ell \bigg|_{R_c = 0} = \frac{V_t}{R_s} \qquad (7.133)$$

$$I_a \bigg|_{R_c = 0} = -\frac{E_c}{R_a} = -\frac{K_1 K_2 I_\ell \omega}{R_a} = -\frac{K_1 K_2 V_t \omega}{R_a R_s} \qquad (7.134)$$

If the armature and series field resistances are neglected, the torque is approximately given by

$$T_a \simeq \frac{V_t^2}{K_1 K_2} \left(\frac{1}{\omega^2} - \frac{K_1 K_2}{\omega R_c} \right) \qquad (7.135)$$

It is clear that reducing R_c reduces the torque for a constant speed. Conversely, with a constant torque, a reduction in R_c reduces the speed ω. These effects are shown in Figure 7.34.

Example 7.20   Assume that the speed of the series motor of Example 7.19 is controlled to the same specifications, but now we use an armature shunt resistance control scheme. Find the required value of R_c, given that $R_a = 0.1\ \Omega$.

Solution

Our specifications call for

$$T_\ell = 387.89\ \text{N} \cdot \text{m}$$

$$\omega = 55\ \text{rad/s}$$

We are given that

$$R_a = 0.1\ \Omega$$

Thus

$$R_s = 0.05\ \Omega$$

We use the torque expression

$$T_a = \frac{K_1 K_2 V_t^2 (R_c - K_1 K_2 \omega)}{(R_a + R_c)\{R_s + [R_c/(R_a + R_c)](R_a + K_1 K_2 \omega)\}^2}$$

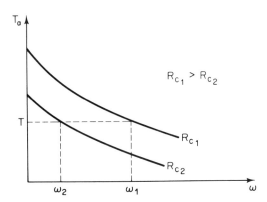

FIGURE 7.34   Effect of armature shunt resistance R_c on torque–speed characteristic of dc series motor.

We thus have

$$387.89 = \frac{2.7765 \times 10^3(R_c - 2.4433)}{(0.1 + R_c)\{0.05 + [R_c/(0.1 + R_c)](2.5433)\}^2}$$

After a few manipulations, we obtain

$$R_c^2 - 38.799R_c - 4.0393 = 0$$

The required answer is obtained as

$$R_c = 38.903 \ \Omega$$

Series and Shunt Armature Resistance Control

In this method a combination of methods 1 and 2 is employed. As shown in Figure 7.35, a resistance R_b is placed in series with the series field winding while a shunt resistance R_c is placed in parallel with the armature circuit.

It is clear that the torque–speed relation for this case can be obtained from the previous analysis simply by replacing R_s by $(R_s + R_b)$. The result is

$$T_a = \frac{K_1K_2V_t^2(R_c - K_1K_2\omega)}{(R_a + R_c)\{R_s + R_b + [R_c/(R_a + R_c)](R_a + K_1K_2\omega)\}^2} \quad (7.136)$$

For a constant torque, increasing R_b reduces the speed, while decreasing R_c reduces the speed.

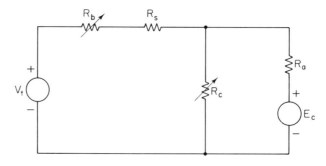

FIGURE 7.35 Series and shunt armature resistance control of a dc series motor.

Example 7.21   For the motor of Example 7.19, suppose that a resistance $R_c = 50\ \Omega$ is available. It is now decided to use series and shunt armature resistance control to achieve the required specifications. Find the value of the required R_b.

Solution

With $R_c = 50\ \Omega$, our torque equation becomes

$$387.89 = \frac{2.7765 \times 10^3 (50 - 2.4433)}{(50.1)[0.05 + R_b + (50/50.1)\,(2.5433)]^2}$$

We solve for R_b to obtain

$$R_b = 18.419 \times 10^{-3}\ \Omega$$

Shunt Diverter

In this method a diverter resistance R_d is placed in shunt with the series field winding as shown in Figure 7.36. The line current is equal to the armature current in this case, but the field current is less.

   The developed torque is given by

$$T_a = K_1 K_2 I_s I_a \qquad (7.137)$$

But from the circuit, we have

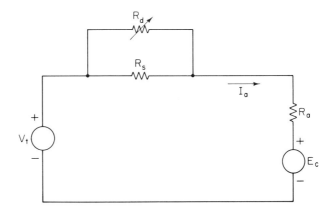

FIGURE 7.36   Shunt diverter speed control of dc series motor.

$$I_s = \frac{R_d}{R_s + R_d} I_a \tag{7.138}$$

Thus

$$T_a = K_1 K_2 \frac{R_d}{R_s + R_d} I_a^2 \tag{7.139}$$

From the circuit we can show that

$$V_t = E_c + I_a \left(R_a + \frac{R_d R_s}{R_d + R_s} \right) \tag{7.140}$$

But

$$E_c = K_1 K_2 I_a \frac{R_d}{R_d + R_s} \omega \tag{7.141}$$

Therefore,

$$I_a = \frac{V_t}{R_a + [R_d R_s/(R_d + R_s)] + [K_1 K_2 R_d \omega/(R_d + R_s)]} \tag{7.142}$$

As a result, we conclude that

$$T_a = \frac{K_1 K_2 \alpha V_t^2}{[R_a + \alpha(R_s + K_1 K_2 \omega)]^2} \tag{7.143}$$

where

$$\alpha = \frac{R_d}{R_s + R_d} \tag{7.144}$$

If we assume that R_a and R_s are negligible, we can write

$$T_a \simeq \frac{V_t^2}{\alpha K_1 K_2 \omega^2} \tag{7.145}$$

Since $\alpha < 1$, we realize that for the same speed, the torque is increased by use of the diverter.

Example 7.22   Assume that a diverter is used with the motor of Example 7.19, such that α = 0.9. For the speed of 55 rad/s, find the resulting torque.

Solution

The torque developed is obtained as

$$T_a = \frac{2.7765 \times 10^3(0.9)}{[0.1 + 0.9(0.05 + 2.5433)]^2}$$

As a result, we conclude that

$$T_a = 421.8 \text{ N} \cdot \text{m}$$

Note that this torque is higher than that without the diverter.

Line Voltage Control

The torque–speed characteristic can be changed simply by adjusting the supply voltage V_t. This can be facilitated by using solid-state drives, as discussed later.

Shunt Motor Speed Control

There is a number of methods to control the speed of a dc shunt motor. Among these, we have (1) field resistance control, (2) shunt field and series armature resistance control, (3) series and shunt armature resistance control, and (4) line voltage control. Each of these methods is discussed in turn.

Field Resistance Control

In this method a variable resistance R_e is inserted in series with the field winding. The effect of an increasing R_e is to decrease the current in the field. Figure 7.37 shows the shunt motor's equivalent circuit in this case. The following analysis provides us with the torque–speed characteristic with R_e included.

The back EMF is given by

$$E_c = K_1 K_2 \frac{V_t}{R_e + R_f} \omega \tag{7.146}$$

Recall that without R_e, the constant K_{sh} was defined as

$$K_{sh} = K_1 K_2 \frac{V_t}{R_f} \tag{7.147}$$

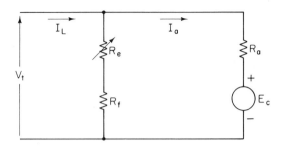

FIGURE 7.37 Field resistance control for a shunt dc motor.

As a result, we write

$$E_c = K_{sh} \frac{R_f}{R_e + R_f} \omega \qquad (7.148)$$

The developed torque is thus

$$T_a = K_{sh} \frac{R_f}{R_e + R_f} I_a \qquad (7.149)$$

The armature current is obtained from

$$E_c = V_t - I_a R_a$$

Thus

$$I_a = \frac{V_t - (K_{sh}R_f\omega/R_e + R_f)}{R_a}$$

The torque–speed relation can thus be obtained as

$$\omega = \omega_0\left(1 + \frac{R_e}{R_f}\right) - \left(1 + \frac{R_e}{R_f}\right)^2 \frac{R_a}{K_{sh}^2} T_a \qquad (7.150)$$

where

$$\omega_0 = \frac{V_t}{K_{sh}} \qquad (7.151)$$

   The result of the foregoing development is that we can conclude that the torque–speed characteristic is still a straight line, as shown in Figure 7.38. The effect of the additional field resistance R_e is to increase the speed at zero torque and increase the magnitude of the slope of the line.

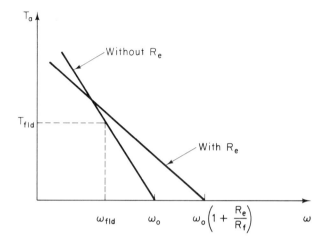

FIGURE 7.38   Torque–speed characteristic of a dc shunt motor with field resistance control.

Example 7.23   The armature resistance of a dc shunt motor is 0.1 Ω and its shunt field resistance is 120 Ω. The value of ω_0 is 127.8 rad/s and the constant K_{sh} is 0.93897. The motor torque is 65 N · m. Find the value of R_e for the motor to run at 1100 r/min carrying this load.

Solution

We use

$$\omega = \omega_0 x - x^2 \left(\frac{R_a}{K_{sh}^2} T_a \right)$$

where

$$x = 1 + \frac{R_e}{R_f}$$

Thus

$$\frac{2\pi(1100)}{60} = 127.8x - \frac{0.1(65)}{(0.93897)^2} x^2$$

As a result,

$$x = 16.381 \quad \text{or} \quad 0.954$$

We take

$$x = 1 + \frac{R_e}{R_f} = 16.381$$

Thus

$$\frac{R_e}{R_f} = 15.381$$

As a result,

$$R_e = 15.381(120) = 1.846 \times 10^3 \ \Omega$$

Shunt Field and Series Armature Resistance Control

As indicated in Figure 7.39, a variable resistance R_b is inserted in series with the armature circuit in addition to the resistance R_e inserted in series with the field winding. The torque–speed relation is given by

$$\omega = \omega_0 \left(1 + \frac{R_e}{R_f}\right) - \left(1 + \frac{R_e}{R_f}\right)^2 \frac{R_a + R_b}{K_{sh}^2} T_a \qquad (7.152)$$

This is the same expression as that obtained for case 1, but with R_a replaced by $(R_b + R_a)$. The effect of additional R_b is to reduce the speed as shown in Figure 7.40.

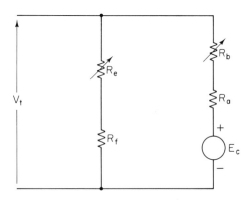

Figure 7.39 Shunt field and series armature resistance control of a dc shunt motor.

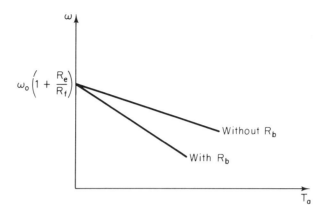

FIGURE 7.40 Speed–torque characteristic for shunt field and series armature resistance control.

Example 7.24 For the motor of Example 7.23, suppose that $R_e = 1.5 \times 10^3 \; \Omega$. Now we like to achieve the same results but with shunt field and series armature resistance control. Find the required value of R_b.

Solution

We use

$$\omega = \omega_o\left(1 + \frac{R_e}{R_f}\right) - \left(1 + \frac{R_e}{R_f}\right)^2 \frac{R_a + R_b}{K_{sh}^2} T_a$$

Thus

$$115.19 = 127.8\left(1 + \frac{1500}{120}\right) - \left(1 + \frac{1500}{120}\right)^2 \frac{(R_a + R_b)65}{(0.93897)^2}$$

As a result,

$$R_a + R_b = 0.11983$$

The required value of R_b is then

$$R_b \simeq 0.02 \; \Omega$$

Series and Shunt Armature Resistance Control

As shown in Figure 7.41, the armature circuit is shunted by the resistance R_c in addition to the resistance R_b inserted in series with the armature. The back EMF E_c is given by

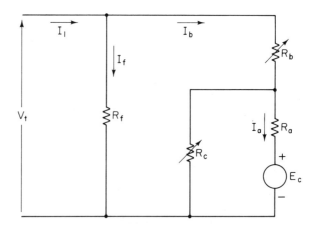

FIGURE 7.41 Series and shunt armature resistance speed control of a dc shunt motor.

$$E_c = K_{\text{sh}}\omega \tag{7.153}$$

The torque developed by the armature is therefore given by

$$T_a = K_{\text{sh}}I_a \tag{7.154}$$

The armature current I_a is obtained by solving the following two loop equations:

$$V_t - E_c = R_a I_a + R_b I_b \tag{7.155}$$

$$V_t = -R_c I_a + (R_b + R_c)I_b \tag{7.156}$$

The solution for I_a is

$$I_a = \frac{R_c V_t - (R_b + R_c)E_c}{R_a(R_b + R_c) + R_b R_c} \tag{7.157}$$

As a result, we have

$$T_a = \frac{K_{\text{sh}}}{R_a(R_b + R_c) + R_b R_c}[R_c V_t - K_{\text{sh}}\omega(R_b + R_c)] \tag{7.158}$$

This can be rearranged to

$$\omega = \omega_o \frac{1}{1 + R_b/R_c} - \frac{R_a + [R_b R_c/(R_b + R_c)]}{K_{\text{sh}}^2}T_a \tag{7.159}$$

where

$$\omega_0 = \frac{V_t}{K_{sh}} \qquad (7.160)$$

Clearly, increasing R_b and R_c reduces the motor speed.

Example 7.25 For the motor of Example 7.23, assume that series and shunt armature resistance control is used with $R_b = 0.03 \; \Omega$. Find the value of R_c to obtain a torque of 65 N · m at a speed of 1100 r/min.

Solution

Let

$$y = \frac{1}{1 + (R_b/R_c)}$$

Thus

$$\omega = \omega_0 y - \frac{R_a + R_b y}{K_{sh}^2} T_a$$

As a result,

$$115.19 = 127.8y - \frac{(0.1 + 0.03y)65}{(0.93897)^2}$$

The solution for y is

$$y = 0.97591$$

Thus

$$R_c = 1.2152 \; \Omega$$

Line Voltage Control

In this case solid-state drives are used to provide adjustable voltage levels for the armature circuit as well as the field circuit, if desired.

Compound Motor Speed Control

The mechanisms for speed control of a compound dc motor are similar in basic concept to those employed with a shunt motor. The following methods will be discussed briefly: (1) shunt field resistance control, (2) shunt field and series armature resistance control, (3) series and shunt armature resistance control, and (4) line voltage control.

Shunt Field Series Resistance Control

A variable resistance R_e is inserted in series with the shunt field winding as shown in the circuit of Figure 7.42. It is clear that for this case

$$I_{sh} = \frac{V_t}{R_f + R_e} \tag{7.161}$$

Increasing R_e decreases I_{sh} and hence the field flux,

$$\phi_f = K_3 I_{sh} + K_4 I_a \tag{7.162}$$

The compound motor's torque–speed relation still applies:

$$T_a = K_1 \left[K_3 I_{sh} \frac{V_t - K_1 K_3 I_{sh} \omega}{R_a + R_s + K_1 K_4 \omega} \right.$$
$$\left. + K_4 \frac{(V_t - K_1 K_3 I_{sh} \omega)^2}{(R_a + R_s + K_1 K_4 \omega)^2} \right] \tag{7.163}$$

Shunt Field and Series Armature Resistance Control

This arrangement is shown in Figure 7.43. As was done in the case of a shunt motor in addition to the series resistance in the shunt field circuit denoted by R_e, we also include the variable resistance R_b in series with the

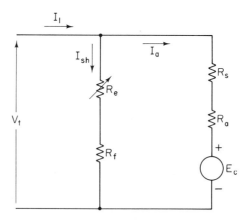

FIGURE 7.42 Compound motor speed control through a series resistance in shunt field.

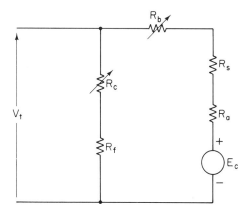

FIGURE 7.43   Shunt field and series armature resistance control of a compound motor.

armature circuit. It is clear that

$$I_{sh} = \frac{V_t}{R_e + R_f} \tag{7.164}$$

The torque–speed characteristic will be modified by replacing $R_s + R_a$ by $R_s + R_a + R_b$ in the expression derived for the straight circuit. As a result, we write

$$T_a = K_1 \left[K_3 I_{sh} \frac{V_t - K_1 K_3 I_{sh}\omega}{R_a + R_s + R_b + K_1 K_4\omega} \right.$$
$$\left. + K_4 \frac{(V_t - K_1 K_3 I_{sh}\omega)^2}{(R_a + R_s + R_b + K_1 K_4\omega)^2} \right] \tag{7.165}$$

Series and Shunt Armature Resistance Control

In this case we include R_b and R_c as shown in Figure 7.44. We note the similarity of this circuit to the corresponding circuit in use with the shunt motor. We can therefore assert that

$$I_a = \frac{R_c V_t - (R_b + R_s + R_c)E_c}{R_a(R_b + R_s + R_c) + (R_b + R_s)R_c} \tag{7.166}$$

This is obtained by replacing R_b in Equation (7.157) by $(R_s + R_b)$. We have also that

$$E_c = K_1(K_3 I_{sh} + K_4 I_b)\omega \tag{7.167}$$

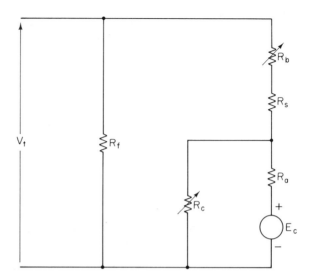

FIGURE 7.44   Series and shunt armature resistance control of a dc compound motor.

The current I_b is given by

$$I_b = I_a[1 + (R_a/R_c)] + E_c/R_c \qquad (7.168)$$

Clearly, we have

$$I_{sh} = \frac{V_t}{R_f} \qquad (7.169)$$

The torque is given by

$$T_a = K_1(K_3 I_{sh} + K_4 I_b)I_a \qquad (7.170)$$

The torque–speed characteristic can thus be obtained by substituting for I_a and I_b in Eq. (7.170).

Line Voltage Control

This is similar in principle to the shunt motor case, with adjustable voltages being applied to the shunt field as well as the series field and armature circuits.

7.12 **REVERSAL OF DIRECTION OF ROTATION**

The requirement of reversing the direction of rotation of a direct current motor is met by reversing the direction of armature current with respect to the direction of the magnetic field, or vice versa. Reversal of both circuits will produce the same direction of rotation. As a general rule, it is preferred to reverse the armature current rather than field current, since the field windings are highly inductive, making it necessary to avoid switching before the field stored energy has been fully dissipated.

The mechanism for reversing a dc motor is shown in Figure 7.45, where reversal of the direction of rotation is achieved through the reversal of the polarity of the voltage applied to the armature circuit using a switch.

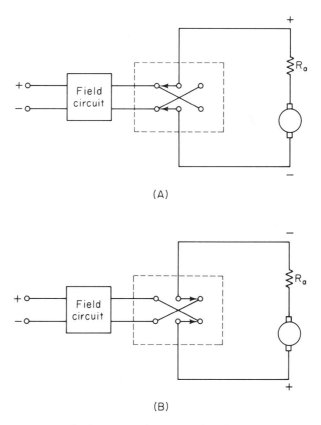

FIGURE 7.45 Reversal of rotation direction of a dc motor: (a) forward; (b) backward.

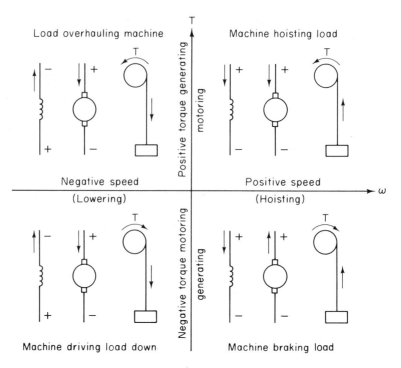

FIGURE 7.46   Four-quadrant operation of dc machine.

Note that in the figure, the block labeled "field circuit" serves as a general-purpose convention where the appropriate circuit (series, shunt, or compound) for field excitation is placed.

It is one of the advantages of dc motors that they can operate in a number of modes according to the four quadrant operation scheme shown in Figure 7.46. A summary of the prevailing conditions in the four modes is given in Table 7.1.

7.13 STARTING DC MOTORS

The armature of a dc motor is designed to have a low resistance to minimize the losses and hence improve efficiency. At starting, the speed of the motor is zero and thus the back EMF E_c is zero. The armature current at starting will therefore be high, as can be seen from

$$I_{a_{st}} = \frac{V_a}{R_a}$$

TABLE (7.1)

Summary of Four-Quadrant Operation of a DC motor

	Quadrant			
	FIRST	*SECOND*	*THIRD*	*FOURTH*
Mode	Hoisting	Overhauling	Down Drive	Braking
Action	Motoring	Generating	Motoring	Generating
$(V_t - E_c)$ and I_a	+	−	+	−
P_a	+	−	+	−
T	+	+	−	−
ω	+	−	−	+

The supply voltage to the dc motor is fixed in many installations and mechanisms for reducing the voltage applied to the armature at starting must be sought. If a variable dc supply is available, the motor is started at a low voltage which is then gradually increased as the motor picks up speed. The conventional method for starting, however, involves inserting a large resistance value in the armature circuit at starting and gradually reducing its value as the motor picks up speed.

An interesting design problem is to find the values of the resistance sections to limit the current to acceptable values as the motor is started. Consider the manual starting arrangement shown in Figure 7.47. The starter is arranged in resistance sections, $r_1, r_2, r_3, \ldots, r_{n-1}$, with the nth section being the resistance of the armature and series field resistance. The total resistance available in the circuit at starting ($t = 0$) is denoted by R_1 and is given by

$$R_1 = r_1 + r_2 + r_3 + \ldots + r_n$$

At $t = 0$, the motor is at standstill and there is no back EMF to oppose the applied voltage V_t. The armature current at $t = 0$ will therefore be given by

$$i_a(0) = \frac{V_t}{R_1}$$

This current value will be assumed to be the maximum allowable, I_{max}. As a result,

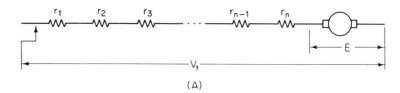

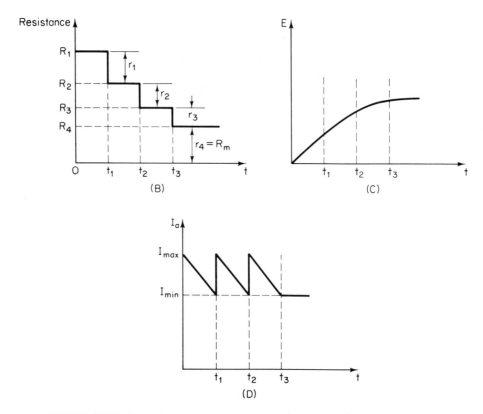

FIGURE 7.47 Dc series motor starter: (a) schematic; (b) resistance–time diagram; (c) EMF buildup; (d) armature current–time diagram.

$$I_{\max} = \frac{V_t}{R_1} \tag{7.171}$$

Immediately following the switching, the rotating motor will develop a back EMF that opposes the applied voltage. At $t = t_1^-$, *the back EMF is assumed to have reached a value* E_{1f} *while the circuit resistance is still* R_1. The armature current at that instant is

$$i_a(t_1^-) = \frac{V_t - E_{1f}}{R_1}$$

This current is smaller than $i_a(0)$, as shown in Figure 7.47. The value of armature current at t_1^- will be assumed to be the specified minimum

$$I_{\min} = i_a(t_1^-) = \frac{V_t - E_{1f}}{R_1}$$

It is now possible to remove one section r_1 from the circuit by notching up one step. The resistance available now is R_2:

$$R_2 = R_1 - r_1$$

At $t = t_1^+$, the back EMF is E_{2i} with the subscript 2 referring to the second time interval and the subscript i referring to initiating the interval. The armature current will rise to the value

$$i_a(t_1^+) = \frac{V_t - E_{2i}}{R_2}$$

This again will be deemed to be the maximum allowable:

$$I_{\max} = \frac{V_t - E_{2i}}{R_2}$$

The back EMF will increase in the time interval t_1 to t_2^- with an increase in motor speed. As a result, at t_2^- the current will drop to

$$I_{\min} = i_a(t_2^-) = \frac{V_t - E_{2f}}{R_2}$$

At t_2^+, the section r_2 is removed, and the process is continued.

We can now consider the problem of designing resistance steps for a dc motor starter. At the time instant t_k, we have immediately before notching up

$$I_{\min} = \frac{V_t - E_{kf}}{R_k} \qquad (7.172)$$

After notching up, we have

$$I_{max} = \frac{V_t - E_{(k+1)i}}{R_{(k+1)}} \tag{7.173}$$

Recall that the back EMF is proportional to the product of speed and flux, and since we cannot expect a change in speed instantaneously, we can conclude that

$$\frac{E_{(k+1)i}}{E_{kf}} = \frac{\phi_{max}}{\phi_{min}}$$

Let us assume that

$$\frac{\phi_{max}}{\phi_{min}} = a$$

Thus

$$\frac{E_{(k+1)i}}{E_{kf}} = a \tag{7.174}$$

We also define the ratio of maximum and minimum current as K:

$$K = \frac{I_{max}}{I_{min}} \tag{7.175}$$

Equation (7.172) can be written using Eq. (7.171) as

$$E_{kf} = I_{max}R_1 - I_{min}R_k \tag{7.176}$$

Equation (7.173) can be written as

$$E_{(k+1)i} = I_{max}R_1 - I_{max}R_{k+1}$$

Using Eq. (7.174), we have

$$aE_{kf} = I_{max}(R_1 - R_{k+1}) \tag{7.177}$$

Eliminating E_{kf} between Eqs. (7.176) and (7.177) and using (7.175), we obtain

$$a\left(R_1 - \frac{R_k}{K}\right) = R_1 - R_{k+1}$$

Thus by a slight rearrangement, we get

$$R_{k+1} = R_1(1 - a) + R_k \frac{a}{K} \tag{7.178}$$

Replacing $(k + 1)$ by (k) and (k) by $(k - 1)$ in the above, we get

$$R_k = R_1(1 - a) + R_{k-1} \frac{a}{K} \tag{7.179}$$

The resistance of the kth section, r_k, is given by

$$r_k = R_k - R_{k+1}$$

Using Eqs. (7.178) and (7.179), we get

$$r_k = \frac{a}{K}(R_{k-1} - R_k)$$

$$= \frac{a}{K} r_{k-1}$$

Define

$$b = \frac{a}{K} \tag{7.180}$$

As a result,

$$r_k = b r_{k-1} \tag{7.181}$$

The resistance sections therefore form a geometric progression with ratio b. The total resistance R_1 at starting is given by

$$R_1 = r_1 + r_2 + \ldots + r_{n-1} + r_n$$

Using Eq. (7.180), we get

$$R_1 = r_1(1 + b + b^2 + \ldots + b^{n-2}) + r_n$$

Performing the required summation, we obtain

$$R_1 = r_1 \frac{1 - b^{n-1}}{1 - b} + r_n \tag{7.182}$$

Recall that

$$r_1 = R_1 - R_2 \qquad (7.183)$$

Using Eqs. (7.178) and (7.180) in (7.183), we get

$$r_1 = R_1 - R_1(1 - a) - bR_1$$

Thus

$$r_1 = (a - b)R_1 \qquad (7.184)$$

Using Eq. (7.180) in (7.184), we get

$$r_1 = b(K - 1)R_1 \qquad (7.185)$$

Substituting Eq. (7.185) in (7.182), we get

$$R_1 = b(K - 1)\frac{1 - b^{n-1}}{1 - b}R_1 + r_n \qquad (7.186)$$

Dividing by R_1 and recalling that

$$r_n = R_m = R_a + R_s$$

we obtain, after some rearranging,

$$b^{n-1} = 1 - \frac{1 - b}{b(K - 1)}\left(1 - \frac{R_m}{R_1}\right)$$

Taking logarithms of both sides, we conclude that

$$n = 1 + \frac{log\{1 - [(1 - b)/b(K - 1)](1 - R_m/R_1)\}}{log\ b} \qquad (7.187)$$

Equation (7.187) specifies the required number of sections n.

Example 7.26 A starter for a 220-V dc series motor is required such that the maximum current be 270 A and the minimum current not be less than 162 A. The armature and series field resistances add up to 0.12 Ω. The magnetization characteristic of the motor is as follows:

Motor Current (A)	162	180	198	216	234	252	270
EMF at 750 r/min	193	200	206	211.2	215.8	220	224

Find the required number of resistance steps n.

Solution

With the given information, we have

$$I_{max} = 270 \text{ A}$$

$$I_{min} = 162 \text{ A}$$

Thus

$$K = \frac{I_{max}}{I_{min}} = \frac{270}{162} = 1.66667$$

From the magnetization characteristic

$$E_{max} = 224$$

$$E_{min} = 193$$

Thus

$$a = \frac{E_{max}}{E_{min}} = \frac{224}{193} = 1.16062$$

We can now find b:

$$b = \frac{a}{K} = \frac{1.16062}{1.66667} = 0.696373$$

The resistance R_1 is obtained from

$$R_1 = \frac{V_t}{I_{max}} = \frac{220}{270} = 0.814815 \ \Omega$$

We also have

$$R_m = 0.12 \ \Omega$$

All the necessary ingredients of Eq. (7.187) are now available and we thus have

$$n = 1 + \frac{\log\{1 - [(1 - 0.696)/0.696(1.667 - 1)]\,(1 - 0.12/0.8148)\}}{\log 0.696373}$$

$$= 3.2543$$

We note that n is noninteger. We therefore take the closest integer of higher value than the calculated value. Thus take

$$n = 4$$

Equation (7.186) will not be satisfied with the adjustment of n to the next larger integer unless a change is made in one of the other parameters. Two possible remedies are to increase I_{min} or decrease I_{max}. The object is to satisfy Eq. (7.186) rewritten as

$$1 - \frac{R_m}{R_1} = b(K - 1)\frac{1 - b^{n-1}}{1 - b} \qquad (7.188)$$

Note that if I_{max} is left unchanged, the left-hand side of (7.188) is also unchanged. A change in I_{min} will change the values of K, a, and b. A simple procedure to find the required value of I_{min} requires a plot of y versus I_{min}, where

$$y = b(K - 1)\frac{1 - b^{n-1}}{1 - b}$$

The intersection of the characteristic with the constant y_0 line provides the required value of I_{min}. Here y_0 is given by

$$y_0 = 1 - \frac{R_m}{R_1}$$

We can now continue with our example to find the required value of I_{min} for the choice $n = 4$.

Example 7.26 (*Continued*) The task here is to find I_{min}. We start by noting that

$$y_0 = 1 - \frac{R_m}{R_1} = 1 - \frac{0.12}{0.8148} = 0.8527$$

Since we had $I_{min} = 162$ A already, we move up to the next point on the magnetization characteristic. Thus take

$$I_{min} = 180 \text{ A}$$

$$E_{min} = 200 \text{ V}$$

Thus

$$K = \frac{270}{180} = 1.5$$

$$a = \frac{224}{200} = 1.12$$

$$b = \frac{a}{K} = 0.74667$$

We now calculate

$$y = b(K - 1)\frac{1 - b^3}{1 - b}$$

$$= 0.86023$$

The process is repeated for $I_{min} = 198$ A to obtain

$$y = 0.70558$$

In Figure 7.48 we plot the two points in the $y - I_{min}$ plane. A simple linear approximation to solve for I^*_{min} requires solving

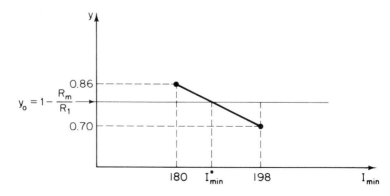

FIGURE 7.48 Finding I^*_{min}.

$$\frac{198 - I_{min}^*}{I_{min}^* - 180} = \frac{0.8527 - 0.70}{0.86 - 0.8527}$$

As a result, we conclude that

$$I_{min}^* = 180.8 \text{ A}$$

Having obtained the proper values of n and I_{min}, we can now conclude our design example by finding the resistance steps.

Example 7.26 For $I_{min} = 180.8$ A, we obtain
(Conclusion)

$$K = \frac{270}{180.8} = 1.4934$$

$$a = \frac{224}{200.3} = 1.1183$$

$$b = \frac{1.1183}{1.4934} = 0.7488$$

Now we have

$$R_1 = 0.8148 \ \Omega$$

$$r_1 = b(K - 1)R_1 = 0.7488(1.4934 - 1)(0.8148) = 0.301 \ \Omega$$

$$r_2 = br_1 = 0.22544 \ \Omega$$

$$r_3 = br_2 = 0.16882 \ \Omega$$

This concludes the starter design.

SOME SOLVED PROBLEMS

Problem 7.A.1

The magnetization curve of a dc generator at a speed of 1500 r/min is given in the following table:

E_g (V)	7	52	136	205	231	248
I_f (A)	0	2	4	6	8	10

a. Approximate the curve by straight-line segments of the form

$$E_g = \frac{n}{1500}(a + bI_f)$$

by finding a and b for each region.
b. Find the generated voltage E_g for a field current of 6.5 A at a speed of 1750 r/min.
c. Find the field current required to set up a generated voltage of 200 V at a speed of 1600 r/min.
d. Find the required speed for a generated voltage of 245 V when the field current is 5.75 A.

Solution

(a) Since we have data for $n = 1500$, we then have to fit

$$E_g = a + bI_f$$

Take the region $0 \le I_f < 2$; for $I_f = 0$, we have

$$E_g = 7 = a + b(0)$$

Hence

$$a = 7$$

For $I_f = 2$, we then have

$$52 = 7 + 2b$$

Hence

$$b = 22.5$$

As a result,

$$E_g = (7 + 22.5I_f)\frac{n}{1500} \qquad 0 \le I_f < 2$$

In a similar manner, we get

$$E_g = (-32 + 42I_f)\frac{n}{1500} \qquad 2 \le I_f < 4$$

$$E_g = (-2 + 34.5I_f)\frac{n}{1500} \qquad 4 \le I_f < 6$$

$$E_g = (127 + 13I_f)\frac{n}{1500} \qquad 6 \le I_f < 8$$

$$E_g = (163 + 8.5I_f)\frac{n}{1500} \qquad 8 \le I_f < 10$$

(b) For $I_f = 6.5$ A, we have, at $n = 1750$ r/min,

$$E_g = \frac{1750}{1500}[127 + 13(6.5)] = 246.75 \text{ V}$$

(c) For a speed of 1500 r/min, a generated voltage of 200 V is obtained in the region $4 \le I_f < 6$. We check to see if 200 V is obtained in this region at 1600 r/min:

$$E_g = \frac{1600}{1500}(-2 + 34.5I_f)$$

For $I_f = 6$, we get

$$E_g = 218.667 \text{ V}$$

For $I_f = 4$, we get

$$E_g = 145.067 \text{ V}$$

Thus 200 V is obtained in this region.

$$200 = \frac{1600}{1500}(-2 + 34.5I_f)$$

As a result,

$$I_f = 5.493 \text{ A}$$

(d) For a field current of 5.75 A, the applicable formula is

$$E_g = \frac{n}{1500}(-2 + 34.5I_f)$$

Thus

$$245 = \frac{n}{1500} [-2 + (34.5)(5.75)]$$

As a result,

$$n = 1871.42 \text{ r/min}$$

Problem 7.A.2

A dc series motor is operated at full load from the 240-V mains at a speed of 600 r/min. The armature back EMF is found to be 217.20 V at a line current of 38 A.

   a. Find the armature resistance assuming that the series field resistance is 0.2 Ω.
   b. Find the no-load speed given that the no-load line current is 1 A.

Solution

(a) We have

$$V_t = E_c + I_a(R_a + R_s)$$
$$240 = 217.2 + 38(R_a + R_s)$$

Thus

$$R_a + R_s = 0.6$$

As a result, the armature resistance is

$$R_a = 0.4 \ \Omega$$

(b) At no load,

$$E_c = 240 - (1)(0.6) = 239.6 \text{ V}$$

We use the speed relation

$$\frac{E_{c_2}}{E_{c_1}} = \frac{n_2 I_{a_2}}{n_1 I_{a_1}}$$

$$\frac{239.6}{217.2} = \frac{n_2}{600}\left(\frac{1}{38}\right)$$

Thus

$$n_2 = 25,151.38 \text{ r/min}$$

Problem 7.A.3

A 10-hp 600-r/min 240-V dc series motor has a rated line current of 38 A, an armature resistance of 0.4 Ω, and a series field resistance of 0.2 Ω. It is required to provide a starting torque of 225% of this rated load torque. Find the value of the required starting resistor.

Solution

Since torque in a series motor is proportional to the square of the current, we have

$$\frac{T_{st}}{T_{fld}} = \left(\frac{I_{st}}{I_{fld}}\right)^2$$

Thus

$$I_{st} = \sqrt{2.25}\, I_{fld}$$

$$= 57 \text{ A}$$

As a result,

$$240 = 57(0.4 + 0.2 + R_{st})$$

The value of the starting resistor is thus obtained as

$$R_{st} = 3.61 \ \Omega$$

Problem 7.A.4

A 50-hp 230-V 1300-r/min dc shunt motor delivers its rated load at rated speed when taking 180.5 A from a 230-V mains. The resistance of the shunt field circuit is 115 Ω and the resistance of the armature circuit is 0.05 Ω. It is desired to operate the motor at 600 r/min to deliver 120% of the full-load torque. This is arranged by inserting a series resistance R_n in the armature circuit. Find the required value of R_n.

Solution

The full-load line current is given by

$$I_L = I_{a_1} + I_f$$

$$180.5 = I_{a_1} + \frac{230}{115}$$

Thus

$$I_{a_1} = 178.5 \text{ A}$$

As a result we can calculate the armature back EMF as

$$E_{c_1} = V_t - I_{a_1}R_a$$

$$= 230 - 178.5(0.05)$$

$$= 221.075 \text{ V}$$

The corresponding speed is

$$n_1 = 1300 \text{ r/min}$$

The torque is proportional to armature current and thus

$$\frac{T_2}{T_1} = \frac{I_{a2}}{I_{a1}}$$

$$1.2 = \frac{I_{a2}}{178.5}$$

As a result,

$$I_{a2} = 214.2$$

The back EMF is proportional to the speed and thus

$$\frac{E_{c2}}{E_{c1}} = \frac{n_2}{n_1}$$

$$E_{c2} = (221.075)\left(\frac{600}{1300}\right)$$

$$= 102.035 \text{ V}$$

We can now calculate the required resistance R_n from

$$V_t = E_{c2} + I_{a2}(R_a + R_n)$$

$$230 = 102.035 + (214.2)(R_a + R_n)$$

Thus

$$R_a + R_n = 0.597 \ \Omega$$

As a result,

$$R_n = 0.547 \ \Omega$$

Problem 7.A.5

A 50-hp 230-V 1300-r/min dc shunt motor has an armature resistance of 0.05 Ω. At rated speed and output, the armature current is 178.5 A and the field current is 2 A. It is required to calculate:

a. The field circuit resistance.
b. The armature back EMF.
c. The power input to the motor, the rotational losses, and the efficiency.
d. The output torque.

Assume that the field flux is reduced to 70% of its original value by using a field rheostat, while the motor torque load remains unchanged. Find the new speed and efficiency. Assume that rotational losses vary with the square of speed.

Solution

(a) The field circuit resistance is obtained as

$$R_f = \frac{230}{2} = 115 \ \Omega$$

(b) The back EMF is obtained from

$$E_c = V_t - I_a R_a$$

$$= 230 - 178.5(0.05)$$

$$= 221.075 \text{ V}$$

(c) The line current is

$$I_L = I_a + I_f = 178.5 + 2 = 180.5 \text{ A}$$

The power input is thus obtained as

$$P_{\text{in}} = V_t I_L = 230(180.5)$$
$$= 41,515 \text{ W}$$

The rated output is

$$P_o = 50 \times 746 = 37,300 \text{ W}$$

The power developed by the armature is

$$P_a = E_c I_a = 221.075(178.5) = 39,461.9 \text{ W}$$

Thus the rotational losses are obtained as

$$P_{\text{rot}} = P_o - P_a$$
$$= 2161.9 \text{ W}$$

The efficiency is now obtained as

$$\eta = \frac{P_o}{P_{\text{in}}} = \frac{37,300}{41,515} \doteq 0.898$$

(d) The output torque is obtained as

$$T_o = \frac{P_o}{\omega}$$
$$= \frac{37,300}{(2\pi/60)(1300)}$$
$$= 273.99 \text{ N} \cdot \text{m}$$

The torque is proportional to armature current and field flux and thus for fixed torque,

$$\phi_1 I_{a_1} = \phi_2 I_{a_2}$$

Thus

$$I_{a2} = \frac{178.5}{0.7} = 255 \text{ A}$$

The new back EMF is

$$E_{c2} = 230 - 255(0.05)$$

$$= 217.25 \text{ V}$$

The back EMF is proportional to speed and flux and thus

$$\frac{E_{c2}}{E_{c1}} = \frac{\phi_2}{\phi_1} \frac{n_2}{n_1}$$

$$\frac{217.25}{221.075} = 0.7\left(\frac{n_2}{1300}\right)$$

As a result,

$$n_2 = 1825 \text{ r/min}$$

To compute the efficiency, we need the input power. The field current is

$$I_{f2} = 0.7I_{f1} = 0.7(2) = 1.4 \text{ A}$$

The line current is thus

$$I_{L2} = I_{a2} + I_{f2}$$

$$= 255 + 1.4 = 256.4 \text{ A}$$

As a result,

$$P_{\text{in}} = V_t I_{L2} = 230(256.4) = 58,972 \text{ W}$$

The armature developed power is

$$P_a = E_{c2}I_{a2} = 217.25(255) = 55,398.75 \text{ W}$$

The new rotational losses are

$$P_{\text{rot}} = 2161.9\left(\frac{1825}{1300}\right)^2$$

$$= 4260.63 \text{ W}$$

Thus the power output is

$$P_o = P_a - P_{rot} = 51{,}137.36 \text{ W}$$

As a result, we compute the efficiency as

$$\eta = \frac{51{,}137.36}{58{,}972} = 0.867$$

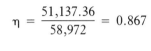

Problem 7.A.6

A 600-V dc motor has the following parameters:

$$R_a = 0.1 \ \Omega \qquad R_s = 0.06 \ \Omega$$
$$R_f = 450 \ \Omega$$

The motor is connected as a series motor and draws an armature current of 200 A at a speed of 600 r/min when driving a constant-torque load. Neglect rotational losses.

    a. Find the back EMF, internal armature power, and load torque.
    b. Find the motor's efficiency.

The motor is now connected as a shunt motor and draws a line current of 200 A when driving the same load.

    c. Find the speed of the motor.
    d. Find the motor efficiency.

Solution

As a series motor, we have

$$R_a + R_s = 0.16\Omega$$
$$I_a = 200 \text{ A} \qquad n = 600 \text{ r/min}$$
$$V_t = 600 \text{ V}$$

(a) The back EMF is obtained as

$$E_c = V_t - I_a(R_a + R_s)$$

$$= 600 - 200(0.16) = 568 \text{ V}$$

The armature power is

$$P_a = E_c I_a$$
$$= 568(200) = 113.6 \times 10^3 \text{ W}$$

The torque is

$$T_a = \frac{P_a}{\omega} = \frac{113.6 \times 10^3}{(2\pi/60)(600)} = 1808 \text{ N} \cdot \text{m}$$

(b) The input power is

$$P_{\text{in}} = V_t I_a = 600(200) = 120 \times 10^3 \text{ W}$$

$$\eta = \frac{113.6}{120} = 0.9466$$

As a shunt motor, we have

$$I_L = 200 \text{ A}$$

$$I_a = 200 - \frac{600}{450} = 198.667 \text{ A}$$

(c) The back EMF is

$$E_c = 600 - 198.667(0.1) = 580.133 \text{ V}$$

Thus the armature power is

$$P_a = 580.133(198.667) = 115.25 \times 10^3 \text{ W}$$

The torque is

$$T_a = 1808 = \frac{115.25 \times 10^3}{\omega}$$

Thus

$$\omega = 63.746 \text{ rad/s}$$

$$n = \frac{60\omega}{2\pi} = 608.73 \text{ r/min}$$

(d) The input power is

$$P_{\text{in}} = V_t I_L = 600(200) = 120 \times 10^3$$

Thus

$$\eta = \frac{115.25}{120} = 0.96042$$

Problem 7.A.7

A 240-V dc motor has a combined armature and series field resistance of 0.6 Ω and a shunt field resistance of 180 Ω. The motor is operated as a series motor with the armature current being 40 A at a speed of 595 r/min. When operated as a compound motor, the line current is found to be 42 A. Assuming that the flux is increased by 50% due to compounding, find the speed of the motor when running as a compound motor.

Solution

When operating as a series motor, we have

$$E_c = 240 - 40(0.6) = 216 \text{ V}$$

But

$$E_c = K_1 \phi_f \omega$$

Thus

$$K_1 \phi_f \bigg|_s = \frac{E_c}{\omega} = \frac{216}{(2\pi/60)(595)} = 3.4666$$

When operating as a compound motor, we have

$$I_a = I_L - I_f = 42 - \frac{240}{180} = 40.667 \text{ A}$$

Thus

$$E_c = 240 - 0.6(40.667) = 215.6 \text{ V}$$

Now

$$K_1\phi_f\Big|_c = K_1\phi_f\Big|_s (1.5) = 5.2$$

As a result, the compound motor speed is

$$\omega = \frac{215.6}{5.2} = 41.462 \text{ rad/s}$$

Problem 7.A.8

For the motor of Problem 7.A.7, the series field resistance is 0.1 Ω. The motor is operated as a shunt motor drawing an armature current of 48 A when delivering a torque of 188.5 N · m. The motor is operated as a compound motor delivering the same torque. Find the motor's speed assuming that the flux is increased by 20% due to compounding.

Solution

When operating as a shunt motor,

$$E_c = 240 - 48(0.5) = 216 \text{ V}$$

$$\omega = \frac{E_c I_a}{T} = \frac{216(48)}{188.5} = 55.003 \text{ rad/s}$$

$$K_1\phi_f = \frac{216}{55.003} = 3.9271$$

When operating as a compound motor,

$$T = K_1\phi_f I_a$$
$$188.5 = 3.9271(1.2)I_a$$
$$I_a = 40 \text{ A}$$

Thus

$$E_c = 240 - 40(0.5) = 220 \text{ V}$$

$$\omega = \frac{220}{3.9271(1.2)} = 46.684 \text{ rad/s}$$

Problem 7.A.9

The combined armature and series field resistance of a 600-V 150-hp dc series motor is 0.16 Ω. The motor speed is 600 r/min for an armature current of 200 A. The motor is to drive a load with the following torque–speed characteristic:

$$T_L = 105 \sqrt{\omega}$$

where ω is in rad/s and the torque is in N · m.

   a. Find the speed of the motor when driving this load.
   b. Compute the torque, horsepower output, and armature current corresponding to this load.

Solution

(a) The back EMF E_c is obtained for 200-A armature current as

$$E_c = 600 - 200(0.16) = 568 \text{ V}$$

But for a series motor,

$$E_c = K_1 K_2 I_a \omega$$

Thus

$$K_1 K_2 = \frac{568/200}{(2\pi/60)(600)}$$

$$= 45.2 \times 10^{-3}$$

The armature developed torque is related to the speed by the following relation for a series motor:

$$T_a = \frac{K_1 K_2 V_t^2}{[K_1 K_2 \omega + (R_a + R_s)]^2}$$

Thus

$$T_a = \frac{16.272 \times 10^3}{(0.16 + 45.2 \times 10^{-3}\omega)^2}$$

   The speed of the motor is obtained by matching the T_a and T_L expressions

$$T_L = 105 \sqrt{\omega} = \frac{16.272 \times 10^3}{(0.16 + 45.2 \times 10^{-3}\omega)^2} = T_a$$

The solution by trial and error is obtained as

$$\omega = 86.714 \text{ rad/s}$$

(b) The corresponding torque is

$$T_L = 977.764 \text{ N} \cdot \text{m}$$

The horsepower delivered is found to be

$$\text{hp} = 113.654$$

The armature current is obtained from

$$T = K_1 K_2 I_a^2$$

This turns out to be

$$I_a = 147.078 \text{ A}$$

PROBLEMS

Problem 7.B.1

The generated armature voltage of a separately excited dc generator is 230 V at a speed of 1000 r/min when the field current is 1.25 A.

   a. Find the generated armature voltage for a field current of 1.1 A at 1200 r/min.
   b. Find the speed in r/min when the generated armature voltage is 210 V for a field current of 1.2 A.
   c. Find the field current at 1200 r/min for a generated armature voltage of 205 V.

Problem 7.B.2

The generated armature voltage of a separately excited dc generator is originally 120 V. Find the voltage if the speed is decreased to 95% and the field current is increased to 110% of their original values.

Problem 7.B.3

For the generator of Example 7.1, it is required to:

a. Find the generated voltage for a field current of 2.7 A at 1100 r/min.
b. Find the field current when the generated voltage is 162 V at 1400 r/min.
c. Find the speed when the generated voltage is 145 V at a field current of 2.6 A.

Problem 7.B.4

For the generator of Problem 7.A.1, it is required to:

a. Find the field current that results in a generated voltage of 240 V at a speed of 1500 r/min.
b. Find the speed for a generated voltage of 250 V at a field current of 9 A.
c. Find the generated voltage for a field current of 8.5 A at a speed of 1700 r/min.

Problem 7.B.5

A 10-kW, 125-V separately excited dc generator is driven at a speed of 1000 r/min. The field current is 2 A when the generator delivers rated armature current. Find the generated armature voltage at full load. Find the required field current if the armature current is 75% of its rated value with the speed remaining constant. The armature resistance is 0.15 Ω.

Problem 7.B.6

a. A 30-kW 250-V separately excited dc generator has an armature resistance of 0.125 Ω. Find the armature current for rated terminal voltage and a generated armature voltage of 265 V.
b. If the armature current drops to 105 A with the generated armature voltage kept at its original value, find the power delivered to the load.

Problem 7.B.7

The generator of Problem 7.B.6 delivers an output power of 20 kW when the generated armature voltage is 265 V. Find the armature current and terminal voltage.

Problem 7.B.8

A dc series-connected generator has a combined armature and series field resistance of 0.16 Ω. The generator is rated at 8 kW and 125-V terminal voltage.

a. Find the generated armature voltage if the armature current is 40 A and the terminal voltage is kept at its rated value.
b. The generator delivers 6 kW when the generated armature voltage is 132 V. Find the corresponding terminal voltage.
c. Find the load power if the generated armature voltage is 127 V and the terminal voltage is 124.5 V.

Problem 7.B.9

The power input to a dc series generator is 27 kW while the output is 25 kW at a terminal voltage of 250 V. Find the combined armature and series field resistance as well as the generated armature voltage.

Problem 7.B.10

A 30-kW 250-V dc shunt generator has an armature resistance of 0.124 Ω and a field resistance of 160 Ω. Find the generated armature voltage and efficiency when the generator delivers rated power output.

Problem 7.B.11

The generator of Problem 7.B.10 operates at an efficiency of 92% with a terminal voltage of 250 V. Find the armature current, input, and output power of the machine.

Problem 7.B.12

A 50-kW 250-V dc shunt generator delivers rated load at rated voltage for a generated armature voltage of 255.1 V. Assume that the armature resistance is 0.025 Ω. Find the shunt field resistance. Find the generated armature voltage when the generator is delivering half-rated load at rated terminal voltage.

Problem 7.B.13

A 25-kW 250-V dc long shunt compound generator has an armature resistance of 0.1 Ω and a series field resistance of 0.05 Ω. The shunt field resistance is 125 Ω. When the machine is delivering rated load at rated voltage, find the generator's efficiency.

Problem 7.B.14

For the generator of Example 7.6, find the efficiency at 80% of rated load at a 600-V terminal voltage.

Problem 7.B.15

A 240-V dc motor has an armature resistance of 0.4 Ω. The armature current is 40 A at a steady speed of 1200 r/min.

a. Find the armature's back EMF.
b. Find the internal power of the armature and the developed torque.
c. Find the speed of the motor if the field flux is reduced to 75% of its value.
d. Find the torque of the motor if the field flux is reduced to 80% of its value corresponding to the conditions of part (b).

Problem 7.B.16

A 600-V 150-hp 600-r/min dc series motor has an armature and series field resistance of 0.16 Ω. The full-load current is 210 A.

a. Find the back EMF at full load.
b. Find the armature developed power and torque at full load.
c. Compute the efficiency of the motor, neglecting rotational losses.

Problem 7.B.17

A 230-V dc series motor has a line current of 80 A and a rated speed of 750 r/min. The armature circuit and series field resistances are 0.14 Ω and 0.11 Ω, respectively. It is required to calculate

a. The speed when the line current drops to 40 A.
b. The no-load speed for a line current of 2 A.
c. The speed at 150% rated load when the line current is 120 A and the series field flux is 125% of the full-load flux due to saturation.

Problem 7.B.18

The power developed by the armature of a 230-V dc series motor at a speed of 750 r/min is 16,800 W. The combined armature and field resistance is 0.25 Ω. Find the armature current and back EMF under these conditions. If the line current is reduced to half its previous value, find the new motor speed assuming that the field flux is proportional to the field current.

Problem 7.B.19

A 250-V 25-hp dc series motor takes a current of 85 A when its speed is 62.83 rad/s. The armature circuit resistance is 0.1 Ω and the series field resistance is 0.05 Ω.

  a. Find the speed when the line current is 100 A.
  b. Find the torque at a speed of 100 rad/s.

Problem 7.B.20

A 20-hp 1150-r/min 230-V dc shunt motor has an armature resistance of 0.188 Ω and a shunt field resistance of 144 Ω. It is required to compute:

  a. The line current if the motor is connected directly across the 230-V supply without a starting resistance.
  b. The resistance of a starting resistor which limits starting current to 150% of the full rated current of 74.6 A.
  c. Full-load back EMF.

Problem 7.B.21

The armature resistance of a 100-hp 600-V shunt motor is 0.1 Ω. The shunt field resistance is 600 Ω. The motor draws a line current of 11 A on no load at a speed of 1200 r/min. At full load, the line current is 138.148 A.

  a. Find the armature current and armature developed power on no load.
  b. Find the full-load speed and efficiency of the motor.

Problem 7.B.22

A 20-hp 230-V 1150-r/min dc shunt motor has an armature resistance of 0.3 Ω. The armature current is 36.7 A for a given load. As the mechanical load is increased, the field flux is increased by 15% with a corresponding increase in armature current to 75 A. It is required to find:

  a. The armature back EMF at the 36.7-A load.
  b. The armature back EMF and speed at the 75-A load.
  c. The armature power developed at both loads.

Problem 7.B.23

A 10-hp 230-V dc shunt motor has an armature circuit resistance of 0.30 Ω. The armature current is 36.7 A at a speed of 1150 r/min. An additional resistance of 1 Ω is inserted in series with the armature while the shunt field resistance of 128 Ω and torque output remain unchanged.

a. Find the resulting change in line current.
b. Find the speed at which the motor will run.

Problem 7.B.24

The rotational losses of a 50-hp 230-V dc shunt motor at full rated load are given by 1445 W. Assume that the armature resistance is 0.2 Ω and that the full-load speed is 1750 r/min. Find the armature current at rated load.

Problem 7.B.25

A 1000-hp 500-V dc shunt motor has an armature resistance of 0.007 Ω. At full-load power, the armature current is 1550 A at 750 r/min. Calculate the full-load rotational losses. Find the output power and motor speed if the armature current is 1200 A.

Problem 7.B.26

A 230-V 25-hp dc shunt motor has an armature resistance of 0.2 Ω. Find the maximum horsepower developed by the armature and the corresponding armature current. Find the full-load armature current. Neglect rotational losses.

Problem 7.B.27

The following data are available for a dc series motor:

$$V_t = 240 \text{ V} \qquad E_c = 217.2 \text{ V at 600 r/min}$$

$$R_a + R_s = 0.6 \ \Omega$$

Find the torque and speed of the motor when driving a load with the following torque–speed characteristic:

$$T_\ell = 3.9536\omega$$

where T_ℓ is in N $\cdot$ m and ω is in rad/s.

Problem 7.B.28

Repeat Problem 7.B.27 for a load with the following torque–speed characteristic:

$$T_\ell = 182.27 \times 10^{-3}\omega^2$$

Problem 7.B.29

A 600-V dc series motor has a combined armature and series field resistance of 0.16 Ω. The motor draws an armature current of 210 A when running at 600 r/min. Establish the torque-speed characteristic of the motor. The motor drives a load with the characteristic

$$T_\ell = A\omega^2$$

The speed is found to be 600 r/min. Find the constant A.

Problem 7.B.30

A 230-V dc series motor has a combined armature and series field resistance of 0.25 Ω. The armature current is 80 A at 750 r/min. Find the torque–speed characteristic of the motor. The motor drives a load with the following characteristic:

$$T_\ell = 100e^{a\omega}$$

Find a, given that the motor's speed is 725 r/min when driving this load.

Problem 7.B.31

The armature resistance of a dc shunt motor is 0.05 Ω. The armature current is found to be 180.5 A at a speed of 1300 r/min. Assume that the motor is operating off a 230-V supply. Find the torque and speed when the motor drives a load with the following torque–speed characteristic:

$$T_\ell = 638.25 \times 10^{-3}\omega$$

Where T_ℓ is in N $\cdot$ m and ω is in rad/s.

Problem 7.B.32

Repeat Problem 7.B.31 for a load characteristic given by

$$T_\ell = 78.69 \sqrt{\omega}$$

Problem 7.B.33

The armature resistance of a dc shunt motor is 0.3 Ω. The motor operates from a 230-V mains and draws an armature current of 36.7 A at 1150 r/min. Establish the torque-speed characteristic of the motor. Find the torque and speed of the motor when it drives a load with the characteristic

$$T_\ell = 0.5\omega$$

Problem 7.B.34

A 600-V dc shunt motor has a resistance of the armature circuit of 0.1 Ω. The motor draws an armature current of 137.148 A at 1174.528 r/min. Establish the torque–speed characteristic of the motor. The motor is driving a load with the following characteristic

$$T_\ell = A \sqrt{\omega}$$

Find A assuming that the motor's speed is 122 rad/s when driving this load.

Problem 7.B.35

The power delivered by a 230-V dc shunt motor is 20 hp at 1150 r/min. The armature resistance is 0.188 Ω. Establish the torque–speed characteristic of the motor. Assume that the motor is driving a load with the torque–speed characteristic

$$T_\ell = a + 0.2\omega$$

Find a such that the motor runs at a speed of 125 rad/s when driving this load.

Problem 7.B.36

For the motor of Problem 7.B.16, find the additional series resistance required to produce full-load torque at a speed of 580 r/min.

Problem 7.B.37

For the motor of Problem 7.B.27, find the additional series resistance required to carry the load at a speed of 48 rad/s.

Problem 7.B.38

Repeat Problem 7.B.36 using armature shunt resistance control assuming that armature resistance is 0.1 Ω.

Problem 7.B.39

Repeat Problem 7.B.37 using armature shunt resistance control assuming that the armature resistance is 0.4 Ω.

Problem 7.B.40

For the motor of Problems 7.B.36 and 7.B.38, assume that $R_c = 150 \ \Omega$ is available. Find the required value of R_b in a series shunt resistance control scheme to achieve the required specification.

Problem 7.B.41

For the motor of Problems 7.B.37 and 7.B.39 assume that $R_c = 50\ \Omega$ is available. Find the required value of R_b in a series shunt resistance control scheme to achieve the required specifications.

Problem 7.B.42

Assume that a shunt diverter with $\alpha = 0.9$ is used for the motor of Problem 7.B.16. Find the resulting torque at a speed of 580 r/min. Assume that $R_a = 0.1\ \Omega$.

Problem 7.B.43

Assume that a shunt diverter with $\alpha = 0.85$ is used for the motor of Problem 7.B.27. Find the resulting torque at 48 rad/s assuming that the armature resistance is 0.4 Ω.

Problem 7.B.44

Consider the shunt motor of Problem 7.B.20. Find the full-load torque. Assume that field resistance control is used with $R_e = 10\ \Omega$. Find the speed in r/min for the same torque as in the full-load case.

Problem 7.B.45

Assume for the motor of Problem 7.B.44 that an armature resistance $R_b = 0.05\ \Omega$ is used in addition to the extra field resistance of 10 Ω. Find the resulting speed in r/min for the same torque.

Problem 7.B.46

For the motor of Problem 7.B.44, series and shunt armature resistance control is used with $R_b = 0.05\ \Omega$ and $R_c = 2\ \Omega$. Find the speed in r/min for full-load torque as obtained in Problem 7.B.44.

Problem 7.B.47

Find R_e for the motor of Problem 7.B.44 such that the motor carries full-load torque at a speed of 1200 r/min.

Problem 7.B.48

Find R_b for the motor of Problem 7.B.45 such that the speed is 1180 r/min for the same torque.

Problem 7.B.49

Assume that the motor of Problem 7.B.46 is controlled with the same values of R_c and R_b. Find the load torque if the speed is 1080 r/min.

Problem 7.B.50

For the motor of Problem 7.A.7, evaluate the constants K_1K_4 and K_1K_3 and hence determine the torque developed for the given conditions using Eq. (7.94).

Problem 7.B.51

For the motor of Problems 7.A.7 and 7.B.50, assume that shunt field resistance control with $R_e = 20 \ \Omega$ is used. Find the torque for the speed given in Problem 7.A.7.

Problem 7.B.52

For the motor of Problem 7.B.51, assume that additional series armature resistance of 0.05 Ω is used in conjunction with R_e. Find the new value of the torque for the same speed.

Problem 7.B.53

For the motor of Problem 7.B.51, assume that series and shunt resistance control is used with $R_c = 100 \ \Omega$ and $R_b = 0.05 \ \Omega$. Assume that $R_s = 0.4 \ \Omega$ and $R_a = 0.2 \ \Omega$. Find the armature current and the torque for the speed of 41.462 rad/s.

Problem 7.B.54

For the motor of Problem 7.B.51, find the additional shunt field resistance R_e required to drive a load with torque of 215 N · m at the same specified speed.

Problem 7.B.55

For the motor of Problem 7.B.52, find R_b such that a load torque of 210 N · m is carried at the same specified speed.

Problem 7.B.56

For the motor of Example 7.26, find the required number of starter steps to obtain

$$I_{max} = 198 \text{ A} \qquad I_{min} = 100 \text{ A}$$

Assume that the magnetization characteristic of the motor is linear for $I <$ 162 A.

Problem 7.B.57

For the design of Problem 7.B.56, find the actual minimum current by interpolating between 110 and 130 A. Find the required starter steps.

E
I
G
H
T

Fractional-Horsepower
AC Motors

8.1 INTRODUCTION

In this chapter we discuss motors of the fractional-horsepower class which are in common use for applications requiring low power output, small size, and reliability. Standard ratings for this class range from $\frac{1}{20}$ to 1 hp. Motors rated for less than $\frac{1}{20}$ hp are called subfractional-horsepower motors and are rated in millihorsepower and range from 1 to 35 mhp. These small motors provide power for all types of equipment in the home, office, and commercial installations. The majority are of the induction-motor type and operate from a single-phase supply. Our discussion begins with an examination of rotating magnetic fields in single-phase induction motors.

8.2 ROTATING MAGNETIC FIELDS IN SINGLE-PHASE INDUCTION MOTORS

In studying the polyphase induction motor in Chapter 5, we discussed the production of a rotating magnetic field due to three-phase stator windings that are displaced by 120° in space as well as in time. To understand the operation of common single-phase induction motors, it is necessary to start by discussing two-phase induction machines. In a true two-phase machine two stator windings, labeled *AA' and BB'*, are placed at 90° spatial displacement as shown in Figure 8.1. The voltages v_A and v_B form a set of balanced two-phase voltages with a 90° time (or phase) displacement. Assuming that

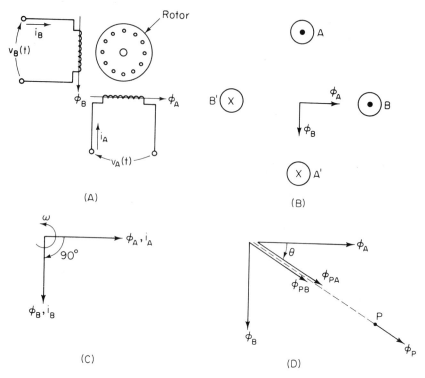

FIGURE 8.1 Rotating magnetic field in a balanced two-phase stator: (a) winding schematic;   (b) flux orientation;   (c) phasor diagram;   (d) space phasor diagram.

the two windings are identical, then the resulting flux ϕ_A and ϕ_B are given by

$$\phi_A = \phi_M \cos \omega t \tag{8.1}$$

$$\phi_B = \phi_M \cos (\omega t - 90°) = \phi_M \sin \omega t \tag{8.2}$$

where ϕ_M is the peak value of the flux. In Figure 8.1(b) the flux ϕ_A is shown to be at right angles to ϕ_B in space. It is clear that because of Eqs. (8.1) and (8.2), the phasor relation between ϕ_A and ϕ_B is shown in Figure 8.1(c) with ϕ_A taken as the reference phasor.

The resultant flux ϕ_P at a point P displaced by a spatial angle θ from the reference is given by

$$\phi_P = \phi_{PA} + \phi_{PB}$$

where ϕ_{PA} is the component of ϕ_A along the OP axis and ϕ_{PB} is the component of ϕ_B along the OP axis, as shown in Figure 8.1(d). Here we have

$$\phi_{PA} = \phi_A \cos \theta$$

$$\phi_{PB} = \phi_B \sin \theta$$

As a result, we have

$$\phi_P = \phi_M(\cos \omega t \cos \theta + \sin \omega t \sin \theta)$$

The relationship above can be written alternatively as

$$\phi_P = \phi_M \cos (\theta - \omega t) \qquad (8.3)$$

The flux at point P is a function of time and the spatial angle θ, and has a constant amplitude ϕ_M. This result is similar to that obtained in Chapter 5 for the balanced three-phase induction motor.

The flux ϕ_P can be represented by a phasor ϕ_M that is coincident with the axis of phase a at $t = 0$. The value of ϕ_P is $\phi_M \cos \theta$ at that instant as shown in Figure 8.2(a). At the instant $t = t_1$, the phasor ϕ_M has rotated an angle ωt_1 in the positive direction of θ, as shown in Figure 8.2(b). The value of ϕ_P is seen to be $\phi_M \cos (\theta - \omega t_1)$ at that instant. It is thus clear that the flux waveform is a rotating field that travels at an angular velocity ω in the forward direction of increase in θ.

The result obtained here for a two-phase stator winding set and for a three-phase stator winding set can be extended to an N-phase system. In this case the N windings are placed at spatial angles of $2\pi/N$ and excited by sinusoidal voltages of time displacement $2\pi/N$. Our analysis proceeds as follows. The flux waveforms are given by

$$\phi_1 = \phi_M \cos \omega t$$

$$\phi_2 = \phi_M \cos\left(\omega t - \frac{2\pi}{N}\right)$$

$$\cdot$$
$$\cdot$$
$$\cdot$$

$$\phi_i = \phi_M \cos\left[\omega t - (i - 1)\frac{2\pi}{N}\right]$$

The resultant flux at a point P is given by the sum

$$\phi_P = \sum_{i=1}^{N} \phi_{Pi}$$

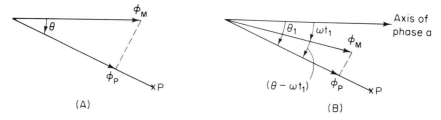

FIGURE 8.2   Illustrating of forward rotating magnetic field: (a) $t = 0$;   (b) $t = t_1$.

Each of the individual components is given by

$$\phi_{Pi} = \phi_i \cos\left[\theta - (i - 1)\frac{2\pi}{N}\right]$$

Thus we have

$$\phi_{Pi} = \phi_M \cos\left[\theta - (i - 1)\frac{2\pi}{N}\right] \cos\left[\omega t - (i - 1)\frac{2\pi}{N}\right]$$

This reduces to

$$\phi_{Pi} = \phi_M \cos(\theta - \alpha_i) \cos(\omega t - \alpha_i)$$

or

$$\phi_{pi} = \frac{\phi_M}{2}\left[\cos(\omega t - \theta) - \cos(\omega t + \theta - 2\alpha_i)\right]$$

where we define

$$\alpha_i = (i - 1)\frac{2\pi}{N}$$

It is thus clear that the resultant flux is

$$\phi_P = \frac{\phi_M}{2}\left[\sum_{i=1}^{N} \cos(\omega t - \theta) - \sum_{i=1}^{N} \cos(\omega t + \theta - 2\alpha_i)\right]$$

The second sum is zero, and we conclude that

$$\phi_P = \frac{N\phi_M}{2}\cos(\theta - \omega t) \tag{8.4}$$

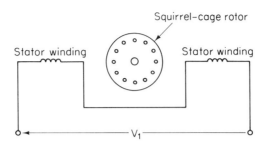

FIGURE 8.3   Schematic of a single-phase induction motor.

A rotating magnetic field of constant magnitude will be produced by an N-phase winding excited by balanced N-phase currents when each phase is displaced $2\pi/N$ electrical degrees from the next phase in space.

In order to understand the operation of a single-phase induction motor, we consider the configuration shown in Figure 8.3. The stator carries a single-phase winding and the rotor is of the squirrel-cage type. This configuration corresponds to a motor that has been brought up to speed, as will be discussed presently.

Let us now consider a single-phase stator winding as shown in Figure 8.4(a). The flux ϕ_A is given by

$$\phi_A = \phi_M \cos \omega t \tag{8.5}$$

The flux at point P displaced by angle θ from the axis of phase a is clearly given by

$$\phi_P = \phi_A \cos \theta$$

Using Eq. (8.5), we obtain

$$\phi_P = \phi_M \cos \theta \cos \omega t \tag{8.6}$$

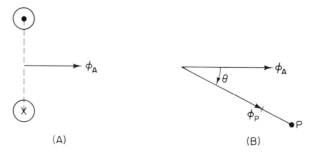

FIGURE 8.4   (a) Single-phase winding;   (b) the flux at a point P.

An alternative form of Eq. (8.6) is given by

$$\phi_P = \frac{\phi_M}{2} [\cos(\theta - \omega t) + \cos(\theta + \omega t)] \qquad (8.7)$$

The flux at point P can therefore be seen to be the sum of two waveforms ϕ_f and ϕ_b given by

$$\phi_f = \frac{\phi_M}{2} \cos(\theta - \omega t) \qquad (8.8)$$

$$\phi_b = \frac{\phi_M}{2} \cos(\theta + \omega t) \qquad (8.9)$$

The waveform ϕ_f is of the same form as that obtained in Eq. (8.3), which was shown to be rotating in the forward direction (increase in θ from the axis of phase a). The only difference between Eqs. (8.8) and (8.3) is that the amplitude of ϕ_f is half of that of ϕ_P in Eq. (8.3). The subscript f in Eq. (8.8) signifies the fact that $\cos(\theta - \omega t)$ is forward rotating wave.

Consider now the waveform ϕ_b of Eq. (8.9). At $t = 0$, the value of ϕ_b is $(\phi_M/2) \cos \theta$ and is represented by the phasor $(\phi_M/2)$, which is coincident with the axis of phase a as shown in Figure 8.5(a). Note that at $t = 0$, both ϕ_f and ϕ_b are equal in value. At a time instant $t = t_1$, the phasor $(\phi_M/2)$ is seen to be at angle ωt_1 with the axis of phase a, as shown in Figure 8.4(b). The waveform ϕ_b can therefore be seen to be rotating at an angular velocity ω in a direction opposite to that of ϕ_f and we refer to ϕ_b as a backward-rotating magnetic field. The subscript (b) in Eq. (8.9) signifies the fact that $\cos(\theta + \omega t)$ is a backward-rotating wave.

The result of our discussion is that in a single-phase induction machine there are two magnetic fields rotating in opposite directions. Each field produces an induction-motor torque in a direction opposite to the other. If the rotor is at rest, the forward torque is equal and opposite to the

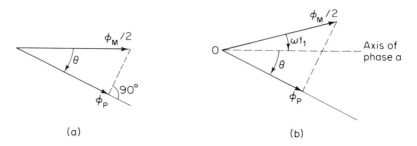

FIGURE 8.5   Showing that ϕ_b is a backward-rotating wave: (a) $t = 0$;   (b) $t = t_1$.

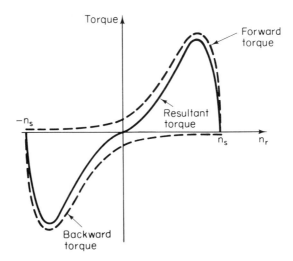

FIGURE 8.6   Torque–speed characteristics of a single-phase induction motor.

backward torque and the resulting torque is zero. What we have just discovered is that a single-phase induction motor is incapable of producing a
torque at rest and is therefore not a self-starting machine. If the rotor is
made to rotate by an external means, each of the two fields would produce
a torque–speed characteristic similar to a balanced three-phase (or two-
phase) induction motor, as shown in Figure 8.6 in the dashed curves. The
resultant torque–speed characteristic is shown in a solid line. The foregoing
argument will be confirmed once we develop an equivalent circuit for the
single-phase induction motor.

8.3   EQUIVALENT CIRCUITS FOR SINGLE-PHASE INDUCTION MOTORS

In Section 8.2 we have seen that in a single-phase induction motor, the
pulsating flux wave resulting from the single winding stator MMF is equal
to the sum of two rotating flux components. The first component is referred
to as the forward field and has a constant amplitude equal to half of that of
the stator waveform. The forward field rotates at synchronous speed. The
second component, referred to as the backward field, is of the same constant
amplitude but rotates in the opposite (or backward) direction at synchronous speed. Each component induces its own rotor current and creates induction motor action in the same manner as in a balanced three-phase induction motor. It is on this basis that we conceive of the circuit model of

Figure 8.7(a). Note that R_1 and X_1 are the stator resistance and leakage reactance, respectively, and V_1 is the stator input voltage. the EMF E_1 is assumed to be the sum of two components, E_{1f} and E_{1b}, corresponding to the forward and backward field waves, respectively. Note that since the two waves have the same amplitude, we have

$$E_{1f} = E_{1b} = \frac{E_1}{2} \tag{8.10}$$

The rotor circuit is modeled as the two blocks shown in Figure 8.7(a), representing the rotor forward circuit model Z_f and the rotor backward circuit model Z_b, respectively.

The model of the rotor circuit for the forward rotating wave Z_f is simple since we are essentially dealing with induction-motor action and the rotor is set in motion in the same direction as the stator synchronous speed. The model of Z_f is shown in Figure 8.7(b) and is similar to that of the rotor of a balanced three-phase induction motor. The impedances dealt with are half of the actual values to account for the division of E_1 into two equal voltages. In this model X_m is the magnetizing reactance, R'_2 and X'_2 are rotor resistance and leakage reactance, both referred to the stator side. The slip s_f is given by

$$s_f = \frac{n_s - n_r}{n_s} \tag{8.11}$$

This is the standard definition of slip as the rotor is revolving in the same direction as that of the forward flux wave.

The model of the rotor circuit for the backward-rotating wave Z_b is shown in Figure 8.7(c) and is similar to that of Z_f, with the exception of the backward slip, denoted by s_b. The backward wave is rotating at a speed of $-n_s$ and the rotor is rotating at n_r. We thus have

$$s_b = \frac{(-n_s) - n_r}{-n_s} = 1 + \frac{n_r}{n_s} \tag{8.12}$$

Using Eq. (8.11), we have

$$s_f = 1 - \frac{n_r}{n_s} \tag{8.13}$$

As a result, we conclude that the slip of the rotor with respect to the backward wave is related to its slip with respect to the forward wave by

$$s_b = 2 - s_f \tag{8.14}$$

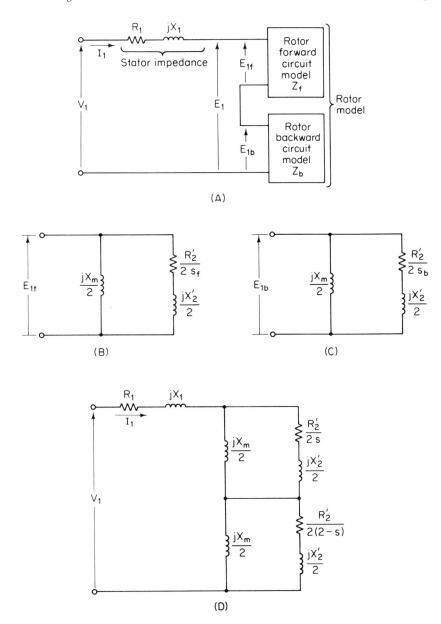

FIGURE 8.7   Developing an equivalent circuit of for single-phase induction motors: (a) basic concept;   (b) forward model;   (c) backward model;   (d) complete equivalent circuit.

We now let s be the forward slip,

$$s_f = s \tag{8.15}$$

and thus

$$s_b = 2 - s \tag{8.16}$$

On the basis of Eqs. (8.15) and (8.16), a complete equivalent circuit as shown in Figure 8.7(d) is now available. The core losses in the present model are treated separately in the same manner as the rotational losses.

The forward impedance Z_f is obtained as the parallel combination of $(jX_m/2)$ and $[(R'_2/2s) + j(X'_2/2)]$, given by

$$Z_f = \frac{j(X_m/2) \, [(R'_2/2s) + j(X'_2/2)]}{(R'_2/2s) + j[(X_m + X'_2)/2]} \tag{8.17}$$

Similarly, for the backward impedance, we get

$$Z_b = \frac{j(X_m/2) \, \{[R'_2/2(2 - s)] + j(X'_2/2)]\}}{[R'_2/2(2 - s)] + j[(X_m + X'_2)/2]} \tag{8.18}$$

Note that with the rotor at rest, $n_r = 0$, and thus with $s = 1$, we get $Z_f = Z_b$. An example is appropriate now.

Example 8.1   The following parameters are available for a 60-Hz four-pole single-phase 110-V ½-hp induction motor:

$$R_1 = 1.5 \ \Omega \qquad R'_2 = 3 \ \Omega$$

$$X_1 = 2.4 \ \Omega \qquad X'_2 = 2.4 \ \Omega$$

$$X_m = 73.4 \ \Omega$$

Calculate Z_f, Z_b, and the input impedance of the motor at a slip of 0.05.

Solution

$$Z_f = \frac{j36.7(30 + j1.2)}{30 + j37.9} = 22.796 \, \underline{/40.654°}$$

$$= 17.294 + j14.851 \ \Omega$$

The result above is a direct application of Eq. (8.17). Similarly, using Eq. (8.18), we get

$$Z_b = \frac{j36.7[(1.5/1.95) + j1.2]}{(1.5/1.95) + j37.9} = 1.38 \underline{/58.502°}\ \Omega$$

$$= 0.721 + j1.1766\ \Omega$$

We observe here that $|Z_f|$ is much larger than $|Z_b|$ at this slip, in contrast to the situation at starting $(s = 1)$, for which $Z_f = Z_b$.

The input impedance Z_i is obtained as

$$Z_i = Z_1 + Z_f + Z_b = 19.515 + j18.428$$

$$= 26.841 \underline{/43.36°}\ \Omega$$

Equations (8.17) and (8.18) yield the forward and backward imped-ances on the basis of complex number arithmetic. The results can be written in the rectangular forms

$$Z_f = R_f + jX_f \tag{8.19}$$

and

$$Z_b = R_b + jX_b \tag{8.20}$$

Using Eq. (8.17), we can write

$$2R_f = \frac{a_f X_m^2}{a_f^2 + X_t^2} \tag{8.21}$$

and

$$X_f = \frac{R_f}{a_f X_m}(a_f^2 + X_t X_2') \tag{8.22}$$

where

$$a_f = \frac{R_2'}{s} \tag{8.23}$$

$$X_t = X_2' + X_m \tag{8.24}$$

In a similar manner we have, using Eq. (8.18),

$$2R_b = \frac{a_b X_m^2}{a_b^2 + X_t^2} \tag{8.25}$$

$$X_b = \frac{R_b}{a_b X_m}(a_b^2 + X_t X_2') \qquad (8.26)$$

where

$$a_b = \frac{R_2'}{2 - s} \qquad (8.27)$$

It is often desirable to introduce some approximations in the formulas just derived. As is the usual case, for $X_t > 10\, a_b$, we can write an approximation to Eq. (8.25) as

$$2R_b \simeq a_b\left(\frac{X_m}{X_t}\right)^2 \qquad (8.28)$$

As a result, by substitution in Eq. (8.26), we get

$$X_b \simeq \frac{X_2' X_m}{2X_t} + \frac{a_b R_b}{X_m} \qquad (8.29)$$

We can introduce further simplifications by assuming that $X_m/X_t \simeq 1$, to obtain from Eq. (8.28)

$$2R_b \simeq a_b = \frac{R_2'}{2 - s} \qquad (8.30)$$

Equation (8.29) reduces to the approximate form

$$X_b \simeq \frac{X_2'}{2} + \frac{a_b^2}{2X_m} \qquad (8.31)$$

Of course, if we neglect the second term in Eq. (8.31), we obtain the most simplified representation of the backward impedance as

$$R_b = \frac{R_2'}{2(2 - s)} \qquad (8.32)$$

$$X_b = \frac{X_2'}{2} \qquad (8.33)$$

Equations (8.32) and (8.33) imply that $X_m/2$ is considered an open circuit in the backward field circuit, as shown in Figure 8.8.

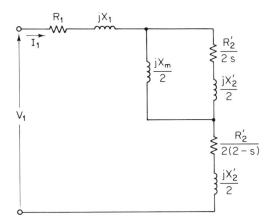

FIGURE 8.8   Approximate equivalent circuit of a single-phase induction motor.

8.4   POWER AND TORQUE RELATIONS

The development of an equivalent-circuit model of a running single-phase induction motor enables us to quantify power and torque relations in a simple way. The power input to the stator P_i is given by

$$P_i = V_1 I_1 \cos \phi_1 \tag{8.34}$$

where ϕ_1 is the phase angle between V_1 and I_1. Part of this power will be dissipated in stator ohmic losses, P_{ℓ_s}, given by

$$P_{\ell_s} = I_1^2 R_1 \tag{8.35}$$

The core losses will be accounted for as a fixed loss and is treated in the same manner as the rotational losses at the end of the analysis. The air-gap power P_g is thus given by

$$P_g = P_i - P_{\ell_s} \tag{8.36}$$

The air-gap power is the power input to the rotor circuit and can be visualized to be made up of two components. The first component is the power taken up by the forward field and is denoted by P_{gf}, and the second is the backward field power denoted by P_{gb}. Thus we have

$$P_g = P_{gf} + P_{gb} \tag{8.37}$$

As we have modeled the forward field circuit by an impedance Z_f, it is natural to write

$$P_{gf} = |I_1|^2 R_f \tag{8.38}$$

Similarly, we write

$$P_{gb} = |I_1|^2 R_b \tag{8.39}$$

The ohmic losses in the rotor circuit are treated in a similar manner. The losses in the rotor circuit due to the forward field $P_{\ell_{rf}}$ can be written as

$$P_{\ell_{rf}} = s_f P_{gf} \tag{8.40}$$

Similarly, the losses in the rotor circuit due to the backward field $P_{\ell_{rb}}$ are written as

$$P_{\ell_{rb}} = s_b P_{gb} \tag{8.41}$$

Equations (8.40) and (8.41) are based on arguments similar to those used with the balanced three-phase induction motor. Specifically, the total rotor equivalent resistance in the forward circuit is given by

$$R_{rf} = \frac{R_2'}{2s_f} \tag{8.42}$$

This is written as

$$R_{rf} = \frac{R_2'}{2} + \frac{R_2'(1 - s_f)}{2s_f} \tag{8.43}$$

The first term corresponds to the rotor ohmic loss due to the forward field and the second represents the power to mechanical load and fixed losses. It is clear from Figure 8.9 that

$$P_{\ell_{rf}} = |I_{rf}|^2 \frac{R_2'}{2} \tag{8.44}$$

and

$$P_{gf} = |I_{rf}|^2 \frac{R_2'}{2s_f} \tag{8.45}$$

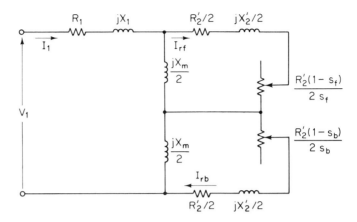

FIGURE 8.9   Equivalent circuit of single-phase induction motor showing rotor loss components in the forward and backward circuits.

Combining Eqs. (8.44) and (8.45), we get (8.40). A similar argument leads to Eq. (8.41). It is noted here that Eqs. (8.38) and (8.45) are equivalent, since the active power to the rotor circuit is consumed only in the right-hand branch, with $jX_m/2$ being a reactive element.

The net power from the rotor circuit is denoted by P_m and is given by

$$P_m = P_{mf} + P_{mb} \qquad (8.46)$$

The component P_{mf} is due to the forward circuit and is given by

$$P_{mf} = P_{gf} - P_{\ell rf} \qquad (8.47)$$

Using Eq. (8.40), we get

$$P_{mf} = (1 - s_f)P_{gf} \qquad (8.48)$$

Similarly, P_{mb} is due to the backward circuit and is given by

$$P_{mb} = P_{gb} - P_{\ell rb} \qquad (8.49)$$

Using Eq. (8.41), we get

$$P_{mb} = (1 - s_b)P_{gb} \qquad (8.50)$$

Recall that

$$s_f = s$$

$$s_b = 2 - s$$

As a result,

$$P_{m_f} = (1 - s)P_{g_f} \tag{8.51}$$

$$P_{m_b} = (s - 1)P_{g_b} \tag{8.52}$$

We now substitute Eqs. (8.51) and (8.52) into (8.46), to obtain

$$P_m = (1 - s)(P_{g_f} - P_{g_b}) \tag{8.53}$$

The shaft power output P_o can now be written as

$$P_o = P_m - P_{\text{rot}} - P_{\text{core}} \tag{8.54}$$

The rotational losses are denoted by P_{rot} and the core losses are denoted by P_{core}.

The output torque T_o is obtained as

$$T_o = \frac{P_o}{\omega_r} \tag{8.55}$$

If fixed losses are neglected, then

$$T_m = \frac{P_m}{\omega_s(1 - s)} \tag{8.56}$$

As a result, using Eq. (8.53), we get

$$T_m = \frac{1}{\omega_s}(P_{g_f} - P_{g_b}) \tag{8.57}$$

The torque due to the forward field is

$$T_{m_f} = \frac{P_{m_f}}{\omega_r} = \frac{P_{g_f}}{\omega_s} \tag{8.58}$$

The torque due to the backward field is

$$T_{m_b} = \frac{P_{m_b}}{\omega_r} = -\frac{P_{g_b}}{\omega_s} \tag{8.59}$$

It is thus clear that the net mechanical torque is the algebraic sum of a forward torque T_{m_f} (positive) and a backward torque T_{m_b} (negative). Note that at starting, $s = 1$ and $R_f = R_b$, and as a result $P_{gf} = P_{gb}$, giving zero output torque. This confirms our earlier statements about the need for starting mechanisms for a single-phase induction motor. This is treated in Section 8.5. At the present time we deal with an example.

Example 8.2 For the single-phase induction motor of Example 8.1, it is required to find the power and torque output and the efficiency when running at a slip of 5%. Neglect core and rotational losses.

Solution

In example 8.1 we obtained

$$Z_i = 26.841 \underline{/43.36°}$$

As a result, with $V_1 = 110 \underline{/0}$, we obtain

$$I_1 = \frac{110 \underline{/0}}{26.841 \underline{/43.36°}} = 4.098 \underline{/-43.36°} \text{ A}$$

The power factor is thus

$$\cos \phi_1 = \cos 43.36° = 0.727$$

The power input is

$$P_1 = V_1 I_1 \cos \phi_1 = 327.76 \text{ W}$$

We have from Example 8.1 for $s = 0.05$,

$$R_f = 17.294 \text{ } \Omega \qquad R_b = 0.721 \text{ } \Omega$$

Thus we have

$$P_{gf} = |I_1|^2 R_f = (4.098)^2(17.294) = 290.46 \text{ W}$$

$$P_{gb} = |I_1|^2 R_b = (4.098)^2(0.721) = 12.109 \text{ W}$$

The output power is thus obtained as

$$P_m = (1 - s)(P_{gf} - P_{gb})$$

$$= 0.95(290.46 - 12.109) = 264.43 \text{ W}$$

As we have a four-pole machine, we get

$$n_s = \frac{120(60)}{4} = 1800 \text{ r/min}$$

$$\omega_s = \frac{2\pi n_s}{60} = 188.5 \text{ rad/s}$$

The output torque is therefore obtained as

$$T_m = \frac{1}{\omega_s}(P_{gf} - P_{gb})$$

$$= \frac{290.46 - 12.109}{188.5} = 1.4767 \text{ N} \cdot \text{m}$$

The efficiency is now calculated as

$$\eta = \frac{P_m}{P_1} = \frac{264.43}{327.76} = 0.8068$$

It is instructive to account for the losses in the motor. Here we have the static ohmic losses obtained as

$$P_{\ell_s} = |I_1|^2 R_1 = (4.098)^2(1.5) = 25.193 \text{ W}$$

The forward rotor losses are

$$P_{\ell_{rf}} = sP_{gf} = 0.05(290.46) = 14.523 \text{ W}$$

The backward rotor losses are

$$P_{\ell_{rb}} = (2 - s)\, P_{gb} = 1.95(12.109) = 23.613 \text{ W}$$

The sum of the losses is

$$P_\ell = 25.193 + 14.523 + 23.613 = 63.329 \text{ W}$$

The power output and losses should match the power input:

$$P_m + P_\ell = 264.43 + 63.33 = 327.76 \text{ W}$$

which is indeed the case. This concludes the solution to the example.

8.5   STARTING SINGLE-PHASE INDUCTION MOTORS

We have shown earlier that a single-phase induction motor with one stator winding is not capable of producing a torque at starting [see, for example, Eq. (8.53) with $s = 1$]. Once the motor is running, it will continue to do so, since the forward field torque dominates that of the backward field component. We have also seen that with two stator windings that are displaced by 90° in space and with two-phase excitation a purely forward rotating field is produced, and this form of a motor (like the balanced three-phase motor) is self-starting.

Methods of starting a single-phase induction motor rely on the fact that given two stator windings displaced by 90° in space, a starting torque will result if the flux in one of the windings lags that of the other by a certain phase angle ψ. To verify this, we consider the situation shown in Figure 8.10. Assume that

$$\phi_A = \phi_M \cos \omega t \tag{8.60}$$

$$\phi_B = \phi_M \cos(\omega t - \psi) \tag{8.61}$$

Clearly, the flux at P is given by the sum of ϕ_{P_A} and ϕ_{P_B}, where

$$\phi_{P_A} = \phi_A \cos \theta \tag{8.62}$$

$$\phi_{P_B} = \phi_B \sin \theta \tag{8.63}$$

As a result, we have

$$\phi_{P_A} = \frac{\phi_M}{2} \left[\cos(\theta - \omega t) + \cos(\theta + \omega t) \right] \tag{8.64}$$

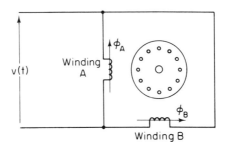

FIGURE 8.10   Two stator windings to explain the starting mechanism of single-phase induction motors.

$$\phi_{PB} = \frac{\phi_M}{2} \{\cos \psi [\sin(\theta + \omega t) + \sin(\theta - \omega t)]$$

$$+ \sin \psi [\cos(\theta - \omega t) - \cos(\theta + \omega t)]\} \qquad (8.65)$$

The flux at P is therefore obtained as

$$\phi_P = \frac{\phi_M}{2} [a_{f_r} \cos(\theta - \omega t) + a_{f_i} \sin(\theta - \omega t)$$

$$+ a_{b_r} \cos(\theta + \omega t) + a_{b_i} \sin(\theta + \omega t)] \qquad (8.66)$$

where we have

$$a_{f_r} = 1 + \sin \psi \qquad (8.67)$$

$$a_{f_i} = \cos \psi \qquad (8.68)$$

$$a_{b_r} = 1 - \sin \psi \qquad (8.69)$$

$$a_{b_i} = \cos \psi \qquad (8.70)$$

Note that we can also define the magnitudes a_f and a_b by

$$a_f^2 = a_{f_r}^2 + a_{f_i}^2 = 2(1 + \sin \psi) \qquad (8.71)$$

$$a_b^2 = a_{b_r}^2 + a_{b_i}^2 = 2(1 - \sin \psi) \qquad (8.72)$$

The angles α_f and α_b are defined next by

$$\cos \alpha_f = \frac{a_{f_r}}{a_f} = \sqrt{\frac{1 + \sin \psi}{2}} \qquad (8.73)$$

$$\cos \alpha_b = \frac{a_{b_r}}{a_b} = \sqrt{\frac{1 - \sin \psi}{2}} \qquad (8.74)$$

$$\sin \alpha_f = \frac{a_{f_i}}{a_f} = \frac{\cos \psi}{\sqrt{2(1 + \sin \psi)}} \qquad (8.75)$$

$$\sin \alpha_b = \frac{a_{b_i}}{a_b} = \frac{\cos \psi}{\sqrt{2(1 - \sin \psi)}} \qquad (8.76)$$

We can now write the flux ϕ_P as

$$\phi_P = \frac{\phi_M}{2} [a_f \cos(\theta - \omega t + \alpha_f) + a_b \cos(\theta + \omega t - \alpha_b)] \qquad (8.77)$$

It is clear that ϕ_P is the sum of a forward rotating component ϕ_f and a backward rotating component ϕ_b given by

$$\phi_P = \phi_f(t) + \phi_b(t) \tag{8.78}$$

where

$$\phi_f(t) = \frac{a_f \phi_M}{2} \cos(\theta - \omega t + \alpha_f) \tag{8.79}$$

$$\phi_b(t) = \frac{a_b \phi_M}{2} \cos(\theta + \omega t - \alpha_b) \tag{8.80}$$

Let us note here that from Eqs. (8.71) and (8.72), we can see that

$$a_f > a_b \tag{8.81}$$

As a result, the magnitude of the forward rotating wave is larger than that of the backward rotating wave. Equation (8.3) for a two-phase machine is a special case of Eq. (8.77) with $\psi = 90°$, and hence, $a_b = 0$, as can be seen from Eq. (8.72). It is now clear that for the arrangement of Figure 8.10, a starting torque should result. This is the basis of the starting mechanisms for single-phase induction motors discussed in this section.

Single-phase induction motors are referred to by names that describe the method of starting. A number of types of single-phase induction motors are now discussed.

Split-Phase Motors

A single-phase induction motor with two distinct windings on the stator that are displaced in space by 90 electrical degrees is called a split-phase motor. The main (or running) winding has a lower R/X ratio than the auxiliary (or starting) winding. A starting switch disconnects the auxiliary winding when the motor is running at approximately 75 to 80% of synchronous speed. The switch is centrifugally operated. The rotor of a split-phase motor is of the squirrel-cage type. At starting, the two windings are connected in parallel across the line as shown in Figure 8.11.

The split-phase design is one of the oldest single-phase motors and is most widely used in the ratings of 0.05 to 0.33 hp. A split-phase motor is used in machine tools, washing machines, oil burners, and blowers, to name just a few of its applications.

The torque–speed characteristic of a typical split-phase induction motor is shown in Figure 8.12. At starting the torque is about 150% of its full-

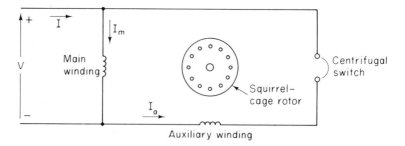

FIGURE 8.11 Schematic diagram of a split-phase induction motor.

load value. As the motor speed picks up, the torque is increased (except for a slight decrease at low speed) and may reach higher than 250% of full-load value. The switch is opened and the motor runs on its main winding alone and the motor reaches its equilibrium speed when the torque developed is matched by the load.

Capacitor-Start Motors

The class of single-phase induction motors in which the auxiliary winding is connected in series with a capacitor is referred to as that of capacitor motors. The auxiliary winding is placed 90 electrical degrees from the main winding. There are three distinct types of capacitor motors in common practice. The first type, which we discuss presently, employs the auxiliary winding and capacitor only during starting and is thus called a capacitor-start

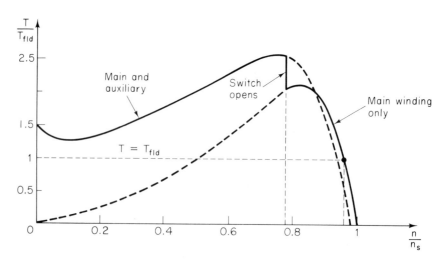

FIGURE 8.12 Torque–speed characteristic of a split-phase induction motor.

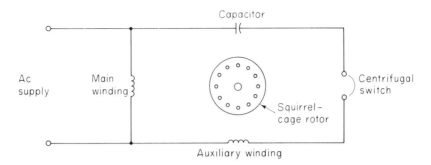

FIGURE 8.13   Capacitor-start motor.

motor. It is thus clear that a centrifugal switch that opens at 75 to 80% of synchronous speed is used in the auxiliary winding circuit (sometimes called the capacitor phase). A sketch of the capacitor-start motor connection is shown in Figure 8.13. A commercial capacitor-start motor is not simply a split-phase motor with a capacitor inserted in the auxiliary circuit but is a specially designed motor that produces higher torque than the corresponding split-phase version.

Capacitor-start motors are extremely popular and are available in all ratings from 0.125 hp up. For ratings at ⅓ hp and above, capacitor-start motors are wound as dual-voltage so that they can be operated on either a 115- or a 230-V supply. In this case, the main winding is made of two sections that are connected in series for 230-V operation or in parallel for 115-V operation. The auxiliary winding in a dual-voltage motor is made of one section which is connected in parallel with one section of the main winding for 230-V operation.

It is important to realize that the capacitor voltage increases rapidly above the switch-open speed and the capacitor can be damaged if the centrifugal switch fails to open at the designed speed. It is also important that switches not flutter, as this causes a dangerous rise in the voltage across the capacitor.

A typical torque–speed characteristic for a capacitor-start single-phase induction motor is shown in Figure 8.14. The starting torque is very high, which is a desirable feature of this type of motor.

Permanent-Split Capacitor Motors

The second type of capacitor motors is referred to as the permanent-split capacitor motor, where the auxiliary winding and the capacitor are retained at normal running speed. This motor is used for special-purpose applications requiring high torque and is available in ratings from 10^{-3} to ⅓–¾

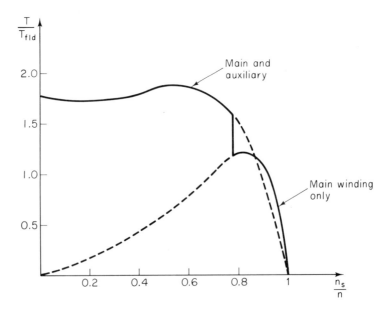

FIGURE 8.14   Torque–speed characteristic of a capacitor-start motor.

hp. A schematic of the permanent-split capacitor motor is shown in Figure 8.15.

A typical torque–speed characteristic for a permanent-split capacitor motor is shown in Figure 8.16. The starting torque is noticeably low since the capacitance is a compromise between best running and starting conditions. The next type of motor overcomes this difficulty.

Two-Value Capacitor Motors

A two-value capacitor motor is the third type of the class of capacitor motors and starts with one value of capacitors in series with the auxiliary winding and runs with a different capacitance value. This change can be done either

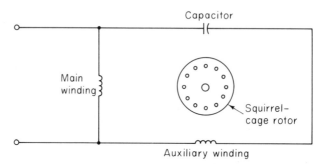

FIGURE 8.15   Permanent-split capacitor motor.

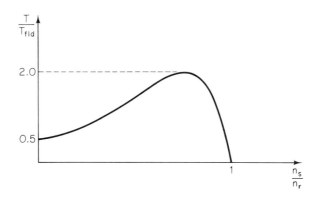

FIGURE 8.16 Torque–speed characteristic of a permanent-split single-phase induction motor.

using two separate capacitors or through the use of an autotransformer. This motor has been replaced by the capacitor-start motor for applications such as refrigerators and compressors.

For the motor using an autotransformer, a transfer switch is used to change the tap on the autotransformer, as shown in Figure 8.17(a). This arrangement appears to be obsolete now and the two-capacitor mechanism illustrated in Figure 8.17(b) is used.

A typical torque–speed characteristic for a two-value capacitor motor is shown in Figure 8.18. Note that optimum starting and running conditions can be accomplished in this type of motor.

Repulsion-Type Motors

A repulsion motor is a single-phase motor with power connected to the stator winding and a rotor whose winding is connected to a commutator. The brushes on the commutator are short-circuited and are positioned such that there is an angle of 20 to 30° between the magnetic axis of the stator winding and the magnetic axis of the rotor winding. A representative torque–speed characteristic for a repulsion motor is shown in Figure 8.19. A repulsion motor is a variable-speed motor.

If in addition to the repulsion winding, a squirrel-cage type of winding is embedded in the rotor, we have a repulsion-induction motor. The torque–speed characteristic for a repulsion-induction motor is shown in Figure 8.20 and can be thought of as a combination of the characteristics of a single-phase induction motor and that of a straight repulsion motor.

A repulsion-start induction motor is a single-phase motor with the same windings as a repulsion motor, but at a certain speed the rotor winding is short circuited to give the equivalent of a squirrel-cage winding. The repulsion-start motor is the first type of single-phase motors that gained

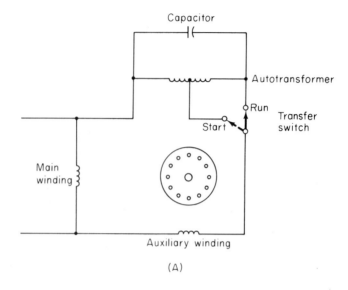

(A)

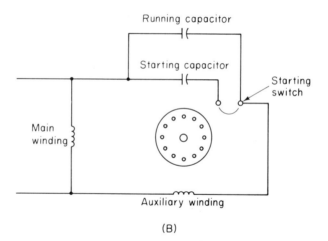

(B)

FIGURE 8.17 Two-value capacitor motor: (a) autotransformer type;   (b) two-capacitor type.

wide acceptance. In recent years, however, it has been replaced by capacitor-type motors. A typical torque–speed characteristic of a repulsion-start induction motor is shown in Figure 8.21.

Shaded-Pole Induction Motors

For applications requiring low power of ¼ hp or less, a shaded-pole induction motor is the standard general-purpose device for constant-speed appli-

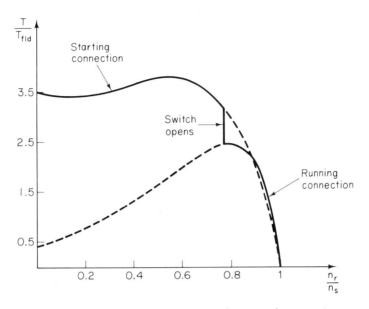

FIGURE 8.18 Torque–speed characteristic of a two-value capacitor motor.

cations. The torque characteristics of a shaded-pole motor are similar to those of a permanent-split capacitor motor as shown in Figure 8.22. It is important to understand the action of pole shading, which is discussed next.

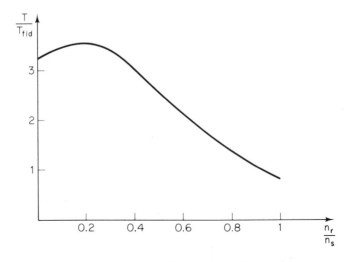

FIGURE 8.19 Torque–speed characteristic of a repulsion motor.

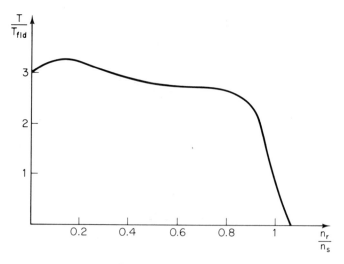

FIGURE 8.20   Torque–speed characteristic of a repulsion-induction motor.

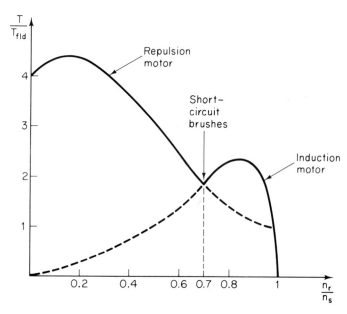

FIGURE 8.21   Torque–speed characteristic of a repulsion-start single-phase induction motor.

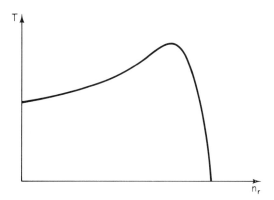

FIGURE 8.22   Torque–speed characteristic of a shaded-pole induction motor.

Shaded-Pole Action

Consider the magnetic structure shown in Figure 8.23(a). If we neglect the reluctance of the iron path, an equivalent circuit as shown in Figure 8.23(b) is obtained. The flux ϕ_s is seen to be equal to the flux ϕ_m if the two air gaps are identical. The MMF of the main coil $F_1 = N_1 I_1$ is sinusoidal if the coil is connected to an ac source, and thus ϕ_s and ϕ_m are also sinusoidal of equal magnitude and phase.

The magnetic structure of Figure 8.23(c) is similar to that of Figure 8.23(a), except that a coil of N_s turns is placed on the rightmost portion. The coil is called a shadding coil and its two terminals are short-circuited. As a result of the sinusoidal variation of ϕ_s, an induced EMF E_s is produced in the shading coil and it lags ϕ_s by 90°. A current I_s is set up in the shading coil and lags E_s by the angle $\theta_s = \arctan(\omega L_s/R_s)$, where ω is due to the supply and L_s and R_s are the inductance and resistance of the shading coil winding, respectively. A phasor diagram showing ϕ_s, E_s, and I_s is given in Figure 8.23(d).

The shading coil is thus seen to produce an MMF $F_2 = N_s I_s$, which is proportional to and in phase with I_s. The magnetic equivalent circuit of Figure 8.23(e) shows the effect of the shading pole. The MMF of the right-hand reluctance is $\mathscr{R}_s$ and is given by

$$\phi_s \mathscr{R}_s = F_1 + F_2$$

It is clear by inspection of Figure 8.23(f) that F_1 is obtained by subtracting F_2 from $\phi_s \mathscr{R}_s$. Observing that F_1 is impressed on $\mathscr{R}_m$ leads us to conclude that ϕ_m is in phase with F_1.

It is clear now that the effect of the shading coil is to cause the flux ϕ_s to lag ϕ_m by a certain angle ψ. This is shown in Figure 8.23(g) and is the basis of explaining the operation of shaded-pole motors.

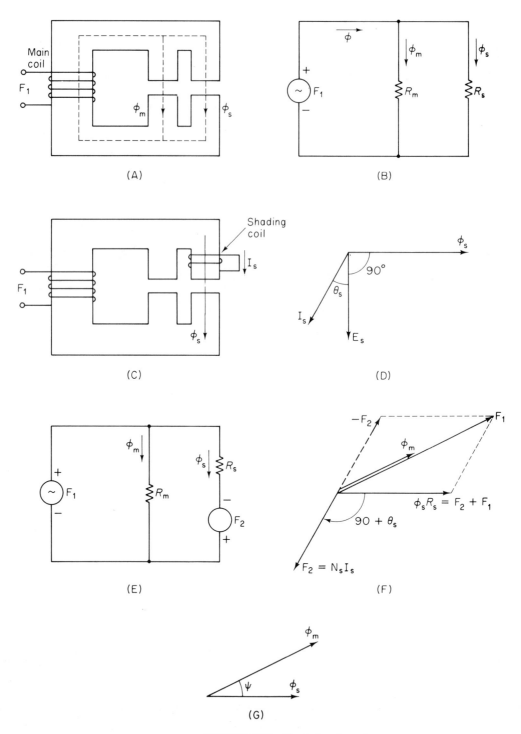

FIGURE 8.23 Shaded-pole action.

8.6   UNBALANCED OPERATION OF TWO-PHASE MOTORS: BASICS

We have seen earlier in this chapter that a two-phase induction motor with identical stator windings and a balanced two-phase voltage will produce induction-motor action in the same manner as a three-phase induction motor. It is now our intention to study the performance of a two-phase motor when operating from an unbalanced supply. The schematic diagram of Figure 8.24 shows the main winding denoted by phase A and the auxiliary winding, which is called phase B. We will assume that V_A and V_B do not form a balanced two-phase set. Under this assumption the study of the motor can be facilitated by use of the concept of symmetrical components treated in this section.

Symmetrical Components for Two-Phase Systems

Problems involving unbalanced operation of two-phase systems can be conveniently analyzed using the concept of symmetrical components. The general idea is that a phasor can be resolved into a number of other phasors, the sum of which yields the original phasor. In the analysis of unbalanced three-phase systems, this implies that each phasor is resolved into three constituent phasors. In the case of an unbalanced two-phase system, each phasor is resolved into the sum of two constituent phasors. It is clear that there is an infinite number of ways of doing this. In the symmetrical components method this is used to advantage by placing further requirements on the resulting phasors to obtain a unique solution to the problem.

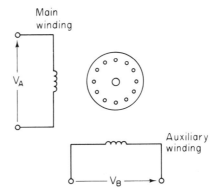

FIGURE 8.24   Two-phase induction motor.

Consider the two voltage phasors V_A and V_B shown in Figure 8.25. The two phasors do not constitute a balanced two-phase system. We can express each phasor as

$$V_A = V_{A+} + V_{A-} \qquad (8.82)$$

$$V_B = V_{B+} + V_{B-} \qquad (8.83)$$

The components V_{A+}, V_{A-}, V_{B+}, and V_{B-}, are not determined yet. Let us require that V_{A+} and V_{B+} form a balanced two-phase voltage set related by

$$V_{B+} = -jV_{A+} \qquad (8.84)$$

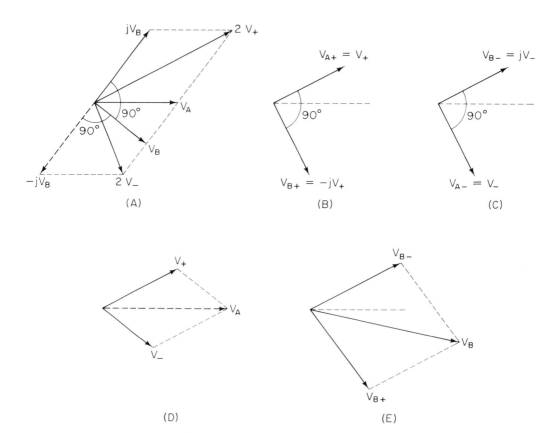

FIGURE 8.25 Symmetrical components for two-phase systems: (a) obtaining V_+ and V_- from V_A and V_B;   (b) positive sequence phasors;   (c) negative sequence phasors;   (d) reconstructing V_A;   (e) reconstructing V_B.

Note that V_{A+} leads V_{B+} by 90° and we call this set a positive sequence or forward voltage set. We also require that

$$V_{B-} = jV_{A-} \tag{8.85}$$

In this case V_{A-} lags V_{B-} by 90° and we call this set a negative sequence or backward voltage set.

With the foregoing requirements, we obtain

$$V_A = V_{A+} + V_{A-} \tag{8.86}$$

$$V_B = -jV_{A+} + jV_{A-} \tag{8.87}$$

For simplicity of notation, let us introduce

$$V_+ = V_{A+} \tag{8.88}$$

$$V_- = V_{A-} \tag{8.89}$$

Thus we have

$$V_A = V_+ + V_- \tag{8.90}$$

$$V_B = -jV_+ + jV_- \tag{8.91}$$

If V_A and V_B are given, the solution for V_+ and V_- is given by

$$V_+ = 0.5(V_A + jV_B) \tag{8.92}$$

$$V_- = 0.5(V_A - jV_B) \tag{8.93}$$

This is obtained by multiplying Eq. (8.91) by j and adding the resulting equation to Eq. (8.90) to obtain (8.92). Similarly, we multiply Eq. (8.91) by $-j$ and add the result to Eq. (8.90) to obtain (8.93). The phasor diagram of Figure 8.25(a) illustrates the procedure of applying Eqs. (8.92) and (8.93) to obtain V_+ and V_-. Note that

$$V_{B+} = -jV_+ \tag{8.94}$$

$$V_{B-} = jV_- \tag{8.95}$$

The positive (forward) sequence voltage set associated with V_A and V_B is shown in Figure 8.25(b), whereas the corresponding negative (backward) sequence voltage set is shown in Figure 8.25(c). The inverse relation of reconstructing V_A from V_+ and V_- is shown in Figure 8.25(d) and that of reconstructing V_B from its components is also shown in Figure 8.25(e).

Example 8.3   Find the positive (forward) and negative (backward) sequence components of the unbalanced voltage set

$$V_A = V_B = V$$

This set clearly results from application of the same single-phase voltage to a two-phase device.

Solution

We have, by application of Eq. (8.92),

$$V_+ = 0.5(1 + j1)V = \frac{V}{\sqrt{2}} \underline{/45°}$$

We also have, by application of Eq. (8.93),

$$V_- = 0.5(1 - j1)V = \frac{V}{\sqrt{2}} \underline{/-45°}$$

As a result, we have for phase A

$$V_{A+} = V_+ = \frac{V}{\sqrt{2}} \underline{/45°}$$

$$V_{A-} = V_- = \frac{V}{\sqrt{2}} \underline{/-45°}$$

For phase B, we get

$$V_{B+} = -jV_+ = \frac{V}{\sqrt{2}} \underline{/-45°}$$

$$V_{B-} = jV_- = \frac{V}{\sqrt{2}} \underline{/45°}$$

To verify our results, we have

$$V_A = V_+ + V_- = \frac{V}{\sqrt{2}} (1 \underline{/45°} + 1 \underline{/-45°}) = V$$

$$V_B = V_{B+} + V_{B-} = V$$

The following example illustrates the inverse problem.

Example 8.4   Assume that the following sequence voltages are given:

$$V_+ = 0.924 \underline{/22.5°}$$

$$V_- = 0.3827 \underline{/-67.5°}$$

Find the corresponding V_A and V_B.

Solution

We have

$$V_A = V_+ + V_- = 0.924 \underline{/22.5°} + 0.3827 \underline{/-67.5°}$$

As a result,

$$V_A = 1 \underline{/0}$$

We also have

$$V_B = -jV_+ + jV_- = 0.924 \underline{/-67.5°} + 0.3827 \underline{/22.5°}$$

$$= 1 \underline{/-45°}$$

It is important to realize that the concept of symmetrical components is also applicable to current phasors. We thus have the resolution

$$I_A = I_+ + I_- \tag{8.96}$$

$$I_B = -jI_+ + jI_- \tag{8.97}$$

With the inverse relations

$$I_+ = 0.5(I_A + jI_B) \tag{8.98}$$

$$I_- = 0.5(I_A - jI_B) \tag{8.99}$$

The apparent power into phase A is obtained from

$$S_A = V_A I_A^* \tag{8.100}$$

In terms of sequence values, we get

$$S_A = (V_+ + V_-)(I_+^* + I_-^*)$$
$$= V_+I_+^* + V_-I_-^* + V_+I_-^* + V_-I_+^*$$

(8.101)

Similarly,

$$S_B = V_B I_B^*$$
$$= V_+I_+^* + V_-I_-^* - V_+I_-^* - V_-I_+^*$$

(8.102)

As a result, we conclude that the total apparent power is

$$S_t = S_A + S_B = 2(S_+ + S_-)$$

(8.103)

where

$$S_+ = V_+I_+^*$$

(8.104)

$$S_- = V_-I_-^*$$

(8.105)

This is the principle of power invariance.

Example 8.5   Given the following voltage and current phasors in an unbalanced two-phase system, find the positive and negative sequence components and verify the principle of power invariance in both forms.

$$V_A = V \qquad\qquad V_B = V\,\underline{/-45°}$$

$$I_A = I\,\underline{/-30°} \qquad I_B = \frac{I}{2}\,\underline{/-60°}$$

Solution

For the voltage phasors, we obtain

$$V_+ = \tfrac{1}{2}(V + V\,\underline{/45}) = \frac{V}{2}(1 + \cos 45° + j \sin 45°)$$

$$= 0.924V\,\underline{/22.5°}$$

$$V_- = \tfrac{1}{2}(V + V\,\underline{/-135°}) = \frac{V}{2}(1 - \cos 45° - j \sin 45°)$$

$$= 0.3827V\,\underline{/-67.5°}$$

For the current phasors, we obtain

$$I_+ = \tfrac{1}{2}\left(I\,\underline{/-30°} + \frac{I}{2}\,\underline{/30°}\right) = 0.6614I\,\underline{/-10.89°}$$

$$I_- = \tfrac{1}{2}\left(I\,\underline{/-30°} + \frac{I}{2}\,\underline{/-150°}\right) = 0.433I\,\underline{/-60°}$$

The apparent power in the phase form is

$$S = V_A I_A^\star + V_B I_B^\star = VI(1\,\underline{/30} + \tfrac{1}{2}\,\underline{/15°})$$

$$= (1.4886\,\underline{/25.013°})VI$$

In the sequence form, we have

$$S = 2(V_+ I_+^\star + V_- I_-^\star)$$

$$= 2VI[(0.924)(0.6614)\,\underline{/22.5° + 10.89°} + (0.3827)(0.433)\,\underline{/-67.5° + 60°}]$$

$$= (1.4887\,\underline{/25.013°})VI$$

This satisfies the verification requirement.

8.7   UNBALANCED OPERATION OF TWO-PHASE MOTORS: ANALYSIS

The concept of symmetrical components gives us a convenient means of analyzing the operation of a two-phase induction motor in the case of unbalanced two-phase voltages applied to the stator winding. The basic ideas is best illustrated as shown in Figure 8.26. The voltage V_A is the sum of the two components V_{A+} and V_{A-}, and similarly, V_B is the sum of V_{B+} and V_{B-} as shown in the figure. The behavior of the motor can be studied in terms of its response to V_{A+} impressed on phase A and V_{B+} impressed on phase B as shown in Figure 8.27(a) in addition to its response to V_{A-} impressed on phase A and V_{B-} impressed on phase B as shown in Figure 8.27(b). It is therefore clear that results of the analysis of the two sets can be combined to yield the required result as dictated by the principle of superposition.

Let us look at the situation with the positive sequence set shown in Figure 8.27(a). It is assumed that this results in a forward field with a balanced two-phase voltage applied. The equivalent circuit of phase A is thus

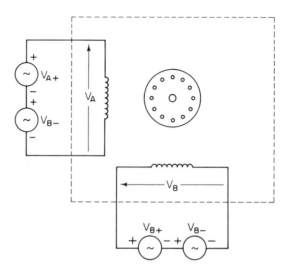

FIGURE 8.26 Application of the symmetrical component method to the unbalanced operation of a two-phase induction motor.

the usual induction motor equivalent circuit as shown in Figure 8.27(c). Here we define

$$Z_+ = \frac{jX_m[(R'_2/s) + jX'_2]}{(R'_2/s) + j(X_m + X'_2)} \qquad (8.106)$$

The impedance Z_+ is the parallel combination of $(R'_2/s + jX'_2)$ and jX_m. All quantities are specified in their phase value. The phase A input impedance is thus given by

$$Z_{i+} = Z_1 + Z_+ \qquad (8.107)$$

where Z_1 is the stator winding impedance given by

$$Z_1 = R_1 + jX_1 \qquad (8.108)$$

For the negative sequence set shown in Figure 8.27(b), the resulting field moves in the backward direction and hence we obtain the equivalent circuit of phase A for the negative sequence as shown in Figure 8.27(d). Note that we replace s by $(2 - s)$ to account for the backward field effect. We now have

$$Z_- = \frac{jX_m \{[R'_2/(2 - s)] + jX'_2\}}{[R'_2/(2 - s)] + j(X_m + X'_2)} \qquad (8.109)$$

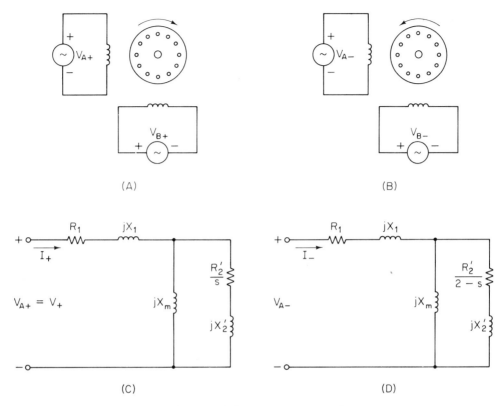

FIGURE 8.27   Application of the superposition to the analysis of a two-phase induction motor: (a) positive sequence effect;   (b) negative sequence effect;   (c) positive sequence equivalent circuit;   (d) negative sequence equivalent circuit.

Furthermore, we get

$$Z_{i-} = Z_1 + Z_- \tag{8.110}$$

The positive sequence current into phase A is thus given by

$$I_+ = \frac{V_+}{Z_{i+}} \tag{8.111}$$

The negative sequence current in phase A is given by

$$I_- = \frac{V_-}{Z_{i-}} \tag{8.112}$$

If we desire to reconstruct I_A, we use

$$I_A = I_+ + I_- \qquad (8.113)$$

For phase B we use

$$I_B = -jI_+ + jI_- \qquad (8.114)$$

To proceed with the analysis we have the air-gap power delivered by the positive sequence voltage obtained as

$$P_{g+} = 2|I_+|^2 R_+ \qquad (8.115)$$

Note that the factor 2 accounts for the two phases and R_+ is the real part of Z_+. Similarly, for the negative sequence air-gap power

$$P_{g-} = 2|I_-|^2 R_- \qquad (8.116)$$

Again R_- is the real part of Z_-. It is clear now that the remaining steps are familiar to us based on the analysis of the single-phase motor. The mechanical power is obtained as

$$P_m = (1 - s)(P_{g+} - P_{g-}) \qquad (8.117)$$

The net output power is obtained by subtracting the core and rotational losses from P_m:

$$P_0 = P_m - P_{\text{rotational}} - P_{\text{core}} \qquad (8.118)$$

The torque output can be computed from

$$T_o = \frac{P_o}{\omega_r} \qquad (8.119)$$

where

$$\omega_r = \omega_s(1 - s) \qquad (8.120)$$

As usual, ω_s is the synchronous speed in radians per second. The efficiency is computed from

$$\eta = \frac{P_o}{P_{\text{in}}} \qquad (8.121)$$

The power input to the motor can be computed as the sum of output and losses in various parts of the motor. An example should illustrate the procedures involved.

Example 8.6   The stator windings of a two-phase 220-V 60-Hz squirrel-cage induction motor are supplied with an unbalanced set of voltages given by

$$V_A = 240 \underline{/0}$$

$$V_B = 220 \underline{/-85°}$$

The parameters of the motor are given by

$$R_1 = 0.6\ \Omega \qquad X_1 = 2.4\ \Omega$$

$$R_2' = 1\ \Omega \qquad X_2' = 3\ \Omega$$

$$X_m = 72\ \Omega$$

Core and rotational losses are given by 150 W. The motor is running at a slip of 0.05. It is required to find the output power under these conditions.

Solution

The positive sequence voltage is obtained using Eq. (8.92) as

$$V_+ = \frac{V_A + jV_B}{2} = \frac{240 \underline{/0} + 220 \underline{/5°}}{2}$$

$$= 229.782 \underline{/2.391°}\ V$$

The negative sequence voltage is obtained using Eq. (8.93) as

$$V_- = \frac{V_A - jV_B}{2} = \frac{240 \underline{/0} + 220 \underline{/-175°}}{2}$$

$$= 14.158 \underline{/-42.62°}\ V$$

The impedance Z_+ is obtained as

$$Z_+ = \frac{j72[(1/0.05) + j3]}{(1/0.05) + j(72 + 3)} = 17.208 + j7.469\ \Omega$$

Thus the input impedance to positive sequence voltage is

$$Z_{i+} = Z_1 + Z_+$$

$$= 17.208 + j7.469 + 0.6 + j2.4$$

$$= 17.808 + j9.869$$

$$= 20.36 \underline{/28.994°}\ \Omega$$

As a result, the positive sequence current is obtained as

$$I_+ = \frac{V_+}{Z_{i+}} = \frac{229.782 \,\underline{/2.391°}}{20.36 \,\underline{/28.994°}}$$

$$= 11.286 \,\underline{/-26.603°}$$

The impedance Z_- is obtained as

$$Z_- = \frac{j72[(1/1.95) + j3]}{(1/1.95) + j(72 + 3)} = 0.473 + j2.883$$

The input impedance to negative sequence voltage is

$$Z_{i-} = Z_1 + Z_- = 0.473 + j2.883 + 0.6 + j2.4$$

$$= 1.073 + j5.283$$

$$= 5.391 \,\underline{/78.524°}$$

As a result, the negative sequence current is obtained as

$$I_- = \frac{V_-}{Z_{i-}} = \frac{14.158 \,\underline{/-42.62°}}{5.391 \,\underline{/78.524°}} = 2.625 \,\underline{/-121.144°}$$

The positive sequence air-gap power is now computed as

$$P_{g+} = 2|I_+|^2 R_+$$

$$= 2(11.286)^2(17.208)$$

$$= 4383.634 \text{ W}$$

The negative sequence air-gap power is

$$P_{g-} = 2|I_-|^2 R_-$$

$$= 2(2.625)^2(0.473)$$

$$= 6.525 \text{ W}$$

The mechanical power is

$$P_m = (1 - s)(P_{g+} - P_{g-})$$

$$= 0.95(4383.634 - 6.525)$$

$$= 4158.25 \text{ W}$$

We thus have

$$P_0 = P_m - P_{rot} - P_{core}$$

$$= 4158.25 - 150 = 4008.254 \text{ W}$$

The following example illustrates the various quantities involved in determining the losses and efficiency for the motor of Example 8.6.

Example 8.7 Find the rotor losses, stator losses, and power input for the motor of Example 8.6.

Solution

The rotor losses are given by

$$P_{\ell_r} = sP_{g+} + (2 - s)P_{g-}$$

$$= 0.05(4383.634) + (1.95)(6.525)$$

$$= 219.182 + 12.724 = 231.905 \text{ W}$$

Note that we can also obtain this result using

$$P_{\ell_r} = P_{g+} + P_{g-} - P_m$$

$$= 4383.634 + 6.525 - 4158.25$$

$$= 231.909 \text{ W}$$

The stator losses are given by

$$P_{\ell_s} = 2|I_+|^2R_1 + 2|I_-|^2R_1$$

$$= 2(0.6)[(11.286)^2 + (2.625)^2]$$

$$= 161.117 \text{ W}$$

As a result, the input power is

$$P_i = P_{\ell_s} + P_{\ell_r} + P_m$$

$$= 161.117 + 231.905 + 4158.25$$

$$= 4551.276 \text{ W}$$

We can alternatively calculate the input power using

$$S_i = 2(V_+ I_+^* + V_- I_-^*)$$
$$= 2[(229.782 \underline{/2.391°})(11.286 \underline{/26.603°})$$
$$+ (14.158 \underline{/-42.62°})(2.625 \underline{/121.144°})]$$
$$= 2(2268.3 + j1257.029 + 7.394 + j36.422)$$

The real part gives P_{in} as

$$P_{\text{in}} = 2(2268.3 + 7.394) = 4551.389 \text{ W}$$

The efficiency is computed as

$$\eta = \frac{P_o}{P_{\text{in}}} = \frac{4008.254}{4551.389} = 0.881$$

Note that we can compute S_i from the phase currents as well. Here we have

$$I_A = I_+ + I_- = (11.286 \underline{/-26.603°}) + (2.625 \underline{/-121.144°})$$
$$= 11.383 \underline{/-39.893°}$$
$$I_B = -jI_+ + jI_- = 11.286 \underline{/-116.603°} + 2.625 \underline{/-31.144°}$$
$$= 11.788 \underline{/-103.777°}$$

Now

$$S_i = V_A I_A^* + V_B I_B^*$$
$$= (240 \underline{/0})(11.383 \underline{/39.893°}) + (220 \underline{/-85°})(11.788 \underline{/103.777°})$$
$$= 4551.369 + j2586.916$$

Clearly, we get the same result.

An important area of application of two-phase induction motors operating under unbalanced conditions is as a two-phase control motor. In this case, phase A of the motor is the fixed or reference phase, whereas phase B is the control phase. The voltages V_A and V_B are at 90° phase angle with V_A fixed at a rated value. The voltage V_B is supplied from an amplifier that responds to the controller's command signal. The magnitude of V_B is proportional to the controller's command signal. Both voltages are of the same frequency. Changing the magnitude of V_B changes the developed torque of

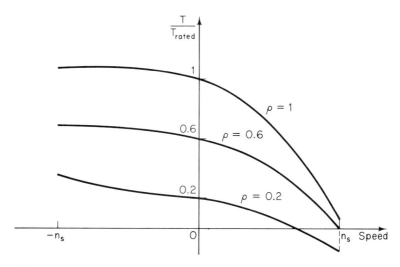

FIGURE 8.28 Typical torque–speed characteristic of a two-phase control motor.

the motor. The torque should be high at low speeds, and the torque–speed characteristics should have a negative slope in the normal operating speed range. These features can be obtained by use of a high-resistance rotor. A typical torque–speed characteristic is shown in Figure 8.28 with the ratio of magnitude V_B to that V_A denoted by ρ.

SOME SOLVED PROBLEMS

Problem 8.A.1

A single-phase induction motor takes an input power of 490 W at a power factor of 0.57 lagging from a 110-V supply when running at a slip of 5%. Assume that the rotor resistance and reactance are 1.78 Ω and 1.28 Ω, respectively, and that the magnetizing reactance is 25 Ω. Find the resistance and reactance of the stator.

Solution

The equivalent circuit of the motor yields

$$Z_f = \frac{\{[1.78/2(0.05)] + j0.64\}(j12.5)}{[1.78/2(0.05)] + j(0.64 + 12.5)} = 5.6818 + j8.3057$$

$$Z_b = \frac{\{[1.78/2(1.95)] + j0.64\}(j12.5)}{[1.78/2(1.95)] + j(0.64 + 12.5)} = 0.4125 + j0.6232$$

As a result of the problem specifications:

$$P_i = 490 \text{ W} \qquad \cos \phi = 0.57 \qquad V = 110$$

$$I_i = \frac{P_i}{V \cos \phi} = \frac{590}{110(0.57)} = 7.815 \,\underline{/-55.2488°}$$

Thus the input impedance is

$$Z_i = \frac{V}{I_i} = \frac{110}{7.815} \,\underline{/55.2498°} = 8.023 + j11.5651$$

The stator impedance is obtained as

$$Z_1 = Z_i - (Z_f + Z_b) = 1.9287 + j2.636 \ \Omega$$

Problem 8.A.2

The forward field impedance of a $\frac{1}{4}$-hp four-pole 110-V 60-Hz single-phase induction motor for a slip of 0.05 is given by

$$Z_f = 12.4 + j16.98 \ \Omega$$

Assume that

$$X_m = 53.5 \ \Omega$$

Find the values of the rotor resistance and reactance.

Solution

We have

$$Z_f = \frac{j0.5X_m[0.5(R_2'/s) + j0.5X_2']}{0.5(R_2'/s) + j0.5(X_2' + X_m)}$$

Thus

$$12.4 + j16.98 = \frac{j26.75[(R_2'/s) + jX_2']}{(R_2'/s) + j(X_2' + 53.5)}$$

As a result, cross-multiplying and separating real and imaginary parts, we obtain

$$0.3652 \frac{R_2'}{s} - 0.4636 X_2' = 24.8$$

$$0.4636 \frac{R_2'}{s} + 0.3652 X_2' = 33.96$$

Solving the two equations, we get

$$X_2' = 2.605 \ \Omega$$

$$\frac{R_2'}{s} = 71.208$$

Thus

$$R_2' = 3.5604 \ \Omega$$

Problem 8.A.3

For the motor of Problem 8.A.2, assume that the stator impedance is given by

$$Z_1 = 1.86 + j2.56 \ \Omega$$

Find the internal mechanical power, output power, power factor, input power, developed torque, and efficiency, assuming that friction losses are 15 W.

Solution

We need the backward field impedance

$$Z_b = \frac{j0.5 X_m'[0.5(R_2'/(2 - s)) + j0.5 X_2']}{0.5(R_2'/(2 - s)) + j0.5(X_2' + X_m')}$$

$$= \frac{j26.75[(1.78/1.95) + j1.3]}{(1.78/1.95) + j(1.3 + 26.75)}$$

$$= 0.8293 + j1.2667 \ \Omega$$

As a result,

$$Z_i = Z_1 + Z_f + Z_b$$
$$= 15.089 + j20.8067 = 25.7023 \underline{/54.05°} \; \Omega$$

The input current is thus

$$I_1 = \frac{V}{Z_i} = \frac{110 \underline{/0}}{25.7023 \underline{/54.05}} = 4.2798 \underline{/-54.05°} \; A$$

The forward gap power is

$$P_{gf} = |I_1|^2 R_f = 227.125 \; W$$

The backward gap power is

$$P_{gb} = |I_1|^2 R_b = 15.1899 \; W$$

As a result, the internal mechanical power is

$$P_m = (1 - s)(P_{gf} - P_{gb})$$
$$= 201.338 \; W$$

The output power is thus

$$P_o = 201.338 - 15 = 186.338 \; W$$

The power factor is given by

$$PF = \cos 54.05° = 0.5871$$

The input power is calculated as

$$P_{in} = 110(4.2798)(0.5871) = 276.383 \; W$$

As a result, the efficiency is

$$\eta = \frac{P_o}{P_{in}} = \frac{186.338}{276.383} = 0.6742$$

The synchronous speed is

$$n_s = \frac{120f}{P} = 1800 \text{ r/min}$$

The rotor speed in rad/s is thus

$$\omega_r = \frac{2\pi n_s}{60}(1 - s) = 57\pi$$

The torque output is given by

$$T_o = \frac{P_o}{\omega_r} = \frac{186.338}{57\pi} = 1.0406 \text{ N} \cdot \text{m}$$

Problem 8.A.4

A two-phase induction motor is connected to an unbalanced voltage source with

$$V_A = 115 \underline{/0} \qquad V_B = 75 \underline{/-90°}$$

The power input to phase A is 464 W at 0.39 PF lagging and the power input to phase B is 123 W at 0.267 PF lagging for a slip of 0.2. Find the following.

    a. The current in phase A and that of phase B.
    b. The impedance of the phase A winding and that of phase B.
    c. The positive and negative sequence currents and voltages.
    d. The motor's input impedance to positive and negative sequence voltages.

Solution

(a) We have for the phase currents

$$I_A = \frac{464}{115(0.39)} \underline{/-\cos^{-1}0.39} = 10.346 \underline{/-67.046°}$$

$$I_B = \frac{123}{75(0.267)} \underline{/-90 - \cos^{-1}0.267} = 6.142 \underline{/-164.514°}$$

(b) The impedances required are obtained as

$$Z_A = \frac{V_A}{I_A} = \frac{115 \underline{/-90°}}{10.346 \underline{/-67.046°}} = 11.116 \underline{/67.046°}$$

$$Z_B = \frac{V_B}{I_B} = \frac{75 \underline{/-90°}}{6.142 \underline{/-164.514°}} = 12.21 \underline{/74.514°}$$

(c) The sequence currents are obtained as

$$I_+ = \frac{I_A + jI_B}{2} = 8.228 \underline{/-69.827°}$$

$$I_- = \frac{I_A - jI_B}{2} = 2.165 \underline{/-56.423}$$

The sequence voltages are

$$V_+ = \frac{V_A + jV_B}{2} = 95 \underline{/0°}$$

$$V_- = \frac{V_A - jV_B}{2} = 20 \underline{/0°}$$

(d) The input impedances are therefore obtained as

$$Z_{i+} = \frac{V_+}{I_+} = \frac{95}{8.228 \underline{/-69.827°}} = 11.546 \underline{/69.827°} \; \Omega$$

$$Z_{i-} = \frac{V_-}{I_-} = \frac{20}{2.165 \underline{/-56.423°}} = 9.238 \underline{/56.423°} \; \Omega$$

Problem 8.A.5

Assume that the stator impedance of the motor of Problem 8.A.4 is given by

$$Z_i = 3.2 + j4$$

Find the output power and efficiency under the specified conditions, neglecting core and rotational losses.

Solution

From the information available, we obtain

$$Z_+ = Z_{i+} - Z_1 = 0.782 + j6.838$$

$$Z_- = Z_{i-} - Z_1 = 1.909 + j3.697$$

As a result,

$$P_{g+} = 2|I_+|^2 R_+ = 2(8.228)^2(0.782) = 105.883 \text{ W}$$

$$P_{g-} = 2|I_-|^2 R_- = 2(2.165)^2(1.909) = 17.896 \text{ W}$$

The output mechanical power is thus obtained using

$$P_m = (1 - s)(P_{g+} - P_{g-})$$

$$= (1 - 0.2)(105.883 - 17.896) = 70.39 \text{ W}$$

The input power from the specifications of Problem 8.A.4 is

$$P_{\text{in}} = P_A + P_B = 464 + 123 = 587 \text{ W}$$

We can thus obtain

$$\eta = \frac{P_m}{P_{\text{in}}} = \frac{70.39}{587} = 0.12$$

PROBLEMS

Problem 8.B.1

The following parameters are available for a single-phase induction motor

$$R_1 = 1.4 \ \Omega \qquad\qquad R_2' = 3.34 \ \Omega$$

$$X_1 = X_2' = 3.2 \ \Omega \qquad X_m = 98 \ \Omega$$

Calculate Z_f, Z_b, and the input impedance of the motor for a slip of 0.06.

Problem 8.B.2

The induction motor of Problem 8.B.1 is a 60-Hz 110-V four-pole machine. Find the output power and torque under the conditions of Problem 8.B.1 assuming that the core losses are 66 W. Neglect rotational losses.

Problem 8.B.3

A four-pole 110-V 60-Hz single-phase induction motor has the following parameters:

$$R_1 = 0.83 \ \Omega \qquad\qquad R_2' = 0.94 \ \Omega$$

$$X_1 = X_2' = 1.92 \ \Omega \qquad X_m = 42 \ \Omega$$

The core losses are equal to the rotational losses, which are given by 40 W. Find the output power and efficiency at a slip of 0.05.

Problem 8.B.4

The following parameters are available for a single-phase 110-V induction motor:

$$R_1 = R_2' = 2.65 \ \Omega$$

$$X_1 = X_2' = 2.7 \ \Omega$$

$$X_m = 70.8 \ \Omega$$

The core losses are 18.5 W and rotational losses are 17 W. Assume that the machine has four poles and operates on a 60-Hz supply. Find the rotor ohmic losses, output power, and torque for a slip of 5%.

Problem 8.B.5

The stator resistance of a single-phase induction motor is 1.95 Ω and the rotor resistance referred to the stator is 3.4 Ω. The motor takes a current of 4.2 A from the 110-V supply at a power factor of 0.624 when running at slip of 0.05. Assume that the core loss is 36 W and that the approximation of Eq. (8.32) is applicable. Find the motor's output power and efficiency neglecting rotational losses.

Problem 8.B.6

A single-phase induction motor takes an input power of 280 W at a power factor of 0.6 lagging from a 110-V supply when running at a slip of 5%. Assume that the rotor resistance and reactance are 3.38 and 2.6 Ω, respectively, and that the magnetizing reactance is 60 Ω. Find the resistance and the reactance of the stator.

Problem 8.B.7

For the motor of Problem 8.B.6, assume that the core losses are 35 W and the rotational losses are 14 W. Find the output power and efficiency when running at a slip of 5%.

Problem 8.B.8

The output torque of a single-phase induction motor is 0.82 N · m at a speed of 1710 rpm. The efficiency is 60% and the fixed losses are 37 W. Assume that motor operates on a 110-V supply and that the stator resistance is 2 Ω. Find the input power factor and input impedance. Assume that the rotor ohmic losses are 35.26 W. Find the forward and backward gap power and the values of R_f and R_b. Assume a four-pole machine.

Problem 8.B.9

The phase voltages in an unbalanced two-phase system are given by

$$V_A = 230 \underline{/0} \qquad V_B = 210 \underline{/-100°}$$

Find the positive and negative sequence voltage components for both phases.

Problem 8.B.10

The positive and negative sequence components of an unbalanced two-phase current set are given by

$$I_+ = 11.26 \underline{/-34.2°}$$
$$I_- = 4 \underline{/-21.8°}$$

Find the phase currents I_A and I_B.

Problem 8.B.11

A two-phase machine is supplied by unbalanced voltages given by

$$V_A = 210 \underline{/80°} \qquad V_B = 230 \underline{/0°}$$

Find the positive and negative sequence voltages V_+ and V_-. The input impedance to positive sequence voltage is given by

$$Z_{i+} = 20 \underline{/30°}$$

The input impedance to negative sequence voltage is

$$Z_{i-} = 5 \underline{/75°}$$

Find the positive and negative sequence currents and hence the phase currents I_A and I_B. Find the apparent power taken by the machine in terms of phase quantities as well as sequence quantities.

Problem 8.B.12

The voltage supply to a two-pole 60-Hz 115-V two-phase induction motor is given by

$$V_A = 115 \underline{/0} \qquad V_B = 75 \underline{/95°}$$

The following are the motor parameters:

$$R_1 = 320 \ \Omega \qquad X_1 = 400 \ \Omega$$
$$R_2 = 1200 \ \Omega \qquad X_2' = 400 \ \Omega$$
$$X_m = 700 \ \Omega$$

Find the output power (neglecting core and rotational losses), input power, efficiency, and torque at a slip of 0.2.

Problem 8.B.13

Find the starting torque of the motor of Problem 8.B.12.

Problem 8.B.14

The voltage supply to a two-pole 60-Hz 115-V two-phase induction motor is given by

$$V_A = 115 \underline{/0} \qquad V_B = 75 \underline{/-90°}$$

The motor parameters are

$$R_1 = 3.2 \ \Omega \qquad X_1 = 4 \ \Omega$$
$$R_2' = 12 \ \Omega \qquad X_2' = 4 \ \Omega$$
$$X_m = 7 \ \Omega$$

Find the output power (neglecting core and rotational losses), input power, efficiency, and torque at a slip of 0.2.

Problem 8.B.15

Repeat Problem 8.B.14 for a slip of 0.6.

Problem 8.B.16

Find the starting torque of the motor of Problem 8.B.14

Problem 8.B.17

A four-pole 220-V 60-Hz two-phase induction motor has the following parameters:

$$R_1 = 3 \ \Omega \qquad\qquad R_2' = 2.4 \ \Omega$$

$$X_1 = X_2' = 3 \ \Omega \qquad X_m = 100 \ \Omega$$

Assume that the applied voltages are

$$V_A = 220 \ \underline{/0} \qquad V_B = 200 \ \underline{/-90°}$$

Core and rotational losses are given by 150 W. For a slip of 4%, find the output power, torque, and efficiency.

Problem 8.B.18

Find the starting torque of the motor of Problem 8.B.17. Neglect core and rotational losses.

Problem 8.B.19

The input impedance of phase A and that of phase B of a four-pole 60-Hz two-phase induction motor running at a slip of 4% are measured as

$$Z_{iA} = 33 + j27 \ \Omega$$

$$Z_{iB} = 71 + j30 \ \Omega$$

The stator impedance per phase is given by

$$Z_1 = 3 + j3 \ \Omega$$

The phase voltages are given by

$$V_A = 220 \ \underline{/0} \qquad V_B = 200 \ \underline{/-90°}$$

It is required to find:

    a. The phase currents I_A and I_B.
    b. The sequence voltages, and currents.
    c. The power input to the motor.
    d. The output power assuming that core and rotational losses are 150 W.
    e. The efficiency of the motor.
    f. The output torque.

Introduction to Power Semiconductor Devices and Systems

9.1 INTRODUCTION

Our coverage of electric machines emphasized the importance of speed control in electric motors. Conventional techniques of motor starting and speed control rely on the use of switched resistive elements, motor-generator sets, and/or variable transformer voltage input. These methods invariably involve devices that have moving parts. The advent of the thyristor following World War II, a solid-state device capable of controlled switching at power levels, heralded a new era of static control circuits and systems. The circuits and systems provide excellent means for control of electric machines in reliable and efficient configurations.

In this chapter we discuss the foundations of solid-state power devices, their properties and ratings, as well as a brief overview of systems resulting from use of such devices.

9.2 SOLID-STATE POWER COMPONENTS

There are a number of high-power semiconductor electronic components that are employed in motor control circuits and systems. An understanding of the characteristics of each component is important as a prelude to the analysis of the basic circuits of solid-state motor drives.

535

Diodes

The diode is a semiconductor device consisting of a *PN* junction wafer designed to conduct current in one direction (from anode to cathode) but offers a high resistance in the reverse direction. The graphic symbol for a diode is shown in Figure 9.1(a) together with the direction of positive diode current i_D and voltage across the diode v_D. Note that according to this convention the diode voltage v_D is positive when the anode (A) terminal $(P$ junction) is higher in potential than the cathode (K) terminal $(N$ junction). In this case the diode is called forward biased, and conduction takes place with i_D having a finite value defined by the conditions in the rest of the circuit. For an ideal diode the voltage drop across the diode is zero in this case, as it acts as a short circuit. When the diode voltage v_D is negative, the diode is called reverse biased, as shown in Figure 9.1(b), and in this case the current $i_D = 0$. The ideal diode characteristic is shown in Figure 9.1(c).

In actual practice, the diode characteristic is more like the one displayed in Figure 9.2. In the forward region $(v_D > 0)$, the diode offers a small resistance to the flow of current. In the reverse region $(v_D < 0)$ a small amount of current called the reverse leakage current exists. If a large enough reverse voltage is applied to the diode, it will break down and allow current to flow in the reverse direction.

SCRs

Over 20 power semiconductor devices constructed of alternate layers of *P*-and *N*-type silicon semiconductor material are described by the generic term "thyristor." The term is derived from thyratron and transistor, the thyratron is a gas-filled tube that has the same characteristics as a thyristor. An important member of the thyristor group is the silicon controlled rectifier (SCR), which combines the properties of a switch and a diode in one power

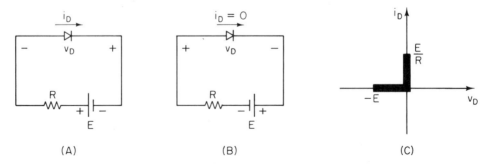

(A)   (B)   (C)

FIGURE 9.1 Simple diode circuits and ideal characteristics: (a) forward-biased state; (b) reverse-biased state; (c) ideal i_D–v_D characteristics.

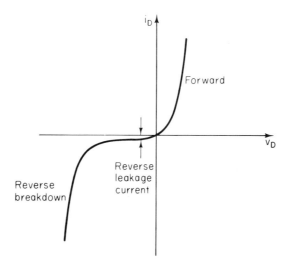

FIGURE 9.2 Actual i_D–v_D characteristic of a diode.

device. An SCR is a four-layer ($PNPN$) semiconductor device with three terminals and three junctions, as shown in Figure 9.3. The anode and cathode are power-level terminals, whereas the gate is a control terminal.

The SCR is a two-state device, acting as a high-impedance circuit element in the off-state, and presenting a low impedance in the on-state. Forward bias is a term describing the application of an external voltage to the anode terminal higher than that at the cathode (i.e., for $v_{AK} > 0$). Reverse bias is the term used when $v_{AK} < 0$. An SCR is similar to a diode when it is reverse biased, as it does not conduct and remains in the off-state for $v_{AK} < 0$. The main difference between an SCR and a diode is that when the SCR is forward biased, it conducts only if the cathode-to-gate current i_G is given a positive value for a short duration. As a result of the positive gate current pulse, the SCR's resistance decreases to a small value close to zero and the SCR remains in the on-state until the anode current ceases to flow.

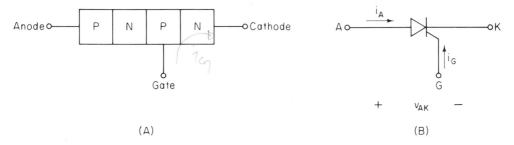

(A)

(B)

FIGURE 9.3 SCR: (a) construction; (b) graphic symbol.

With gate current maintained at zero for some time, the SCR returns to the off-state signified by a high anode-to-cathode resistance. A reverse-biased SCR can be damaged by the application of a positive gate current pulse.

SCR Performance Characteristics

The performance characteristics of the SCR are described in terms of the plot of the anode current i_A versus the anode-to-cathode voltage v_{AK}. First-quadrant performance in the i_A–v_{AK} plane corresponds to forward bias. The effect of the magnitude of v_{AK} on the state of the SCR with zero gate current is examined first.

As discussed earlier, with zero gate current the SCR does not conduct for low values of v_{AK}. As the voltage across the SCR is increased, conduction occurs at a voltage level known as the forward breakover voltage V_{FBO}. At that point the SCR conducts and the voltage between its terminals drops. The anode current increases to a value limited only by the impedance of the external load circuit. The i_A–v_{AK} characteristic for an SCR without gate current is shown in Figure 9.4. The SCR is not designed for turn-on by forward breakover voltage.

With an applied gate current, conduction does not start unless the anode-to-cathode voltage is at a certain level as indicated in Figure 9.5. The higher the value of the gate current, the lower is the anode-to-cathode voltage required for turn-on. The voltage scale in the diagram is much amplified to show the phenomenon clearly.

When the SCR is subject to negative v_{AK}, it is said to be reverse biased. In this case leakage current is small for negative voltages up to V_{RB}, the repetitive peak reverse voltage. At V_{RB} the SCR switches to the on-state with negative anode current. Operating in the reverse mode is in the third

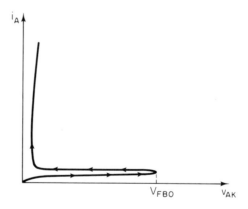

FIGURE 9.4 SCR performance characteristic with zero gate current.

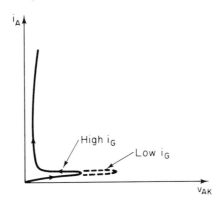

FIGURE 9.5   SCR performance characteristic with applied gate current.

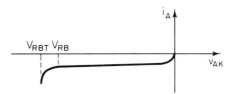

FIGURE 9.6   SCR characteristic in the reverse-blocking mode.

quadrant of the i_A–v_{AK} plane, but should be avoided. Figure 9.6 shows the i_A–v_{AK} characteristic for a reverse-blocking thyristor. Note that in the figure V_{RBT}, the nonrepetitive peak reverse voltage is slightly lower than V_{RB}.

Triacs and Diacs

The SCR has a four-layer ($PNPN$) structure and conducts in one direction only. Connecting two four-layer structures back to back as shown in Figure 9.7 results in a bidirectional device. The same effect is obtained in a five-layer device with a gate and is referred to as a triac. If the five-layer device

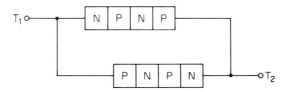

FIGURE 9.7   Bidirectional device.

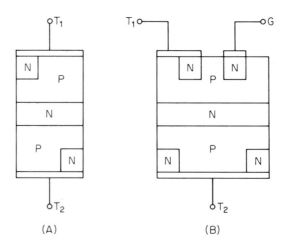

FIGURE 9.8 (a) Diac structure; (b) triac structure.

is not equipped with a gate, it is called a diac. The structure of a diac is shown in Figure 9.8(a) and that of a triac is given in Figure 9.8(b). The graphic symbols for a diac and a triac are shown in Figure 9.9.

The characteristic of a diac is shown in Figure 9.10. When terminal T_1 is at a higher potential than T_2, a small leakage current exists, up to a breakover voltage value V_{BO}, beyond which the diac conducts. This is shown as first-quadrant operation in Figure 9.10. When terminal T_2 is at a higher potential than T_1, similar behavior takes place, as shown in the third-quadrant operation in Figure 9.10.

The triac has a characteristic similar to that of the thyristor but can be made to breakover at a lower voltage by injecting a gate current in the same manner as with the SCR.

Unijunction Transistors

A unijunction transistor is a three-terminal solid-state device used for generating trigger pulses for thyristors. The UJT has an emitter (E) and two bases B_1 and B_2, as shown in Figure 9.11. The emitter is reverse biased when its voltage V_E is less than the emitter peak voltage V_P and a small reverse leakage current exists (Figure 9.12). The UJT turns on for $V_E \geq$

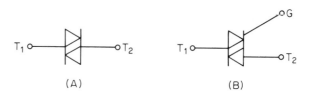

FIGURE 9.9 Graphic symbols: (a) diac; (b) triac.

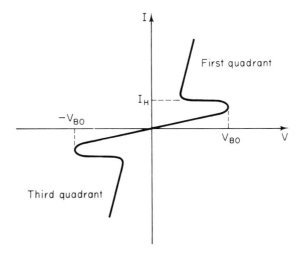

FIGURE 9.10 Diac characteristic.

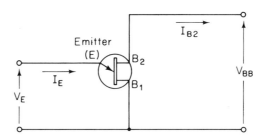

FIGURE 9.11 Unijunction transistor.

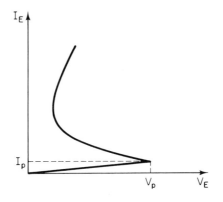

FIGURE 9.12 Current–voltage characteristic of a UJT.

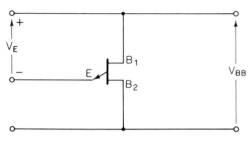

FIGURE 9.13   CUJT circuit.

V_P and I_E greater than I_P. The device resets itself to the off-state when the current from emitter to base 1 is low enough.

Complementary Unijunction Transistors

A complementary UJT (CUJT) is a semiconductor device that is similar to UJT except that the currents and voltages applied are of opposite polarity, as shown in Figure 9.13. These devices offer better stability than a UJT. The CUJT and UJT as well as the programmable unijunction transistor (PUJT) are important elements in SCR triggering circuits.

9.3   POWER ELECTRONIC SYSTEMS

Solid-state power components are able to control the switching of voltages and currents of high power levels. These components are used in power electronic systems that perform specific energy conversion functions based on this switching ability. Power electronic systems are made by suitable interconnections of thyristor devices and other components into a static energy converter. The term *static* is used to emphasize the fact that there is no moving parts in these systems. The function of a static energy converter is to convert energy at its input side from an available form (alternating current or direct current) into another form at the output side (again ac or dc). In this section we introduce a number of these systems in terms of their functions, classification, and the associated terminology. In Chapter 10 we deal with some specific configurations.

AC Voltage Controllers

Given a power source with fixed ac voltage level at the input side, an ac voltage controller converts the energy into ac but with a different value of the rms voltage at the output side, which is connected to the load. The variable-voltage ac output is used in control of lighting, heating, and induction heating equipment as well as fractional-horsepower and three-phase induction motors.

Voltage control is achieved using one of two major techniques:

1. *On-off control:* The ac source is connected to the load for a number of cycles and then disconnects it for a similar interval. The procedure is termed pulse burst modulation, where a fixed time period T is defined corresponding to a number of cycles of the ac source frequency. The period T is divided into an on-interval T_{oN} and an off-interval T_{oFF}. The duration T_{oN} is determined to correspond to speed requirements in the controlled device. The concept of on–off control is illustrated in Figure 9.14(a).

2. *Phase control:* This technique uses the switching characteristics of the SCR to connect the ac source to the load for only a portion of each cycle. The delay angle α is the control parameter in this case.

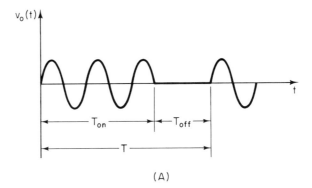

(A)

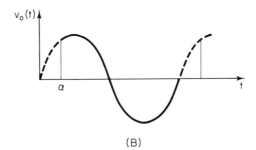

(B)

FIGURE 9.14 Principles of ac voltage control: (a) on–off or pulse burst modulation; (b) phase control.

Controlled Rectifiers

A controlled rectifier is used to vary the average value of the direct voltage applied to the load circuit. The input is a constant-voltage ac power source. The thyristors are phase controlled in this application. Controlled rectifier systems are widely used in dc motor speed-control schemes.

Controlled rectifiers are classified either according to the number of phases of the ac source or according to the number of pulses of current through the load circuit during one cycle of the source voltage. Available configurations vary from the simple single-phase half-wave rectifier (one pulse converter) to the more involved three-phase full-wave converter. An increase in the number of phases of ac power supply results in less fluctuations (power ripple) in the output side.

The term *controlled rectifier* is used whenever gate control is employed by use of thyristors. An uncontrolled rectifier is a device that uses power diodes in its function. A controlled rectifier is called a full converter or two-quadrant converter if all its active components are thyristors, since in this case the inverse process, called synchronous inversion, is possible. Synchronous inversion is the term used when the converter delivers ac from a dc input. Many dc motor control applications require four-quadrant operation, that is, both polarities of dc voltage and current. A four-quadrant converter is obtained by placing two two-quadrant converters in a back-to-back arrangement called the dual converter.

Choppers (DC-to-DC Converters)

The dc-to-dc converter or chopper is a power electronic system that converts power from a constant dc voltage source to direct voltage with variable

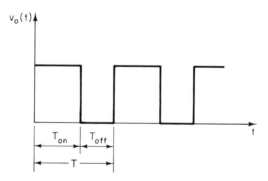

FIGURE 9.15   Principle of chopper operation.

average value. The main function of a chopper is to alternately switch the source dc voltage on and off in one of the following ways:

1. *Pulse width modulation (PWM):* In this case, using terms shown in Figure 9.15, the period T is maintained constant while T_{oN} or T_{oFF} are varied.
2. *Frequency modulation (FM):* Here T_{oN} is held constant while T_{oFF} is varied and hence T is varied.
3. *Combined modulation:* In this arrangement both T_{oN} and T_{oFF} are varied and T is varied as well. This is a combination of the PWM and FM techniques.

Inverters

The term *inverter* is used to describe a power electronic system with fixed dc input and fixed- or variable-frequency ac voltage (single-phase or three-phase) output. Control of the output voltage is achieved by either controlling the dc voltage input to the inverter or by controlling the voltage within the inverter.

Controlling the dc voltage input is done through the use of a chopper. If the source voltage is ac, we may use a phase-controlled converter or an uncontrolled rectifier with a variable dc voltage output. This latter requirement is obtained either using an autotransformer to supply the rectifier or by cascading the rectifier to a chopper.

Controlling the voltage within the inverter involves pulse width modulation, either by controlling the width of the output pulse or by rapidly switching the output of the inverter on and off a number of times during each half-cycle to produce a train of constant-amplitude pulses.

AC-to-AC Converters

This application area requires converting a fixed ac source power of a given frequency and voltage into different frequency and voltage ac output. The most obvious means of achieving this involves an intermediate dc link. In this case the given ac is converted to dc through rectification, which is followed by inversion to the desired ac voltage and frequency. A more efficient system is termed the cycloconverter and is a frequency changer that produces a low output frequency (less than one-third of the input frequency) in one stage. It is noted that the dc link converter concept is applicable for all applications in contrast to the output frequency limitations of a cycloconverter.

9.4   THE SCR: MODEL, TURN-ON, AND DYNAMIC CHARACTERISTICS

The main building block of power electronic systems is the SCR. The present section is intended to offer a brief discussion of pertinent characteristics and ratings of the SCR that should be taken into consideration in designing power electronic circuits. A first step is to consider a circuit model for this device.

Two-Transistor Model

The four-layer ($PNPN$) structure of Figure 9.16(a) can be represented as two complementary three-layer structures as shown in Figure 9.16(b). The first structure is modeled as a PNP transistor T_1 and the second is an NPN transistor T_2, as shown in Figure 9.16(c). The base of T_1 is the collector of T_2, whereas the collector of T_1 is the base of T_2. A gate voltage source V_G is connected through a switch to the gate terminal. The circuit of Figure 9.16(c) is the two-transistor model of the SCR.

Assuming that the gate current is zero $i_G = 0$, then

$$i_A = i_{E_1} = i_{E_2} \tag{9.1}$$

From the circuit, we have

$$i_{B_1} = i_{C_2} \tag{9.2}$$
$$i_{C_1} = i_{B_2} \tag{9.3}$$

From Kirchhoff's current law, neglecting leakage,

$$i_E = i_B + i_C$$

The current transfer ratio β of the transistor is defined by

$$\beta = \frac{i_C}{i_B}$$

As a result we conclude by eliminating i_B that

$$i_C = \frac{\beta}{1 + \beta} i_E$$

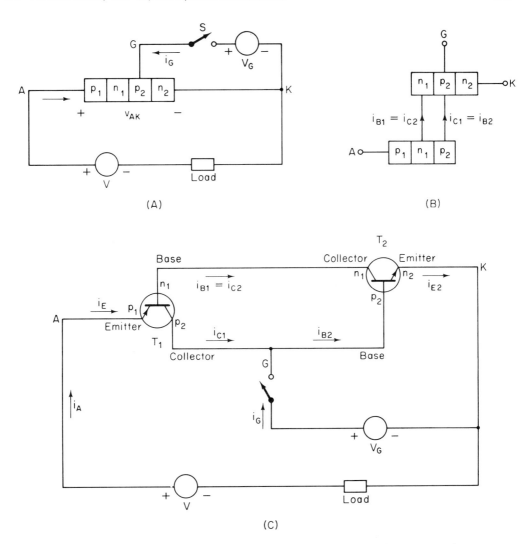

FIGURE 9.16 Two transistor model of SCR: (a) original structure; (b) two-structure model; (c) equivalent circuit.

The relation above applies for transistors 1 and 2 in the model of the thyristor. Thus we have

$$i_{C1} = \frac{\beta_1}{1 + \beta_1} i_{E1} \tag{9.4}$$

$$i_{C2} = \frac{\beta_2}{1 + \beta_2} i_{E2} \tag{9.5}$$

The anode current i_A is equal to the emitter current i_{E_1}. This is the sum of the base current i_{B_1} and collector current i_{C_1}, as well as the leakage current i_{C_0} crossing the common junction n_1–p_2:

$$i_A = i_{C_1} + i_{C_2} + i_{C_0}$$

Substituting for i_{C_1} and i_{C_2} in terms of i_{E_1} and i_{E_2}, we obtain

$$i_A = \frac{\beta_1}{1 + \beta_1} i_{E_1} + \frac{\beta_2}{1 + \beta_2} i_{E_2} + i_{C_0}$$

Note that since $i_A = i_{E_1} = i_{E_2}$, we write

$$i_A\left(1 - \frac{\beta_1}{1 + \beta_1} - \frac{\beta_2}{1 + \beta_2}\right) = i_{C_0}$$

A few simplification steps result in

$$i_A = \frac{(1 + \beta_1)(1 + \beta_2)}{1 - \beta_1\beta_2} i_{C_0} \tag{9.6}$$

The leakage current i_{C_0} is sufficiently small under normal conditions so that it can be neglected. The anode current is very small for $\beta_1\beta_2 \ll 1$, and this corresponds to the off-state (or forward blocking). The product $\beta_1\beta_2$ can approach unity under certain conditions corresponding to the turn-on state, where the anode current will be large. SCR turn-on can be due to a number of causes, as discussed next.

SCR Turn-On

The turning-on of the SCR is associated with an increase in the emitter current. Factors that cause an increase in emitter current are:

1. *Injection of gate current:* The closing of the switch connecting the gate voltage source to the SCR for a short duration allows the flow of current i_G, which increases the base current i_{B_2}. This results in an increase in i_{C_2} and β_2, and thus i_{C_1} and β_1 increase. The product $\beta_1\beta_2$ is thus increased to close to unity. As a consequence, the anode current i_A is increased to an amount that is limited only by the impedance of the external load circuit.

    Injection of a short gate current pulse is the means normally used for SCR turn-on. The effect of the magnitude of grid current i_G on conduction is discussed together with the SCR characteristics.

2. *Optical turn-on:* A beam of light directed at an unshielded SCR's junction can produce sufficient energy to initiate turn-on. In a light-activated SCR this is the normal means for device turn-on.

3. *Thermal turn-on:* An increase in the SCR's temperature results in an increase in the leakage current i_{C_0}. Collector currents and hence the product $\beta_1\beta_2$ increase and turn-on is initiated.

4. *Breakover voltage turn-on:* An increase in the forward anode–cathode voltage above a certain value V_{FB_0}, the forward breakover voltage causes a sufficient increase in the product $\beta_1\beta_2$ and hence causes SCR turn-on without an applied gate current.

5. *Rate of change of voltage:* The n_1–p_2 junction acts as a capacitor, and thus a large rate of change of voltage can produce sufficient charging current to initiate SCR turn-on.

6. *Radiation:* When permitted to strike an SCR, radiation such as X-rays and gamma rays cause a thyristor turn-on.

The initiation of an SCR turn-on results in a dynamic process that involves a number of important parameters which are discussed next.

SCR Dynamic Characteristics

At the instant of gate current application, there is no significant increase in anode current. This lasts for an interval of time referred to as the *delay time* t_d. This is defined either in terms of anode current rise or in terms of anode-to-cathode voltage drop. The time elapsed from initiation of gate current until the anode current reaches 10% of its final value is defined as the delay time t_d. The corresponding definition in terms of anode-to-cathode voltage measures t_d as the time elapsed between the instant of gate current initiation to the instant at which the value of v_{AK} is 90% of V_{FB}, the repetitive forward blocking voltage. Here V_{FB} is defined as the maximum instantaneous value of forward voltage that occurs across the thyristor, including all repetitive voltages.

The *rise time* t_r is the time interval between the instant at which v_{AK} is 90% to the instant when it is 10% of its final value. The turn-on time t_{on} is equal to $(t_d + t_r)$. The time variations of v_{AK}, i_A, and i_G are shown in Figure 9.17, where definitions of t_d, t_r, and t_{on} are clearly marked.

At the time of application of a gate current pulse to the SCR, conduction commences in a relatively small area in the neighborhood of the gate. In many designs, the area of conduction increases at the rate of about 1 cm each 10^{-4} s. The rate of increase of anode current with respect to time in circuits with relatively low inductance to resistance ratio can be higher than the rate of spread of the conduction area. As a result, high power density areas develop that are associated with local hot spots within the device. The hot spots can lead to permanent damage to the SCR.

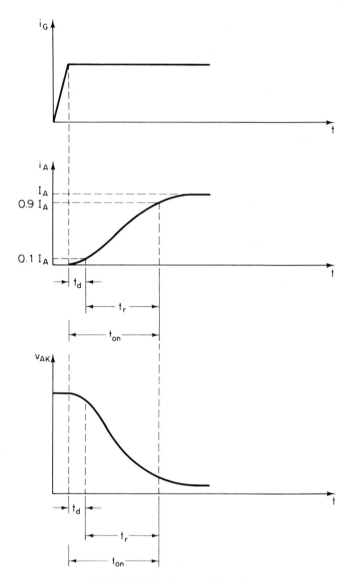

FIGURE 9.17   Turn-on dynamics.

It is clear that rates of change of currents in an SCR are of importance. The term *rate of change of anode current* refers in practice to the critical rate of rise of the on-state current. The maximum allowable rate of increase of anode current is commonly referred to as *di/dt*. To avoid *di/dt* failures two obvious solutions are available. In the first, a high value of gate current in the order of three to five times I_{gt} is applied. The second is directed at improved geometric designs of the cathode to ensure a large initial conduction area.

9.5   THE SCR: RATINGS AND PARAMETERS

A number of additional specification parameters and ratings of the SCR are discussed in this section. One of the most important aspects is that corresponding to heat dissipation in an SCR.

Junction Temperature and Power Dissipation

The operating junction temperature range in thyristor devices depends on the type and design of the unit. The lower temperature limit corresponds to thermal stress due to the difference in the thermal expansion characteristics of materials used in fabricating the device. The upper operating temperature limit is due to the dependence of the breakdown voltage and turn-off time on temperature. A storage temperature is also specified which is related to stability of junction coating compounds.

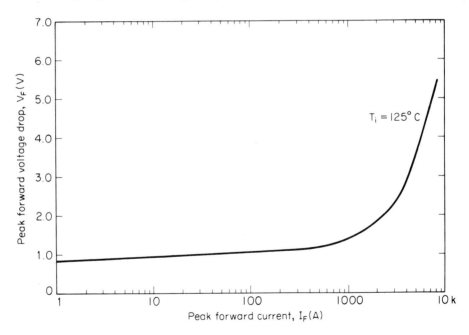

FIGURE 9.18   Forward voltage versus forward current characteristic for Westinghouse Type 282 SCR.

The rated maximum operating junction temperature is used to determine the steady state, transient, and recurrent overload capability of the device for a given heat sink system and maximum ambient temperature. Conversely, for a given loading, the required heat-sink system can be determined using the thermal impedance approach. It should be noted that the device may actually operate beyond the specified maximum operating junction temperature for transient periods and still be applied within its rating.

The forward current rating is a function of the maximum junction temperature, total device losses, and its thermal impedance. For a given junction temperature the on-state forward voltage drop V_F varies with the peak forward current i_F as shown in the typical characteristic shown in Figure 9.18. The device's power dissipation versus average forward current $I_{F(AV)}$ for the same device for different conduction angles is shown in Figure 9.19. Points on the characteristic are obtained from the given V_F–i_F characteristic and the assumed anode current waveform. Power is simply the product of current and voltage. The dependence of power dissipation on waveform is verified by comparing Figure 9.19 and 9.20, with the latter showing the characteristic for a square waveform.

The power dissipation in the junction region is due to a number of components. These are turn-on, switching, conduction, turn-off, blocking,

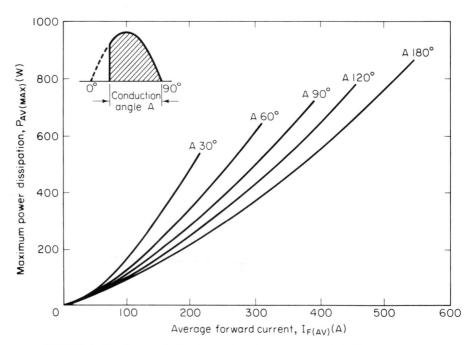

FIGURE 9.19 Power dissipation versus forward current, half-wave sinusoid, for Westinghouse Type 282 SCR.

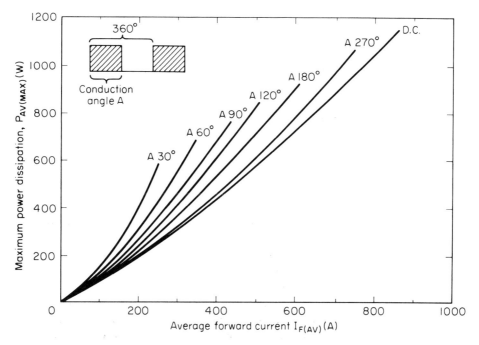

FIGURE 9.20 Power dissipation versus forward current, rectangular wave for Westinghouse Type 282 SCR.

and triggering. For normal duty cycle in the range 50 to 400 Hz and low (*di/dt*) current waveforms, on-state conduction losses are major components of junction heating.

Recurrent and Nonrecurrent SCR Current Ratings

A major requirement in SCR applications is that the junction temperature must not exceed its maximum allowed value. When this requirement is met, the device can be applied on a recurrent (repetitive) basis for normal operating conditions. The design of the circuit is thus based on the available recurrent SCR current ratings.

It is recognized, however, that the SCR may be subject to some abnormal overload conditions, such as surges during its operating life. Protective circuits are designed to intervene when this occurs. The SCR is provided with an overcurrent capability allowing for coordination with protective circuits. This capability is due to a second group of SCR current ratings which are termed nonrecurrent current ratings and allow the maximum allowable junction temperature to be exceeded for a brief duration. Nonrecurrent ratings are understood to apply to load conditions that will not occur for more than a limited number of times in the life of the SCR.

The Joint Electron Device Engineering Council (JEDEC) defines this number to be not less than 100. Nonrecurrent current ratings apply only when the overload is not repeated before the junction temperature has returned to its maximum rated value or less. The length of the interval between surges does not change the rating.

Recurrent Current Ratings

There are two ratings for the SCR in terms of current for normal recurrent operation. The first is based on the average forward current, while the second is in terms of root-mean-square (rms) current rating.

1. *Average current rating (recurrent):* Manufacturer-supplied data provide the average current ratings for a given SCR corresponding to the applied recurrent waveform and conduction angles. The variation of maximum case temperature with average anode current provides guidelines for the circuit designer in choosing the required system configuration.

2. *Rms current rating (recurrent):* Waveforms encountered in SCR applications are characterized by form factors larger than 1 (form factor = rms value/average value). The rms value of the current exceeds the average value considerably. For a fixed average current, the rms value of the current increases as the delay angle increases (conduction angle decrease). It should be recalled that heat generation is a function of the rms value of the current. Thus the rating curves are cut off at a value of average current corresponding to the limiting value of rms current for the circuit.

Nonrecurrent Current Ratings

We stated earlier in this section that nonrecurrent ratings are intended for abnormal conditions that may take place during the operation of the SCR. Presently, we discuss two nonrecurrent current ratings commonly used.

1. I^2t *rating:* During overloads, line faults, and short circuits, the SCR must withstand conditions leading to high junction temperatures. The I^2t rating is used for protective circuit coordination and is intended for emergency operation for less than one half cycle. The current I is the rms value of the fault current and t is the time in seconds.

2. *Surge current rating:* The surge current rating is based on peak current values and 180° conduction and defines a particular capability used under fault conditions. This is a nonrepetitive rating and the SCR is designed to withstand a maximum of 100 surges during its lifetime. The surge current rating is commonly denoted by I_{FM}.

Critical Rate of Rise of Forward Voltage

The application of a fast-rising forward voltage wavefront results in the flow of current through the junctions. The steeper the wavefront, the higher the likelihood for the device to turn on below its forward voltage rating. The resulting current is due to the charging effect and is a function of the junction temperature, the blocking voltage, and the rate dv/dt. Causes of high rate of rise of forward voltage include ac switching transients and surges and high-frequency inverter and cycloconverter applications.

The standard dv/dt definition relates to either of the waveforms shown in Figure 9.21. For the exponential wave, the slope dv/dt is obtained for the intersection at $t = \tau$, where τ is the time constant of the exponential waveform. Note that

$$\tau = \frac{0.632V_{FB}}{dv/dt}$$

The static dv/dt capability for a typical SCR is shown in Figure 9.22. The characteristic can be seen to depend on the bias voltage and the magnitude of voltage step. It should be noted that reverse biasing increases the dv/dt withstanding capability.

To reduce the value of dv/dt at switching, an inductor may be used. This is impractical, however, and the use of snubber circuits is recommended. In its simplest form a snubber is a series combination of a capacitor C_s and a resistance R_s connected across the SCR as shown in Figure

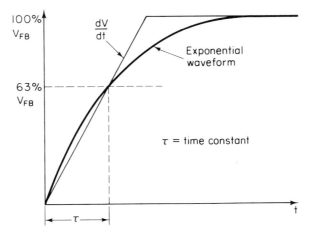

FIGURE 9.21   Standard waveforms for dv/dt definition.

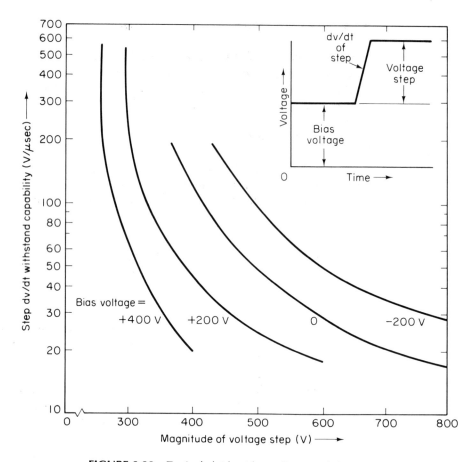

FIGURE 9.22   Typical *dv/dt* withstanding capability of SCR.

9.23. During the negative half cycle of the source voltage, the capacitor C_s will be charged in a positive direction up to the instant of SCR turn-on. At that instant there will be a step increase in current as a result of discharging C_s, and the resulting *di/dt* will be high but limited by R_s.

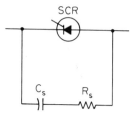

FIGURE 9.23   *dv/dt* suppression, *RC* snubber network.

9.6   INDUCTIVE LOAD SWITCHING AND FREE-WHEELERS

Our development thus far has dealt with an SCR in the conduction state. The following section deals with the process of turning off the SCR. The power electronic systems of interest are used mainly in control of electric machines which are characterized by the presence of significant inductance. The energy-storing ability of these machines presents unique switching problems that need special treatment.

Consider the circuit of Figure 9.24. The load is assumed to be represented by a series combination of a resistance R and an inductance L. The switch S is assumed initially open and hence no current exists in the circuit prior to the instant $t = 0$, when the switch is closed. For $t > 0$, the current $i(t)$ in the circuit is related to the source voltage V by Kirchhoff's voltage equation:

$$L \frac{di}{dt} + Ri = V$$

The solution is given by

$$i(t) = \frac{V}{R} (1 - e^{-t/\tau}) \tag{9.7}$$

where τ is the time constant given by

$$\tau = \frac{L}{R}$$

The current waveform $i(t)$ is shown in Figure 9.24(b). The voltage across the inductor is given by

$$V_L(t) = V e^{-t/\tau} \tag{9.8}$$

This voltage decays to zero as time approaches infinity, as shown in Figure 9.24(c).

The power fed into the inductor is obtained as

$$P_L(t) = \frac{V^2}{R} e^{-t/\tau} (1 - e^{-t/\tau}) \tag{9.9}$$

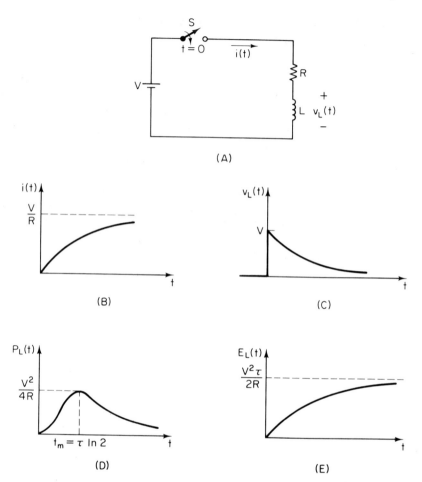

FIGURE 9.24 *RL* series load and associated waveforms: (a) circuit diagram;   (b) current waveform;   (c) inductor voltage;   (d) power into the inductor;   (e) energy into the conductor.

This is shown in Figure 9.24(d). The energy stored in the inductor is found to be

$$E_L(t) = \frac{V^2 L}{2R^2}(1 - e^{-t/\tau})^2 \qquad (9.10)$$

The energy $E(t)$ is shown in Figure 9.24(e). Note that the energy stored in the inductor reaches a steady-state value of

$$E_{ss} = \frac{V^2 L}{2R^2}$$

Assume now that the switch S is reopened at time $t = t_1$. This action demands that the circuit current be reduced to zero instantaneously and thus requires that the rate of change of current be of a large negative value approaching minus infinity. As a result, the voltage across the inductance will approach minus infinity. Associated with this is a release of energy stored in the inductor $E(t_1)$ that tends to maintain the current in the circuit from changing. In the case of power switches such as circuit breakers, the large voltage appearing between the switch's terminals ionizes the environment (air or oil) and causes it to conduct forming a high-resistance arc through which the stored energy is dissipated.

The development above points to a problem that arises when switching is done by using a power semiconductor device. The energy stored in the inductor requires a path to follow that does not include the switching device, which may otherwise be destroyed. This requirement is met by inserting a diode in parallel with the load circuit, as shown in Figure 9.25(a). The diode is sometimes referred to as a "free-wheeling" diode. An example is taken up at this point.

Example 9.1   Assume that in the circuit shown in Figure 9.24, the value of $V = 100$ V and the load parameters are

$$R = 4 \ \Omega \qquad L = 20 \ \text{mH}$$

The switch S is closed at $t = 0$. Find the following as functions of time.

a. The current $i(t)$ in the circuit.
b. The voltages across the resistance and inductance.
c. The power supplied by the voltage source, the power into the inductance, and the power into the resistance.
d. The energy supplied by the voltage source, the energy stored in the inductance, and the energy dissipated in the resistance.

Solution

The circuit time constant is

$$\tau = \frac{L}{R} = \frac{20 \times 10^{-3}}{4} = 5 \times 10^{-3} \ \text{s}$$

(a) Using Eq. (9.7), we get the circuit current as

$$i(t) = 25(1 - e^{-200t})$$

(b) Using Eq. (9.8), we get

$$v_L(t) = 100e^{-200t}$$

The voltage across the resistor is

$$v_R(t) = Ri(t) = 100(1 - e^{-200t})$$

(c) The power supplied by the source is

$$P_s(t) = Vi(t)$$
$$= 2500(1 - e^{-200t})$$

The power into the inductance using Eq. (9.9) is

$$P_L(t) = 2500e^{-200t}(1 - e^{-200t})$$

The power into the resistance is

$$P_R(t) = Ri^2(t) = 2500(1 - e^{-200t})^2$$

Note that the prinicple of conservation of power holds true in our result.
(d) The energy supplied by the source is

$$E_s(t) = \int_0^t P_s(t)\, dt$$
$$= 2500[t + 0.005(e^{-200t} - 1)]$$

The energy stored in the inductance according to Eq. (9.10) is

$$E_L(t) = 6.25(1 - e^{-200t})^2$$

The energy dissipated in the resistor is

$$E_R(t) = 2500(t + 0.01e^{-200t} - 0.0025e^{-400t} - 0.0075)$$

Note again that the sum of $E_L(t)$ and $E_R(t)$ is equal to $E_s(t)$.

Switching with Free-Wheeling Diodes

To provide a path of flow of energy stored in an inductive branch following the opening of the switch S, a diode is inserted as shown in Figure 9.25(a). With the switch S closed, the voltage across the diode's terminals is the source voltage V. The diode does not conduct in this case and the circuit behaves in the same manner as if the diode were an open circuit. As soon as the switch S is opened, a large negative voltage results at the inductor and the diode is now in a conducting state. The diode now appears as a short circuit across the RL branch. It is thus evident that with the switch opened, a mesh including the RL branch and the diode is available for the

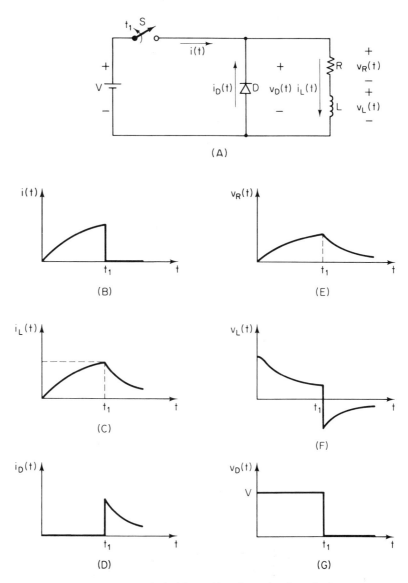

FIGURE 9.25   Switching with a free-wheeling diode.

energy trapped in the inductor to flow. An analysis of the circuit performance is offered next.

Let us now consider the circuit of Figure 9.25(a) where the free-wheeling diode D is placed in parallel with the RL load of the circuit of Figure 9.24(a). The switch S is closed at $t = 0$ and remains closed up to $t = t_1$ when it is re-opened. It is clear that the current in the diode $i_D(t)$ is zero while the switch is closed. The voltage across the diode $v_D(t)$ is that of the supply voltage

$$i_D(t) = 0 \qquad 0 < t < t_1 \tag{9.11}$$

$$v_D(t) = V \qquad 0 < t < t_1 \tag{9.12}$$

The load current $i_L(t)$ is equal to the source current:

$$i_L(t) = i(t) = \frac{V}{R}(1 - e^{-t/\tau}) \qquad 0 < t < t_1 \tag{9.13}$$

The voltages across the resistor R and the inductor L are given by

$$v_R(t) = V(1 - e^{-t/\tau}) \qquad 0 < t < t_1 \tag{9.14}$$

$$v_L(t) = Ve^{-t/\tau} \qquad 0 < t < t_1 \tag{9.15}$$

The expressions given above are simply restatements of the results obtained in the absence of the diode.

The action of opening the switch S at $t = t_1$ will result in the requirement that the current in the line $i(t)$ be zero:

$$i(t) = 0 \qquad t > t_1 \tag{9.16}$$

The load current $i_L(t)$, however, will circulate through the diode and thus

$$i_D(t) = i_L(t) \qquad t > t_1 \tag{9.17}$$

The diode is conducting for $t > t_1$ and hence the voltage across its terminals is zero (assuming ideal diode):

$$v_D(t) = 0 \qquad t > t_1 \tag{9.18}$$

Applying Kirchhoff's voltage equation around the loop formed by the RL series combination and the diode, we obtain

$$v_L(t) + v_R(t) - v_D(t) = 0 \tag{9.19}$$

Thus

$$L \frac{di_L}{dt} + Ri_L(t) = 0 \qquad t > t_1 \tag{9.20}$$

The solution is

$$i_L(t) = i_L(t_1)e^{-(t-t_1)/\tau} \qquad t > t_1 \tag{9.21}$$

The current $i_L(t_1)$ is given by

$$i_L(t_1) = \frac{V}{R}(1 - e^{-t_1/\tau}) \tag{9.22}$$

The voltage across the resistor is thus

$$v_R(t) = Ri_L(t) = Ri_L(t_1)e^{-(t-t_1)/\tau} \qquad t > t_1 \tag{9.23}$$

The voltage across the inductor is given by

$$v_L(t) = -v_R(t) \tag{9.24}$$

In Figure 9.25(b), the line current $i(t)$ is shown to increase according to Eq. (9.13) exponentially up to $t = t_1$. Beyond $t = t_1$ with the switch open the current is zero. The current in the inductor $i_L(t)$ is the same as the line current $i(t)$ for $0 < t < t_1$. With the switch opened, the current $i(t)$ decays according to Eq. (9.21), as shown in Figure 9.25(c). The diode current $i_D(t)$ is shown in Figure 9.25(d) as zero up to $t = t_1$ and is equal to $i_L(t)$ for $t > t_1$. Figure 9.25(e) shows the voltage across the resistor $v_R(t)$, and Figure 9.25(f) shows the voltage across the inductor $v_L(t)$. The diode voltage $v_D(t)$ is shown in Figure 9.25(g). Note that from Figures 9.25(e) to (g) we can verify that $v_D(t) = v_R(t) + v_L(t)$.

It is thus clear from the development above that inserting the diode D in parallel with the load enables the energy stored in the inductor to be dissipated through the resistance R following the opening of the switch. A numerical example using the results of Example 9.1 is considered now.

Example 9.2   Assume that the circuit of Example 9.1 is equipped with a free-wheeling diode as shown in Figure 9.25. The switch is now opened at $t_1 = 0.01$ s. Find the energy stored in the inductor as a function of time for $t > t_1$.

Solution

At $t_1 = 0.01$, we have

$$i_L(t_1) = i(t_1) = 25(1 - e^{-200 \times 0.01}) = 21.617 \text{ A}$$

The energy stored in the inductor at t_1 is

$$E_L(t_1) = 6.25(1 - e^{-200 \times 0.01})^2 = 4.6728 \text{ J}$$

For $t > t_1$, we have, using Eq. (9.21),

$$i_L(t) = 21.617e^{-200(t - 0.01)}$$

The voltage across the inductor is

$$v_L(t) = -v_R(t) = -Ri_L(t) = -86.467e^{-200(t - 0.01)}$$

Thus the power into the inductor is given by

$$P_L(t) = v_L(t)i_L(t)$$
$$= -1869.1127e^{-400(t - 0.01)}$$

Note that the energy stored in the inductor is now being transferred to the resistor via the diode. The energy in the inductor is

$$E_L(t) = E_L(t_1) + \int_{t_1}^{t} P_L(t) \, dt$$
$$= 4.6728 + \int_{t_1}^{t} - 1869.1127e^{-400(t - 0.01)} \, dt$$
$$= 4.6728e^{-400(t - 0.01)}$$

Note that this decays to zero as t increases.

The method discussed above provides a protection mechanism for the switch (read SCR instead of switch for our purposes) by dissipating the energy stored in the inductor through the load resistance. The amounts of heat generation in practical circuits makes this method impractical. Moreover, the circuit efficiency is low. If the source is capable of absorbing the energy stored in the inductor, an energy recovery scheme which is a modification of our present circuit will prove useful, as discussed next.

Trapped Energy Recovery Circuit

The basic idea in the trapped energy recovery circuit is to introduce a transformer in the arrangement. The primary winding is in the RL branch and if possible would be the inductor coil L itself. The secondary winding of the transformer is connected to the source through the free-wheeling diode, as shown in the circuit of Figure 9.26(a). The diode conducts when the switch is opened; energy stored in the inductor L is returned to the source through the transformer secondary and the diode.

To analyze the performance of the circuit, it is convenient to use equivalent circuits such as those shown in Figure 9.26(b) and (c). In the circuit of Figure 9.26(b), the transformer is modeled as an ideal transformer with transformation ratio $N_1:N_2$ and a magnetizing branch consisting of the series combination of the R and L. The magnetizing current is denoted by

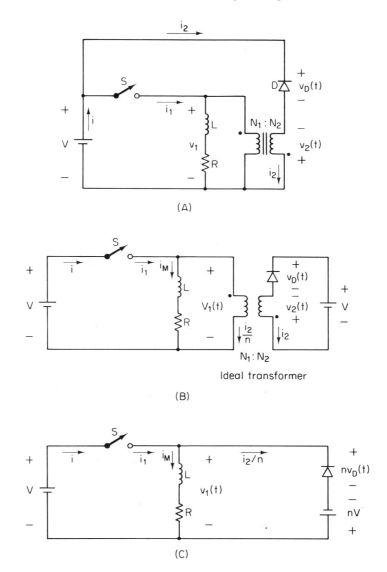

FIGURE 9.26 Trapped energy recovery circuit and reduced models.

i_M. The voltage source in the secondary circuit is included in the circuit as a convenient means of representing the fact that the diode and secondary winding are connected across the source V.

In the circuit of Figure 9.26(c), the secondary circuit is referred to the primary side. The transformation ratio n is defined by

$$n = \frac{N_1}{N_2} \tag{9.25}$$

Note the polarity of the transformer in inspecting the equivalent circuit.

The switch S is closed at $t = 0$ and remains closed up to time $t = t_1$. It is clear that the diode does not conduct in this time interval. The analysis of the circuit is the same as before and we can write

$$\frac{i_2(t)}{n} = 0 \qquad 0 \leqslant t < t_1 \tag{9.26}$$

Thus

$$i_M(t) = i_1(t) \qquad 0 \leqslant t \leqslant t_1 \tag{9.27}$$

With zero initial conditions we can write

$$i_1(t) = \frac{V}{R}(1 - e^{-t/\tau}) \qquad 0 \leqslant t \leqslant t_1 \tag{9.28}$$

The voltage $v_1(t)$ can be expressed in terms of the source voltage as

$$v_1(t) = V \tag{9.29}$$

On the other hand, we can write

$$v_1(t) = n[v_D(t) - V] \tag{9.30}$$

Thus we conclude that the voltage across the diode is given by

$$v_D(t) = V\left(1 + \frac{1}{n}\right) \qquad 0 < t \leqslant t_1 \tag{9.31}$$

The switch S is opened at $t = t_1$. As a result, we have

$$i_1(t) = 0 \qquad t > t_1 \tag{9.32}$$

Thus

$$i_M(t) = \frac{-i_2(t)}{n} \qquad t > t_1 \tag{9.33}$$

The diode is now conducting

$$nv_D(t) = 0 \tag{9.34}$$

We can thus write

$$v_1(t) = -nV \qquad t > t_1 \tag{9.35}$$

We also have

$$v_1(t) = -\frac{1}{n}\left(L\frac{di_2}{dt} + Ri_2\right) \tag{9.36}$$

Thus

$$L\frac{di_2}{dt} + Ri_2(t) = n^2V \tag{9.37}$$

The solution to this equation is

$$i_2(t) = Ae^{-t/\tau} + \frac{n^2V}{R} \tag{9.38}$$

The constant A has to be determined on the basis of boundary conditions at $t = t_1$. Thus we have

$$A = \left[i_2(t_1) - \frac{n^2V}{R}\right]e^{t_1/\tau} \tag{9.39}$$

The solution to the differential equation is written as

$$i_2(t) = i_2(t_1)e^{-(t-t_1)/\tau}$$
$$+ \frac{n^2V}{R}\left[1 - e^{-(t-t_1)/\tau}\right] \qquad t \geqslant t_1 \tag{9.40}$$

The initial condition for the equation above is obtained from the fact that the current in the inductor cannot change instantaneously when the switch is opened at $t = t_1$. Thus

$$i_M(t_1^-) = -\frac{i_2(t_1^+)}{n} = i_1(t_1^-) \qquad (9.41)$$

In the above, the superscript $(-)$ denotes time immediately before t_1 and the superscript $(+)$ denotes time immediately after t_1. The current $i_1(t_1^-)$ is obtained from Eq. (9.28) as

$$i_1(t_1^-) = \frac{V}{R}(1 - e^{-t_1/\tau}) \qquad (9.42)$$

Thus we have

$$i_2(t_1^+) = -\frac{nV}{R}(1 - e^{-t_1/\tau}) \qquad (9.43)$$

Substituting in Eq. (9.40) and collecting terms, we obtain

$$i_2(t) = \frac{n^2V}{R}\left\{1 + \frac{1}{n}e^{-t/\tau}[1 - (1 + n)e^{t_1/\tau}]\right\} \qquad (9.44)$$

The expression above is applicable for $t \geq t_1$. Note that at $t = t_1$, the current $i_2(t_1)$ is negative. The steady-state value of the current $i_2(t)$ according to the above expression is positive. It is thus clear that there is a time instant denoted by t_c, at which i_2 is zero. This is obtained as the solution to

$$1 + \frac{1}{n}e^{-t_c/\tau}[1 - (1 + n)e^{t_1/\tau}] = 0 \qquad (9.45)$$

A few reduction steps result in

$$t_c = \tau \, ln \frac{(n + 1)e^{t_1/\tau} - 1}{n} \qquad (9.46)$$

The value of the voltage $v_1(t)$ is zero at $t = t_c$ and the forward current in the diode i_2 is zero. Thus the diode voltage rises to V, that is,

$$v_D(t_c) = V \qquad (9.47)$$

The diode ceases to conduct and all variables in the circuit are at zero values. The energy stored in the inductor has all been returned to the source and the switching process is completed.

It is instructive to summarize our findings by showing time variations of the variables considered. The current $i_1(t)$ flowing from the switch side to the load is given for $0 \leq t < t_1$ by

$$i_1(t) = \frac{V}{R}(1 - e^{-t/\tau}) \qquad (9.48)$$

Beyond t_1, its value is

$$i_1(t) = 0 \qquad t \geq t_1 \qquad (9.49)$$

This is shown in Figure 9.27(a). The current in the secondary circuit $i_2(t)$ is given by

$$i_2(t) = \begin{cases} 0 & 0 \leq t < t_1 \\ \dfrac{n^2 V}{R}\left\{1 + \dfrac{1}{n}e^{-t/\tau}[1 - (1+n)e^{t_1/\tau}]\right\} & t_1 \leq t < t_c \quad (9.50) \\ 0 & t \geq t_c \end{cases}$$

The time variation of $i_2(t)$ is shown in Figure 9.27(b).

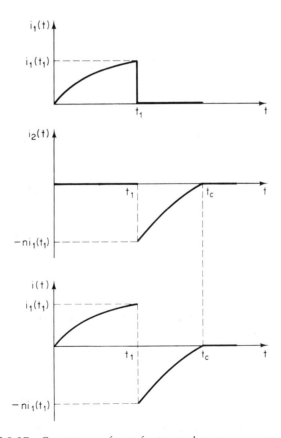

FIGURE 9.27 Current waveforms for trapped energy recovery circuit.

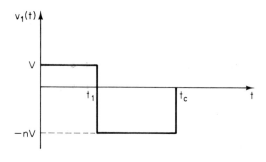

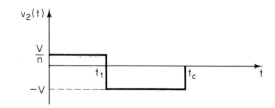

FIGURE 9.28   Voltage waveforms for trapped energy recovery circuit.

The source current $i(t)$ can be seen from inspection of the circuit of Figure 9.26(a) to be

$$i(t) = \begin{cases} i_1(t) & 0 \leqslant t < t_1 \\ i_2(t) & t_1 \leqslant t < t_c \\ 0 & t \geqslant t_c \end{cases} \qquad (9.51)$$

The voltage $v_1(t)$ on the primary side is given by

$$v_1(t) = \begin{cases} V & 0 \leqslant t < t_1 \\ -nV & t_1 < t < t_c \\ 0 & t > t_c \end{cases} \qquad (9.52)$$

The time variation $v_1(t)$ is as shown in Figure 9.28(a). The secondary voltage $v_2(t)$ varies according to

$$
v_2(t) = \begin{cases} \dfrac{V}{n} & 0 \leqslant t < t_1 \\[2mm] -V & t_1 \leqslant t < t_c \\[2mm] 0 & t > t_c \end{cases} \tag{9.53}
$$

The time variation $v_2(t)$ is as shown in Figure 9.28(b). Finally, the diode voltage $v_D(t)$ is given by

$$
v_D(t) = \begin{cases} V\left(1 + \dfrac{1}{n}\right) & 0 \leqslant t < t_1 \\[2mm] 0 & t_1 \leqslant t < t_c \\[2mm] V & t > t_c \end{cases} \tag{9.54}
$$

The voltage function $v_D(t)$ is shown in Figure 9.28(c).

Example 9.3   An energy recovery circuit like that of Figure 9.26 has a transformer ratio $n = 2$. Assume that t_1, L, R, and V have the same values as specified in Examples 9.1 and 9.2. Find the value of t_c. Calculate the value of energy returned to the source, that absorbed by R over the time interval t_1 to t_c. Show that the energy stored in the inductor during the interval $t = 0$ to t_1 is completely discharged in the interval t_1 to t_c.

Solution

Using Eq. (9.46), we get

$$
t_c = 200 \ln\frac{3e^2 - 1}{2}
$$

$$
= 0.0118 \text{ s}
$$

Using Eq. (9.44), we get

$$
i_2(t) = \frac{(2)^2(100)}{4}\left[1 + \tfrac{1}{2}e^{-200t}(1 - 3e^2)\right]
$$

$$
= 100(1 - 10.5836e^{-200t})
$$

The energy supplied by the source is

$$
E_s = \int_{t_1}^{t_c} V i_2(t)\, dt
$$

$$
= -3.6515 \text{ J}
$$

Note that this is negative, indicating that energy is being recovered by the source.

The energy absorbed by the resistor is

$$E_r = \int_{t_1}^{t_c} \frac{R}{n^2} i_2^2(t)\, dt$$

$$= 1.0214 \text{ J}$$

As a result, the sum of energy taken by source and resistor is

$$E_r - E_s = 4.673 \text{ J}$$

The energy stored in the inductor at t_1 was found to be 4.6729 J and hence

$$E_L(t_1) = E_r - E_s$$

9.7 THE SCR: TURN-OFF AND COMMUTATION

When the SCR is conducting, each of the three junctions is forward biased and the base regions are saturated. Conduction is maintained without a gate signal. To achieve turn-off, the load current must be diverted or brought to zero for a specific length of time to allow recombination of charge carriers. Once recombination takes place, the device regains its forward blocking capability. Normally, the anode-to-cathode voltage is reduced until the holding current can no longer be supported to achieve turn-off.

The process of turn-off is known as commutation. In ac applications the required conditions for turn-off can be satisfied. For dc applications, it is necessary to introduce a reverse bias by external means to reduce the anode current. The process is then called forced commutation.

The time interval t_{off} which must elapse after the forward current has ceased before forward voltage may again be applied without turn-on is called "turn-off time." Typical turn-off times are specified by the manufacturer. To measure the required interval, the device is operated with current and voltage waveforms as shown in Figure 9.29. The interval between t_2 and t_7 is then decreased until the SCR can just support reapplied forward voltage.

The waveforms of Figure 9.29 correspond to the conventional rating procedure where a current pulse of sufficient width is applied to ensure that the total cathode region is in conduction prior to the turn-off cycle. A second procedure is termed "concurrent testing" and involves a narrow current

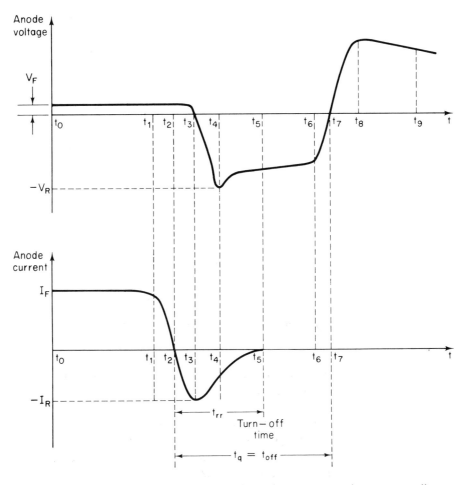

FIGURE 9.29   Voltage and current waveforms for conventional rating turn-off.

pulse as shown in Figure 9.30. In this case forward losses enter into the rating.

In Figure 9.29 the anode current is I_F in the interval t_0 to t_1 and the corresponding anode voltage is V_F. At the time instant t_1, the anode current is reduced at the rate di_R/dt, becomes zero at t_2, and continues decreasing up to t_3, where it attains its peak value I_R, called reverse recovery current. The anode voltage is positive up to t_3. At t_3, the anode voltage becomes reverse biased and decreases to a peak $(-V_R)$ at t_4. Between t_3 and t_5 the negative anode current is decreasing to zero at t_5. The interval t_2 to t_5 is called the reverse recovery time t_{rr}. The reappearance of a forward voltage at t_7 should not cause a turn-on without a gate signal.

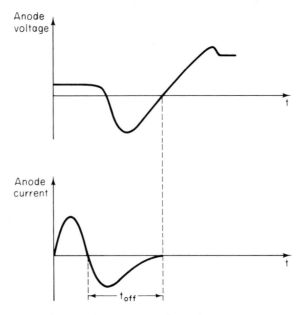

FIGURE 9.30   Voltage and current waveforms for concurrent rating turn-off.

The turn-off time interval t_{off} is not a constant for a given device. The following factors affect the length t_{off}:

1. *Junction temperature:* The required turn-off time increases as the junction temperature increases.
2. *Forward current magnitude:* The value of t_{off} increases with an increase in the magnitude of the peak forward current.
3. *Rate of decay of forward current* (t_1 to t_2): An increase in di_R/dt increases t_{off}.
4. *Peak reverse current* (t_3): A decrease in I_R increases t_{off}.
5. *Reverse voltage* (t_4 to t_6): A decrease in this parameter will increase t_{off}.
6. An increase in the rate of reapplication of forward blocking voltage will increase t_{off}.
7. An increase in forward blocking voltage will increase t_{off}.

Reapplied dv/dt

The specification of the reapplied *dv/dt* is part of the SCR's turn-off specification. It is defined as the maximum allowable rate of reapplication of off-state blocking voltage, while the device is regaining its rated off-state blocking voltage, following the device's turn-off time t_q under given circuit and temperature conditions.

There are many methods to produce commutation in power electronic systems. The chief methods are discussed briefly next.

Method A: Line Commutation

This commutation method applies in the case of circuits with an alternating voltage source in series with the SCR and load such as that shown in Figure 9.31. Load current will flow in the circuit during the positive half-cycle. The current in the circuit is zero at an instant of time, following which the forward voltage on the SCR will be negative. With zero gate signal, the thyristor will turn off provided that the duration of the half cycle is longer than the turn-off time of the device. Figure 9.32 shows the waveforms involved.

Method B: Series Capacitor Commutation

The simplest form of this commutation method involves the insertion of capacitor C in series with the inductive load (RL) as shown in Figure 9.33. The free-wheeling diode D is used to provide means for dissipating the energy stored in L at the time of switching as discussed earlier.

Application of Kirchhoff's voltage law around the loop results in

$$L\frac{di}{dt} + \frac{1}{C}\int_0^t i\, dt + v_c(0) + Ri = V_d \tag{9.55}$$

Note that the thyristor is assumed to conduct and thus $v_{AK} = 0$. The initial voltage on the capacitor is denoted by $v_c(0)$. The equation above is differentiated to yield

$$\frac{d^2i}{dt^2} + \frac{R}{L}\frac{di}{dt} + \frac{l}{LC}i = 0$$

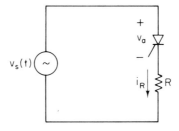

FIGURE 9.31   Circuit to illustrate ac line commutation.

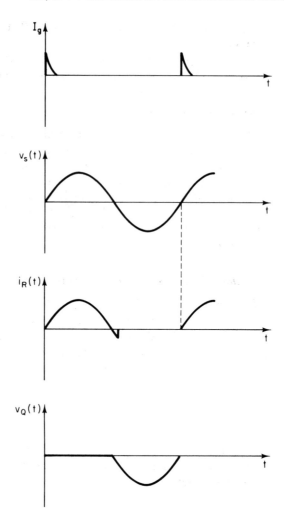

FIGURE 9.32   Waveforms for ac line commutation.

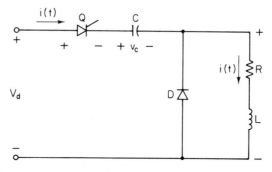

FIGURE 9.33   Basic circuit for series capacitor series commutation.

In dealing with second-order systems such as the present circuit, it is customary to formulate the solution in terms of a canonical (standard) form by defining the resonant frequency ω_n and damping ratio ξ by

$$\omega_n = \frac{1}{\sqrt{LC}} \tag{9.56}$$

$$2\xi\omega_n = \frac{R}{L} \tag{9.57}$$

$$\sigma = \xi\omega_n \tag{9.58}$$

$$\omega_d = \omega_n \sqrt{1 - \xi^2} \tag{9.59}$$

The radian frequency ω_d is termed the damped natural frequency or the ringing radian frequency. It is assumed here that $\xi < 1$, or that the circuit is underdamped. As a result, the current $i(t)$ is given by

$$i(t) = e^{-\sigma t}(Ae^{j\omega_d t} + Be^{-j\omega_d t})$$

Using Euler's equation, we obtain

$$i(t) = e^{-\sigma t}[(A + B) \cos \omega_d t + j(A - B) \sin \omega_d t]$$

The constants A and B are determined from the initial conditions $i(0) = 0$ and

$$v_L(0) + v_c(0) = V_d$$

As a result, we get

$$i(t) = \frac{V_d - v_c(0)}{\omega_d L} e^{-\sigma t} \sin \omega_d t \tag{9.60}$$

It is thus clear that the current in the circuit is a decaying sinusoid, as shown in Figure 9.34.

The capacitor voltage $v_c(t)$ can be obtained as

$$v_c(t) = V_d - \frac{V_d - v_c(0)}{\omega_d} e^{-\sigma t}(\sigma \sin \omega_d t + \omega_d \cos \omega_d t) \tag{9.61}$$

It is clear that the current $i(t)$ in the circuit becomes zero at the time instant t_1 given by

$$\omega_d t_1 = \pi \tag{9.62}$$

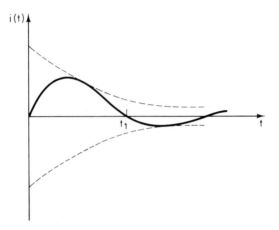

FIGURE 9.34   Current in *RLC* circuit.

At that instant of time the thyristor is turned off. The voltage on the capacitor at that time instant is given by

$$v_c(t_1) = V_d + [V_d - v_c(0)]e^{-\sigma\pi/\omega_d}$$

Since the current $i(t_1)$ is zero, and in accordance with Kirchhoff's voltage law, the voltage across the thyristor is obtained as

$$v_{AK}(t_1) = V_d - v_c(t_1)$$

$$= [v_c(0) - V_d]e^{-\sigma\pi/\omega_d} \qquad (9.63)$$

Note that since $v_c(0) < V_d$, the voltage v_{AK} is negative. The variation of $i(t)$, $v_c(t)$, and $v_{AK}(t)$ with time for this circuit is shown in Figure 9.35.

It is thus possible to turn off the thyristor as the current goes through zero in a finite time. The negative value of v_{AK} following turn-off indicates that some method of discharging the capacitor must be used before the thyristor can conduct again.

Example 9.4   The load of Example 9.1 is commutated at $t_1 = 0.01$ using series capacitor commutation. Find the value of the required capacitance.

Solution

Combining Eqs. (9.62), (9.59), and (9.57), we have

$$\frac{2\xi}{\sqrt{1 - \zeta^2}} = \frac{4(0.01)}{20 \times 10^{-3}\pi}$$

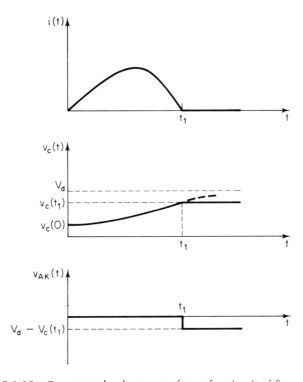

FIGURE 9.35 Current and voltage waveforms for circuit of figure 9.33.

As a result

$$\xi = 0.3033$$

Using Eq.(9.57), we get

$$\omega_n = \frac{R}{2L\xi} = \frac{4}{(40 \times 10^{-3})(0.3033)} = 329.691$$

But from Eq. (9.56),

$$C = \frac{1}{\omega_n^2 L} = \frac{1}{(329.691)^2(20 \times 10^{-3})}$$

As a result, the required capacitance value is

$$C = 460 \times 10^{-6} \text{ F}$$

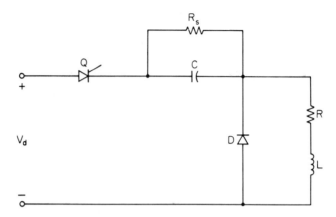

FIGURE 9.36   Series capacitor commutation with shunt resistor for capacitor discharge.

One possible means of discharging the capacitor is to employ a resistor in parallel with the capacitor, as shown in Figure 9.36. The time constant of the R_sC combination determines the minimum time to elapse before a gate pulse is applied to the thyristor to commence conduction.

An improved scheme of discharging the capacitor is shown in Figure 9.37. In this case an inductor L_s in series with a thyristor Q_2 are connected in parallel with the capacitor. While Q_1 is conducting, thyristor Q_2 is off and the capacitor is charged. The thyristor Q_2 is turned on at a time later than the turn-off time of Q_1, allowing discharge of the capacitor through the inductor L_s and the thyristor Q_2.

The principle of load commutation employing a series capacitor is

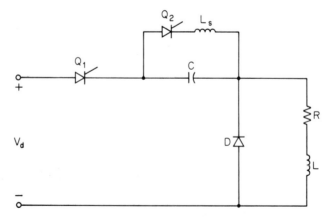

FIGURE 9.37   Series capacitor commutation with capacitor discharge through inductor L_s and thyristor Q_2.

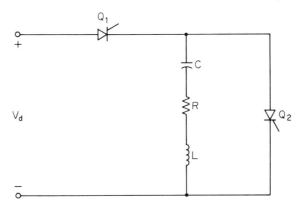

FIGURE 9.38 Basic series inverter circuit.

used in series inverters for high output frequency in the range 200 Hz to 100 kHz. Figure 9.38 shows a basic series inverter circuit. Note that the thyristor Q_2 is placed in parallel with the load and series capacitor. The thyristors are turned on alternately by short pulses of gate currents i_{G_1} and i_{G_2} applied at intervals matching the required load frequency. It should be noted that in this circuit arrangement the source may be short-circuited through Q_1 and Q_2. This occurs when Q_1 has not had enough time to turn off, and simultaneously Q_2 is turned on by the high dv/dt resulting from the capacitor discharge. It is for this reason that Q_1 and Q_2 are provided with protective elements. The protection is in the form of snubber r_1c_1 and inductor L_1 for thyristor Q_1 and similarly for Q_2, as shown in Figure 9.39.

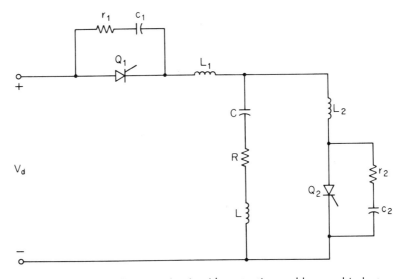

FIGURE 9.39 Series inverter circuit with protective snubbers and inductors.

Method C: Parallel-Capacitor Commutation

The basic concept in parallel-capacitor commutation is to switch a pre-charged capacitor across the thyristor at the beginning of commutation. The application of a reverse bias across the thyristor causes turn-off. As shown in Figure 9.40, the capacitor is switched across the thyristor by the application of a gate pulse i_{G_2} to the second thyristor Q_2, causing it to conduct and therefore creating the extinction loop made of C, Q_2, and Q_1.

Initially Q_1 is conducting and the load current flows through the loop made by the source, thyristor Q_1, and load RL. The capacitor is initially charged to a voltage $(-V_{c0})$ by external means and Q_2 is not conducting. At the time Q_2 is fired (through the application of gate pulse i_{G_2}), the load current I_L is diverted from Q_1 to the branch Q_2–C. We thus have

$$I_L = C \frac{dv_c}{dt}$$

Assuming I_L to be constant in the interval of interest, we conclude that

$$v_c(t) = v_c(0) + \frac{I_L t}{C}$$

$$= -V_{c0} + \frac{I_L t}{C}$$

Thus the capacitor voltage rises linearly from $(-V_{c0})$ to positive values. The capacitor voltage becomes zero at t_0 obtained from

$$0 = -V_{c0} + \frac{I_L t_0}{C}$$

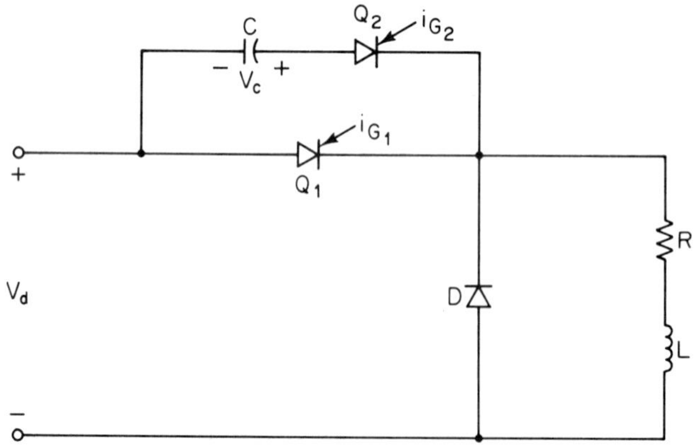

FIGURE 9.40 Basic concept of parallel-capacitor commutation.

Thus

$$t_0 = \frac{CV_{c0}}{I_L} \qquad (9.64)$$

The time t_0 is shown as the available turn-off time.

Example 9.5   Assume that the load of Example 9.1 is commutated using parallel capacitor commutation of the circuit shown in Figure 9.40. Find the available turn-off time for the following combinations:

     a. $C = 10 \times 10^{-3}$ F and $V_{c0} = 10$ V
     b. $C = 200 \times 10^{-6}$ F and $V_{c0} = 50$ V

Assume that commutation commences at $t = 0.01$ s.

Solution

From Example 9.1 at $t = 0.01$, we have

$$i_L(0.01) = 21.617 \text{ A}$$

Using Eq. (9.64), we have

$$t_o = \frac{CV_{c0}}{21.617}$$

For the combination of (a), we get

$$t_0 = 4.626 \times 10^{-3} \text{ s}$$

For the combination of (b), we get

$$t_0 = 0.462 \times 10^{-3} \text{ s}$$

     The circuit of Figure 9.41 is a practical implementation of the concept of parallel-capacitor commutation. When Q_1 is conducting, the capacitor C will charge to potential V_{c0} through R_1. As a result of triggering Q_2, a reverse bias voltage will be applied across Q_1 causing turn-off in a manner similar to that in the circuit of Figure 9.40. Note that for the circuit shown in Figure 9.41, the capacitor charging current does not flow through the load. The initial capacitor voltage value in this configuration is limited to the source voltage. This is disadvantageous in high-frequency and high-cur-

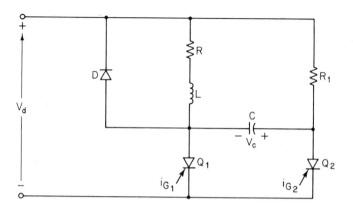

FIGURE 9.41   Basic parallel-capacitor commutation circuit.

rent applications, as this voltage value may not be enough to achieve turn-off of Q_1.

In the circuit shown in Figure 9.42, the resistor R_1 of Figure 9.41 is replaced by the series combination of an inductor L_1 and auxiliary thyristor Q_3. This modification results in resonant charging of the capacitor. The main thyristor Q_1 is turned on and the auxiliary thyristor Q_3 is turned on simultaneously. As a result, C is charged with the polarity indicated through the loop L_1, Q_3, C, Q_1, and the source. The capacitor potential V_c is higher than the source voltage V_d. It should be noted that this arrangement requires a higher rating for Q_1 as it will have to support both load current and the capacitor charging current. The possibility that Q_2 and Q_3 can be on simultaneously due to a fault condition can lead to loss of devices.

To overcome the difficulties encountered with the preceding circuit, an alternative circuit (in-line chopper switch) is used as shown in Figure

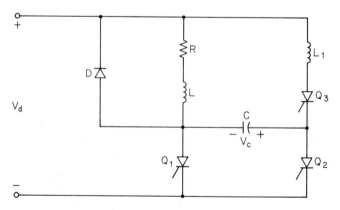

FIGURE 9.42   Parallel-capacitor commutation circuit using LC charging.

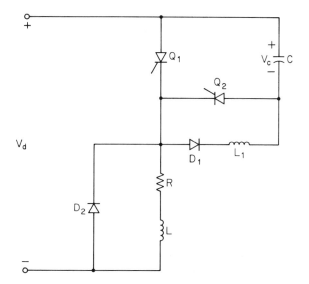

FIGURE 9.43 Parallel-capacitor commutation circuit with capacitor charging flowing into the load (in-line chopper switch).

9.43. The operation of the circuit is as follows. First Q_2 must be triggered to charge the capacitor in the polarity shown. As soon as C is charged, Q_2 will turn off due to lack of current. When Q_1 is triggered, the current flows into two paths: the load current flows in RL; the commutating current flows through Q_1, D_1, L_1, and C, and the charge on C is reversed and held with the hold-off diode D_1. At any desired time Q_2 may be triggered, which then places C across Q_1 via Q_2 and Q_1 is thus turned off.

Method D: Parallel LC Commutation

In parallel LC commutation, reverse voltage is applied to the load carrying thyristor Q_1 from the overshoot of a series LC resonant circuit connected across the SCR as shown in Figure 9.44. The series capacitor approach relies on the load circuit to form part of the tuned path. This provides a limited control range and results in reduced effectiveness of the commutation circuit due to variations in the load impedance. The parallel LC commutation concept avoids these limitations.

Before the gate pulse is applied to Q_1, the capacitor C charges up to the polarity indicated. When Q_1 is triggered, current flows into two paths. The load current I_L flows through the load; a pulse of current flows through the resonant LC circuit and charges C up in the reverse polarity. The resonant circuit current will reverse and attempt to flow through Q_1 in opposition to the load current. The thyristor Q_1 will turn off when the reverse resonant circuit current is greater than the load current. Note that a nonlin-

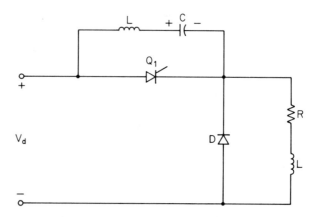

FIGURE 9.44 Parallel capacitor-inductor or *LC* free-running commutation circuit.

ear inductor may be used in place of L. As soon as Q_1 is turned off, C will recharge with its original polarity through L and the load.

The Morgan Circuit

The Morgan circuit is an improved version of the parallel capacitor-inductor circuit and employs a saturable reactor, as shown in Figure 9.45. From the previous cycle of operation, the capacitor is charged in the polarity shown. The reactor core is in positive saturation at this time. The polarity of a positively saturated core is shown in the circuit. The triggering of Q_1 and subsequent conduction closes the loop Q_1L_2C. The following sequence of events takes place:

1. As a result of the polarity of voltage applied to the reactor, it is pulled out of saturation. The unsaturated reactor presents a high inductance to the circuit. The capacitor is then discharging into L_2.

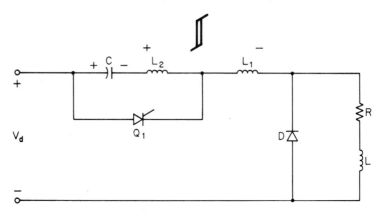

FIGURE 9.45 Morgan circuit.

2. As soon as full-load current flows, the capacitor current decreases and the voltage induced across L_2 will reverse in polarity. The reactor core then goes into negative saturation and the inductance of L_2 changes from the high value to the low saturated value.

3. The resonant charging of C now proceeds more rapidly to a peak value. This is followed by a decrease in current and the voltage across L_2 reverses. The reversal of voltage pulls the reactor from saturation. The inductance of L_2 rises to the high value and recharging of C proceeds again.

4. The voltage across the inductor is held for a period of time and the positive saturation occurs.

5. The capacitor is switched directly across Q_1 via the saturated inductance of L_2. If the reverse current exceeds the load current, Q_1 will be turned off. The remaining charge in C is then dissipated in the load and then C is charged up and ready for the next cycle.

The Jones Circuit

The Jones commutation circuit is used extensively in electric vehicle control schemes. The circuit is shown in Figure 9.46 and does not require that the capacitor be charged before commutation can be achieved.

Let us assume that C is not charged initially. When Q_1 is triggered, the load current flows through Q_1, L_1, and load. As a result of the close coupling between L_1 and L_2, an induced voltage appears in L_2. The capacitor is therefore charged to the polarity indicated in the figure. The charge on C is trapped by the diode D_1. When Q_2 is triggered, turn-off of Q_1 commences and C gets charged in a polarity opposite to that indicated in the figure. At the next time Q_1 is triggered, C discharges through Q_1, L_2 with a reverse polarity and is now ready for the next commutation. The value of the voltage on C is dependent on whether the voltage induced in L_2 is greater than the reversal of positive charge built up while Q_2 was

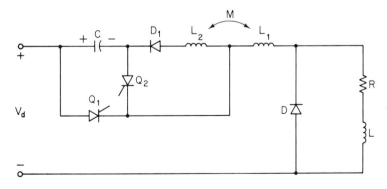

FIGURE 9.46 Jones circuit.

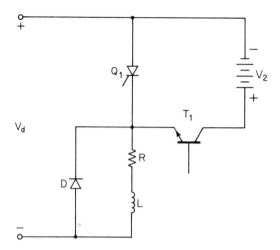

FIGURE 9.47   Auxiliary transistor, external pulse commutation circuit.

conducting. It is emphasized here that the periods of conduction of Q_1 guarantee the charging of C for commutation requirements.

Method E: External Pulse Commutation

In this class of commutation circuits, the commutation energy is obtained from an external source in the form of a pulse. The duration of the commutation pulse must be greater than the turn-off time of the device. The simplest circuit is shown in Figure 9.47 and employs an auxiliary voltage source V_2 with the polarity shown. At the desired instant of commutation a current pulse is applied to the auxiliary transistor T_1. As a result, V_2 is

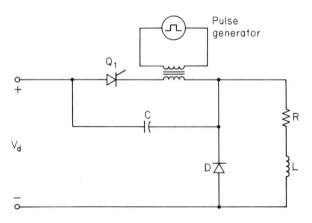

FIGURE 9.48   External pulse commutation circuit.

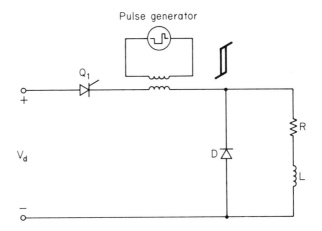

FIGURE 9.49   External pulse commutation circuit.

connected across Q_1 turning it off. The base drive pulse is maintained until complete turn-off has been achieved.

Another example of this class is shown in Figure 9.48. In this case the external pulse generator energy is injected in series with Q_1 through a pulse transformer. The transformer is designed so that no saturating takes place under loading conditions. The voltage drop across the transformer is negligible. When Q_1 is triggered, current flows through the load and pulse transformer. To turn Q_1 off a positive pulse is applied to the cathode of Q_1 from the generator through the transformer. Capacitor C is charged only to about 1V and can be considered a short circuit for the duration of commutation.

Another variant of pulse commutation circuits is shown in Figure 9.49. When Q_1 is turned on, the pulse transformer saturates and offers a low impedance in the load current path. In preparation for turn-off, the pulse transformer is desaturated by the application of a negative pulse. This temporarily increases the voltage across the load and the current through it. As soon as the transformer is desaturated, a pulse in the reverse polarity is applied to turn-off the thyristor. The pulse should be held for the required turn-off time.

SOME SOLVED PROBLEMS

Problem 9.A.1

Consider the circuit of Figure 9.25 with circuit parameters as specified in Examples 9.1 and 9.2. Find the current i_L and the energy stored in the inductor at $t_2 = 0.02$ s. If the switch is closed at t_2, find the energy stored in the inductor as a function of time.

Solution

From Example 9.2 with $t = t_2$, we obtain

$$i_L(0.02) = 21.617e^{-2} = 2.926 \text{ A}$$
$$E_L(0.02) = 4.6728e^{-4} = 85.6 \times 10^{-3} \text{ J}$$

These values indicate a significant decrease in current and stored energy in 0.02 s.

When the switch is closed, the diode ceases to conduct and the voltage source is impressed on the *RL* series combination. The circuit equation is

$$L\frac{di}{dt} + Ri = V$$

The solution is

$$i_L(t) = \frac{V}{R} + Ae^{-(t-t_2)/\tau}$$

The constant *A* is found as

$$A = i_L(t_2) - \frac{V}{R}$$

As a result,

$$i_L(t) = i_L(t_2)e^{-(t-t_2)/\tau} + \frac{V}{R}(1 - e^{-(t-t_2)/\tau})$$

Substituting numerical values, we get

$$i_L(t) = 25 - 22.074e^{-200(t-0.02)}$$

The voltage across the inductor is obtained using

$$v_L(t) = L\frac{di_L}{dt} = 88.296e^{-200(t-0.02)}$$

The power into the inductor is thus

$$p_L(t) = v_L(t)i_L(t)$$
$$= 2207.4e^{-200(t-0.02)} - 1949.05e^{-400(t-0.02)}$$

The energy stored in the inductor is given by

$$E_L(t) = E_L(t_2) + \int_{t_2}^{t} p_L(t)\, dt$$

As a result,

$$E_L(t) = 85.6 \times 10^{-3} + 11.037(1 - e^{-200(t - 0.02)}) - 4.8726(1 - e^{-400(t - 0.02)})$$

The value of energy stored in the inductor is therefore given by

$$E_L(t) = 6.25 - 11.037e^{-200(t - 0.02)} + 4.8726e^{-400(t - 0.02)}$$

Problem 9.A.2

The analysis of the trapped energy recovery circuit of Figure 9.26 was carried out assuming that the load resistance R is present. Find an expression for t_c similar to that defined by Eq. (9.46), for $R = 0$.

Solution

The value of t_c is obtained to correspond to $i_2(t) = 0$. Substitution of $R = 0$ in either Eq. (9.44) or (9.46) will result in an undetermined answer. By setting $R = 0$, the expression for $i_1(t)$ is from basic principles

$$i(t) = \frac{Vt}{L} \qquad 0 \leqslant t \leqslant t_1$$

Equation (9.37) with $R = 0$ yields

$$L \frac{di_2}{dt} = n^2 V$$

Thus

$$i_2(t) = i_2(t_1^+) + \frac{n^2 V}{L}(t - t_1)$$

Here we have, from Eq. (9.41),

$$i_2(t_1^+) = -n i_1(t_1^-)$$

As a result, we have the equivalent of Eq. (9.44) given by

$$i_2(t) = \frac{n^2 V}{L}\left[t - t_1\left(1 + \frac{1}{n}\right)\right] \qquad t_1 < t$$

Now we obtain t_c from

$$i_2(t_c) = 0$$

The result is

$$t_c = t_1\left(1 + \frac{1}{n}\right)$$

The same answer for i_2 can be obtained by finding the limit of i_2 in Eq. (9.44) as R tends to zero.

Problem 9.A.3

Assume that a 500µF capacitor is used for commutation in the circuit of Example 9.4. Find the time t_1.

Solution

Using Eq. (9.56) we get

$$\omega_n^2 = \frac{1}{20 \times 10^{-3} \times 500 \times 10^{-6}} = 10^5$$

Thus

$$\omega_n = 316.2$$

Using Eq. (9.57), we get

$$\xi = \frac{R}{2L\omega_n} = \frac{4}{40 \times 10^{-3} \times 316.2} = 0.3162$$

From Eq. (9.59), we calculate

$$\omega_d = 316.2\sqrt{1 - (0.3162)^2} = 300$$

As a result from Eq. (9.62), we obtain

$$t_1 = \frac{\pi}{\omega_d} = 10.47 \times 10^{-3} \text{ s}$$

Problem 9.A.4

Show the series capacitor commutation circuit of Figure 9.33 that the value of C for commutation is limited by

$$C < \frac{4L}{R^2}$$

Solution

Commutation depends on finding t_1 such that the current is zero at that finite time. This takes place if the circuit is underdamped, $\xi < 1$. To see this, note that we can write

$$t_1^2 = \frac{\pi^2}{\omega_n^2(1 - \xi^2)}$$

Hence for t_1 to be real, we need

$$\xi < 1$$

Now we have

$$\xi = \frac{R}{2L\omega_n}$$

Thus

$$\xi^2 = \frac{R^2 C}{4L}$$

As a result,

$$\frac{R^2 C}{4L} < 1$$

or

$$C < \frac{4L}{R^2}$$

Problem 9.A.5

Assume in the parallel-capacitor commutation circuit of Figure 9.41 that

$$V_d = 100 \text{ V} \qquad R_1 = 100 \text{ }\Omega \qquad C = 10^{-4} \text{ F}$$

Find the capacitor voltage at $t = 0.01$ s. Assume that the capacitor is initially not charged.

Solution

The voltage V_d is impressed on the series combination of R_1 and C, with D and Q_2 not conducting. As a result, we have

$$v_c(t) = V_d (1 - e^{-t/\tau_1})$$

where

$$\tau_1 = RC$$

Here we have

$$\tau_1 = 100(10^{-4}) = 10^{-2}$$
$$V_d = 100$$

As a result,

$$v_c(t) = 100(1 - e^{-100t})$$

For $t = 0.01$, we obtain

$$v_c(0.01) = 63.21 \text{ V}$$

PROBLEMS

Problem 9.B.1

Assume that in the circuit shown in Figure 9.24, the value of V is 100 V and the load parameters are

$$R = 80 \text{ }\Omega \qquad L = 10 \text{ mH}$$

The switch S is closed at $t = 0$. Find the current and energy stored in the inductor as a function of time.

Problem 9.B.2

Repeat Problem 9.B.1 for $V = 120$ V, $R = 1\ \Omega$, and $L = 0.03$ H.

Problem 9.B.3

The current and energy stored in the inductor of Figure 9.24 at $t_1 = 4$ ms are 278 A and 38.7 J, respectively. The resistance is 0.25 Ω. Find the inductance L, the time constant τ, and the applied voltage as well as the current and energy stored in the inductor as a function of time.

Problem 9.B.4

The voltage across the inductor in Figure 9.24 is 23 V, at $t = 1250$ μs. The power into the inductor at that time is 8000 W. The voltage source is 110 V dc. Find the circuit time constant and the resistance and inductance of the circuit. Find the current as a function of time.

Problem 9.B.5

Assume that the circuit of Problem 9.B.1 is equipped with a free-wheeling diode, as shown in Figure 9.25. The switch is opened at $t_1 = 125$ μs. Find the energy stored in the inductor as a function of time for $t > t_1$.

Problem 9.B.6

Repeat Problem 9.B.5 for the circuit of Problem 9.B.2 with $t_1 = 30$ ms.

Problem 9.B.7

Assume that in the circuit shown in Figure 9.25 the voltage V is 110 V, the inductance is $L = 10^{-3}$ H, and the time constant $\tau = 4$ ms. The switch is closed at $t = 0$ and then opened at $t_1 = 4$ ms. Find the inductor current, voltage, and power at the time instant $t_2 = 8$ ms.

Problem 9.B.8

Assume that in the circuit shown in Figure 9.25, the voltage V is 110 V and the resistance $R = 0.25\ \Omega$. The current in the inductor at $t_1 = 1250$ μs is 348 A, with the switch closed. Find the inductance. If the switch is opened at t_1, find the current and energy in the inductor as a function of time for $t > t_1$.

Problem 9.B.9

Consider the circuit of Problems 9.B.1 and 9.B.5. Find the current in the inductor at $t = 250$ μs. If the switch is closed at t_2, find the current in the inductor as a function of time. (Use the formula in Problem 9.A.1.)

Problem 9.B.10

Consider the circuit of Problems 9.B.2 and 9.B.6. Find the current in the inductor at $t_2 = 0.06$ s. If the switch is closed at t_2, find the current and energy in the inductor as functions of time.

Problem 9.B.11

For the circuit of Problem 9.B.7, the switch is closed at t_2. Find the current and power in the inductor as functions of time.

Problem 9.B.12

For the circuit of Problem 9.B.8, the switch is closed at $t_2 = 2500$ μs. Find the current through and voltage across the inductor as a function of time.

Problem 9.B.13

For the trapped energy recovery circuit of Figure 9.26, prove from Eq. (9.44) and (9.45) that an equivalent form of Eq. (9.46) is given by

$$i_2(t) = \frac{n^2 V}{R} (1 - t^{-(t - t_c)/\tau})$$

Show that the energy returned to the source is given by

$$E_s = \frac{n^2 V^2}{R} [\Delta_c + \tau (1 - e^{\Delta_c/\tau})]$$

where

$$\Delta_c = t_c - t_1$$

Verify the results of Example 9.3 using the equations just derived.

Problem 9.B.14

An energy recovery circuit with a transformer ratio of $n = 2$ is used for the load of Problems 9.B.1 and 9.B.5. Find t_c and the energy returned to the source during the interval t_1 to t_c.

Problem 9.B.15

An energy recovery circuit with transformer ratio of $n = 2$ is used for the load of Problems 9.B.2 and 9.B.6. Find t_c and the energy returned to the source during the interval t_1 to t_c.

Problem 9.B.16

An energy recovery circuit with transformer ratio of $n = 4$ is used for the load of Problems 9.B.3 and 9.B.7. Find t_c, the energy returned to the source, and that dissipated in the resistance in the interval t_1 to t_c.

Problem 9.B.17

An energy recovery circuit with transformer ratio of $n = 4$ is used for the load of Problems 9.B.4 and 9.B.8. Find t_c and the energy returned to the source during the interval t_1 to t_c.

Problem 9.B.18

Show that the turns ratio of the transformer in the trapped energy recovery circuit is given (in terms of the various time parameters involved) by

$$ n = \frac{e^{t_1/\tau} - 1}{e^{t_c/\tau} - e^{t_1/\tau}} $$

Problem 9.B.19

Assume for the circuit of Problem 9.B.14 that the time t_c is to be less than 150 μs. Find the required transformer turns ratio (use the formula derived in Problem 9.B.18).

Problem 9.B.20

Repeat Problem 9.B.19 for the circuit of Problem 9.B.15 with t_c less than 35×10^{-3} s.

Problem 9.B.21

Repeat Problem 9.B.19 for the circuit of Problem 9.B.16 with t_c less than 5×10^{-3} s.

Problem 9.B.22

Repeat Problem 9.B.19 for the circuit of Problem 9.B.17 with t_c less than 1750 μs.

Problem 9.B.23

Show for the series capacitor comutation method that the capacitance value
is related to L, R, and the commutation time by

$$C = \frac{4Lt_1^2}{4\pi^2 L^2 + R^2 t_1^2}$$

Use this formula to verify the results of Example 9.4.

Problem 9.B.24

Find the value of the capacitance C required to commutate the load of Problem 9.B.1 at $t_1 = 125$ μs using series capacitance commutation (use the
formula given in Problem 9.B.23).

Problem 9.B.25

Repeat Problem 9.B.24 for the load of Problem 9.B.2 for $t_1 = 30 \times 10^{-3}$ s.

Problem 9.B.26

Repeat Problem 9.B.24 for the load of Problem 9.B.3 for $t_1 = 4$ ms.

Problem 9.B.27

Repeat Problem 9.B.24 for the load of Problem 9.B.4 for $t_1 = 1250$ μs.

Problem 9.B.28

Assume that a 400-μF capacitor is used for the series capacitor commutation
circuit of Example 9.4. Find the value of t_1 (use the formula derived in
Problem 9.B.23) to show that

$$t_1 = \frac{2\pi L}{\sqrt{4L/C - R^2}}$$

Problem 9.B.29

Assume that a 0.2-μF capacitor is used for series capacitor commutation of
the circuit of Problem 9.B.1. Find the value of t_1 (use the formula given in
statement of Problem 9.B.28).

Problem 9.B.30

Repeat Problem 9.B.29 for the circuit of Problem 9.B.2 with $C = 2 \times 10^{-3}$ F.

Problem 9.B.31

Problem 9.B.29 for the circuit of Problem 9.B.3 with $C = 10^{-3}$ F.

Problem 9.B.32

Repeat Problem 9.B.29 for the circuit of Problem 9.B.4 with $C = 500$ μF.

Problem 9.B.33

Parallel-capacitor commutation is used for the load of Problems 9.B.1 and 9.B.5 with $C = 0.2 \times 10^{-6}$. Find the required capacitor voltage for an available turn-off time of 125 μs. Repeat for $C = 1$ μF.

Problem 9.B.34

Parallel-capacitor commutation is used for the load of Problems 9.B.2 and 9.B.6 with $C = 4 \times 10^{-3}$ F and $V_{c0} = 50$ V. Find the available turn-off time. Suppose that the turn-off time is to be 30 ms. What value of capacitance is required for $V_{c0} = 50$ V?

Problem 9.B.35

Parallel-capacitor commutation is used for the load of Problems 9.B.3 and 9.B.7 with $C = 2 \times 10^{-3}$ F. Find the capacitor voltage for $t_0 = 4$ ms. Find the capacitance value required with $V_{c0} = 80$ V to achieve the same value of t_0.

Problem 9.B.36

Parallel-capacitor commutation is used for the load of Problems 9.B.4 and 9.B.8, with $C = 10^{-3}$ F and $V_{c0} = 75$ V. Find the available turn-off time.

T
E
N

Power Electronic Systems

10.1 INTRODUCTION

Power semiconductor or power electronic devices can be interconnected in
a number of system configurations as highlighted briefly in Section 9.3. In
this chapter we give further details on circuit configurations and the input–
output relations prevailing for load circuit configurations typical of those
employed in adjustable speed drives found in practice.

10.2 AC VOLTAGE CONTROLLERS

As discussed in Section 9.3, ac voltage controllers produce an ac voltage
output with variable rms value from a fixed-voltage ac source input. A number of circuit arrangements are possible for single-phase and three-phase
applications.

Single-Phase Controller Circuits

The circuit shown in Figure 10.1(a) is a single-phase half-wave controller
circuit suitable for resistive loads. The thyristor Q_1 is triggered at $\omega t = \alpha$
and conducts for the remainder of the positive half-cycle. At $\omega t = \pi$, the
diode conducts for the negative half-cycle as it is connected in antiparallel
with Q. The output voltage waveform $v_o(t)$ is shown in Figure 10.1(b). The
average voltage output is obtained as

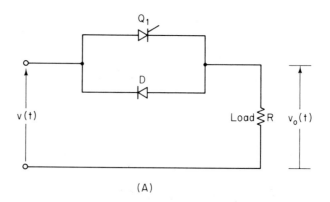

(A)

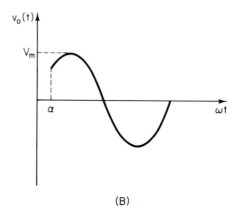

(B)

FIGURE 10.1 Single-phase half-wave ac voltage controller.

$$V_{o_{av}} = \frac{1}{2\pi} \int_{\alpha}^{2\pi} V_m \sin \omega t \, d\omega t$$

Thus

$$V_{o_{av}} = \frac{V_m}{2\pi} (\cos \alpha - 1) \qquad (10.1)$$

The rms value of $v_o(t)$ is given by

$$V_{o_{rms}}^2 = \frac{1}{2\pi} \int_{\alpha}^{2\pi} V_m^2 \sin^2 \omega t \, d\omega t$$

$$= \frac{V_m^2}{4\pi} \left(2\pi - \alpha + \frac{1}{2} \sin 2\alpha \right) \qquad (10.2)$$

From Eq. (10.1), we note that the average (dc component) of the output voltage is not zero, for $\alpha > 0$. As a result, a dc current component will pass through the ac source. From Eq. (10.2) we note that the output voltage (in rms) is controlled from 100% of the input voltage (in rms) at $\alpha = 0$ to 70.7% $(1/\sqrt{2})$ of the input voltage (in rms) at $\alpha = \pi$.

A single-phase full-wave ac voltage controller circuit is obtained if the diode D of Figure 10.1(a) is replaced by a thyristor Q_2 connected in anti-parallel with Q_1, as shown in Figure 10.2(a). Thyristor Q_1 is triggered at $\omega t = \alpha$, whereas Q_2 is triggered half a cycle later at $\omega t = \pi + \alpha$. In applications up to 400 Hz, the triac is used in place of Q_1 and Q_2. With inductive loads, the gate pulses are made longer to attain current latching. Figure 10.2(b) shows the output voltage waveform for a resistive load.

It is observed in the circuit of Figure 10.2(a) that the cathodes K_1 and K_2 are not connected in common. The gating signals for Q_1 and Q_2 must therefore be isolated from each other. Otherwise, the cathodes will be con-

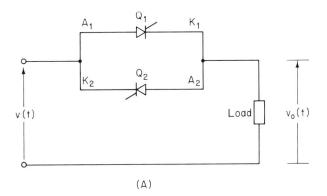

(A)

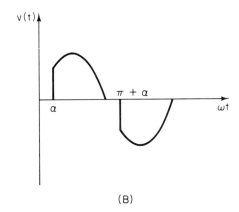

(B)

FIGURE 10.2   Single-phase full-wave ac voltage controller.

nected at gating times and the thyristors will be shorted out. An alternative arrangement of the full-wave controller with cathodes of Q_1 and Q_2 connected to a common point is shown in Figure 10.3. Isolation of the gate circuits is not required at the cost of adding two diodes D_1 and D_2 and having both a thyristor and a diode conducting at the same time.

An alternative arrangement consisting of one thyristor and four diodes provides full-wave single-phase ac voltage control is shown in Figure 10.4. Note that in this arrangement the thyristor conducts in both half-cycles along with two diodes.

On load tap changing is used in the circuit of Figure 10.5 with two antiparallel pairs of thyristors. Tap N_1 is selected to provide slightly less than the minimum required power; then phase control adds the necessary incremental power to obtain fine control. Additional taps may be introduced between N_1 and N_2 provided that an antiparallel pair of thyristors is added for each tap.

Analysis: The single-phase full-wave ac voltage controller with an *RL* load circuit is analyzed now. The presence of the inductance results in some constraints on the angle α at which the SCR is triggered.

It is important to define the angle α in the following manner. In Figure 10.6(a), a diode full-wave circuit with a resistive load is shown, with $v(t) = V_m \sin \omega t$. Clearly, diode 1 begins to conduct at $\omega t = 0$. The angle α by which Q_1 is retarded is shown in Figure 10.6(b) and is amount of angular displacement between the instant at which D_1 conducts and that at which Q_1 begins conduction. The antiparallel thyristor Q_2 is triggered at

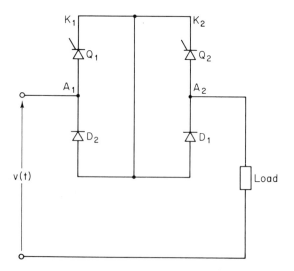

FIGURE 10.3 Full-wave ac voltage controller with common-cathode thyristors.

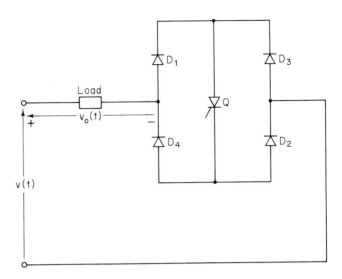

FIGURE 10.4 Full-wave single-phase ac voltage controller using one thyristor and four diodes.

angle $\omega t = \pi$, that is, half a cycle displacement from the instant of conduction of Q_1.

The following analysis is based on the single-phase full-wave ac voltage controller of Figure 10.2 repeated in Figure 10.7(a) with an RL load circuit.

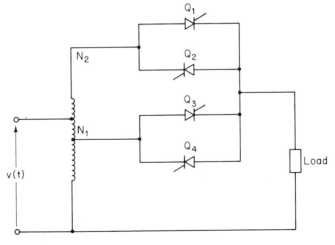

FIGURE 10.5 Single-phase tap changer.

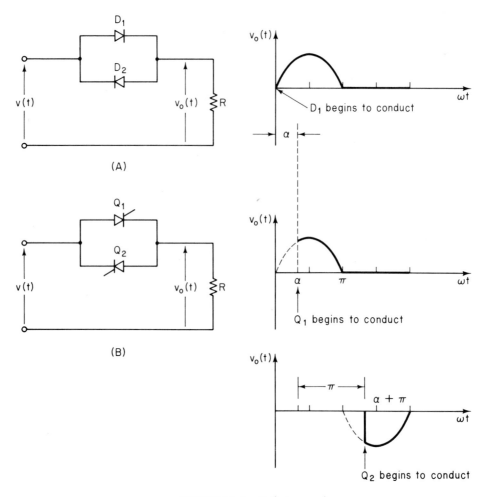

FIGURE 10.6   Defining angle α.

The ac source has a voltage waveform shown in Figure 10.7(b) given by

$$v(t) = V_m \sin \omega t \tag{10.3}$$

Thyristor 1 is fired at $\omega t = \alpha$ and conducts while thyristor 2 is blocking, as it is reverse biased in the first half cycle. Current $i_1(t)$ flows in thyristor 1 to the load circuit and consists of a sinusoidal steady-state term and a transient term:

$$i_1(t) = \frac{V_m}{|Z|} \sin(\omega t - \phi) + A e^{-t/\tau} \tag{10.4}$$

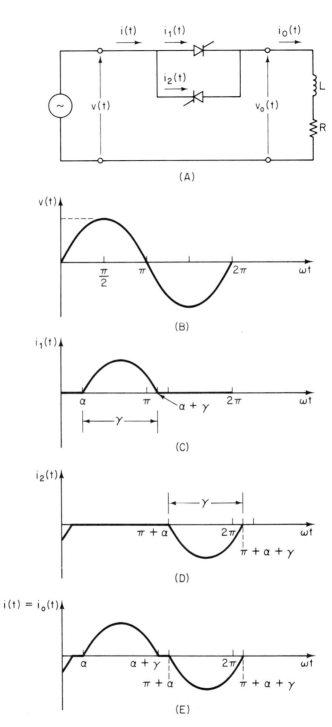

FIGURE 10.7 Single-phase full-wave ac voltage controller and associated voltage and current waveforms.

The constant A is obtained from the initial condition that current is zero at $\omega t = \alpha$. The conduction angle γ defines the interval during which thyristor 1 conducts by $\alpha \leqslant \omega t < \alpha + \gamma$. The current $i_1(t)$ is thus given by

$$i_1(t) = \begin{cases} \dfrac{V_m}{|Z|} \left[\sin(\omega t - \phi) - \sin(\alpha - \phi)e^{(\alpha - \omega t)/\omega \tau}\right] \\ \qquad\qquad\qquad\qquad\qquad\qquad \alpha \leqslant \omega t < \alpha + \gamma \\ 0 \qquad\qquad\qquad\qquad\qquad\quad \alpha + \gamma \leqslant \omega t \leqslant 2\pi + \alpha \end{cases} \qquad (10.5)$$

Here we have

$$|Z|^2 = \omega^2 L^2 + R^2 \qquad (10.6)$$

$$\tan \phi = \frac{\omega L}{R} \qquad (10.7)$$

$$\tau = \frac{L}{R} \qquad (10.8)$$

At $\omega t = \alpha + \gamma$, the current in thyristor 1 is zero and thyristor 1 ceases to conduct. The conduction angle γ can therefore be obtained from the requirement $i_1(\alpha + \gamma) = 0$, to yield

$$\sin(\alpha + \gamma - \phi) = \sin(\alpha - \phi)\,e^{-\gamma/\tan\phi} \qquad (10.9)$$

An alternative form to determine γ is obtained from Eq. (10.9) as

$$\frac{\sin \gamma}{e^{-\gamma/\tan\phi} - \cos \gamma} = \tan(\alpha - \phi) \qquad (10.10)$$

The conduction angle γ is required to be less than π, so that thyristor 1 ceases to conduct prior to firing of thyristor 2 at $\omega t = \pi + \alpha$. From Eq. (10.10), for $\gamma > \pi$, we find $\alpha < \phi$. We thus conclude that α should be greater or equal to ϕ.

$$\alpha_{\min} = \phi \qquad (10.11)$$

If α is chosen to be equal to ϕ, the current $i_1(t)$ is purely sinusoidal as the transient term drops out and the conduction angle $\gamma = \pi$.

Equation (10.10) is used to determine γ for a given α and ϕ. Two limiting cases are of interest.

1. *Resistive load:* Here $\phi = 0$ and the current $i_1(t)$ is sinusoidal:

$$i_1(t) = \frac{V_m}{|Z|} \sin \omega t \tag{10.12}$$

The conduction angle γ is given by the relation

$$\gamma = \pi - \alpha \tag{10.13}$$

This relation is a straight line labeled A in Figure 10.8.

2. *Inductive load:* Here $\phi = \pi/2$ and the current $i_1(t)$ is given by

$$i_1(t) = \frac{V_m}{\omega L} (\cos \alpha - \cos \omega t) \tag{10.14}$$

The conduction angle is

$$\gamma = 2(\pi - \alpha) \tag{10.15}$$

The relation between α and γ is the straight line labeled B in Figure 10.8.

The output voltage $v_o(t)$ at the load terminals is equal to the source voltage $v(t)$ in the time intervals during which either thyristor conducts as shown in Figure 10.9. It is clear that the average value of $v_o(t)$ is zero over one cycle of the source voltage. The effective value of the output voltage is obtained from

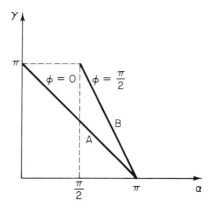

FIGURE 10.8   Variation of conduction angle γ with angle α for resistive and inductive loads in a single-phase full-wave ac voltage controller.

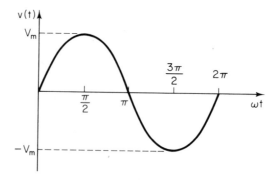

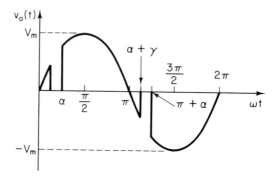

FIGURE 10.9 Voltage waveforms in a single-phase full-wave ac voltage controller.

$$V^2_{o\text{rms}} = \frac{V_m^2}{\pi} \int_\alpha^{\alpha + \gamma} \sin^2 \theta \, d\theta$$

As a result,

$$V^2_{o\text{rms}} = \frac{V_m^2}{4\pi} [\sin 2\alpha - \sin 2(\alpha + \gamma) + 2\gamma] \qquad (10.16)$$

The ratio of the effective values of output voltage and source voltage is therefore given by

$$\frac{V_{o\text{rms}}}{V} = \sqrt{\frac{[\sin 2\alpha - \sin 2(\alpha + \gamma)] + 2\gamma}{2\pi}} \qquad (10.17)$$

The source current $i(t)$ is equal to the output current and the average value of $i(t)$ is zero over one cycle of the source voltage. The current flowing through each of the thyristors has an average value given by

$$I_1 = \frac{V_m}{2\pi \,|Z|} \,[\cos(\alpha - \phi) - \cos(\alpha + \gamma - \phi)$$
$$+ \tan\phi \,\sin(\alpha - \phi)(e^{-\gamma/\tan\phi} - 1)] \quad (10.18)$$

An example will be appropriate now.

Example 10.1   A single-phase full-wave ac voltage controller supplies a load circuit with $\phi = \pi/4$. Find the required angle α and the ratio of effective (rms) output voltage to source voltage for the following values of conduction angle γ.

   a. $\gamma = 150°$.
   b. $\gamma = 175°$.

Solution

We use Eq. (10.10) to obtain α in both cases, then use Eq. (10.17) to obtain the required ratio.
   (a) We have

$$\tan(\alpha - \phi) = \frac{\sin\gamma}{e^{-\gamma/\tan\phi} - \cos\gamma}$$

$$\tan\!\left(\alpha - \frac{\pi}{4}\right) = \frac{\sin 150°}{e^{-[(150\,\times\,\pi)/180]} - \cos 150°}$$

$$= 0.533$$

As a result, we get

$$\alpha = 73.04°$$

The required ratio is obtained as

$$\frac{V_o}{V} = \sqrt{\frac{\sin[2(73.04)] - \sin[2(73.04 + 150)] + (2 \times 150 \times \pi)/180}{2\pi}}$$

$$= 0.874$$

   (b) We now have for $\gamma = 175°$,

$$\tan(\alpha - \phi) = \frac{\sin 175°}{e^{-[(175/180)\,\times\,\pi]} - \cos 175°} = 0.8353$$

$$\alpha = 45 + 4.775 = 49.78°$$

$$\frac{V_o}{V} = \sqrt{\frac{\sin[2(49.78)] - \sin[2(49.78 + 175)] + (2 \times 175 \times \pi)/180}{2\pi}}$$

$$= 0.985$$

We note here that the controller acts as a step-down transformer. The effect of reducing α is to increase γ and hence increase the ratio of V_o to V.

Three-Phase Voltage Controllers

Controlling the three-phase voltage supply to a three-phase load involves a number of possible combinations depending on the nature of the load connection. A number of circuits are discussed next where the load is simply referred to as a motor.

Circuit 1: Motor in Y, Thyristors in Y

A simple extension of the single-phase controller circuit of Figure 10.2 to the case when only the motor line terminals A', B', and C' are available results in the circuit shown in Figure 10.10. For motors connected in Δ, a

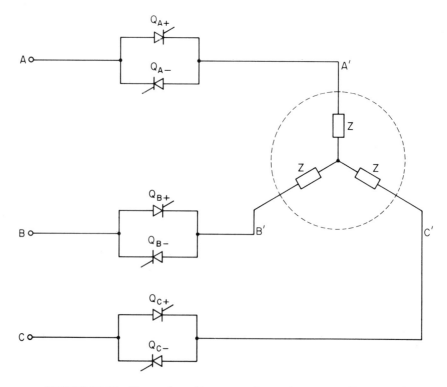

FIGURE 10.10   Three-phase Y-connected ac voltage controller circuit.

simple Δ–Y transformation on the winding impedances results in the circuit of Figure 10.10. Three pairs of antiparallel connected thyristors are used in this configuration which may have its neutral point connected to the neutral of a four-wire system. Observe that one can replace three thyristors Q_{A-}, Q_{B-}, and Q_{C-} by diodes to obtain a half-wave controller. In the latter case it is not permissible to connect the neutral of the load to the neutral of a four-wire system.

The antiparallel thyristor pairs can each be replaced by a bridge such as that shown in Figure 10.4 to obtain the alternative circuit shown in Figure 10.11. As we have noted earlier, this circuit is inefficient.

The three-phase Y circuit will operate with full control in a three-wire system with only two antiparallel thyristor sets or bridge thyristor sets. When a four-wire system is necessary, a third set of either type is necessary.

If each individual phase winding of the load is accessible, it is possible to connect the three antiparallel thyristors (or bridge thyristor) in the Y configuration shown in Figure 10.12.

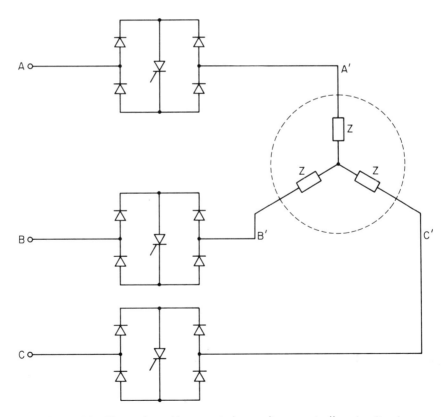

FIGURE 10.11 Three-phase Y-connected ac voltage controller circuit using one thyristor and four diodes in each line.

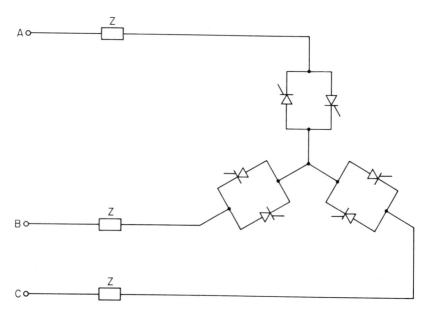

FIGURE 10.12   Three-phase Y-connected ac voltage controller with thyristor in the open Y of the load.

Unsymmetrical arrangements with only one or two phases controlled by thyristors or with one thyristor replaced by a diode for low-cost applications in three-wire systems are possible, but their performance is too poor for large system applications.

Circuit 2: Motor in Y, Thyristors in Δ

The circuit of Figure 10.12 can be modified to the form shown in Figure 10.13, where the three thyristor sets are connected in Δ. The motor and line currents in circuit 2 are the same as in circuit 1. Note that only thyristors' currents are different between the two arrangements.

Circuit 3: Motor in Δ, Thyristors in Δ

For motors connected in Δ, the arrangement of Figure 10.14 can be used. Here a pair of thyristors connected in back to back (i.e., antiparallel) is inserted in series with each winding Z.

10.3   CONTROLLED RECTIFIERS

A controlled rectifier is used to vary the average value of the direct voltage applied to a load circuit. The system introduces phase-controlled thyristors between the constant-voltage ac source and the load. Controlled rectifiers

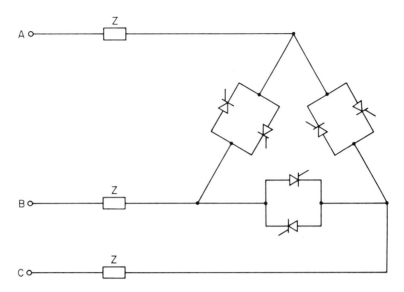

FIGURE 10.13   Three-phase ac voltage controller circuit 2, motor in Y, thyristors in Δ.

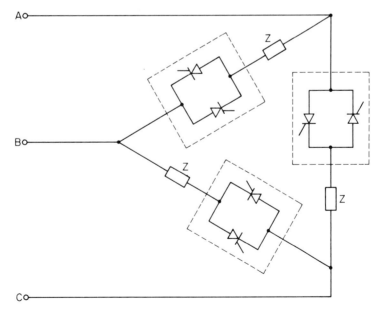

FIGURE 10.14   Three-phase ac voltage controller circuit 3, motor in Δ, thyristors in Δ.

are widely used in dc motor speed control, high-voltage dc transmission, and other application areas requiring direct voltage from an available ac source.

Many configurations of controlled-rectifier circuits can be built. The circuits may be classified according to the number of current pulses in the load circuit during one cycle of the source voltage waveform.

The most elementary controlled-rectifier circuit is that of a single-phase half-wave controlled rectifier. This circuit is impractical as it requires an ideal ac voltage source. A study of the circuit, however, is necessary to understand the performance of practical circuits based on this elementary circuit.

We begin this section by a discussion of the single-phase half-wave controlled rectifier when the load circuit contains a dc voltage source. This load circuit is a model of the armature circuit of a dc motor. The output current and voltage waveforms are obtained, and this is followed by expressions for the average values of output voltage and current.

Controlled Half-Wave Rectifiers

The basic circuit of a controlled half-wave rectifier is shown in Figure 10.15. The voltage source $v(t)$ is connected to the load through a thyristor Q. The load circuit consists of a constant EMF, E_c, in series with a resistance R and an inductance L. The ac voltage is assumed to be of the form

$$v(t) = V_m \sin \omega t$$

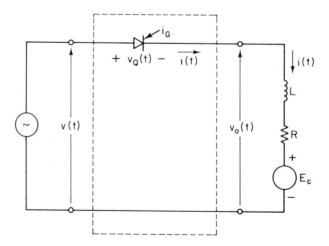

FIGURE 10.15   Basic controlled half-wave rectifier circuit.

As a result, the voltage across the thyristor terminals is given by

$$v_Q(t) = v(t) - E_c - Ri(t) - L\frac{di}{dt} \tag{10.19}$$

Our primary unknown is the output current $i(t)$.

If the thyristor is not conducting, $i(t) = 0$ and the thyristor voltage is given by

$$v_Q(t) = V_m \sin \omega t - E_c \tag{10.20}$$

When a gate current i_G is applied, the thyristor conducts only if it is forward biased, so that

$$v_Q(t) > 0$$

Thus we see that this requires

$$\sin \omega t \geq \frac{E_c}{V_m} \tag{10.21}$$

Let α be the angle corresponding to gate current application time t_g; thus

$$\alpha = \omega t_g \tag{10.22}$$

Clearly, the minimum value of α required so that the thyristor conducts is given by the expression

$$\sin \alpha_{min} = \frac{E_c}{V_m} \tag{10.23}$$

Application of gate current at an angle less than the minimum α will fail to fire the thyristor since it is reverse biased ($v_Q < 0$).

Assume now that the thyristor's gate current is applied at $\omega t = \alpha$, with $\alpha \geq \alpha_{min}$. The thyristor will conduct and the circuit conditions are described by

$$L\frac{di}{dt} + Ri = V_m \sin \omega t - E_c$$

The initial condition on the current is

$$i\left(\frac{\alpha}{\omega}\right) = 0$$

The thyristor is assumed to conduct up to $\omega t = \alpha + \gamma$, where γ is the conduction angle.

The solution of the circuit differential equation is given by

$$i(t) = \frac{V_m}{|Z|} \sin(\omega t - \phi) - \frac{E_c}{R} + Ae^{-t/\tau} \qquad \alpha \le \omega t \le \alpha + \gamma \quad (10.24)$$

The first term is the steady-state solution due to the sinusoidal voltage input, where as usual, the load impedance is expressed as

$$Z = R + j\omega L = |Z| \underline{/\phi} \tag{10.25}$$

Thus

$$|Z|^2 = R^2 + \omega^2 L^2 \tag{10.26}$$

$$\tan \phi = \frac{\omega L}{R} \tag{10.27}$$

The second term in the current expression is due to the dc voltage E_c. The third term corresponds to the transient response, with A being a constant and τ the circuit time constant:

$$\tau = \frac{L}{R} \tag{10.28}$$

The constant A is determined using the initial condition on the current:

$$i\left(\frac{\alpha}{\omega}\right) = 0 = \frac{V_m}{|Z|} \sin(\alpha - \phi) - \frac{E_c}{R} + Ae^{-\alpha/\omega\tau}$$

Note that $\omega\tau = \tan \phi$ and $R = |Z| \cos \phi$; thus

$$A = \frac{V_m}{|Z|} e^{\alpha/\tan\phi} \left[\frac{E_c}{V_m \cos \phi} - \sin(\alpha - \phi) \right] \tag{10.29}$$

We therefore have the output current during the conduction period as

$$i(t) = \frac{V_m}{|Z|} \left\{ \sin(\omega t - \phi) - \frac{E_c}{V_m \cos \phi} + \left[\frac{E_c}{V_m \cos \phi} \right.\right.$$
$$\left.\left. - \sin(\alpha - \phi) \right] e^{-(t - \alpha/\omega)/\tau} \right\} \qquad \alpha \le \omega t \le \alpha + \gamma \quad (10.30)$$

Equation (10.30) is valid up to $\omega t = \alpha + \gamma$ when $i(t)$ is zero, as shown in Figure 10.16. The value of γ can be determined from the condition

$$i\left(\frac{\alpha + \gamma}{\omega}\right) = 0$$

As a result, we find that

$$\frac{(E_c/V_m \cos \phi) - \sin(\alpha + \gamma - \phi)}{(E_c/V_m \cos \phi) - \sin(\alpha - \phi)} = e^{-\gamma/\tan\phi} \qquad (10.31)$$

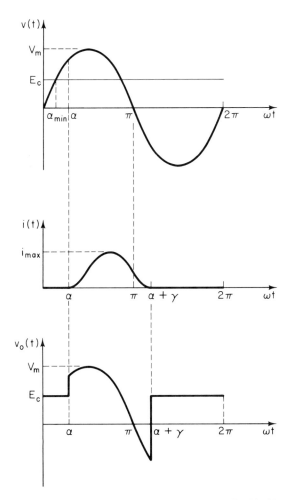

FIGURE 10.16 Voltage and current waveforms for controlled half-wave rectifier.

For a given circuit, E_c, V_m, and ϕ are available. Specifying α leads us to solve Eq. (10.31) for the conduction angle γ. Note that

$$i(t) = \begin{cases} 0 & 0 \leq \omega t < \alpha \\ 0 & \alpha + \gamma < \omega t \leq 2\pi \end{cases} \qquad (10.32)$$

The voltage output $v_0(t)$ is given by

$$v_0(t) = \begin{cases} E_c & 0 \leq \omega t < \alpha \\ V_m \sin \omega t & \alpha \leq \omega t \leq \alpha + \gamma \\ E_c & \alpha + \gamma < \omega t \leq 2\pi \end{cases} \qquad (10.33)$$

This concludes the discussion of output voltage and current waveforms in a single-phase controlled half-wave rectifier.

Example 10.2 A controlled half-wave rectifier supplies a load circuit with $\phi = \pi/4$, and the ratio of the back EMF E_c to the maximum source voltage V_m is 0.707. Assume that α is set to its minimum permissible value. Find the conduction angle γ.

Solution

We have, by Eq. (10.23),

$$\sin \alpha_{\min} = 0.707$$

Therefore,

$$\alpha = \alpha_{\min} = 45°$$

Substituting in Eq. (10.31), we obtain

$$1 - \sin \gamma = e^{-\gamma}$$

This is perhaps the simplest possible relation obtained based on Eq. (10.31), but we still need trial and error to obtain the solution for γ. In this case it is easy to verify that

$$\gamma = 119°$$

Average Output Voltage and Current

The average of the output voltage over one cycle of the supply frequency is given by

$$
V_{o_{av}} = \frac{1}{2\pi} \left(\int_0^\alpha E_c \, d\theta + \int_\alpha^{\alpha+\gamma} V_m \sin\theta \, d\theta + \int_{\alpha+\gamma}^{2\pi} E_c \, d\theta \right)
$$

$$
= \frac{1}{2\pi} \{ V_m[\cos\alpha - \cos(\alpha+\gamma)] + E_c(2\pi - \gamma) \}
$$

This can be rewritten as

$$
V_{o_{av}} = E_c \left(1 - \frac{\gamma}{2\pi} \right) + \frac{V_m}{2\pi} [\cos\alpha - \cos(\alpha+\gamma)] \qquad (10.34)
$$

To obtain the average of the output current, we find the average voltage across the resistor $V_{R_{av}}$ using

$$
V_{o_{av}} = V_{R_{av}} + V_{L_{av}} + E_c \qquad (10.35)
$$

We note first that the average of the voltage across the inductor is zero, as can be seen from

$$
\begin{aligned}
V_{L_{av}} &= \frac{1}{2\pi} \int_0^{2\pi/\omega} v_L(t) \, dt \\
&= \frac{1}{2\pi} \int_0^{2\pi/\omega} L \, di \\
&= \frac{L}{2\pi} \left[i\left(\frac{2\pi}{\omega}\right) - i(0) \right] \\
&= 0
\end{aligned} \qquad (10.36)
$$

As a result, we obtain, by combining Eqs. (10.34) to (10.36),

$$
V_{R_{av}} = \frac{V_m}{2\pi} [\cos\alpha - \cos(\alpha+\gamma)] - \frac{\gamma}{2\pi} E_c \qquad (10.37)
$$

The average value of the output current is therefore obtained as

$$
I_{o_{av}} = \frac{1}{2\pi R} \{ V_m[\cos\alpha - \cos(\alpha+\gamma)] - \gamma E_c \} \qquad (10.38)
$$

Let us observe that if $\alpha = \alpha_{min}$, we obtain

$$I_{o_{av}} = \frac{1}{2\pi R}\left[V_m(1 - \cos \gamma)\sqrt{1 - \left(\frac{E_c}{V_m}\right)^2} + E_c(\sin \gamma - \gamma) \right] \quad (10.39)$$

where γ is the solution of Eq. (10.31). The average output voltage is simply given by

$$V_{o_{av}} = RI_{o_{av}} + E_c \quad (10.40)$$

Example 10.3 Assume for the half-wave controlled rectifier of Example 10.2 that the effective value of the ac source voltage is 120 V and that the load resistance is 5 Ω. Find the average output voltage and load current.

Solution

Here we have

$$V_m = 120\sqrt{2} = 169.71 \text{ V}$$

Thus the back EMF is given by

$$E_c = 0.707V_m = 120 \text{ V}$$

With

$$\alpha = 45° \quad \text{and} \quad \gamma = 119°$$

we obtain using Eq. (10.34) the average output voltage,

$$V_{o_{av}} = 120\left[1 - \frac{119}{360} \right] + \frac{169.71}{2\pi}[\cos 45° - \cos (164°)]$$

$$= 125.4 \text{ V}$$

To obtain the average current in the load circuit, we apply Eq. (10.39) to get

$$I_{o_{av}} = \frac{1}{2\pi(5)}\left[169.71(1 - \cos 119°)\sqrt{1 - \tfrac{1}{2}} \right.$$

$$\left. + 120(\sin 119° - \frac{119}{180} \times \pi \right]$$

$$= 1.08 \text{ A}$$

A less involved procedure uses Eq. (10.40), rewritten as

$$I_{o_{av}} = \frac{V_{o_{av}} - E_c}{R}$$

$$= \frac{125.4 - 120}{5} = 1.08 \text{ A}$$

Special Cases

The analysis of the controlled half-wave rectifier circuit performance with a general load consisting of an R, L, and E_c combination can be used to obtain corresponding expressions for some important special cases in two categories. The first category corresponds to a load circuit with no EMF. The second deals with circuits where either R or L are negligible but E_c is present.

a) RL Load Circuits

Assume that $E_c = 0$; the load circuit is then a series combination of a resistance R and an inductance L as may be encountered in the model of the field circuit of a dc motor. We note immediately that

$$\alpha_{min} = 0$$

Thus gating the thyristor at $t = 0$ is permissible. The current waveform is obtained from Eq. (10.30) with $E_c = 0$, as

$$i(t) = \frac{V_m}{|Z|} [\sin(\omega t - \phi) + \sin(\phi - \alpha)e^{-(t - \alpha/\omega)/\tau}]$$

$$\alpha \leq \omega t \leq \alpha + \gamma \quad (10.41)$$

The conduction angle γ is obtained from Eq. (10.31) as the solution to

$$\frac{\sin(\alpha + \gamma - \phi)}{\sin(\alpha - \phi)} = e^{-\gamma/\tan\phi} \quad (10.42)$$

The voltage and current waveforms of a controlled half-wave rectifier with an RL load are shown in Figure 10.17.

The average value of the current for this circuit is given by Eq. (10.38) with $E_c = 0$ as

$$I_{o_{av}} = \frac{V_m}{2\pi R} [\cos \alpha - \cos (\alpha + \gamma)] \quad (10.43)$$

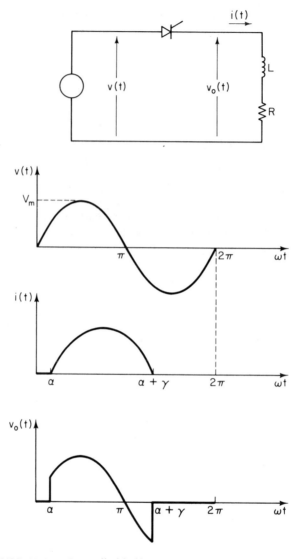

FIGURE 10.17   Controlled half-wave rectifier with *RL* load circuit.

The average output voltage is

$$V_{o\text{av}} = RI_{o\text{av}} \qquad (10.44)$$

Example 10.4   The effective value of the source voltage is 120 V in the controlled half-wave rectifier circuit of Figure 10.17 and the load power factor angle $\phi = 45°$.

Assume that $R = 5 \ \Omega$. Find the angle α, average output voltage, and current corresponding to the following conduction angles.

a. $\gamma = 225°$.
b. $\gamma = 210°$.
c. $\gamma = 150°$.

Solution

We use Eq. (10.42), rewritten as

$$[\cos(\gamma - \phi) - e^{-\gamma/\tan\phi} \cos \phi] \sin \alpha \\ + [\sin(\gamma - \phi) + e^{-\gamma/\tan\phi} \sin \phi] \cos \alpha = 0$$

With $\phi = 45°$, we have

$$\left[\cos(\gamma - 45) - \frac{e^{-\gamma}}{\sqrt{2}} \right] \sin \alpha + \left[\sin(\gamma - 45) + \frac{e^{-\gamma}}{\sqrt{2}} \right] \cos \alpha = 0$$

(a) For $\gamma = 225°$, we obtain

$$\left(\cos 180° - \frac{0.02}{\sqrt{2}} \right) \sin \alpha + \left(\sin 180 + \frac{0.02}{\sqrt{2}} \right) \cos \alpha = 0$$

$$\tan \alpha = 0.014$$
$$\alpha = 0.79°$$

Using Eq. (10.43), we get

$$I_{o\text{av}} = \frac{120 \ \sqrt{2}}{2\pi(5)} [\cos 0.79 - \cos(0.79 + 225)]$$

$$= 9.17 \text{ A}$$

From Eq. (10.44), we obtain

$$V_{o\text{av}} = 5(9.17) = 45.85 \text{ V}$$

(b) For $\gamma = 210°$, we obtain, following the same procedure,

$$\alpha = 15.7°$$
$$I_{o\text{av}} = 8.97 \text{ A}$$
$$V_{o\text{av}} = 44.85 \text{ V}$$

(c) For $\gamma = 150°$, we obtain

$$\alpha = 73°$$

$$I_{o_{av}} = 5.53 \text{ A}$$

$$V_{o_{av}} = 27.65 \text{ V}$$

We note that γ is decreased as a result of an increase in α, with a corresponding reduction in average output voltage and current.

Resistive Load Circuits

If the load circuit has a negligible inductance, we have $L = 0$, and we get

$$\phi = 0 \qquad |Z| = R$$

The transient term becomes zero and the current $i(t)$ is given by

$$i(t) = \frac{V_m}{R} \sin \omega t \qquad \alpha \leqslant \omega t \leqslant \alpha + \gamma \qquad (10.45)$$

The conduction angle γ is obtained directly from the equation above as the solution to

$$\sin(\alpha + \gamma) = 0$$

This leads to

$$\alpha + \gamma = \pi \qquad (10.46)$$

As a result,

$$I_{o_{av}} = \frac{V_m(1 + \cos \alpha)}{2\pi R} \qquad (10.47)$$

$$V_{o_{av}} = \frac{V_m(1 + \cos \alpha)}{2\pi} \qquad (10.48)$$

Pertinent waveforms for this circuit are shown in Figure 10.18.

Example 10.5 The effective value of the source voltage is 120 V for a controlled half-wave rectifier with a load resistance of 5 Ω. Find the average output voltage and current when the delay angle is given by

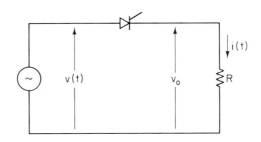

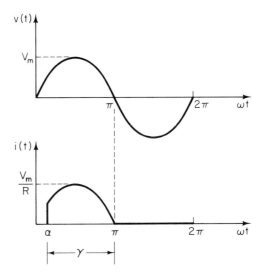

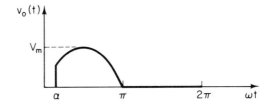

FIGURE 10.18 Controlled half-wave rectifier with resistive load circuit.

a. $\alpha = 15°$.
b. $\alpha = 30°$.
c. $\alpha = 60°$.

Solution

We use Eqs. (10.47) and (10.48) to obtain

$$I_{o_{av}} = \frac{120\sqrt{2}}{2\pi(5)}(1 + \cos\alpha)$$

$$V_{o_{av}} = 5I_{o_{av}}$$

(a) For $\alpha = 15°$, we get

$$I_{o_{av}} = 10.62 \text{ A}$$

$$V_{o_{av}} = 53.1 \text{ V}$$

(b) For $\alpha = 30°$, we get

$$I_{o_{av}} = 10.08 \text{ A}$$

$$V_{o_{av}} = 50.4 \text{ V}$$

(c) For $\alpha = 60°$, we get

$$I_{o_{av}} = 8.10 \text{ A}$$

$$V_{o_{av}} = 40.5 \text{ V}$$

Inductive Load Circuits

If the load circuit has a negligible resistance, then with $R = 0$, we have

$$\phi = \frac{\pi}{2} \qquad |Z| = \omega L$$

As a result, from Eq. (10.30), we get

$$i(t) = \frac{V_m}{\omega L}(\cos\alpha - \cos\omega t) \qquad \alpha \leqslant \omega t \leqslant \alpha + \gamma \qquad (10.49)$$

The conduction angle is obtained by setting $i(t) = 0$ at $\omega t = \alpha + \gamma$. As a result,

$$\cos\alpha = \cos(\alpha + \gamma)$$

The solution is

$$\alpha + \gamma = 2\pi - \alpha \qquad (10.50)$$

The average current is obtained using

$$I_{oav} = \frac{V_m}{\omega L} \left[\frac{1}{2\pi} \int_\alpha^{2\pi - \alpha} (\cos \alpha - \cos \omega t) \, d\omega t \right]$$

As a result,

$$I_{oav} = \frac{V_m}{\pi \omega L} [\sin \alpha + (\pi - \alpha) \cos \alpha] \qquad (10.51)$$

Pertinent waveforms for this case are shown in Figure 10.19.

Example 10.6   Assume for the rectifier of Example 10.5 that the load is inductive with $\omega L = 5 \; \Omega$. Find the average load current for the following values of α.

    a. $\alpha = 15°$.
    b. $\alpha = 30°$.
    c. $\alpha = 60°$.

Solution

We apply Eq. (10.51) to obtain

$$I_{oav} = \frac{120 \sqrt{2}}{5\pi} [\sin \alpha + (\pi - \alpha) \cos \alpha]$$

(a) For $\alpha = 15°$, we get

$$I_{oav} = 32.849 \; A$$

(b) For $\alpha = 30°$, we get

$$I_{oav} = 29.9 A$$

(c) For $\alpha = 60°$, we get

$$I_{oav} = 20.67 \; A$$

As the delay angle is increased, the average current in the load decreases.

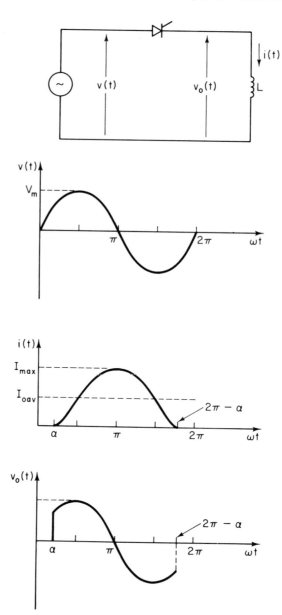

FIGURE 10.19   Voltage and current waveforms for a half-wave controlled rectifier with a purely inductive load.

b) EMF and One Passive Element

The second category of special cases involves E_c and either R or L. We take a look at:

E_c and R Load Circuits

Here we have $L = 0$, and thus

$$\phi = 0 \qquad |Z| = R \qquad \tau = 0$$

As a result, the current is obtained from Eq. (10.30) as

$$i(t) = \frac{V_m}{R} \sin \omega t - \frac{E_c}{R} \qquad \alpha \leq \omega t \leq \alpha + \gamma \qquad (10.52)$$

The angle γ is obtained by setting $i(t) = 0$ to get

$$\frac{E_c}{V_m} = \sin(\alpha + \gamma)$$

Recall that

$$\frac{E_c}{V_m} = \sin \alpha_{min}$$

As a result,

$$\sin(\alpha + \gamma) = \sin \alpha_{min}$$

From the above,

$$\gamma = \pi - \alpha_{min} - \alpha \qquad (10.53)$$

The average current is given by Eq. (10.38) as

$$I_{o_{av}} = \frac{1}{2\pi R} \{V_m[\cos \alpha - \cos(\alpha + \gamma)] - \gamma E_c\} \qquad (10.54)$$

Example 10.7 A half-wave rectifier circuit supplies a load circuit composed of a 5-Ω resistance in series with an EMF $E_c = 80$ V. Assume that the effective value of the ac source voltage is 120 V. Find the average current in the load for $\alpha = 0$ and $\alpha = 30°$.

Solution

We have

$$\alpha_{min} = \sin^{-1} \frac{80}{120 \sqrt{2}} = 28.1°$$

From Eq. (10.53), we get

$$\gamma = 180 - 28.1 - \alpha$$
$$= 151.9° - \alpha$$

Using Eq. (10.54), we obtain

$$I_{o_{av}} = \frac{1}{10\pi} \left[120 \sqrt{2} \, (\cos \alpha - \cos 151.9°) - 80\pi \, \frac{151.9 - \alpha}{180} \right]$$

For $\alpha = 0$, we get

$$I_{o_{av}} = 3.42 \text{ A}$$

For $\alpha = 30°$, we get

$$I_{o_{av}} = 4.03 \text{ A}$$

E_c and L Load Circuits

Here we have $R = 0$, and we have from the circuit equation

$$L \frac{di}{dt} = V_m \sin \omega t - E_c$$

By direct integration, we get

$$i(t) = \frac{1}{L} \int_{\alpha/\omega}^{t} (V_m \sin \omega t - E_c) \, dt$$

As a result,

$$i(t) = \frac{1}{\omega L} [V_m(\cos \alpha - \cos \omega t) - E_c(\omega t - \alpha)]$$

$$\alpha \le \omega t \le \alpha + \gamma \quad (10.55)$$

The conduction angle γ is obtained from the solution to

$$\cos \alpha - \cos(\alpha + \gamma) = \frac{E_c}{V_m} \gamma \tag{10.56}$$

The average current output is obtained from

$$I_{o\mathrm{av}} = \frac{1}{2\pi} \int_{\alpha}^{\alpha + \gamma} i(t) \, d\omega t$$

The result is given by

$$I_{o\mathrm{av}} = \frac{V_m}{2\pi\omega L} \left[\cos \alpha(\gamma - \sin \gamma) + \sin \alpha(1 - \cos \gamma) - \frac{E_c}{V_m} \frac{\gamma^2}{2} \right] \tag{10.57}$$

Example 10.8 A single-phase half-wave controlled rectifier supplies an E_c–L load with $E_c = 80$ V. The effective value of the ac voltage source is 120 V. Find the angle α corresponding to the following conduction angles.

a. $\gamma = 60°$.
b. $\gamma = 120°$.
c. $\gamma = 200°$.

Solution

We rewrite Eq. (10.56) in the form

$$\sin(\alpha + \theta) = \frac{E_c \gamma}{V_m d}$$

where

$$\sin \theta = \frac{a}{d} \qquad \cos \theta = \frac{b}{d}$$

$$a = 1 - \cos \gamma \qquad b = \sin \gamma$$

$$d^2 = a^2 + b^2$$

(a) For $\gamma = 60°$, we obtain

$$a = 0.5 \qquad b = 0.866$$

$$d = 1 \qquad \theta = 30°$$
$$\sin(\alpha + \theta) = 0.494$$
$$\alpha + \theta = 29.6° \quad \text{or} \quad 150.4°$$

Take the second solution to obtain

$$\alpha = 150.4 - 30 = 120.4°$$

(b) For $\gamma = 120°$, we obtain

$$a = 1.5 \qquad b = 0.866$$
$$d = 1.732 \qquad \theta = 60°$$

$$\sin(\alpha + \theta) = 0.57$$
$$\alpha + \theta = 145.3°$$

As a result,

$$\alpha = 85.3°$$

(c) For $\gamma = 200°$, we obtain

$$a = 1.94 \qquad b = -0.342$$
$$d = 1.97 \qquad \theta = 100°$$
$$\sin(\alpha + \theta) = 0.9$$
$$\alpha + \theta = 115.8°$$

As a result,

$$\alpha = 15.8°$$

Single-Phase Full-Wave Controlled Rectifiers

Full-wave rectification of a single-phase alternating source v_1 can be obtained using either the circuit of Figure 10.20(a) or that of Figure 10.20(b). In the single-phase full-wave controlled bridge rectifier circuit of Figure 10.20(a) thyristors Q_{1A} and Q_{1B} conduct the current i_o to the load during part of a half cycle of the source voltage, determined by the delay angle α. At $\omega t = \pi + \alpha$, thyristors Q_{2A} and Q_{2B} conduct the current i_o to the load. As a result, the load current consists of two pulses during one cycle of the

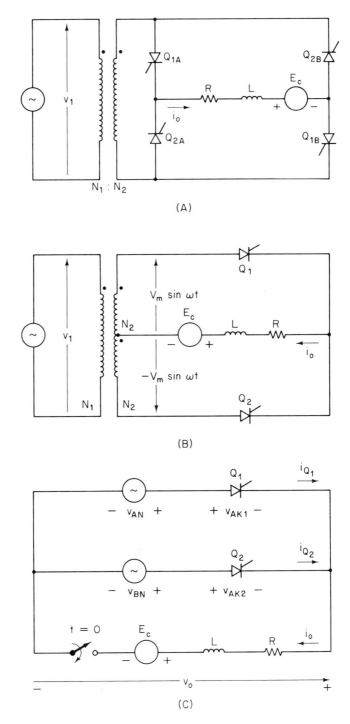

FIGURE 10.20 (a) Single-phase full-wave controlled bridge rectifier; (b) single-phase full-wave controlled rectifier using center-tapped secondary transformer; (c) equivalent circuit of (a) and (b).

635

source voltage. An alternative circuit is shown in Figure 10.20(b), where a transformer with a center-tapped secondary winding is employed. In this case the voltage source is transformed into two single-phase sources of opposite polarity as shown in the figure. Both circuits shown in parts (a) and (b) are equivalent to the circuit shown in part (c). Analysis of the performance of a single-phase full-wave rectifier circuit will be carried out using the equivalent circuit of Figure 10.20(c).

Analysis: Assume that the alternating voltages $v_{AN}(t)$ and $v_{BN}(t)$ are given by

$$v_{AN}(t) = V_m \sin \omega t \qquad\qquad (10.58)$$

$$v_{BN}(t) = V_m \sin(\omega t + \pi) \qquad\qquad (10.59)$$

The two voltage waveforms are shown in Figure 10.21(a). Gate current i_{G_1} is applied to thyristor 1 at $\omega t = \alpha$ as shown in Figure 10.21(b). Gate current i_{G_2} is applied to thyristor 2 at $\omega t = \pi + \alpha$ as shown in Figure 10.21(d). For $\omega t > \alpha$, thyristor 1 and the load circuit function as a single-phase half-wave controlled rectifier and as a result, the current $i_{Q_1}(t)$ in thyristor 1 is as shown in Figure 10.21(c). At $\omega t = \alpha + \gamma$, the current becomes zero and thyristor 1 ceases to conduct for $\omega t > \alpha + \gamma$. The current i_{Q_1} and the conduction angle γ are found using analysis of the half-wave rectifier circuit of this section. The load current $i_0(t)$ will remain at zero in the interval $\alpha + \gamma_t < \omega t < \pi + \alpha$. At $\omega t = \pi + \alpha$, thyristor 2 is fired and its current $i_{Q_2}(t)$ follows the pattern shown in Figure 10.21(e). For the interval $\pi + \alpha < \omega t < \pi + \alpha + \gamma$, thyristor 2 and the load circuit function as a half-wave controlled rectifier circuit independent from thyristor 1 which is blocking. The load current in that interval is supplied by thyristor 2 as shown in Figure 10.21(f).

The load current consists of two distinct pulses in one cycle of the source voltage. The first pulse begins at $\omega t = \alpha$ and continues to $\omega t = \alpha + \gamma$. The second pulse occurs at $\omega t = \pi + \alpha$ and has the same duration as that of the first pulse. The load current is zero for a definite period of time lasting for $\pi - \gamma/\omega$ seconds. This mode of operation is termed the discontinuous current mode, as the load current is not continuously greater than zero. Of course, this requires that $\gamma < \pi$.

Analysis of the single-phase full-wave controlled rectifier operating in the discontinuous mode is simple. Each thyristor operates with the load circuit as a half-wave circuit for the interval during which it conducts. As a result, conclusions drawn earlier are applicable in this case. It should be noted, however, that since now we have two pulses per cycle of the source voltage, the average current of the full-wave case is double that for the corresponding half-wave circuit.

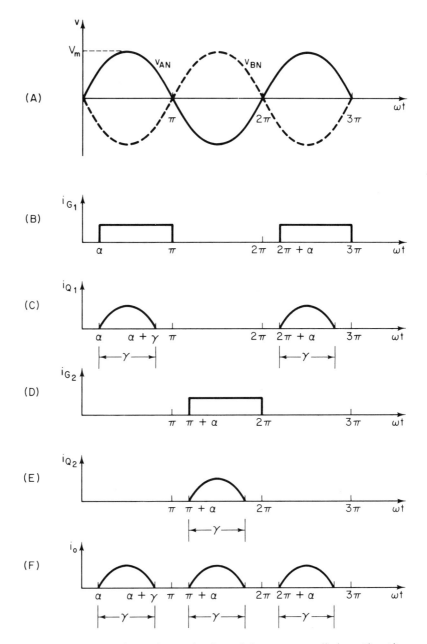

FIGURE 10.21 Waveforms for single-phase full-wave controlled rectifier (discontinuous mode).

Continuous Mode of Operation

Operation in the discontinuous mode requires that $\pi + \alpha > \alpha + \gamma$, that is, thyristor 2 is gated at a time later than that when thyristor 1 ceases to conduct, or $\gamma < \pi$. There are conditions under which the conduction angle is greater than π. To illustrate this for a simple case, consider a load circuit with inductance only. We have, by Eq. (10.50),

$$\alpha + \gamma = 2\pi - \alpha$$

Thus

$$\gamma = 2\pi - 2\alpha$$

For discontinuous current operation, we require that

$$\alpha > \frac{\pi}{2}$$

Any choice of α less than 90° for this circuit will result in operation in the continuous mode, as discussed presently.

For a given load circuit and delay angle α we can calculate the conduction angle for the corresponding half-wave rectifier circuit using Eq. (10.31) or any of its special cases. Let us denote this angle by γ_t, and assume that $\gamma_t > \pi$. At $\omega t = \pi + \alpha$, thyristor 2 conducts due to application of gate current i_{G_2}. The voltage across thyristor 1 denoted by v_{AK1} is then given at that time by

$$v_{AK1}(\pi + \alpha) = v_{AN}(\pi + \alpha) - v_{BN}(\pi + \alpha)$$

$$= -2V_m \sin \alpha$$

Thus thyristor 1 is reverse biased at that point in time and ceases to conduct. Note that since $\pi + \alpha < \alpha + \gamma_t$, the current through thyristor 1 at $\pi + \alpha$ is positive, as shown in Figure 10.22(c). The load current at $\omega t = \pi + \alpha$ is equal to $i_{Q_1}(\pi + \alpha)$, which is positive. It can thus be concluded that the current through thyristor 2 at $\omega t = \pi + \alpha$ is equal to that through thyristor 2, as shown in Figure 10.22(e). The output or load current in the present case is shown in Figure 10.22(f) and is seen to be continuously greater than zero, and hence this mode of operation is called the continuous mode of operation. Note that the actual conduction angle of each thyristor in this mode is $\gamma = \pi$ and is less than the theoretical value γ_t.

Critical Delay Angle

The conduction angle γ_t for a given set of parameters E_c, V_m, R, and L varies with the chosen delay angle α. We have seen that if $\gamma_t < \pi$, the

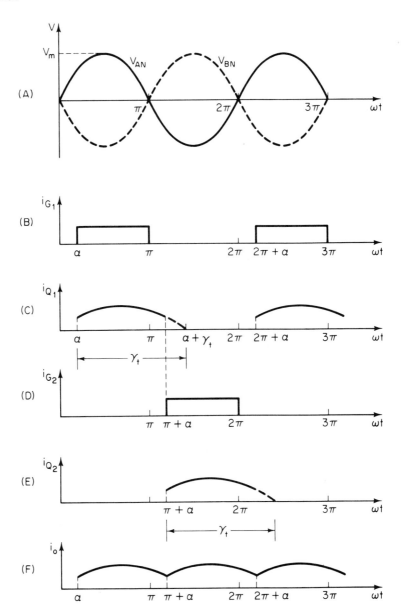

FIGURE 10.22 Waveforms for single-phase full-wave controlled rectifier (continuous mode of operation).

output current of a full-wave rectifier circuit is discontinuous. On the other hand, if $\gamma_t > \pi$, the output current is continuous. The boundary between the two modes is defined by the requirement that $\gamma_t = \pi$. In this case the output current is zero only at $\omega t = \pi + \alpha_c$, where α_c is called the critical delay angle. For example, in the case of a purely inductive load circuit ($E_c = 0$, $R = 0$), we have seen that $\alpha_c = \pi/2$. Obtaining expressions for α_c in terms of E_c, V_m, and the load circuit power factor, angle ϕ is carried out by substituting $\gamma = \pi$ in the equation defining γ.

Three-Phase Controlled Rectifiers

Three-phase ac supply sources are common in practice due to their advantages in providing smooth operating conditions. Now we discuss controlled rectifiers with three-phase ac voltages as the supply. A three-phase half-wave controlled rectifier circuit is shown in Figure 10.23(a). The output current through and the voltage across the load resistance R are shown in Figure 10.23(b). It may be seen that this is a three-pulse circuit, as there are three load current pulses in one period of the source waveform. The circuit is impractical since ideal voltage sources are required.

The circuit most commonly encountered in practice is the three-phase

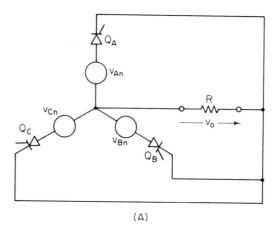

(A)

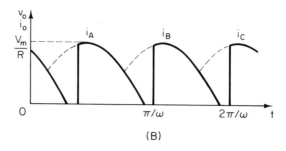

(B)

FIGURE 10.23   Three-phase half-wave controlled rectifier.

full-wave controlled rectifier circuit shown in Figure 10.24(a). A pair of thyristors is associated with each phase of the source, as shown. The sequence in which gate current pulses are applied to the thyristors can best be understood with reference to Figure 10.24(b). Assuming that a phase

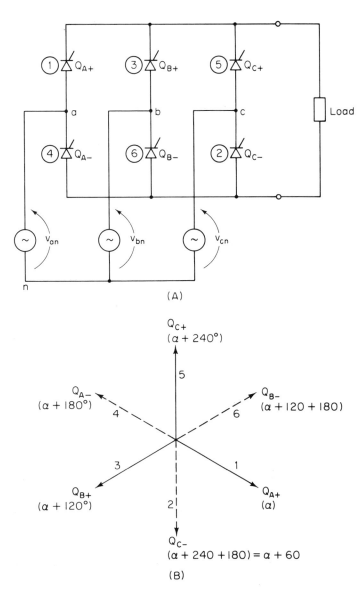

FIGURE 10.24 Three-phase full-wave controlled rectifier: (a) circuit; (b) phasor illustration of firing sequence.

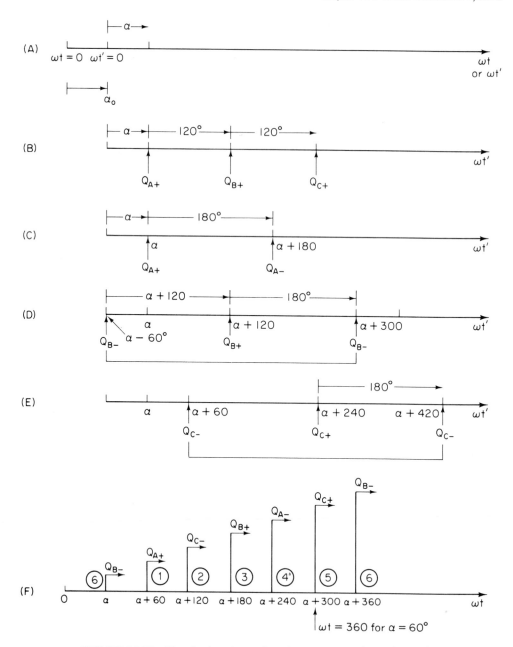

FIGURE 10.25 Developing the gating time sequence for a three-phase full-wave controlled rectifier.

sequence *abc* is employed, the diagram shown in Figure 10.24(b) is constructed. Gate current is applied to thyristor $A+$ at $\omega t' = \alpha$, where t' denotes the time measured from a suitable reference related to the system time reference by $\omega t' = \omega t - \alpha_0$, as shown in Figure 10.25(a). The angle α_0 is chosen such that subsequent analysis can be made easier. For our present case $\alpha_0 = 60°$ is a convenient value. It is thus clear that Q_{B+} is gated (fired) at $\omega t' = \alpha + 120°$ and that Q_{C+} is gated at $\omega t' = \alpha + 240°$. This is shown in Figure 10.25(b). Thyristor Q_{A-} is gated 180° later than Q_{A+}, that is, at $\omega t' = \alpha + 180°$, as shown in Figure 10.25(c). Similarly, Q_{B-} is gated 180° later than Q_{B+} at $\omega t' = \alpha + 300$, as shown in Figure 10.25(d). Figure 10.25(e) pertains to thyristors associated with phase *C*. Note that Q_{B-} is fired at $\omega t' = \alpha + 300$ as well as $\omega t' = \alpha + 300 - 360 = \alpha - 60°$ and that Q_{C-} is fired at $\omega t' = \alpha + 420$ as well as at $\omega t' = \alpha + 60°$, as shown. In part (f) of the figure we show the instants when each thyristor is gated and conclude that the time sequence of gating is as shown in the circled numbers 1 to 6.

Assume that the load circuit is purely resistive and that the thyristors are replaced by diodes. We further assume that the line-to-line voltages of the supply are given by the waveforms shown in Figure 10.26(a) as

$$v_{ab}(t) = V_m \sin \omega t$$

$$v_{bc}(t) = V_m \sin(\omega t - 120°)$$

$$v_{ca}(t) = V_m \sin(\omega t + 120°)$$

At $\omega t = \pi/6$, we have $v_{ab} = 0.5V_m$, $v_{bc} = -V_m$, and $v_{ca} = 0.5V_m$. The largest line-to-line voltage is v_{cb}, and therefore thyristors Q_{C+} and Q_{B-} conduct at that time. Next, take $\omega t = \pi/2$, at which v_{ab} has the largest voltage and therefore Q_{A+} and Q_{B-} conduct. The active current paths in the circuit at these times are shown in Figure 10.26(b) and (c).

It is clear from Figure 10.26 and our present discussion that the operation of the circuit of Figure 10.24 can be described using the following six intervals:

Interval I: $\pi/3 \leq \omega t < 2\pi/3$, v_{ab} is applied to the load through thyristors $A+$ and $B-$. The voltage applied to the load is

$$v_1(t) = v_{ab}(t) = V_m \sin \omega t$$

Interval II: $2\pi/3 < \omega t < \pi$, v_{ac} is applied to the load through thyristors $A+$ and $C-$ with

$$v_2(t) = v_{ac}(t) = V_m \sin(\omega t - 60°)$$

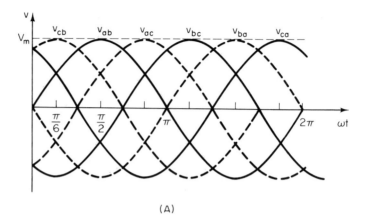

(A)

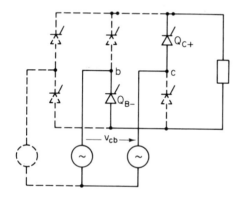
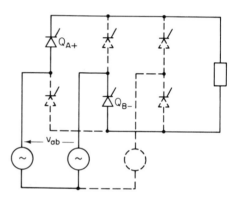

(B) active current path at $\omega t = \dfrac{\pi}{6}$;   (C) active current path at $\omega t = \dfrac{\pi}{2}$.

FIGURE 10.26 Illustrating operation of three-phase rectifier of Figure 10.24(a): (a) line voltage waveforms for circuit of Figure 10.24; (b) active current path at $\omega t = \pi/6$; (c) active current path at $\omega t = \pi/2$.

Interval III: $\pi < \omega t < 4\pi/3$, v_{bc} is applied to the load through thyristors $B+$ and $C-$ and

$$v_3(t) = v_{bc}(t) = V_m \sin(\omega t - 120°)$$

Interval IV: $4\pi/3 < \omega t < 5\pi/3$, v_{ba} is applied to the load through thyristors $B+$ and $A-$, and

$$v_4(t) = v_{ba}(t) = V_m \sin(\omega t - 180°)$$

Interval V: $5\pi/3 < \omega t < 2\pi$, v_{ca} is applied to the load through thyristors $A-$ and $C+$, with

$$v_5(t) = v_{ca}(t) = V_m \sin(\omega t - 240°)$$

Interval VI: $2\pi < \omega t < 7\pi/3$, v_{cb} is applied to the load through $C+$ and $B-$ and

$$v_6(t) = v_{cb}(t) = V_m \sin(\omega t - 300°)$$

It is clear that the rectifier is a six-pulse converter and that its operation can be examined using an equivalent circuit containing a balanced six-phase voltage source and six thyristors, as shown in Figure 10.27.

In the equivalent circuit of Figure 10.27(a), thyristor Q_1 represents the two thyristors active in interval I in the actual circuit and similar statements can be made about the five other thyristors in the equivalent circuit. The reference point for the angle α is at $\alpha_0 = \pi/3$, as mentioned earlier. Inspection of Figure 10.24 leads us to conclude that the operation of a three-phase full-wave rectifier given in the circuit of Figure 10.24(a) can be described in terms of six single-phase half-wave circuits and since the sources are balanced, our performance equations are similar to those obtained for half-wave rectifiers, with some modifications discussed presently.

In the circuit of Figure 10.24 there are six current pulses, and hence the average output current is six times the average current obtained by a single-phase half-wave controlled rectifier. The value of α in the single-phase half-wave rectifier analysis equations must be reduced by 60°, due to the datum from which α is measured in the three-phase circuit.

The range of α for which discontinuous operation takes place in the three-phase case is much smaller than the corresponding range in the single-phase case.

In Figures 10.28, voltage and current waveforms for a three-phase controlled rectifier operating in the continuous current mode are shown. In this case, the average output voltage is obtained using

$$V_o = \frac{3}{\pi} \int_{\alpha+60}^{\alpha+120} V_m \sin\theta\, d\theta = \frac{3V_m}{\pi} \cos\alpha \qquad (10.60)$$

The fundamental frequency of the output is six times that of the source frequency.

A three-phase controlled rectifier circuit with a dc voltage source E_c in its load circuit can operate as an inverter, transferring dc energy from the rectifier output (load) side to the ac source. The criterion to establish the mode of operation as an inverter is simply that the average power output is negative.

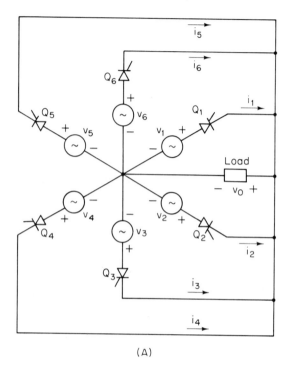

(A)

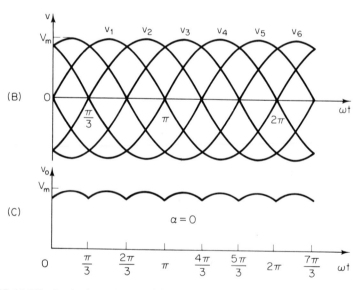

FIGURE 10.27 Equivalent circuit of three-phase full-wave controlled rectifier and voltage waveforms.

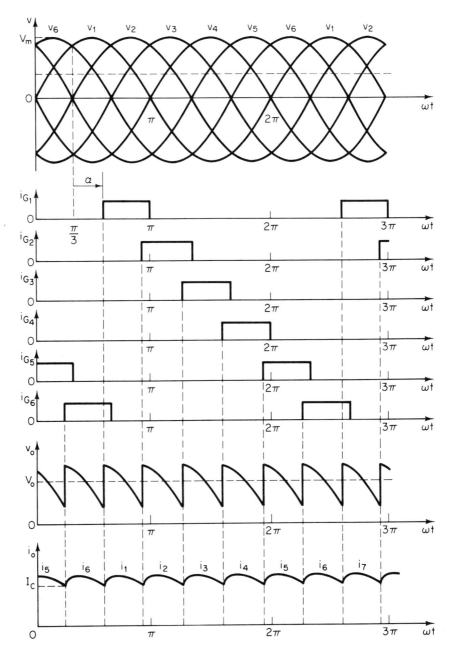

FIGURE 10.28   Voltages and current waveforms for a three-phase controlled rectifier operating in the continuous-current mode.

10.4 DC-TO-DC CONVERTERS (CHOPPERS)

A dc-to-dc converter or chopper provides a dc output from a dc supply. The average of the output is varied relative to the input by varying the porportion of the time during which the output is connected to the input.

The basic circuit of a dc-to-dc converter, or chopper, is shown in Figure 10.29(a) with a resistive load. A thyristor Q is connected in series with the dc voltage source V_i and the load circuit. A free-wheeling diode D is connected in parallel with the load circuit. The thyristor's gate current is a sequence of pulses of duration t_{on}, at a period T as shown in Figure 10.29(b). The input voltage $v_i(t)$ is constant at the value V_i, as shown in Figure 10.29(c). The output current $i_o(t)$ is zero when the thyristor is turned off. When the thyristor is conducting, the voltage appearing at the load terminals is V_i and thus we have for the first period

$$v_o(t) = \begin{cases} V_i & 0 < t \leq t_{on} \\ 0 & t_{on} < t \leq T \end{cases} \tag{10.61}$$

$$i_o(t) = \begin{cases} \dfrac{V_i}{R} & 0 < t \leq t_{on} \\ 0 & t_{on} < t \leq T \end{cases} \tag{10.62}$$

The output current and voltage waveforms are shown in Figure 10.29(d) and (e). Clearly,

$$i_i(t) = i_o(t)$$

as shown in Figure 10.29(f). The waveforms are repeated periodically with period T.

The dc value of the output voltage, denoted by V_o, is the average of $v_o(t)$ over the period

$$V_o = \frac{1}{T} \int_0^T v_o(t)\, dt = \frac{1}{T} \int_0^{t_{oN}} V_i\, dt$$

Thus

$$V_o = \frac{t_{on}}{T} V_i \tag{10.63}$$

The average value of the output current is

$$I_o = \frac{V_o}{R} \tag{10.64}$$

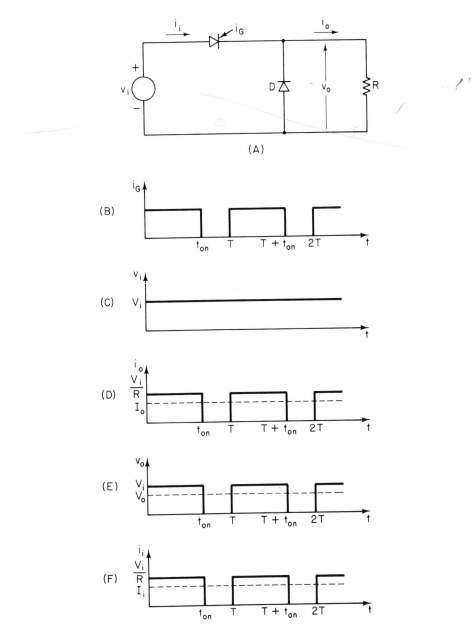

FIGURE 10.29 Basic chopper circuit and pertinent waveforms: (a) circuit current pulses; (b) gate current pulses; (c) input voltage; (d) output current; (e) output voltage; (f) input current.

Note that for this configuration both V_o and I_o are positive, which corresponds to the operation in the first quadrant of the V_o–I_o plane.

Analysis of Basic Chopper Circuit

Choppers are used to control speed of dc motors by varying the applied dc voltage. It is therefore important to study the dynamics of a chopper connected to a load consisting of an EMF E_c, in series with a resistance and an inductance as shown in Figure 10.30. The load circuit corresponds to the armature of a dc machine. It is assumed that E_c is constant in the following analysis.

Application of Kirchhoff's voltage law (KVL) to the output loop yields

$$Ri_o + L\frac{di_o}{dt} = V_o - E_c \tag{10.65}$$

Define the time constant τ by

$$\tau = \frac{L}{R} \tag{10.66}$$

This yields

$$\frac{di_o}{dt} + \tau i_o = \frac{V_o - E_c}{R} \tag{10.67}$$

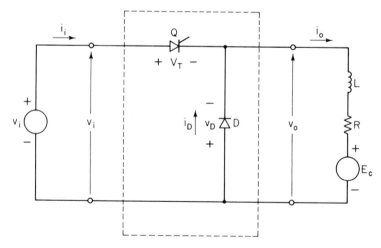

FIGURE 10.30   First-quadrant chopper connected to a load circuit.

Let us assume that at $t = 0^+$, the current i_o is given by the initial value denoted by I_{min}. The thyristor is turned on at $t = 0$, and stays on up to $t = t_{on}$, so that $V_o = V_i$. The current $i_o(t)$ for $t > 0$ is thus given by

$$i_o(t) = I_{min} e^{-t/\tau} + \frac{V_i - E_c}{R}(1 - e^{-t/\tau}) \qquad 0 \le t < t_{on} \quad (10.68)$$

The current i_o increases exponentially as time increases.

Assume now that at $t = t_{on}$, the thyristor ceases to conduct. The value of i_o at $t = t_{on}$ is denoted by I_{max} and is given by

$$i_o(t_{on}) = I_{max} = I_{min} e^{-t_{on}/\tau} + \frac{V_i - E_c}{R}(1 - e^{-t_{on}/\tau}) \qquad (10.69)$$

This expression is obtained by substituting $t = t_{on}$ in Eq. (10.68).

For the time interval $t > t_{on}$, the voltage V_o is zero since the diode conducts. The KVL equation for the output loop is thus

$$\tau \frac{di_o}{dt} + i_o = -\frac{E_c}{R} \qquad (10.70)$$

The solution for $i_o(t)$ is obtained in the usual manner observing continuity of current in the inductor, requiring that

$$i_o(t_{on}) = I_{max} \qquad (10.71)$$

The result is given by

$$i_o(t) = I_{max} e^{-(t - t_{on})/\tau} - \frac{E_c}{R}(1 - e^{-(t - t_{on})/\tau}) \qquad (10.72)$$

The current decreases exponentially as time increases.

Assume now that at $t = T$, the thyristor conducts. The value of i_o at this instant of time is denoted by I_{min} and is given by

$$i_o(T) = I_{min} = I_{max} e^{-(T - t_{on})/\tau} - \frac{E_c}{R}(1 - e^{-(T - t_{on})/\tau}) \qquad (10.73)$$

Note that the values of i_o at $t = 0$ and $t = T$ are equal to I_{min}.

We can now conclude that the current $i_o(t)$ varies from a minimum value of I_{min} given by Eq. (10.73) at $t = 0$ to a maximum value of I_{max} given by Eq. (10.69) at $t = t_{on}$ and then decreases to a value I_{min} at the end of the period T. The values of I_{max} and I_{min} in terms of thyristor conduction time t_{on}, the period T, the circuit's time constant τ, the input voltage V_i, and the counter EMF E_c are obtained by combining Eqs. (10.69) and (10.73) as

$$I_{max} = \frac{V_i}{R}\left(\frac{1 - e^{-t_{on}/\tau}}{1 - e^{-T/\tau}} - \frac{E_c}{V_i}\right) \tag{10.74}$$

$$I_{min} = \frac{V_i}{R}\left(\frac{e^{t_{on}/\tau} - 1}{e^{T/\tau} - 1} - \frac{E_c}{V_i}\right) \tag{10.75}$$

We explore the dependence of I_{max} and I_{min} on the value of the thyristor conduction time t_{on}. An upper limit on t_{on} is given by the period T corresponding to the case where the thyristor is continuously turned on. For $t_{on} = T$, we can see that

$$I_{max} = I_{min} = \frac{V_i - E_c}{R} \tag{10.76}$$

The value of I_{min} is positive for $t_{on} = T$, assuming that $V_i > E_c$. For a small value of t_{on}, according to Eq. (10.75), the value of I_{min} is negative. This cannot take place, since the thyristor is not conducting and a negative i_o cannot pass through the diode. The lowest possible value of I_{min} is therefore zero.

The critical value of t_{on} denoted by t_{on}^* for which I_{min} is zero is obtained from Eq. (10.75) as

$$t_{on}^* = \tau \, \ell n\left[1 + \frac{E_c}{V_i}(e^{T/\tau} - 1)\right] \tag{10.77}$$

We will investigate two possible cases. In the first case t_{on} is larger than the critical value t_{on}^* and the second case corresponds to t_{on} being less than the critical value t_{on}^*.

Assume that $t_{on} > t_{on}^*$, and hence I_{min} is positive. Figure 10.31(a) shows the thyristor's gate current pulses, and the corresponding output current $i_o(t)$ is shown in Figure 10.31(b). Note that for $0 < t < t_{on}$, the current is given by Eq. (10.68), whereas for $t_{on} < t < T$, the current is given by Eq. (10.72). The main point is that the current $i_o(t)$ is always present and this mode of operation is referred to as the continuous current mode of operation.

The second mode of operation corresponds to $t_{on} < t_{on}^*$, for which $I_{min} = 0$. The output current $i_o(t)$ for $0 < t < t_{on}$, is given by Eq. (10.68), rewritten as

$$i_o(t) = \frac{V_i - E_c}{R}(1 - e^{-t/\tau}) \qquad 0 < t < t_{on} \tag{10.78}$$

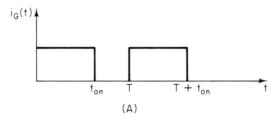

(A)

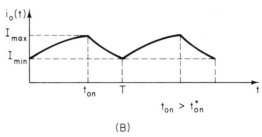

$t_{on} > t^*_{on}$

(B)

FIGURE 10.31 Waveforms for the continuous-current-mode operation of basic chopper circuit: (a) gate current pulses; (b) output current $i_o(t)$.

The maximum value of $i_o(t)$ is obtained from either Eq. (10.78) or from Eq. (10.69) by setting $I_{min} = 0$ as

$$i_o(t_{on}) = I_{max} = \frac{V_i - E_c}{R}(1 - e^{-t_{on}/\tau}) \qquad (10.79)$$

For $t > t_{on}$, we have, from Eqs. (10.72) and (10.79),

$$i_o(t) = \frac{V_i - E_c}{R}(1 - e^{-t_{on}/\tau})e^{-(t - t_{on})/\tau}$$

$$-\frac{E_c}{R}(1 - e^{-(t - t_{on})/\tau}) \qquad t > t_{on} \qquad (10.80)$$

The current $i_o(t)$ decreases according to Eq. (10.80) and is zero at time t_x obtained from

$$0 = \left[\frac{V_i}{R}(1 - e^{-t_{on}/\tau}) + \frac{E_c}{R}e^{-t_{on}/\tau}\right]e^{-(t_x - t_{on})/\tau} - \frac{E_c}{R}$$

As a result, we find t_x as

$$t_x = \tau \ln\left[1 + \frac{V_i}{E_c}(e^{t_{on}/\tau} - 1)\right] \qquad (10.81)$$

The current $i_o(t)$ remains zero for values of t greater than t_x up to the time T, when the thyristor conducts, and we have

$$i_o(t) = 0 \qquad t_x < t \leqslant T$$

The present case is referred to as the discontinuous mode of operation. Figure 10.32(a) shows the gate current pulses and Figure 10.32(b) shows the corresponding output current for this case.

Our analysis of the basic chopper circuit has centered so far on the output current waveforms. The output voltage $v_o(t)$ is treated now. With reference to the circuit of Figure 10.30, it can be seen that the output voltage $v_o(t)$ when the thyristor conducts is given by

$$v_o(t) = V_i \qquad 0 < t < t_{on}$$

Immediately following t_{on}, the diode D conducts up to the time instant t_x. As a result,

$$v_o(t) = 0 \qquad t_{on} < t < t_x$$

since the load is shorted out by D. In the remainder of the period, the output voltage is equal to E_c, since i_o is zero:

$$v_o(t) = E_c \qquad t_x < t < T$$

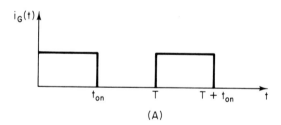

(A)

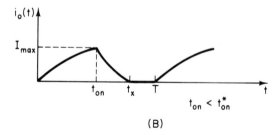

(B)

FIGURE 10.32 Gate current and output current waveforms for the discontinuous mode of operation of a chopper.

The average value of the output voltage V_o can therefore be found using

$$V_o = \frac{1}{T}\left[\int_0^{t_{on}} V_i \, dt + \int_{t_x}^T E_c \, dt\right]$$

As a result,

$$V_o = \frac{t_{on}}{T} V_i + \frac{T - t_x}{T} E_c \qquad (10.82)$$

For continuous operation, note that $t_x = T$. The average value of the output current I_o can thus be obtained simply using

$$I_o = \frac{V_o - E_c}{R} \qquad (10.83)$$

It is important to consider two examples at this time.

Example 10.9 In a basic chopper circuit, the following information is available:

$$
\begin{array}{ll}
V_i = 100 \text{ V} & E_c = 12 \text{ V} \\
L = 0.8 \text{ mH} & R = 0.2 \text{ } \Omega \\
T = 2.400 \text{ ms} & t_{on} = 1 \text{ ms}
\end{array}
$$

a. Determine the mode of operation.
b. Find the average output voltage and current.
c. Find the maximum and minimum values of the output current.

Solution

(a) We have

$$\tau = \frac{L}{R} = \frac{0.8 \times 10^{-3}}{0.2} = 4 \times 10^{-3} \text{ s}$$

We find the critical value of on-time using Eq. (10.77) as

$$t_{on}^\star = 4 \times 10^{-3} \, \ell n \left[1 + \frac{12}{100} \left(e^{2.4/4} - 1\right)\right]$$

$$= 0.376 \times 10^{-3} \text{ s}$$

Since $t_{on}^{\star} < t_{on}$, we conclude that the operation of the chopper is continuous.

(b) The average output voltage is given by Eq. (10.82) with $t_x = T$ as

$$V_o = \frac{t_{on}}{T} V_i = \frac{1}{2.4} (100) = 41.667 \text{ V}$$

The average output current is thus obtained as

$$I_o = \frac{V_o - E_c}{R} = \frac{41.667 - 12}{0.2} = 148.33 \text{ A}$$

(c) Using Eq. (10.74), we have

$$I_{max} = \frac{100}{0.2} \left(\frac{1 - e^{-1/4}}{1 - e^{-2.4/4}} \right) - \frac{12}{0.2}$$

$$= 185.13 \text{ A}$$

Using Eq. (10.75), we have

$$I_{min} = \frac{100}{0.2} \left(\frac{e^{1/4} - 1}{e^{2.4/4} - 1} \right) - \frac{12}{100}$$

$$= 112.74 \text{ A}$$

In the following example we deal with the discontinuous mode of operation.

Example 10.10 Suppose in the chopper circuit of Example 10.9 that $E_c = 50$ V, with all other parameters unchanged. It is required to repeat Example 10.9 under the new conditions.

Solution

(a) In the present case, we obtain

$$t_{on}^{\star} = 4 \times 10^{-3} \ln \left[1 + \frac{50}{100} (e^{2.4/4} - 1) \right]$$

$$= 1.38 \times 10^{-3} \text{ s}$$

The operation is therefore in the discontinuous mode since $t_{on}^{\star} > t_{on}$.

(b) We need to determine t_x using Eq. (10.81):

$$t_x = \tau \, \ell n \left[1 + \frac{V_i}{E_c} (e^{t_{on}/\tau} - 1) \right]$$

$$= 4 \times 10^{-3} \, \ell n \left[1 + \frac{100}{50} (e^{1/4} - 1) \right]$$

$$= 1.8 \times 10^{-3} \text{ s}$$

We now apply Eq. (10.82) to get V_o:

$$V_o = \frac{1}{2.4} (100) + \frac{2.4 - 1.8}{2.4} (50)$$

$$= 54.167 \text{ V}$$

The average output current is therefore obtained as

$$I_o = \frac{54.167 - 50}{0.2} = 20.83 \text{ A}$$

(c) Using Eq. (10.79), we get

$$I_{max} = \frac{V_i - E_c}{R} (1 - e^{-t_{on}/\tau})$$

$$= \frac{100 - 50}{0.2} (1 - e^{-1/4})$$

$$= 55.3 \text{ A}$$

Since the current is discontinuous, we note that $I_{min} = 0$ A.

Two-Quadrant Choppers

The basic circuit of the one-quadrant chopper of Figure 10.30 is modified by introducing a second thyristor Q_2 and a diode D_2 as shown in Figure 10.33(a). This allows for the output current to take on negative values. Gate current pulses i_{G_1} of duration t_{on} at a period T are applied to thyristor Q_1 as shown in Figure 10.33(b) in a manner similar to that in the one-quadrant chopper. Thyristor Q_2 is switched on in the intervals during which Q_1 is turned off by applying the gate current pulses i_{G_2} as shown in Figure 10.33(c). Pulses i_{G_2} have the same period T but are of duration $(T - t_{on})$.

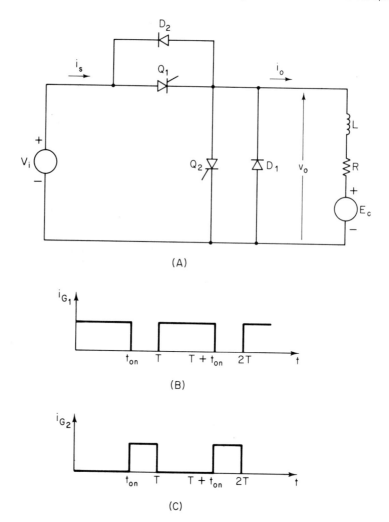

(A)

(B)

(C)

FIGURE 10.33 Two quadrant chopper: (a) basic circuit; (b) thyristor Q_1 gate current pulse; (c) thyristor Q_2 gate current pulse.

The operation of the circuit depends on the sign of i_o and the state of the thyristors Q_1 and Q_2 as follows:

1. *Positive output current:* The circuit functions as a one-quadrant chopper in the following manner:
   (a) *If Q_1 is on,* current i_o is supplied to the load from the source through Q_1. Here we have $i_s = i_o$, since Q_2 is turned off.
   (b) *If Q_1 is off,* current i_o flows through the free-wheeling diode D_1. The source current i_s is zero in this case.

2. *Negative output current:* This corresponds to second quadrant operation, and we have

   (a) *If Q_1 is on,* Q_2 is off and current flows from the load through the
      free-wheeling diode D_2 to the source. Here we have $i_s = i_o$.

   (b) *If Q_1 is off,* Q_2 is on and the load current i_o circulates through the
      thyristor Q_2. The source current is zero in this case.

The two-quadrant chopper circuit is composed essentially of two one-quadrant circuits that operate in the continuous mode of operation and does not
permit discontinuous load currents.

    The operation of the circuit can be analyzed using the same equations
developed for the one-quadrant chopper circuit in the continuous mode.
Recall that the maximum and minimum values of the output current are
given by Eqs. (10.74) and (10.75) as

$$I_{\max} = \frac{V_i}{R}\left(\frac{1 - e^{-t_{on}/\tau}}{1 - e^{-T/\tau}} - \frac{E_c}{V_i}\right)$$

$$I_{\min} = \frac{V_i}{R}\left(\frac{e^{t_{on}/\tau} - 1}{e^{T/\tau} - 1} - \frac{E_c}{V_i}\right)$$

In the present case, $I_{\max}$ and $I_{\min}$ can assume negative values depending on
the circuit parameters. Three cases of interest are examined:

Case 1: Both $I_{\max}$ and $I_{\min}$ are positive. The circuit functions as a one-
      quadrant chopper.

Case 2: The maximum current $I_{\max}$ is positive, whereas $I_{\min}$ is negative.
      The resulting average output current I_o can be positive, as shown in
      Figure 10.34, corresponding to first-quadrant operation, or negative,
      corresponding to second-quadrant operation. The diagrams of Figure
      10.34 show the pertinent waveforms as well as the configuration of
      active elements of the chopper circuit (shown in heavy lines) for first-
      quadrant operation.

Case 3: Both $I_{\max}$ and $I_{\min}$ are negative and thus I_o is negative correspond-
      ing to second-quadrant operation. Pertinent waveforms for this case
      are shown in Figure 10.35.

Four-Quadrant Choppers

The arrangement of Figure 10.36(a) shows a basic circuit of a dc-to-dc converter that is able to operate in any of the four quadrants of the v_o–I_o plane.
It is essentially an interconnection of a pair of two-quadrant choppers with
appropriate control signals.

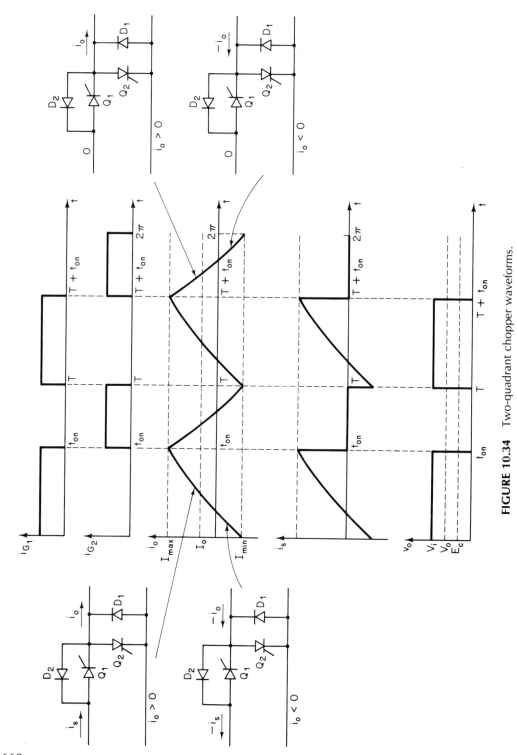

FIGURE 10.34 Two-quadrant chopper waveforms.

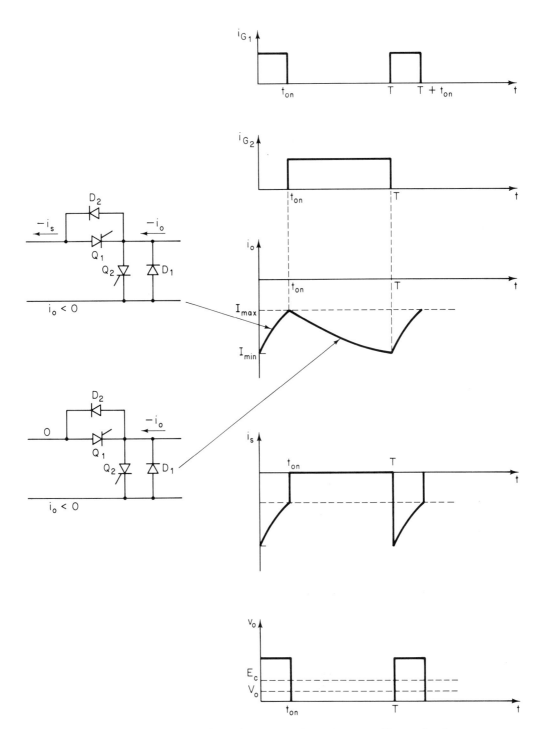

FIGURE 10.35 Waveforms for two-quadrant chopper with negative I_o.

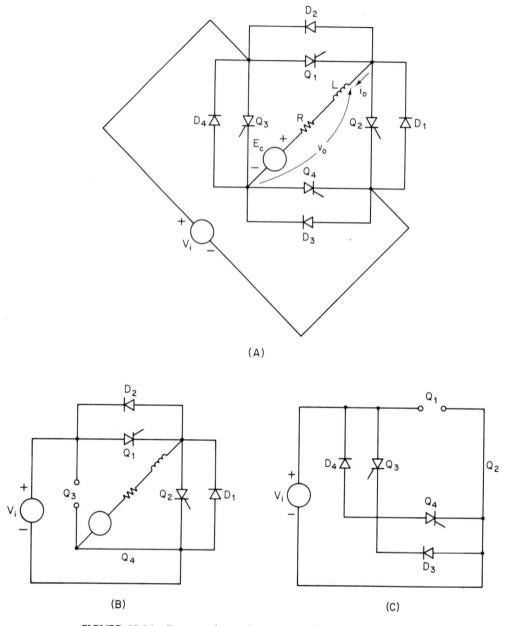

FIGURE 10.36 Four-quadrant chopper: (a) basic circuit; (b) operation in first and second quadrants; (c) operation in third and fourth quadrants.

For operation in the first and second quadrants, Q_4 is turned on continuously and Q_3 is turned off continuously. The circuit now is equivalent to the form shown in Figure 10.36(b), which is clearly a two-quadrant configuration. For operation in the third and fourth quadrants, Q_2 is turned on continuously and Q_1 is turned off continuously. As a result, the equivalent circuit of Figure 10.36(c) applies. This is clearly a two-quadrant configuration with negative values of V_o.

10.5 INVERTERS

An inverter converts dc power to ac power at required output voltage and frequency. The output voltage of an inverter has a periodic waveform which is not sinusoidal but can be made to closely approximate this desired waveform by using a number of techniques. Basic inverter circuits are discussed first.

Half-Bridge Inverter

Figure 10.37(a) shows the basic configuration of a single-phase half-bridge or center-tapped dc source inverter. A three-wire dc source is required for this circuit. Thyristors Q_1 and Q_2 are alternately turned on to connect point a to the positive and negative lines, respectively. The gate currents for the thyristors are shown in Figure 10.37(b) and (c), where T is related to the output frequency by

$$T = \frac{1}{f}$$

The output voltage $v_o(t)$ shown in Figure 10.37(d) is a square wave with frequency f and amplitude $(V_d/2)$. The waveform of the output current $i_o(t)$ depends on the composition of the load circuit. Diodes D_1 and D_4 conduct current from an inductive load back to the source when thyristors are not conducting.

Full-Bridge Single-Phase Inverters

In the full-bridge inverter configuration shown in Figure 10.38(a), the disadvantage of the half-bridge arrangement requiring a three-wire dc supply with center-tapped source is overcome at the expense of adding two additional sets of thyristors and diodes. Thyristors Q_1 and Q_2 are triggered simultaneously, as shown in Figure 10.38(b), whereas thyristors Q_2 and Q_3

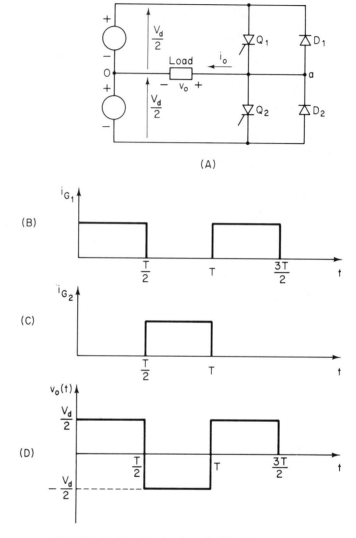

FIGURE 10.37   Single-phase half-bridge inverter.

are triggered simultaneously half a cycle later, as shown in Figure 10.38(c). The output voltage waveform $v_o(t)$ is similar to that of the half-bridge circuit but now has an amplitude of V_d as shown in Figure 10.38(d).

The voltage waveform $v_o(t)$ can be expressed as the sum of sinusoidal waveforms using the Fourier series expansion. The result is

$$v_o(t) = \sum_{n=1}^{\infty} b_n \sin n\omega t \qquad (10.84)$$

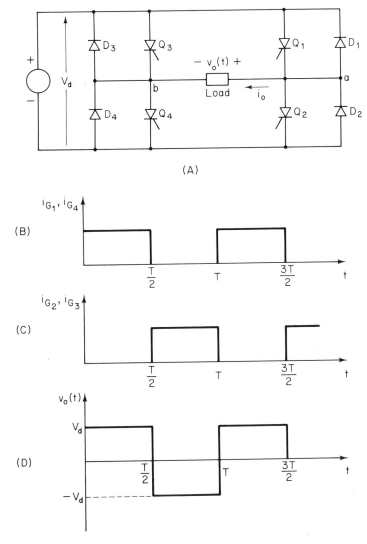

FIGURE 10.38 Single-phase full-bridge inverter.

where

$$\omega = 2\pi f = \frac{2\pi}{T} \tag{10.85}$$

The amplitude coefficients b_n are given by

$$b_n = \begin{cases} \dfrac{4V_d}{n\pi} & n = 1, 3, 5, \ldots \\[2mm] 0 & n = 2, 4, 6, \ldots \end{cases} \tag{10.86}$$

The output voltage contains only odd harmonics. The fundamental component of the output voltage is thus given by

$$v_1(t) = \frac{4V_d}{\pi} \sin \omega t \tag{10.87}$$

The third-harmonic (sinusoidal) component is

$$v_3(t) = \frac{4V_d}{3\pi} \sin 3\omega t \tag{10.88}$$

The third harmonic of the output voltage has an amplitude that is 0.33 times the amplitude of the fundamental.

Assume that the load circuit is a series RLC combination. The steady-state value of the current $i_n(t)$ in the load circuit due to the nth harmonic voltage $v_n(t)$ is obtained by using the complex impedance Z_n defined by

$$Z_n = R + jn\omega L + \frac{1}{jn\omega C} \tag{10.89}$$

In polar form we write

$$Z_n = |Z_n| \, \underline{/\phi_n} \tag{10.90}$$

As a result, we have

$$i_n(t) = \frac{b_n}{|Z_n|} \sin(n\omega t - \phi_n) \tag{10.91}$$

The output current is therefore given by

$$i_o(t) = \sum_{n=1}^{\infty} \frac{b_n}{|Z_n|} \sin(n\omega t - \phi_n) \tag{10.92}$$

Again the output current contains only odd harmonics.

The waveform of the output current depends on the composition of the load circuit as shown in Figure 10.39(a). For a resistive load $i_o(t)$ is a square wave of amplitude $I_c = V_d/R$ as shown in part (b) of the figure. For

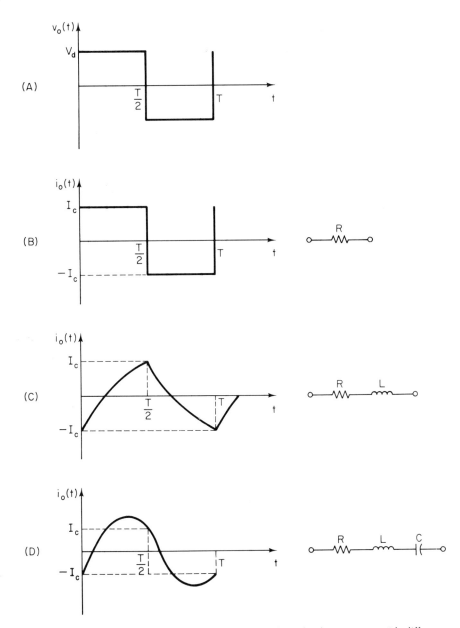

FIGURE 10.39 Current waveforms for a single-phase bridge inverter with different load circuit compositions: (a) output voltage; (b) output current with resistive load; (c) output current with inductive load; (d) output current with overdamped *RLC* load.

an inductive load circuit one obtains the variation shown in part (c). Here
the current varies from a minimum of $-I_c$ at $t = 0, T, 2T, \ldots$, to $+I_c$ at
$t = T/2, 3T/2, \ldots$, assuming, of course, steady operation. For an RLC
load such that the inductive effect dominates the capacitive (overdamped),
we obtain the waveform shown in Figure 10.39(d).

The output current at the instant of commutation $t = T/2$ is denoted
by I_c in the waveforms of Figure 10.39. The value of I_c is obtained by
substituting in Eq. (10.92) with $t = T/2$. Note that in the inductive cases,
$I_c > 0$ and forced commutation should be employed. For capacitive loads
and underdamped RLC circuits, the value of I_c is negative and it may be
possible to rely on load commutation of the thyristors.

Voltage Control Techniques

Controlling the ac output of an inverter to meet operational requirements
can be achieved using a number of techniques that depend on the nature of
the dc source and available hardware. The most desirable arrangement permits varying the ratio of the dc input voltage and the ac output voltage using
pulse-width modulation.

Pulse-width modulation requires means of forced commutation and
can be classified in the following categories: (1) single-pulse modulation, (2)
multiple-pulse modulation, and (3) sinusoidal-pulse modulation.
These basic methods are discussed presently.

1. *Single-pulse modulation:* In single-pulse modulation voltage control, the
   output voltage waveform is in the form of one pulse for each half a
   cycle of the required load voltage. In Figure 10.40(b), the pulse is
   shown centered at $\omega t = \pi/2$ (i.e., $t = T/4$ with pulse width of δ). The
   start of each pulse is thus seen to be retarded by an angle $[(\pi - \delta)/2]$.
   The pulse width δ is the parameter controlling the ac voltage supplied
   to the load, as can be seen from the following Fourier series expansion:

$$v_o(t) = \sum_{n=1}^{\infty} b_n \sin n\omega t$$

with

$$b_n = \begin{cases} \dfrac{4V_d}{n\pi} \sin \dfrac{n\delta}{2} & n = 1, 3, 5, \ldots \\ 0 & n = 2, 4, \ldots \end{cases}$$

The ratio of the amplitude of the third harmonic b_3 to the amplitude of the fundamental component is

$$\frac{b_3}{b_1} = \frac{1}{3}\left(3 - 4\sin^2\frac{\delta}{2}\right) \qquad 0 < \delta < \pi$$

A reduction in δ can be seen to increase the ratio b_3/b_1 and in general in an increase in the harmonic content of the output voltage.

2. *Multiple-pulse modulation:* The general idea of multiple-pulse modulation is illustrated in Figure 10.41. The output voltage waveform consists of N pulses in each half cycle of the required inverter output voltage at specified frequency f. The frequency of the pulses f_p is given by

$$f_p = 2fN$$

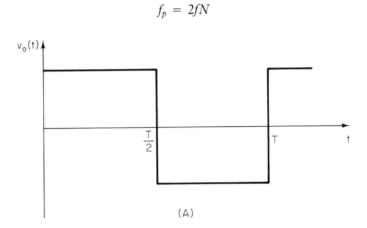

(A)

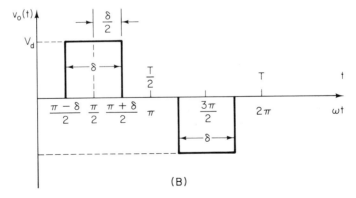

(B)

FIGURE 10.40   Single-pulse modulation waveforms.

Figure 10.41(a) shows the output voltage waveform for $N = 2$. Here the pulse centers are spaced by a time interval of $T/4$ seconds. The pulse width δ_2 should be less than $\omega T/4$ or $\pi/2$. Figure 10.41(b) is for $N = 3$ and it is clear that $\delta_3 < \pi/3$. In general, $\delta_N \leqslant \pi/N$.

3. *Sinusoidal pulse modulation:* Thyristor switching times are obtained in this method as shown in Figure 10.42(a). A reference sinusoidal signal $v_R(t)$ with amplitude V_R and frequency f equal to the desired inverter frequency output is employed:

$$v_R(t) = V_R \sin \omega t$$

A triangular (carrier) wave $v_T(t)$ with amplitude V_T and frequency f_T is compared with the reference sinusoid. The switching times are determined as the points of intersection of $v_T(t)$ and $v_R(t)$ as shown in the figure. Pulse width is determined by the time duration during which $v_T < v_R$ in the positive half cycle of v_R and that for which the converse is true in the negative half cycle of v_R. Two voltage control parameters are the modulation index M defined by

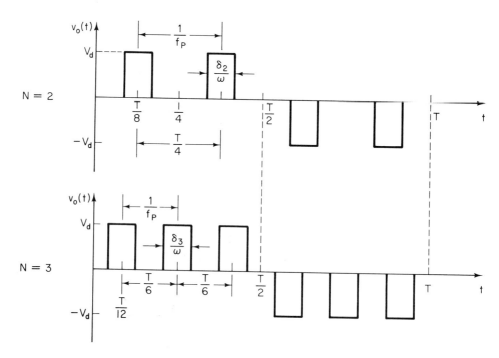

FIGURE 10.41   Multiple-pulse modulation waveforms.

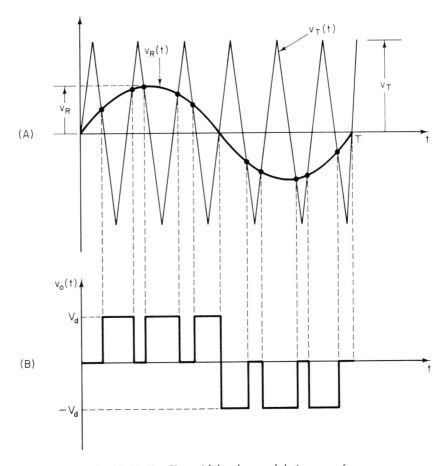

FIGURE 10.42 Sinusoidal pulse-modulation waveforms.

$$M = \frac{V_R}{V_T}$$

and the chopping ratio N given by

$$N = \frac{f_T}{f}$$

In three-phase applications, N is taken as 6 (or multiple of 3), as shown in the figure, which is drawn for $M = 0.5$. This technique is sometimes referred to as the triangle interception or subharmonic method.

Three-Phase Inverters

There are two schemes to obtain a three-phase ac output from a direct voltage source. The first scheme involves three single-phase inverters with the gating signals displaced by 120° of the output frequency interval. The three single-phase voltages are then fed to a three-phase transformer. The three primary windings of the transformer are isolated from each other, but the secondaries can be connected in either Y or Δ.

The second scheme involves a three-phase bridge inverter circuit as shown in Figure 10.43. The gate current pulses are applied at 60° intervals of the output voltage waveforms in the sequence indicated in the figure to yield the balanced voltages v_{AB}, v_{BC}, and v_{CA}. There are six intervals in one cycle of the ac voltage waveform, as indicated in Figure 10.44, showing the gate current pulses for a practical circuit implementation. Each thyristor is turned on for a duration of $2\pi/3$ radians in sequence. Note that an interval of $\pi/3$ radians elapses between the end of the gate signal applied to Q_1 and the beginning of the gate signal to Q_4, which is in series with Q_1. The same comment can be made about the pairs Q_3 and Q_6, as well as Q_5 and Q_2. This arrangement reduces the possibility of shorting-out the dc source.

We note that in interval I, Q_1 and Q_6 conduct and therefore terminals A and B are connected across the source voltage V_i. In the case of a balanced three-phase resistive load, the voltage across each of the two resistors is $V_i/2$ as shown in Figure 10.45(a). It is clear, then, that for interval I, we have

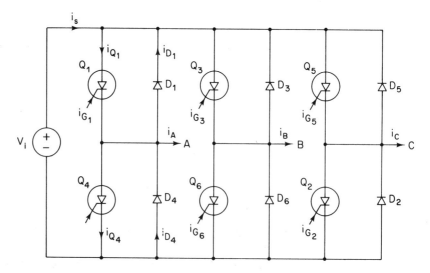

FIGURE 10.43   Three-phase bridge inverter circuit.

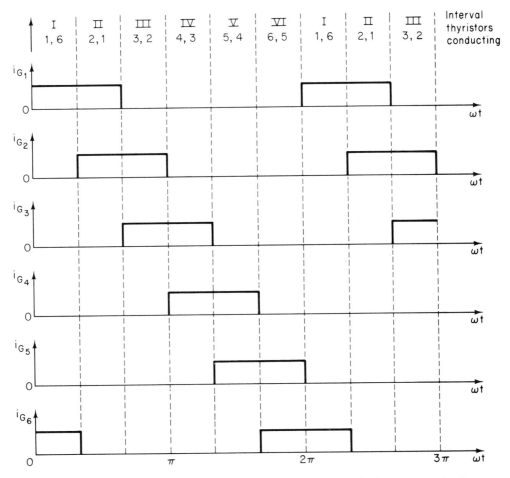

FIGURE 10.44 Gate signals for a three-phase bridge inverter circuit.

$$v_{AN}(t) = \frac{V_i}{2}$$

$$v_{BN}(t) = -\frac{V_i}{2}$$

Similar conclusions can be drawn for each of the remaining intervals. In Figure 10.45(b) we show the circuit conditions for interval II. The voltages v_{AN} and v_{BN} are shown in parts (c) and (d), and the resulting output voltage $v_{AB}(t)$ is shown in Figure 10.45(e). The voltages $v_{BC}(t)$ and $v_{CA}(t)$ can be obtained in a similar manner.

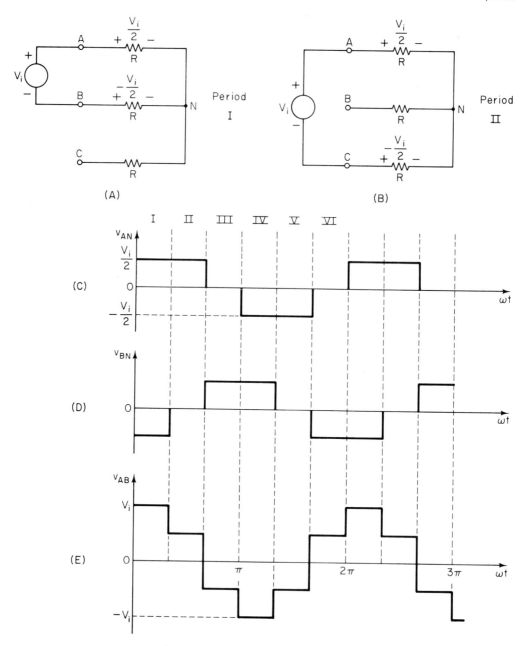

FIGURE 10.45 Determining voltage waveforms in a three-phase bridge inverter.

Example 10.11 A single-phase, full-bridge inverter has a dc supply of 220 V and operates at a period of 2 ms. The load at fundamental frequency is the following series combination:

$$R = 0.4 \ \Omega \qquad \omega L = 8 \ \Omega \qquad \frac{1}{\omega C} = 8.4 \ \Omega$$

Find the fundamental and third-harmonic output voltage and current waveforms.

Solution

The maximum value of the nth-harmonic voltage component is given by Eq. (10.86) as

$$b_n = \frac{4V_d}{n\pi} \qquad n = 1, 3, 5, \ldots$$

Thus we have

$$b_n = \frac{4(220)}{n\pi} = \frac{280.1}{n}$$

As a result, the fundamental voltage component is

$$v_1(t) = 280.11 \sin \omega t$$

where

$$\omega = 2\pi f = \frac{2\pi}{T} = 1000\pi$$

The third harmonic is given by

$$v_3(t) = 93.4 \sin 3\omega t$$

The impedance of the load for the nth harmonic is given by

$$Z_n = 0.4 + j\left(8n - \frac{8.4}{n} \right)$$

Thus

$$Z_1 = 0.4 - j0.4 = 0.57 \ \underline{/-45°}$$
$$Z_3 = 0.4 + j21.2 = 21.2 \ \underline{/88.9°}$$

We can therefore conclude that

$$i_1(t) = 495.2 \sin(\omega t + 45°)$$
$$i_3(t) = 4.4 \sin(3\omega t - 88.9°)$$

Note that the third harmonic of current is inductive with a relatively small magnitude, whereas the fundamental component is capacitive.

SOME SOLVED PROBLEMS

Problem 10.A.1

A single-phase full-wave ac voltage controller supplies a load circuit with $\phi = 80°$. Find the conduction angles γ for $\alpha = 85°$ and $160°$.

Solution

We use Eq. (10.9) to obtain

$$\sin(\alpha + \gamma - 80) = \sin(\alpha - 80) \, e^{-\gamma/\tan 80°}$$

We solve iteratively for γ for given values of α to get

For $\alpha = 160°$, $\gamma = 39.2°$.

For $\alpha = 85°$, $\gamma = 172.06°$.

Problem 10.A.2

A single-phase full-wave ac voltage controller has a firing angle of $70°$. Find the load power factor for the following conduction angles.

a. $\gamma = 130°$.
b. $\gamma = 150°$.
c. $\gamma = 175°$.

Solution

We use Eq. (10.9) and solve iteratively for ϕ in each case to obtain:
  (a) For $\gamma = 130°$, $\phi = 20.01°$ and $\cos \phi = 0.94$.
  (b) For $\gamma = 150°$, $\phi = 41.41°$ and $\cos \phi = 0.75$.
  (c) For $\gamma = 175°$, $\phi = 66.02°$ and $\cos \phi = 0.406$.

Problem 10.A.3

The following information is available about a basic chopper circuit:

$$V_i = 120 \text{ V} \qquad V_o = 40 \text{ V} \qquad E_c = 30 \text{ V}$$
$$t_{on} = 0.5 \text{ ms} \qquad T = 3 \text{ ms}$$

Determine the mode of operation of the chopper and the time constant of the load circuit.

Solution

The ratio of average output voltage to source voltage is given by

$$\frac{V_o}{V_i} = \frac{40}{120} = \frac{1}{3}$$

The ratio of on-time to period is

$$\frac{t_{on}}{T} = \frac{0.5}{3} = \frac{1}{6}$$

Since $V_o/V_i > t_{on}/T$, operation is discontinuous.

$$V_o = \frac{t_{on}}{T} V_i + \frac{T - t_X}{T} E_c$$
$$40 = \frac{1}{6}(120) + \frac{3 - t_X}{3}(30)$$

Thus

$$t_X = 1 \text{ ms}$$

Now use Eq. (10.81):

$$1 = \tau \, \ell n[1 + 4 \, (e^{0.5/\tau} - 1)]$$

Thus we have

$$e^{1/\tau} = 1 + 4(e^{1/2\tau} - 1)$$

Let $x = e^{1/2\tau}$, then solve

$$x^2 - 4x + 3 = 0$$

Thus

$$x = 3 = e^{1/2\tau}$$

As a result, the required time constant is given by

$$\tau = 0.46 \text{ ms}$$

Problem 10.A.4

The load connected to the terminals of a single-phase full-bridge inverter consists of a resistance R connected in series with an inductance with $\omega L = 10\ \Omega$ and a capacitance with $1/\omega C = 10.2\ \Omega$ at fundamental frequency. The maximum value of the third-harmonic voltage is 100 V and the third-harmonic current magnitude is 1% of the magnitude of the fundamental component. Find the load resistance and the power it consumes due to fundamental and third-harmonic components.

Solution

From Eq. (10.86), we obtain

$$100 = \frac{4V_d}{3\pi}$$

As a result

$$V_d = 75\pi\ \text{V}$$

From this we conclude that

$$b_1 = \frac{4V_d}{\pi} = 300\ \text{V}$$

The fundamental component of the current has a maximum value of

$$I_{1\text{max}} = \frac{b_1}{|Z_1|}$$

The third-harmonic component of the current has a maximum value of

$$I_{3\text{max}} = \frac{b_3}{|Z_3|}$$

As a result,

$$\frac{1}{100} = \frac{I_3}{I_1} = \frac{b_3 Z_1}{b_1 Z_3} = \frac{|Z_1|}{3|Z_3|}$$

Therefore,

$$10^4[R^2 + (10 - 10.2)^2] = 9\left[R^2 + \left(3 \times 10 - \frac{10.2}{3}\right)^2\right]$$

We solve to obtain

$$R = 0.77 \ \Omega$$

The effective value of the fundamental component of current is thus obtained as

$$I_1 = \frac{300/\sqrt{2}}{[(0.77)^2 + (10 - 10.2)^2]^{1/2}} = 266.65 \ \text{A}$$

Similarly,

$$I_3 = \frac{100/\sqrt{2}}{[(0.77)^2 + (3 \times 10 - 10.2/3)^2]^{1/2}} = 2.66 \ \text{A}$$

Note that $I_1/I_3 = 100$ could be used to obtain I_3. The fundamental component of power taken by the resistor is thus

$$P_1 = I_1^2 R = 54.75 \times 10^3 \ \text{W}$$

The third-harmonic component is therefore

$$P_3 = I_3^2 R = 10^{-4} P_1 = 5.5 \ \text{W}$$

PROBLEMS

Problem 10.B.1

A single-phase half-wave controller circuit as shown in Figure 10.1(a) supplies power to a load resistance of 12 Ω. The source voltage is 120 V. Find the average voltage output and the rms value of the output voltage and current.

Problem 10.B.2

The average output voltage of a single-phase half-wave controller circuit is -15 V. Assume that the source voltage is 120 V. Find the firing angle α and the rms value of the output voltage.

Problem 10.B.3

The rms value of the voltage output of a single-phase half-wave controller circuit is 110 V. Assume that the source voltage is 120 V. Find the firing angle α and the average value of the output voltage.

Problem 10.B.4

A single-phase full-wave ac voltage controller supplies a load circuit with $\phi = \pi/3$. Find the angle α and the ratio V_o/V for a conduction angle of 160°.

Problem 10.B.5

A single-phase full-wave ac voltage controller supplies a load circuit with $\phi = \pi/3$. Find the angle α and the ratio V_o/V for a conduction angle of 140°.

Problem 10.B.6

A single-phase full-wave ac voltage controller supplies a load circuit with $\phi = \pi/3$. Find the conduction angle γ for $\alpha = 75°$.

Problem 10.B.7

Repeat Problem 10.B.6 for $\alpha = 95°$.

Problem 10.B.8

Repeat Problem 10.B.6 for $\phi = 45°$ and $\alpha = 75°$.

Problem 10.B.9

Repeat Problem 10.B.6 for $\phi = 45°$ and $\alpha = 60°$.

Problem 10.B.10

A single-phase full-wave ac voltage controller has a firing angle of 80°. Find the load angles ϕ for the conduction angles $\gamma = 120°$ and $\gamma = 170°$.

Problem 10.B.11

A single-phase full-wave ac voltage controller has a firing angle of 40° and a conduction angle of 160°. Find the angle ϕ of the load circuit.

Problem 10.B.12

A single-phase full-wave ac voltage controller has a firing angle of 160° and a conduction angle of 25°. Find the angle ϕ of the load circuit.

Problem 10.B.13

A single-phase half-wave controlled rectifier as shown in Figure 10.15 operates with a firing angle of 30° from a 120-V ac supply. The load's power factor angle ϕ is 75° and the dc source is defined by $E_c/V_m = 0.25$. Find the conduction angle γ and the average output current if the load resistance is 10 Ω.

Problem 10.B.14

A single-phase half-wave controlled rectifier as shown in Figure 10.15 operates with a firing angle of 45° from a 120-V ac supply. The angle ϕ is given as 45° and $E_c/V_m = 0.5$. Find the conduction angle and the average output current if the load resistance is 10 Ω.

Problem 10.B.15

Repeat Problem 10.B.14 for $\alpha = 75°$ and $\phi = 30°$.

Problem 10.B.16

Repeat Problem 10.B.14 for $\alpha = 30°$, $\phi = 30°$, and $E_c/V_m = -0.3$.

Problem 10.B.17

Find the ratio E_c/V_m for a single-phase half-wave controlled rectifier supplying a load with $\phi = 30°$ at $\alpha = 75°$ for conduction angles of 60° and 160°.

Problem 10.B.18

Repeat Problem 10.B.17 for $\alpha = 60°$ and conduction angles of 120° and 200°.

Problem 10.B.19

A single-phase half-wave controlled rectifier supplies a load with $\phi = 30°$ and $E_c/V_m = 0.5$. The conduction angle is 100°. Find the firing angle α.

Problem 10.B.20

Repeat Problem 10.B.20 for $E_c/V_m = -0.3$ and $\gamma = 200°$.

Problem 10.B.21

A single-phase half-wave controlled rectifier supplies a load circuit with $E_c/V_m = 0.7$. The firing angle $\alpha = 75°$ and the corresponding conduction angle is 80°. Find the load circuit's power factor angle ϕ.

Problem 10.B.22

Repeat Problem 10.B.21 for $\alpha = 30°$, $\gamma = 210°$, and $E_c/V_m = -0.5$.

Problem 10.B.23

A single-phase half-wave controlled rectifier supplies a load with $\phi = 25°$ and a resistance of 10 Ω ($E_c = 0$). The conduction angle is 175° and the ac source voltage is 120 V. Find the firing angle α and the average output current.

Problem 10.B.24

Repeat Problem 10.B.23 for $\phi = 30°$ and $\gamma = 150°$.

Problem 10.B.25

A single-phase half-wave controlled rectifier has a firing angle of 40° when supplying a load impedance with $\phi = 30°$. Find the conduction angle.

Problem 10.B.26

Repeat Problem 10.B.25 for $\alpha = 15°$ and $\phi = 30°$.

Problem 10.B.27

The firing angle of a single-phase controlled rectifier is 30° at a conduction angle of 185° when supplying a load impedance. Find the power factor of the load.

Problem 10.B.28

Repeat Problem 10.B.27 for $\alpha = 60°$ and $\gamma = 150°$.

Problem 10.B.29

A single-phase half-wave rectifier is connected to a resistive load. The ac supply voltage is 110 V.

   a. Find the average output current if the firing angle is 45° and the load resistance is 5.5 Ω.
   b. Find the load resistance if the average output current is 9 A for a firing angle of 50°.

Problem 10.B.30

The load resistance connected to the output terminals of a single-phase half-wave rectifier is 5 Ω.

   a. Find the firing angle if the average output current is 8 A with an ac supply voltage of 110 V.

b. The average output current is 9 A for a firing angle of 52°. Find the effective value of the ac supply.

Problem 10.B.31

An inductive reactance $\omega L = 10$ is connected to the output terminals of a single-phase half-wave rectifier.

a. Find the average output current for a firing angle of 45° with a 110-V ac supply.
b. Find the supply voltage if the average output current is 12 A with the same firing angle.

Problem 10.B.32

A single-phase half-wave rectifier operates from a 120-V ac supply. The firing angle is 70° with an inductive reactance connected to the output terminals.

a. Find the average output current for a reactance of 5 Ω.
b. If the average output current is 19 A, find the inductive reactance of the load.

Problem 10.B.33

A single-phase half-wave rectifier supplies a load consisting of a dc voltage source E_c and a resistance R. Assume that $E_c/V_m = 0.5$. Find the ratio of average output currents for firing angles of 30° and 60°, respectively.

Problem 10.B.34

A single-phase half-wave controlled rectifier supplies a load circuit consisting of a 5 Ω resistor in series with an EMF $E_c = 60$ V. Assume that the effective value of the ac source voltage is 120 V. Find the average current in the load for $\alpha = 35°$.

Problem 10.B.35

A single-phase half-wave controlled rectifier supplies a load circuit consisting of an inductance in series with an EMF E_c. Find the ratio E_c/V_m for the following combinations of firing angle α and conduction angle γ.

a. $\alpha = 30°$, $\gamma = 200°$.
b. $\alpha = 80°$, $\gamma = 190°$.

Problem 10.B.36

The ratio E_c/V_m for a single-phase half-wave controlled rectifier is 0.6 when it supplies a load circuit with negligible resistance with a conduction angle of 170°. Find the firing angle α.

Problem 10.B.37

The following information is available about a basic chopper circuit:

$$V_i = 120 \text{ V} \quad E_c = 15 \text{ V} \quad T = 3 \text{ m sec}$$
$$L = 0.9 \text{ mH} \quad R = 0.15 \ \Omega$$
$$t_{oN} = 1 \text{ ms}$$

a. Determine the mode of operation of the chopper.
b. Find the average values of the output voltage and current.

Problem 10.B.38

The following information is available about a basic chopper circuit:

$$V_i = 120 \text{ V} \quad E_c = 20 \text{ V} \quad T = 4 \text{ ms}$$

$$L = 9 \text{ mH} \quad R = 20 \ \Omega \quad t_{on} = 3.3 \text{ ms}$$

a. Determine the mode of operation of the chopper.
b. Find the average values of the output voltage and current.

Problem 10.B.39

The following information is available about a basic chopper circuit:

$$V_i = 120 \text{ V} \quad E_c = 30 \text{ V} \quad T = 3 \text{ ms}$$
$$L = 9 \text{ mH} \quad R = 15 \ \Omega \quad t_{on} = 0.5 \text{ ms}$$

Determine the mode of operation of the chopper and find the average value of the output voltage.

Problem 10.B.40

Repeat Problem 10.B.39 with the following specifications:

$$V_i = 120 \text{ V} \quad E_c = 60 \text{ V} \quad T = 4 \text{ ms}$$
$$L = 9 \text{ mH} \quad R = 20 \ \Omega \quad t_{on} = 3 \text{ ms}$$

Problem 10.B.41

The input voltage to a basic chopper is 120 V and its output is 100 V with a load back EMF of 80 V. Assume that the chopping period is 3 ms and that the on-time is 2 ms. Find the time constant of the load circuit and the critical on-time.

Problem 10.B.42

Repeat Problem 10.B.41 for $V_i = 120$ V, $V_o = 45$ V, $E_c = 40$ V, $t_{on} = 1.5$ ms, and $T = 5$ ms.

Problem 10.B.43

A single-phase full-bridge inverter supplies a load consisting of a resistance of 10 Ω, an inductance $\omega L = 10$ Ω, and a capacitance $1/\omega C = 10.2$ Ω, where the reactances are given at fundamental frequency. The voltage of the dc source is 200 V. Find the fundamental and third-harmonic components of the output current as functions of time.

Problem 10.B.44

The dc voltage supply to a single-phase full-bridge inverter is 240 V. The load resistance is 8 Ω, its inductive reactance at fundamental frequency is 12 Ω, and the corresponding capacitive reactance is 12.8 Ω. Find the effective values of the fundamental and third-harmonic components of the load current.

Problem 10.B.45

An inductive load impedance is connected to the terminals of a single-phase full-bridge inverter. The effective value of the fundamental component of the load current is 30 A. Find the value of the dc source voltage and the effective value of the third-harmonic component of the load current. Assume that $Z = 1 + j2$ at fundamental frequency.

Problem 10.B.46

The effective value of the third-harmonic component of the load current in a single-phase full-bridge inverter is 4 A. Assume that the load impedance at fundamental frequency is $(0.8 + j0.8)$ Ω. Find the effective value of the fundamental component of the load current.

Problem 10.B.47

An inductive load is connected to the terminals of a single-phase full-bridge inverter. The power delivered to the load through the fundamental component of current is 50 times that due to the third harmonic. Find the power factor of the load at fundamental frequency.

Problem 10.B.48

An inductive load with a power factor of 0.8 at fundamental frequency is connected to the load terminals of a single-phase full-bridge inverter. Find the ratio of power absorbed by the load resistance at fundamental frequency to that at the third harmonic.

Problem 10.B.49

The third harmonic's power factor of an inductive load is 0.25 lagging. The power taken by the load due to third harmonic is 500 W. Find the power factor of the load and the power absorbed at fundamental frequency.

Problem 10.B.50

The power factor at fundamental frequency for the load of Problem 10.B.49 is improved to 0.8 lagging by inserting a capacitor in series with the load. Find the capacitive reactance of the capacitor at fundamental frequency in terms of load resistance. Find the new load power factor for the third harmonic.

Adjustable-Speed Drives

11.1 INTRODUCTION

In Chapters 9 and 10 we discussed the foundations and operational principles of power semiconductor (power-electronic) devices and systems. The present chapter is intended to highlight the application of power semiconductor systems in static control of electric motors. A combination of a power electronic system such as a voltage controller and an electric motor with the associated control mechanisms is referred to as an adjustable-speed drive.

Adjustable-speed drives can be either of the ac or the dc type. In the ac type, the supply to an induction and/or a synchronous motor is controlled by an ac voltage controller, a cycloconverter, or an inverter to achieve prescribed speed for given load conditions. In the dc type a separately excited or series dc motor is controlled using a chopper or a rectifier circuit to achieve a desired speed characteristic. The choice of the power semiconductor driving system depends on the type of available power supply and load characteristic. We will start with the application to induction-motor drives.

11.2 ADJUSTABLE-SPEED INDUCTION-MOTOR DRIVES

The speed of an induction motor driving a mechanical load with a given torque–speed characteristic can be controlled by controlling either its synchronous speed or the rotor's slip. For a fixed number of stator poles, the synchronous speed is controlled by the supply frequency. The slip of the rotor under load is controlled by regulating either of the voltage or the current applied to the stator. In a wound-rotor induction motor, the rotor slip can be controlled by recovering power from the rotor circuit.

Schemes of adjustable-speed drives for an induction motor can be broadly classified into three main categories:

1. *Variable-voltage constant frequency:* This is also known as stator voltage control. In this scheme voltage applied to the stator is varied using ac voltage controllers of Section 10.2 while maintaining a constant frequency.

2. *Variable-frequency control:* The stator frequency is varied while varying either the applied voltage or current.

3. *Slip power recovery:* The equivalent rotor resistance can be varied electronically by use of a semiconductor power circuit connected to the rotor terminals to recover power at slip frequency and convert it to the supply line.

The variable-frequency category can be further classified into the classes of dc link conversion schemes and that of cycloconverters. The dc link schemes involve rectification of the ac supply, which is followed by inversion to provide the required power drive. The inverters can be either voltage fed or current fed. In the former case the controlled variable is the voltage applied to the stator along with the frequency. The current-fed inverter scheme provides control through the stator current and frequency. Voltage-fed inverters can be either square-wave inverters or pulse-width-modulated (PWM) inverters. Figure 11.1 shows the classification of induction motor drive systems indicated above.

Three-Quadrant Operation

A three-phase induction machine operates normally in the motoring mode, where the developed torque and rotor speed are in the first quadrant of the T–ω diagram, as shown in Figure 11.2. In this case the developed torque is such that the rotor attempts to align itself with the rotating magnetic field set up by the three-phase supply in the three-phase stator windings. Operation in the second quadrant is commonly referred to as plugging, and in this case the motor has a direction of rotation opposite to that which would be produced by its own developed torque. In the fourth quadrant, commonly referred to as overhauling, the load tends to drive the motor in a positive direction to more than synchronous speed. Adjustable-speed drives are designed to recognize the possibility of operation in the plugging and overhauling modes.

It should be noted that plugging involves transfer of energy from the

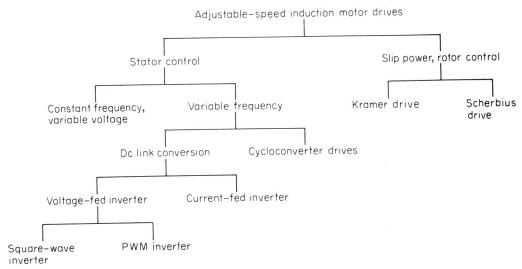

FIGURE 11.1 Classification of adjustable-speed induction-motor drives.

mechanical load to the rotor as well as electrical energy from the stator circuit. As a result, large motor currents can be expected and these are accompanied by large ohmic losses that must be dissipated as heat. When a motor is anticipated to experience plugging conditions, it is preferable to use a wound-rotor induction motor with external resistances connected to the rotor circuit to dissipate the plugging energy.

Note that fourth-quadrant operation is simply reverse motoring. Common terminology refers to second-quadrant operation as reverse regeneration (plugging) and that in the fourth quadrant as forward regeneration (overhauling).

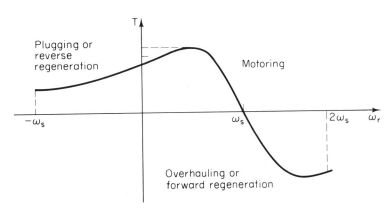

FIGURE 11.2 Complete torque–speed characteristic of an induction motor.

11.3 VARIABLE VOLTAGE–CONSTANT FREQUENCY DRIVES

In this method the voltages applied to the stator windings are controlled using an ac voltage controller while maintaining the frequency of the supply unchanged. A drive shown in Figure 11.3 employs a three-phase Y-connected ac voltage controller of the same configuration as that treated in Section 10.2 and shown in Figure 10.10. A typical application of this drive is at the low to medium power levels, such as pumps and blower-type loads. The same circuit is used as a solid-state starter for medium- to high-horsepower induction motors.

The stator voltage can be varied in this method between zero and full value within the trigger angle range $0 < \alpha < 120$, and as a result a variac-like performance is obtained. This method is simple and economical for control of squirrel-cage class D induction motors with high slip (typically 10 to 15%). Performance of the drive is poor since line currents carry rich harmonics and the drive has a poor power factor.

The torque output of a three-phase induction motor is given by Eq. (5.38) as

$$T = \frac{3|V_1|^2}{\omega_s} \frac{R_2/s}{(R_1 + R_2/s)^2 + X_T^2}$$ (11.1)

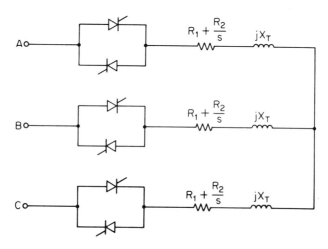

FIGURE 11.3   Induction-motor speed control with a three-phase ac voltage controller.

It is therefore clear that the torque output varies with the square of the applied voltage at a fixed slip. Figure 11.4 shows the torque–speed characteristic of an induction motor with variable stator voltage.

To determine the speed of the motor with a given voltage value, we need the load torque–speed characteristic. Two important types of load characteristics are the constant torque load and the square-law torque typical of blowers and pumps, given by

$$T_\ell = k_\ell \omega_r^2 \tag{11.2}$$

where k_ℓ is a constant, T_ℓ is the load torque, and ω_r is the rotor angular speed. Recall Eq. (5.33):

$$\omega_r = \omega_s(1 - s)$$

We conclude that s, and hence ω_r, can be obtained as the solution of

$$\omega_s^2 k_\ell (1 - s)^2 = \frac{3|V_1|^2}{\omega_s} \frac{R_2/s}{(R_1 + R_2/s)^2 + X_T^2} \tag{11.3}$$

Figure 11.4 shows typical operating points for a three-phase induction motor with a square-law load torque. For a constant-torque load, Eq. (11.1) is solved for s simply by substituting the specified load torque T_ℓ for the torque on the left-hand side.

The motor torque is related to the stator current I_s by Eq. (5.35), where the substitution $I_s \simeq I_r$ is made, rewritten as

$$T = \frac{3|I_s|^2 R_2}{s\omega_s} \tag{11.4}$$

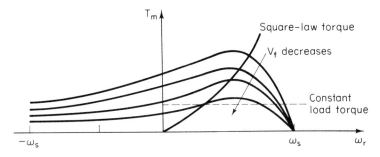

FIGURE 11.4 Torque–speed curves of induction motor with variable stator voltage.

It is clear for a square-law load torque that we get

$$|I_s| = K_s \sqrt{s}\,(1 - s) \tag{11.5}$$

where

$$K_s = \left(\frac{k_\ell \omega_s^3}{3R_2}\right)^{1/2} \tag{11.6}$$

The stator current is a maximum for $s = 0.333$, as may be verified from differentiating the expression for $|I_s|$ with respect to s and setting the result to zero. In this case, we have

$$|I_s|_{\max} = \frac{2}{3\sqrt{3}}\,K_s \tag{11.7}$$

The ratio of maximum stator current to its value at a given slip is therefore given by

$$\frac{|I_s|_{\max}}{|I_s|} = \frac{2}{3\sqrt{3}s\,(1 - s)} \tag{11.8}$$

For a motor with 12% slip at full load, we find that the maximum value of stator current is only 26.3% higher than its rated value.

For a load with constant torque, we have, by Eq. (11.4),

$$|I_s| = K_c \sqrt{s} \tag{11.9}$$

where

$$K_c^2 = \frac{T_\ell \omega_s}{3R_2} \tag{11.10}$$

The maximum value of stator current occurs for $s = 1$ (i.e., at standstill). In this case

$$\frac{|I_s|_{\max}}{|I_s|} = \frac{1}{\sqrt{s}} \tag{11.11}$$

For a motor with 12% slip at full load, the maximum value of stator current is close to three times that at full load.

Example 11.1   A three-phase eight-pole 60-Hz induction motor has the following parameters:

$$R_2 = 0.12 \ \Omega \qquad X_2 = 0.7 \ \Omega$$

Neglect stator resistance and reactance, magnetizing circuit, and rotational losses.

a. Find the slip of the motor when driving a load with characteristics as given by Eq. (11.2) with $k_\ell = 0.0136$, with applied phase voltages of 254 V and 200 V, respectively.

b. Repeat part (a) for $k_\ell = 0.136$ and phase voltages of 300 V and 254 V.

Solution

Using Eqs. (11.3) and (11.6) and rearranging, we obtain a fourth-order equation in s given by

$$s^4 - 2s^3 + \left(1 + \frac{R_2^2}{X_2^2}\right)s^2 - \left(2\frac{R_2^2}{X_2^2} + \frac{|V_1|^2}{X_2^2 K_s^2}\right)s + \frac{R_2^2}{X_2^2} = 0$$

We have

$$R_2 = 0.12 \ \Omega \qquad X_2 = 0.7 \ \Omega$$

$$\omega_s = \frac{4\pi f}{P} = 30\pi \ \text{rad/s}$$

$$\frac{1}{K_s^2} = \frac{3 \ R_2}{k_\ell \omega_s^3} = \frac{3 \times 0.12}{k_\ell (30\pi)^3}$$

(a) For $k_\ell = 0.0136$, we obtain

$$s^4 - 2s^3 + 1.03s^2 - \left(0.0588 + \frac{|V_1|^2}{15.5 \times 10^3}\right)s + 0.0294 = 0$$

With $|V_1| = 254$ V, we solve to obtain

$$s = 6.97 \times 10^{-3}$$

With $|V_1| = 200$ V, we obtain

$$s = 1.12 \times 10^{-2}$$

Note that, as expected, a reduction in voltage results in a decreased speed (increased slip).

(b) For $k_\ell = 0.136$, we obtain

$$s^4 - 2s^3 + 1.03s^2 - \left(0.0588 + \frac{|V_1|^2}{15.5 \times 10^4}\right)s + 0.0294 = 0$$

With $|V_1| = 300$ V, we obtain

$$s = 4.95 \times 10^{-2}$$

With $|V_1| = 254$ V, we obtain

$$s = 7.15 \times 10^{-2}$$

Note that this load results in slower speed with the same applied voltage as the load of part (a).

11.4 VARIABLE VOLTAGE–VARIABLE FREQUENCY DRIVES

It is reasonable to expect improved performance of adjustable-speed drives if a variable-frequency stator supply is employed. It should be noted that the air-gap flux is directly proportional to the stator applied voltage and inversely proportional to the frequency. A reduction in the supply frequency to achieve speed control below rated synchronous speed will be associated with an increase in the air-gap flux if the applied voltage is maintained at rated value. To avoid saturation due to the increased flux, variable-frequency drives employ a variable voltage as well, with the object of maintaining an acceptable air-gap flux level. This concept is generally referred to as constant (V/f) control and is used in drives employing squirrel-cage induction motors of all classifications. In the present section we discuss the theory of variable-frequency operation of the induction motor.

Variable-Frequency Operation

Analysis of operation of a three-phase induction motor with variable frequency requires replacing the fixed reactance X_T used in Chapter 5 by a frequency-dependent term. Let us denote the frequency of the source voltages by f_i; then we have

$$X_T = \omega_i L_T \tag{11.12}$$

where L_T is the combined stator and rotor (referred) inductances and

$$\omega_i = 2\pi f_i \tag{11.13}$$

Equation 5.18 is written as

$$n_s = \frac{120 f_i}{P} \tag{11.14}$$

As a result, the angular synchronous speed ω_s given by Eq. (5.34) can be written in the alternative form

$$\omega_s = \frac{2}{P}\omega_i \tag{11.15}$$

The torque output (neglecting rotational losses) is given by Eq. (5.38) as

$$T = \frac{3|V_1|^2}{\omega_s} \frac{R_2}{s} \frac{1}{(R_1 + R_2/s)^2 + X_T^2} \tag{11.16}$$

We can therefore conclude that an expression for the output torque that recognizes frequency variations is given by

$$T = \frac{3|V_1|^2 P R_2}{2\omega_i s[(R_1 + R_2/s)^2 + \omega_i^2 L_T^2]} \tag{11.17}$$

The torque–speed characteristics of a three-phase induction motor with variable-frequency input can thus be developed on the basis of Eq. (11.17).

The starting (or standstill) torque is obtained by substituting $s = 1$ in Eq. (11.17), to obtain

$$T_{st} = \frac{3|V_1|^2 P R_2}{2\omega_i[(R_1 + R_2)^2 + \omega_i^2 L_T^2]} \tag{11.18}$$

Let us assume that the motor operation is such that the source voltage-to-source frequency ratio is maintained constant. We thus have

$$T_{st} = \frac{K_{st}\omega_i}{(R_1 + R_2)^2 + \omega_i^2 L_T^2} \tag{11.19}$$

where

$$K_{st} = \frac{3|V_1|^2 P R_2}{2\omega_i^2} \qquad (11.20)$$

For large values of ω_i such that $(R_1 + R_2)^2$ is negligible in comparison with $(\omega_i L_T)^2$, we have

$$T_{st} \simeq \frac{K_{st}}{\omega_i L_T^2} \qquad \text{for } \omega_i \gg \frac{R_1 + R_2}{L_T} \qquad (11.21)$$

In this case the starting torque decreases as the input frequency is increased. For small values of ω_i such that $\omega_i^2 L_T^2$ is negligible in comparison with $(R_1 + R_2)^2$, we have

$$T_{st} \simeq \frac{K_{st}\omega_i}{(R_1 + R_2)^2} \qquad \text{for } \omega_i \ll \frac{R_1 + R_2}{L_T} \qquad (11.22)$$

In this case the starting torque increases as the input frequency is increased.

We can conclude from the discussion above that there is an input frequency $\omega_{i_{\max}}^{T_{(st)}}$ for which the starting torque attains a maximum value. This condition is obtained by setting the derivative of T_{st} given by Eq. (11.19) with respect to ω_i to zero. The result is

$$[(R_1 + R_2)^2 + \omega_i^2 L_T^2] - 2\omega_i^2 L_T^2 = 0$$

We thus have

$$\omega_{i_{\max}}^{T_{(st)}} = \frac{R_1 + R_2}{L_T} \qquad (11.23)$$

The value of maximum starting torque is therefore given by

$$T_{st_{\max}} = \frac{K_{st}}{2(R_1 + R_2)L_T} \qquad (11.24)$$

Equation (11.23) determines the lowest possible value of ω_i for proper operation.

The maximum (or breakdown) torque developed by the motor takes place at a slip defined by Eq. (5.41) as

$$s_{\max T_m} = \frac{R_2}{\sqrt{R_1^2 + \omega_i^2 L_T^2}} \qquad (11.25)$$

The corresponding maximum torque is given by Eq. (5.42) as

$$T_{\max_m} = \frac{3|V_1|^2 P}{4\omega_i(R_1 + \sqrt{R_1^2 + \omega_i^2 L_T^2})} \qquad (11.26)$$

If the machine is operating with negative slip, in the regenerative braking mode, then

$$s_{\max_{T_g}} = \frac{-R_2}{\sqrt{R_1^2 + \omega_i^2 L_T^2}} \qquad (11.27)$$

The corresponding maximum regenerative braking torque is given by

$$T_{\max_g} = \frac{-3|V_1|^2 P}{4\omega_i(\sqrt{R_1^2 + \omega_i^2 L_T^2} - R_1)} \qquad (11.28)$$

Note that if the stator resistance is negligible,

$$|T_{\max_m}| = |T_{\max_g}| \simeq \frac{3|V_1|^2 P}{4\omega_i^2 L_T} \qquad (11.29)$$

As a result, if the ratio V_1/ω_i is held constant, the breakdown torque developed by the motor will not change significantly with speed.

In Figure 11.5, a family of torque–speed characteristics of a three-phase induction motor is shown. The angular speed ω_{s0} corresponds to rated frequency operation. The speed at maximum torque for rated voltage and frequency ω_{mT_r} defines the division of the speed range into the constant-volts/Hz region to the left and the constant voltage–variable frequency region to the right. The envelope of the torque–speed characteristics is shown in the figure as constant up to ω_{mT_r} and then follows the curve $T_{\max} \propto (1/\omega_r)$ in the higher-speed range. An example is appropriate now.

Example 11.2   A four-pole 60-Hz 440-V three-phase induction motor is supplied by a variable voltage–variable frequency source. The motor has negligible stator resistance and a rotor resistance of 0.1 Ω. The total inductance is 2×10^{-3} H. Find the lowest possible frequency and voltage of proper operation. Determine the starting torque, the speed at maximum torque, the maximum torque, and the stator's applied voltage when the supply frequency is 8 Hz, 40 Hz, and 60 Hz.

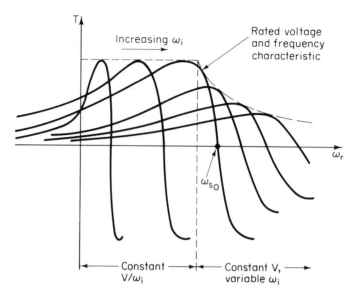

FIGURE 11.5 Torque–speed characteristics with variable-frequency conditions for an induction motor.

Solution

The base voltage and frequency are clearly 440 V (line to line) and 60 Hz. From Eq. (11.23), we get

$$\omega_{i_{max}}^{T_{st}} = \frac{0.1}{2 \times 10^{-3}} - 50 \text{ rad/s}$$

Thus the lowest possible frequency is

$$f_{i_{min}} = \frac{50}{2\pi} = 8 \text{ Hz}$$

The corresponding voltage is obtained by assuming constant volts/Hz operation as

$$V_{i_{min}} = 8\left(\frac{440}{60}\right) = 58.67 \text{ V (line to line)}$$

We use Eq. (11.20), to obtain

$$K_{st} = \frac{3(440/\sqrt{3})^2(4)(0.1)}{2[2\pi(60)]^2} = 0.2724$$

We then use Eq. (11.19) to obtain

$$T_{st} = \frac{0.2724(2\pi f_i)}{(0.1)^2 + (2\pi f_i)^2(2 \times 10^{-3})^2}$$

As a result,

$$T_{st} = \frac{1.7118 f_i}{0.01 + 1.58 \times 10^{-4} f_i^2}$$

From this relation we get the following starting torques:

$$f_i = 8 \text{ Hz} \qquad T_{st} = 681.1 \text{ N} \cdot \text{m}$$
$$f_i = 40 \text{ Hz} \qquad T_{st} = 260.7 \text{ N} \cdot \text{m}$$
$$f_i = 60 \text{ Hz} \qquad T_{st} = 177.55 \text{ N} \cdot \text{m}$$

The slip at maximum torque is given by

$$s_{maxT} = \frac{R_2}{\omega_i L_T} = \frac{25}{\pi f_i}$$

Recall that

$$\omega_{s_i} = \frac{4\pi f_i}{P}$$

Thus we obtain

$$\omega_{r_{maxT}} = \pi f_i - 25$$

As a result, the following angular speeds corresponding to maximum torque are obtained:

$$f_i = 8 \text{ Hz} \qquad \omega_{r_{maxT}} = 0$$
$$f_i = 40 \text{ Hz} \qquad \omega_{r_{maxT}} = 100.66 \text{ rad/s}$$
$$f_i = 60 \text{ Hz} \qquad \omega_{r_{maxT}} = 163.5 \text{ rad/s}$$

The maximum torque remains invariant according to Eq. (11.29) at

$$T_{max} = \frac{3(440/\sqrt{3})^2(4)}{4(120\pi)^2(2 \times 10^{-3})} = 681.1 \text{ N} \cdot \text{m}$$

The last result is the same as the maximum starting torque. The applied voltage is obtained from

$$V_i = f_i\left(\frac{440}{60}\right)$$

As a result, we obtain

$$f_i = 8 \text{ Hz} \qquad V_i = 58.67 \text{ V}$$
$$f_i = 40 \text{ Hz} \qquad V_i = 293.33 \text{ V}$$
$$f_i = 60 \text{ Hz} \qquad V_i = 440 \text{ V}$$

This concludes this example.

11.5 DC LINK CONVERTER DRIVES

Adjustable-speed drives of ac motors requiring variable voltage–variable frequency from a fixed voltage–fixed frequency supply can be designed using an intermediate dc link. Here, ac power is converted to dc using a three-phase rectifier bridge. The dc output of the rectifier is fed through the dc link to a three-phase inverter connected to the motor's stator terminals to provide the required variable-frequency supply.

A number of dc link converter schemes is available, as indicated in Figure 11.6. In schemes (a) and (b), three-phase diode rectifiers are employed. In scheme (a), fixed ac is fed to a transformer whose ratio is controlled as a function of the desired frequency output. The variable-voltage ac is then fed to the rectifier. The rectifier's variable dc voltage output is injected into the inverter. The inverter's output is of the required variable voltage–variable frequency form. Scheme (b) differs from that of (a) in the fact that the output of the inverter is of variable frequency but fixed voltage. The desired variable-voltage feature is achieved by the use of an intermediate transformer between the inverter's output terminals and the motor's stator terminals. These two schemes are not truly static drives, as the variable transformers involve moving parts.

In scheme (c), a phase-controlled thyristor bridge rectifies the incoming ac voltage to variable dc voltage while maintaining the desired voltage-to-frequency ratio. No transformers are involved in this scheme. In scheme (d) the thyristor bridge rectifier of scheme (c) is replaced by a cascade of a diode rectifier bridge and a chopper to provide the variable dc input to the inverter.

In schemes (a) to (d) of Figure 11.6, the inverter provides the required adjustable-frequency feature while adjusting the voltage level is done externally to the inverter. The voltage supply to the ac motor's stator terminals is a series of square voltage waves of adjustable voltage and frequency. A more attractive arrangement uses a pulse-width-modulated inverter (PWM) which provides both frequency and voltage control with the same set of thyristors. The output of the PWM inverter is a series of adjustable-width

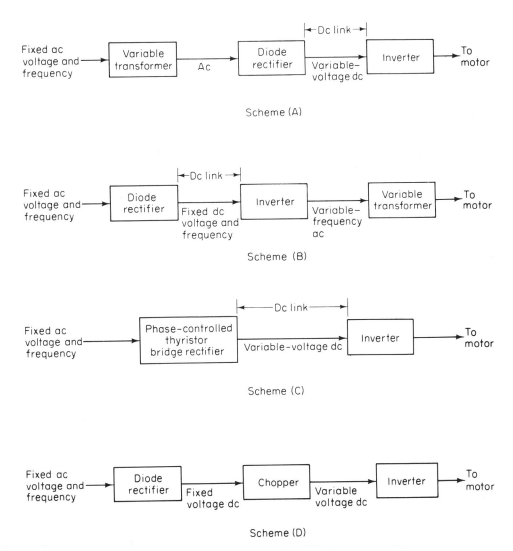

FIGURE 11.6   Square-wave dc link converter drives.

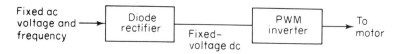

FIGURE 11.7   PWM inverter dc link converter drive.

pulses which provides for an adjustable rms voltage supply. The dc supply to the inverter is furnished by a diode rectifier bridge as shown in Figure 11.7.

Voltage-Fed versus Current-Fed Inverters

In conventional dc link converter drive schemes the inverter is fed with a dc voltage whose magnitude is independent of the load connected to the inverter terminals. In this case the scheme is referred to as a voltage-fed inverter drive. A more recent development is the introduction of current-fed inverter drives where the current rather than the voltage is the independent variable. The major distinction between the two types lies in the Thévenin's impedance seen at the inverter terminals, as discussed next.

Consider an electric system made of a supply network connected to a load Z_L as shown in Figure 11.8(a). It is clear that the supply network can be reduced to the simple Thévenin's equivalent voltage source V_{Th}, and an impedance Z_{Th} connected in series with the source as shown in Figure 11.8(b). It is clear that the load current I_L is given by

$$I_L = \frac{V_{Th}}{Z_{Th} + Z_L}$$

If Z_L is variable, such as is the case with an induction machine, then I_L will vary even though V_{Th} is fixed. If the Thévenin's impedance Z_{Th} is large

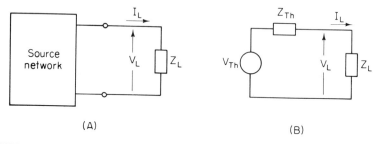

FIGURE 11.8   Thévenin's equivalents to distinguish between current-fed and voltage-fed inverters.

enough, then I_L will remain relatively constant even though Z_L varies. The voltage applied to the load denoted by V_L is given by

$$V_L = \frac{V_{\text{Th}} Z_L}{Z_{\text{Th}} + Z_L}$$

If the Thévenin's impedance Z_{Th} is small relative to Z_L, the voltage applied to the load remains invariant even though the load impedance varies.

We can conclude therefore that if the Thévenin's impedance is relatively high, the source network appears as an independent current source to the load circuit. On the other hand, if the Thévenin's impedance is relatively low, the source network appears as an independent voltage source to the load circuit.

The reader may be wondering why our discussion involves impedances when the input variables to the inverter are referred to as dc. The key to the answer is to recall that the rectifier output is still time varying and harmonics (sinusoids) exist at the rectifier output.

11.6   VOLTAGE-FED INVERTER DRIVES

The theory of variable voltage–variable frequency speed control was discussed in Section 11.4. Motors used in this type are low-slip motors resulting in improved efficiencies. The voltage-frequency and torque-frequency desired characteristics are shown in Figure 11.9. In the subsynchronous region, the voltage/frequency ratio is constant, resulting in constant-torque output. For the frequency range above rated frequency, the voltage is maintained constant, resulting in constant-horsepower operation. It should be pointed out that at low frequency, the resistance effect dominates that of the leakage reactance and additional voltage is impressed to compensate this

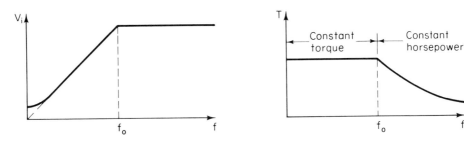

FIGURE 11.9   Desired voltage–frequency and torque–frequency relations for an induction motor.

effect. Note that the *T–f* characteristic of Figure 11.9 is the envelope of the torque–speed characteristics of Figure 11.5.

As discussed earlier, there are two main classes of the voltage-fed inverter drives employing the dc link conversion idea. These are the square-wave inverter drives and the PWM inverter drives discussed now.

Square-Wave Inverter Drive

A typical arrangement of a voltage-fed square-wave inverter ac motor drive is shown in the schematic of Figure 11.10. This is essentially a detail of scheme (c) of Figure 11.6. The inverter is called voltage fed since the capacitor *C* provides a low Thévenin's impedance to the inverter. The inverter voltage waves are not affected by the load.

Voltage-fed square-wave drives are employed in low- to medium-power applications with speed ratio limited to 10:1. This type has been replaced by PWM drives, discussed next.

PWM Inverter Drives

The controlled rectifier of Figure 11.10 can be replaced by an uncontrolled diode rectifier resulting in an uncontrolled dc link voltage. At the same time the inverter is controlled using a pulse-width-modulation strategy to provide for variable frequency–variable voltage output. Common practice calls for use of sinusoidal PWM illustrated in Figure 10.42. PWM voltage control is applicable in the constant-torque region of Figure 11.9, whereas in the constant-power region the inverter operation is identical to a square-wave drive.

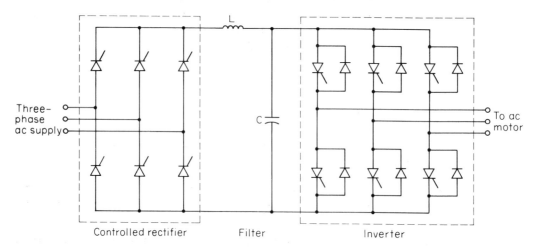

FIGURE 11.10 Voltage-fed square-wave inverter ac motor drive.

An important advantage of PWM inverter drives is that a number of inverter drives can be supplied through a dc bus at the output of a single uncontrolled rectifier. The drive system can be made uninterruptible for possible ac line failure by switching in standby battery in the dc link.

11.7 CURRENT-FED INVERTER DRIVES

A current-fed inverter ac motor drive shown in Figure 11.11 consists of a controlled rectifier bridge, a dc link filter inductor L, a three-phase inverter, and the controlled ac motor. It is thus clear that this drive is a dc link converter drive. The differences between a voltage-fed inverter drive and the current-fed inverter drive are that in the latter, the capacitor C of Figure 11.10 is not present and the inductor is large enough so that the dc link presents a constant regulated dc current to the inverter. The inverter output current consists of six-stepped waves that present a stiff regulated ac current source with adjustable frequency to the ac motor terminals. This type of drive is used with individual motors in medium- to high-horsepower range that have some minimum load present at all times.

The impedance of the motor is low at starting and at low speeds, so that with fixed stator current the terminal and air-gap voltages are relatively small, resulting in low air-gap flux. As the motor speed approaches synchronous speed, the impedance increases and as a result the flux increases to saturation levels. Two torque–speed curves for the motor operating with fixed dc link current corresponding to stator rated current and frequency are shown in Figure 11.12. The dashed curve A corresponds to values computed neglecting saturation. It is characterized by a low starting torque and a high maximum torque occurring close to synchronous speed with a steep decline to zero as the rotor's speed approaches the synchronous speed. For comparison purposes, the torque–speed characteristic of the motor with controlled stator voltage and frequency at rated values marked C in the figure features a higher starting torque and a lower maximum torque.

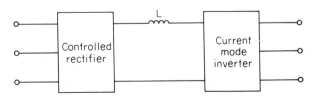

FIGURE 11.11 Variable-current variable-frequency induction motor drive.

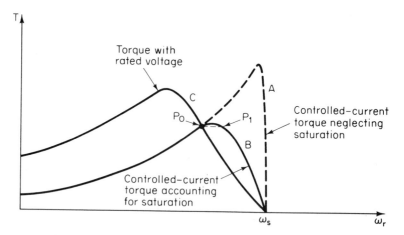

FIGURE 11.12   Torque–speed curves for voltage and current control.

It may be assumed that controlled-current operation provides superior torque than controlled voltage operation for low slip values. This is not true, since curve A is based on a fixed magnetizing reactance and therefore neglects saturation. In the low range of speed, the flux is below its rated value and the torque due to current control is lower than the corresponding torque obtained by rated voltage control. At the point P_0, the intersection of curves A and C, the flux in the current-controlled case corresponds to its rated value. Beyond P_0, saturation effects should be accounted for and the actual torque–speed characteristic with current control is obtained as the curve B, showing a lower maximum torque.

We observe that for the same output torque, the current-fed inverter drive can operate at one of two speeds corresponding to points P_0 and P_1 in Figure 11.12. Operation at point P_0 is preferred since at P_1, higher core losses exist. Closed-loop control to ensure that operation is at P_0 is necessary since P_0 is on the unstable side of the characteristic.

11.8   CYCLOCONVERTER DRIVES

A cycloconverter is a direct frequency changer that converts ac power directly from one frequency level to another frequency level. No dc link is required for a cycloconverter. Normally, cycloconverters have an output frequency range of zero to one-third of the input frequency. A cycloconverter is capable of producing variable voltage–variable frequency output suitable for induction motor drives using the principles of Section 11.4. Cycloconverter drives are used in large-horsepower applications up to 20,000 hp. The system can provide reverse operation and regeneration.

11.9 REGULATION OF SLIP POWER

We have seen in Chapter 5 that speed control can be achieved in a wound-rotor induction motor by inserting passive circuit elements such as external resistors (or capacitors) in the rotor circuit. Speed control using external resistors involves channeling a portion of the air-gap power at slip frequency into the resistors, where it is dissipated as heat. In this speed-control technology, efficiency is sacrificed to attain the required goal. The idea of regulating the slip power by inserting auxiliary, but active devices in the rotor circuit has been around for quite sometime. The conventional Kramer and Scherbius drives employed rotating machines connected to the rotor circuit to channel portions of slip power back to the supply. With the introduction of the new power semiconductor device and system technology, it became clear that these elegant speed control schemes can be realized reliably and at considerably less cost and size than the old technology's implementation.

Static Kramer Drives

A conventional Kramer drive used auxiliary machines connected to the rotor circuit of a wound-rotor induction motor to convert a portion of the power of the rotor circuit from slip frequency to supply frequency and then feeding this power back to the mains. A static Kramer drive employs a diode rectifier connected to the rotor circuit to convert a portion of slip frequency power to dc, which is then converted to ac at supply frequency using a line commutated inverter as shown in Figure 11.13. Speed control is achieved through control of the angle of firing α of the inverter ($90° < \alpha < 180°$).

The torque is proportional to the dc link current under practical simplifying assumptions and the torque–speed characteristics of the system for different values of α are shown in Figure 11.14. Operating at synchronous speed corresponds to $\alpha = 90°$ and the torque is zero at that point, and as a result, both V_d and I_d are zero. Application of the load results in the motor slowing down, increasing the slip and consequently increasing V_d and I_d until the motor's developed torque balances the load torque. For a constant load torque, I_d is constant and the speed can be decreased by increasing cos α such that V_d increases to match V_i.

The static Kramer drive system is used in large-horsepower pump and blower applications requiring a limited range of speed control in the subsynchronous (below ω_s) range.

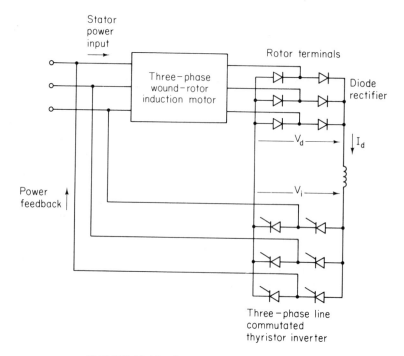

FIGURE 11.13   Static Kramer drive system.

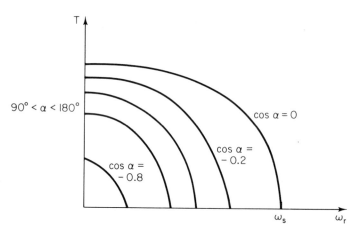

FIGURE 11.14   Torque–speed curves of a static Kramer drive system with different firing angles of inverter.

Static Scherbius Drives

The static Kramer drive permits flow of slip power only from the rotor circuit to the supply lines, and operation is therefore restricted to speeds in the subsynchronous range. In the Scherbius drive, the diode rectifier is replaced by a thyristor bridge, allowing slip power to flow in either direction. The speed of the Scherbius drive can therefore be controlled in both the subsynchronous region and the supersynchronous region. In the latter case, power at slip frequency is injected in the rotor circuit. A line commutated phase-controlled cycloconverter can be used in place of the dc link converter in sophisticated static Scherbius drives.

11.10 ADJUSTABLE-SPEED DC MOTOR DRIVES

Historically, the demand for variable-speed motor drives for industrial applications has been met by dc motors with a fixed field flux and a variable armature voltage for speed control. The variable dc voltage was obtained from a motor-generator set arranged in a Ward–Leonard system introduced in the 1890s. In this arrangement the generator field flux is adjusted to obtain the desired variable dc voltage to provide a wide range of motor speeds. This system gained wide acceptance, and commercially available packaged drives were common beginning immediately before World War II. There are many disadvantages to the Ward–Leonard approach, including low overall efficiency, bulk and size restrictions, and the undesirable requirements for periodic maintenance, inspection, and replacement of generator parts. The advent of the SCR and the associated developments of power electronic systems provided an attractive alternative to the Ward–Leonard approach.

A separately excited direct current motor is a doubly fed device that can be controlled through its armature circuit and/or its field circuit. Depending on the available power source, control is obtained using controlled rectifiers in an ac–dc drive as shown in Figure 11.15(a), or using choppers in a dc–dc drive as shown in Figure 11.15(b). Note that in ac–dc drives a single-phase ac supply may be used in small-horsepower applications.

The strategy adopted in practice for control of dc motors is to divide the speed range of interest into two regions, with the motor's rated (base) speed separating the two regions (Figure 11.16). In the first region, the field excitation is fixed at its rated value, and speed adjustments are made via firing angle control of the motor's armature power circuit. This region is called the constant-torque or armature voltage control region. Above base

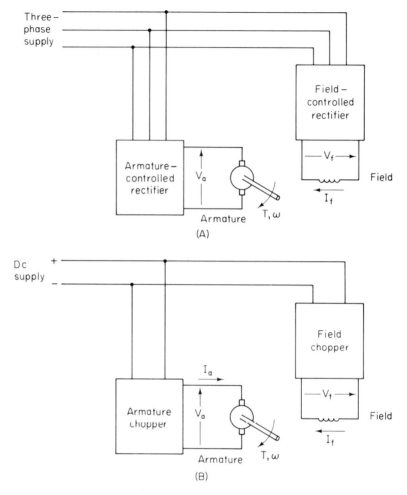

FIGURE 11.15 Separately excited dc motor adjustable-speed drives: (a) ac–dc drive; (b) dc–dc drive.

speed, field weakening is employed, and this is called the field control or constant-horsepower region. Normally, the second region extends to twice the base speed.

Region I: In this region the output torque is required to be maintained constant over the speed range from zero to base motor speed ω_b. The base motor speed is determined by the full-rated armature voltage and full-rated field flux. The field flux is kept constant at its rated value throughout this region. In meeting the constant-torque requirement under these conditions, the armature current should therefore be

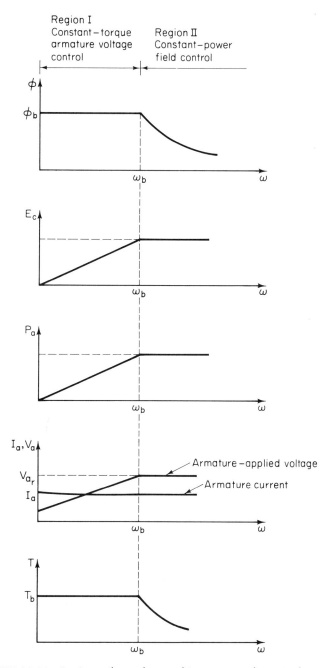

FIGURE 11.16 Regions of speed control in a separately excited dc motor.

maintained constant. This follows since by Eqs. (7.27) and (7.29), we have

$$T_0 = K_1 \phi_f I_a \qquad (11.30)$$

The armature current is related to the applied armature terminal voltage V_a and the motor's back EMF by Eq. (7.23) as

$$I_a = \frac{V_a - E_c}{R_a} \qquad (11.31)$$

Note that the back EMF E_c varies in proportion to the speed according to Eq. (7.24) given by

$$E_c = K_1 \phi_f \omega \qquad (11.32)$$

It is therefore clear that the armature terminal voltage varies with the motor speed according to

$$V_a = K_1 \phi_f \omega + \frac{T_0}{K_1 \phi_f} R_a \qquad (11.33)$$

The power output is obtained using Eq. (7.27) as

$$P_a = E_c I_a = K_1 \phi_f I_a \omega \qquad (11.34)$$

Thus the power output in region I increases in proportion to the motor speed. The torque output is constant at the rated value given by

$$T_o = K_1 \phi_f I_a \qquad (11.35)$$

Region II: For the region where motor speed exceeds the base speed ω_b, armature voltage control cannot be used, since full-rated voltage is applied to the armature terminals. Here the armature voltage V_a is maintained at its rated value V_{a_r}, and field control is applied. In this region, the horsepower output is maintained constant by weakening the field through reducing the voltage applied to the field. To maintain a constant-power output, we require that the product of field flux and speed be kept constant at

$$\phi_f \omega = \frac{P_a}{K_1 I_a} = \text{constant} \qquad (11.36)$$

As a result, the field is weakened as the required speed increases. Note that the back EMF remains at its rated value, and hence the armature current remains constant in this region.

11.11 AC–DC DRIVES FOR DC MOTORS

Controlled rectifiers are used to provide the variable dc voltage to the armature and field circuits in a dc motor. We will focus our attention on the separately excited motor in this section. An important factor to consider is the ripple content in the dc output of the rectifier. In a single-phase half-wave rectifier circuit, one pulse of current is produced every supply cycle, and this provides an output that is rich in harmonics, resulting in excessive heating and torque pulsations. For 60-Hz operation there are 60 pulses per second as opposed to 120 pulses per second for the single-phase full-wave rectifier case. Single-phase sources are used primarily for motors with horsepower output of 5 hp or less, for their simplicity and economy that offsets the poor wave shape.

For larger-horsepower applications the three-phase full-wave bridge configuration is used, and this has 360 pulses per second, which improves the harmonics content. It is often found necessary to employ two parallel phase-shifted three-phase full-wave controlled rectifier circuits to provide lower ripple.

Armature Voltage Control Using Controlled Rectifiers

When the available supply is ac, controlled rectifier circuits are employed to provide control of the voltage applied to the armature circuit. Control is achieved by adjusting the thyristor's firing angles.

Single-Phase AC Supply

In this case a full-wave bridge rectifier circuit such as the one shown in Figure 10.20(a) or one with center tapped secondary as that of Figure 10.20(b) is used. The circuit of Figure 10.20(a) is reproduced with a slight rearrangement in Figure 11.17(a). In the circuit we show instantaneous values of the variables involved, with the motor's back EMF E_c assumed constant. The back EMF actually depends on the motor's speed, which is time varying. A practical approximation is obtained by use of the results of Section 10.3. Our object is to obtain an equivalent circuit as shown in Figure 11.17(b) that represents the average performance of the system over one cycle of the source voltage.

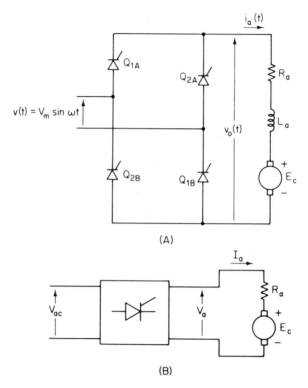

FIGURE 11.17 dc motor armature voltage control using single-phase full-wave controlled rectifier: (a) circuit details; (b) equivalent-circuit model of average performance.

The average voltage applied to the motor's armature circuit is a function of the firing angle α of the thyristors. The required relations are developed using results obtained in Section 10.3.

In Section 10.3, we studied a single-phase half-wave rectifier circuit as shown in Figure 10.15. The load circuit employed in our analysis is clearly a model of the armature circuit of a dc motor. The average current into the load circuit (motor's armature current) is given by Eq. (10.38) as

$$I_{o_{av}} = \frac{1}{2\pi R}\{V_m[\cos\alpha - \cos(\alpha + \gamma)] - \gamma E_c\} \qquad (11.37)$$

The conduction angle γ determines the mode of operation and is obtained from Eq. (10.31) using information available about the load circuit. Let us assume that the armature circuit inductance is chosen large enough to result in continuous operation, and therefore we take $\gamma = \pi$. In this case Eq. (11.37) yields

$$I_{o_{av}} = \frac{V_m \cos \alpha}{\pi R} - \frac{E_c}{2R} \tag{11.38}$$

This is the average value of the armature circuit current over one cycle of the supply frequency if a single-phase half-wave controlled rectifier is used for armature voltage control.

In the more practical case, where a single-phase full-wave controlled rectifier bridge is used as shown in Figure 10.20(a), there are two load current pulses in one cycle of the supply frequency, as shown in Figure 10.22(f). We can therefore conclude that the average value of the motor's armature current over one cycle of the source voltage waveform is obtained by multiplying the right-hand side of Eq. (11.38) by 2 to obtain

$$I_a = \frac{2V_m \cos \alpha}{\pi R} - \frac{E_c}{R} \tag{11.39}$$

As a result, we have

$$E_c = \frac{2V_m \cos \alpha}{\pi} - I_a R_a \tag{11.40}$$

In terms of the dc motor's equivalent circuit shown in Figure 11.17, we have

$$V_a = \frac{2V_m \cos \alpha}{\pi} \tag{11.41}$$

Alternatively, with the rms value of the supply voltage V, we have

$$V_a = \frac{2\sqrt{2}\,V}{\pi} \cos \alpha \tag{11.42}$$

Three-Phase AC Supply

When the available ac supply is a three-phase source, a full-wave controlled rectifier circuit as shown in Figure 10.24 is employed. In this case the average armature applied voltage is given on the basis of Eq. (10.60) as

$$V_a = \frac{3V_m}{\pi} \cos \alpha \tag{11.43}$$

In terms of the effective value of the line-to-line voltage V_{LL}, we have

$$V_a = \frac{3\sqrt{2}}{\pi} V_{LL} \cos \alpha \tag{11.44}$$

Example 11.3 A 230-V three-phase supply is available to drive a separately excited dc motor through a three-phase full-wave bridge rectifier circuit connected to the armature terminals. The armature resistance is 0.2 Ω and the motor draws a current of 205 A when running at 1750 r/min with an armature voltage of 230 V dc.

    a. Find the firing angle α under the specified conditions.
    b. Find the firing angle α required for the motor to run at 875 r/min.
    c. Find the motor's speed for a firing angle of 75°.
    d. Find the motor's speed for a firing angle of zero.

Solution

(a) We apply Eq. (11.44):

$$230 = \frac{3\sqrt{2}}{\pi}(230)\cos\alpha$$

As a result,

$$\alpha = 42.2°$$

(b) For a speed of 1750 r/min, we have

$$E_{c_1} = V_{a_1} - I_a R_a = 230 - 205(0.2) = 189 \text{ V}$$

For a speed of 875 r/min, we require

$$E_{c_2} = E_{c_1}\frac{\omega_2}{\omega_1} = \frac{189}{2} = 94.5 \text{ V}$$

As a result,

$$V_{a_2} = E_{c_2} + I_a R_a = 94.5 + 205(0.2) = 135.5 \text{ V}$$

We can now obtain the required firing angle:

$$V_{a_2} = 135.5 = \frac{3\sqrt{2}}{\pi}(230)\cos\alpha_2$$

$$\alpha_2 = 64.14°$$

(c) With $\alpha_3 = 75°$, we obtain

$$V_{a3} = \frac{3\sqrt{2}}{\pi} (230) \cos 75° = 80.39 \text{ V}$$

Thus

$$E_{c3} = V_{a3} - I_a R_a$$

$$= 80.39 - 205(0.2) = 39.39 \text{ V}$$

$$n_3 = n_1 \frac{E_{c3}}{E_{c1}} = (1750) \frac{39.39}{189} = 364.74 \text{ r/min}$$

(d) With $\alpha_4 = 0°$, we have

$$V_{a4} = \frac{3\sqrt{2}}{\pi} (230) = 310.61 \text{ V}$$

As a result,

$$E_{c4} = 310.61 - 41 = 269.61 \text{ V}$$

$$n_4 = 1750 \left(\frac{269.61}{189} \right) = 2496.4 \text{ r/min}$$

Field Control Using Controlled Rectifiers

With an ac supply available, controlled rectifiers are used to provide variable voltage dc supply to the field circuit. The load on the rectifier is an RL circuit and the rectifier is assumed to operate continuously with $\gamma = \pi$. For a single-phase controlled half-wave rectifier, we have, by Eq. (10.43),

$$I_f = \frac{V_m}{\pi R_f} \cos \alpha \tag{11.45}$$

For a single-phase full-wave controlled rectifier supplying the field circuit, we obtain

$$I_f = \frac{2V_m}{\pi R_f} \cos \alpha \tag{11.46}$$

When a three-phase full-wave controlled rectifier is employed, then from Eq. (10.60) we have the average field voltage given in terms of the maximum value of the line-to-line voltage by

$$V_f = \frac{3V_m}{\pi} \cos \alpha \tag{11.47}$$

Therefore, the average field current is given by

$$I_f = \frac{3V_m}{\pi R_f} \cos \alpha \tag{11.48}$$

It is clear that the average field current is a function of the firing angle α and we write in general

$$I_f = I_{f_b} \cos \alpha \tag{11.49}$$

The base field current I_{f_b} corresponds to rated field flux and is given in terms of the effective value of the ac voltage for the single-phase full-wave rectifier by

$$I_{f_b} = \frac{2\sqrt{2}}{\pi R_f} V \tag{11.50}$$

For a three-phase full-wave rectifier we have

$$I_{f_b} = \frac{3\sqrt{2}}{\pi R_f} V_{LL} \tag{11.51}$$

where V_{LL} is the rms value of the ac line-to-line voltage.

Speed–Firing Angle Relations

From the analysis of the preceding subsections we can develop the relation between motor speed and controlled rectifier firing angles α_a for the armature circuit in region I and α_f for the field in region II.

Region I: It is convenient to write Eqs. (11.42) and (11.44) as

$$V_a = K_a \cos \alpha_a \tag{11.52}$$

where the constant K_a is defined by

$$K_a = \frac{2\sqrt{2}}{\pi} V$$

for a single-phase full-wave rectifier bridge, and

$$K_a = \frac{3\sqrt{2}}{\pi} V_{LL}$$

for a three-phase full-wave rectifier bridge.
Combining Eqs. (11.31) to (11.33), and (11.52), we conclude that

$$\omega = \frac{1}{K_1 \phi_f} \left(K_a \cos \alpha_a - \frac{T_o R_a}{K_1 \phi_f} \right) \tag{11.53}$$

It is clear that an increase in α_a is associated with a decrease in speed in the constant-torque region.

Region II: From Eqs. (11.49), (7.27) and (11.36), we can write

$$\omega = \frac{P_a}{K_1 K_2 I_a I_{f_b}} \frac{1}{\cos \alpha_f} \tag{11.54}$$

It is clear that an increase in α_f increases ω in the constant-horsepower region.

Example 11.4   For the motor of Example 11.3, assume that the speed corresponding to $\alpha = 0$ is the base speed. Find the firing angle of the field rectifier circuit corresponding to a speed of 3000 r/min. What would be the firing angle for a speed that is twice base speed?

Solution

Operation in region II should satisfy Eq. (11.36), requiring that

$$I_{f_b} \omega_b = I_f \omega$$

Invoking Eq. (11.49), we have

$$I_{f_b} \omega_b = I_{f_b} \omega \cos \alpha$$

Thus

$$\cos \alpha = \frac{\omega_b}{\omega} = \frac{2496.38}{3000}$$
$$\alpha = \cos^{-1} 0.83 = 33.7°$$

For $\omega = 2\omega_b$,

$$\cos \alpha = 0.5$$

As a result,

$$\alpha = 60°$$

11.12 DC–DC DRIVES FOR DC MOTORS

A dc supply is used to power motors for rapid transit, electric, railroad trains, battery-operated vehicles, and electric forklift trucks. Dc to dc drives employing chopper control offer an attractive alternative to resistance control schemes that prevailed in this application area. The use of semiconductor power control avoids the excessive heat generated in resistive speed control. This is an important consideration, especially in underground applications such as transportation in tunnels and mining operations.

As discussed in Section 10.4, dc-to-dc converters (choppers) provide a variable direct voltage by varying the duration of thyristor on time t_{on} in relation to the chopping frequency T in the time-ratio-control (TRC) scheme. Choppers are widely employed to control separately excited and series dc motors in traction applications. A typical one-quadrant chopper scheme for armature control of a separately excited dc motor is shown in Figure 10.30 and is repeated in Figure 11.18 with an input LC filter and a smoothing reactor inserted in the circuit to reduce the possible wide variations in supply and motor currents.

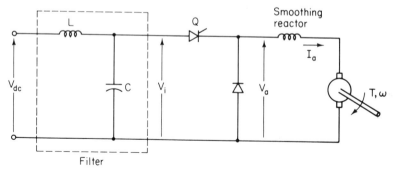

FIGURE 11.18   One-quadrant chopper armature voltage control circuit.

One-Quadrant Chopper Control

In Section 10.4 we found that the current waveform for a load consisting of a dc motor's armature can be either continuous, as shown in Figure 10.31(b), or discontinuous, as shown in Figure 10.32(b). The criterion that determines the mode of operation is the value of the critical on-time $t_{on}^{\star}$ given by Eq. (10.77), repeated here as

$$t_{oN}^{\star} = \tau \, \ell n \left[1 + \frac{E_c}{V_i} (e^{T/\tau} - 1) \right] \tag{11.55}$$

We recall Eq. (11.32) showing that E_c varies with the speed ω, and conclude that the mode of operation depends on speed. Clearly, for $\omega = 0$, the critical on-time is zero and any value of thyristor on-time results in continuous operation of the chopper. For speeds other than zero, definite values of $t_{on}^{\star}$ exist and if $t_{oN} > t_{on}^{\star}$, the output current is continuous, and conversely if $t_{on} < t_{on}^{\star}$, the output current is discontinuous.

The discontinuous current operation mode involves an interval of time between t_x and T when the instantaneous output current is zero for constant V_i. Here the value of t_x is given by Eq. (10.81) as

$$t_x = \tau \, \ell n \left[1 + \frac{V_i}{E_c} (e^{t_{on}/\tau} - 1) \right] \tag{11.56}$$

In this case the average of the output current is given by combining Eqs. (10.82) and (10.83) as

$$I_a = \left(\frac{t_{on}}{T} V_i - \frac{t_x}{T} E_c \right) \frac{1}{R_a} \tag{11.57}$$

In region I, the average output torque is given by Eq. (11.35) as

$$T_0 = K_1 \phi_f I_a \tag{11.58}$$

The field flux is held constant in this region. We also have, by Eq. (11.32),

$$E_c = K_1 \phi_f \omega \tag{11.59}$$

As a result, we conclude, by combining Eqs. (11.57) to (11.59), that the motor's speed is given by

$$\omega = \frac{t_{on}}{t_x} \frac{V_i}{K_1 \phi_f} - \frac{R_a}{(K_1 \phi_f)^2} \frac{T}{t_x} T_o \tag{11.60}$$

In the continuous mode of operation, with $t_{on} > t_{on}^*$, we have $t_x = T$ as a result Eq. (11.60) reduces to

$$\omega = \frac{(t_{on}/T)V_i - (R_a/K_1\phi_f)T_o}{K_1\phi_f} \qquad (11.61)$$

This completes our analysis of armature voltage control using one-quadrant choppers.

In region II, with one-quadrant chopper control of the field circuit, our analysis results are obtained by setting $E_c = 0$, since the field circuit model consists of a series RL combination. From Eq. (11.55), we note that with $E_c = 0$, the field current is continuous for any t_{on}. Therefore, Eqs. (10.82) and (10.83) give for the field circuit average current

$$I_f = \frac{t_{on}}{T} \frac{V_{fi}}{R_f} \qquad (11.62)$$

From Eqs. (11.36), (7.27), and (11.62), we get the desired expression for the average motor speed in region II as

$$\omega = \frac{P_a}{K_1 K_2 I_a I_{fb}} \frac{T}{t_{on}} \qquad (11.63)$$

where the base field current I_{fb} is defined by Eq. (11.62) for $t_{on} = T$ as

$$I_{fb} = \frac{V_{fi}}{R_f} \qquad (11.64)$$

The dc voltage supply to the field circuit is denoted by V_{fi} in the expressions above.

Example 11.5 The armature voltage of a separately excited dc motor is controlled by a one-quadrant chopper with chopping frequency of 200 pulses per second from a 300-V dc source. The motor runs at a speed of 800 r/min when the chopper's time ratio is 0.8. Assume that the armature circuit resistance and inductance are 0.08 Ω and 15 mH, respectively, and that the motor develops a torque of 2.72 N · m per ampere of armature current.

Find the mode of operation of the chopper, the output torque, and horsepower under the specified conditions.

Solution

From the problem specifications at 800 r/min, we get, using Eq. (11.59),

$$E_c = (2.72) \frac{2\pi}{60} (800) = 227.9 \text{ V}$$

The armature circuit time constant is obtained as

$$\tau = \frac{L_a}{R_a} = \frac{15 \times 10^{-3}}{0.08} = 187.5 \times 10^{-3} \text{ s}$$

The chopping period is given by

$$T = \frac{1}{200} = 5 \times 10^{-3} \text{ s}$$

We obtain the critical on-time using Eq. (11.55) as

$$t_{\text{on}}^\star = 187.5 \times 10^{-3} \ell n \left[1 + \frac{227.9}{300} (e^{5/187.5} - 1) \right]$$

$$= 3.8 \times 10^{-3} \text{ s}$$

We know that $t_{\text{on}} = 0.8 \times 5 \times 10^{-3} = 4 \times 10^{-3}$. As a result, we conclude that the chopper output current is continuous.

To obtain the torque output, we use Eq. (11.61) rearranged as

$$T_o = \frac{K_1 \phi_f}{R_a} \left(\frac{t_{oN}}{T} V_i - K_1 \phi_f \omega \right)$$

Thus we obtain

$$T_o = \frac{2.72}{0.08} [0.8(300) - 227.9]$$
$$= 411.4 \text{ N} \cdot \text{m}$$

The power output is obtained as

$$P_o = (411.4) \frac{2\pi}{60} (800) = 34.5 \times 10^3 \text{ W}$$
$$= 46.2 \text{ hp}$$

Problem 11.A.4 deals with the case when the chopper's operation is discontinuous. To illustrate the principle of field control, we have the following example.

Example 11.6 Assume for the motor of Example 11.5 that field chopper control is employed to run the motor at a speed of 1500 r/min while delivering the same power output as obtained at 800 r/min and drawing the same armature current.

Solution

Although we can use Eq. (11.63), we use basic formulas from Chapter 7 instead,

$$E_c = \frac{P_a}{I_a} = \frac{34.5 \times 10^3}{151.3} = 227.9 \text{ V}$$

This is the same back EMF. Recall that

$$E_c = K_1 \phi_f \omega$$

Thus the required field flux is obtained as

$$\phi_{fn} = \phi_{f0} \frac{\omega_0}{\omega_n} = \frac{8}{15} \phi_{f0}$$

where the subscript n denotes the present case and the subscript 0 denotes the field flux for Example 11.5. Assume that ϕ_{f0} corresponds to full applied field flux; then

$$\frac{\phi_{f0}}{\phi_{fn}} = \frac{V_i}{V_o} = \frac{15}{8}$$

The required chopper output voltage is V_o. Now we have

$$\frac{V_o}{V_i} = \frac{t_{on}}{T}$$

Thus

$$\frac{t_{on}}{T} = \frac{8}{15}$$

Assuming that $T = 5 \times 10^{-3}$ s, we get

$$t_{on} = 2.67 \times 10^{-3} \text{ s}$$

SOME SOLVED PROBLEMS

Problem 11.A.1

A three-phase four-pole 60-Hz squirrel-cage induction motor drives a load with a torque–speed characteristic given by

$$T_\ell = 7.2 \times 10^{-3} \, \omega^2$$

The motor parameters are

$$R_1 = 0.06 \, \Omega \qquad R_2 = 0.08 \, \Omega \qquad X_T = 0.6 \, \Omega$$

a. Obtain a formula that enables us to find the motor's slip for a given stator voltage in terms of motor and load parameters.
b. Find the rotor slip, current, torque, and output horsepower for a line-to-line voltage of 440 V.
c. Repeat part (b) for a line-to-line voltage of 330 V.
d. Repeat part (c) for a line-to-line voltage of 220 V.

Solution

(a) Equations (11.3) and (11.6) are combined to give

$$\frac{K_s^2}{|V_1|^2} (1 - s)^2 = \frac{s}{(R_1 s + R_2)^2 + s^2 X_T^2}$$

A few simplification steps result in the following fourth-order polynominal in s:

$$s^4 + a_3 s^3 + a_2 s^2 + a_1 s + a_0 = 0$$

where

$$a_3 = 2\left(\frac{R_1 R_2}{R_1^2 + X_T^2} - 1\right)$$

$$a_2 = 1 + \frac{R_2(R_2 - 4R_1)}{R_1^2 + X_T^2}$$

$$a_1 = \frac{1}{R_1^2 + X_T^2}\left[2R_2(R_1 - R_2) - \frac{|V_1|^2}{K_s^2}\right]$$

$$a_0 = \frac{R_2^2}{R_1^2 + X_T^2}$$

For the motor considered, we calculate the required parameters as

$$\omega_s = 60\pi \text{ rad/s}$$

$$K_s^2 = \frac{k_\ell \omega_s^3}{3R_2} = 200.9 \times 10^3$$

$$a_3 = -1.974$$

$$a_2 = 0.965$$

$$a_1 = -8.8 \times 10^{-3} - \frac{|V_1|^2}{73.06 \times 10^3}$$

$$a_0 = 17.6 \times 10^{-3}$$

Note that only a_1 depends on the stator phase voltage V_1.

(b) For $|V_1| = 440/\sqrt{3}$ V, we have

$$a_1 = -0.8922$$

The slip is obtained by solving the fourth-order polynominal as

$$s = 0.02$$

The current is obtained from

$$|I| = \frac{440/\sqrt{3}}{|(0.06 + 0.08/0.02) + j0.6|} = 61.9 \text{ A}$$

The rotor speed is obtained as

$$\omega_r = 0.98(60\pi) = 184.7 \text{ rad/s}$$

As a result, the load torque is found as

$$T_\ell = 7.2 \times 10^{-3}\omega_r^2 = 245.62 \text{ N} \cdot \text{m}$$

The output horsepower is obtained

$$P_0 = \frac{\omega_r T_\ell}{746} = 60.81 \text{ hp}$$

(c) For $|V_1| = 330/\sqrt{3}$ V, we have

$$a_1 = -0.5057$$

The slip is found as

$$s = 0.0373$$

The current is obtained from

$$|I| = \frac{330/\sqrt{3}}{|(0.06 + 0.08/0.0373) + j0.6|} = 83.38 \text{ A}$$

The rotor speed is obtained as

$$\omega_r = (1 - 0.0373)(60\pi) = 181.47 \text{ rad/s}$$

The load torque is found as

$$T_\ell = 237.1 \text{ N} \cdot \text{m}$$

The output horsepower is

$$P_o = 57.68 \text{ hp}$$

(d) For $|V_1| = 220/\sqrt{3}$ V, we have

$$a_1 = -0.2296$$

The slip is found as

$$s = 0.13$$

The current is obtained as

$$|I| = 140.6 \text{ A}$$

The rotor speed is given by

$$\omega_r = 163.96 \text{ rad/s}$$

The load torque is found as

$$T_\ell = (7.2 \times 10^{-3})(163.96)^2 = 193.57 \text{ N} \cdot \text{m}$$

The output horsepower is obtained as

$$P_o = 42.55 \text{ hp}$$

Problem 11.A.2

A three-phase four-pole induction motor has the following equivalent circuit parameters:

$$R_2 = 0.1 \; \Omega \qquad L_T = 1.6 \times 10^{-3} \; H$$

Neglect stator resistance. The motor's supply provides a constant volts/Hz input to the stator with line-to-line voltage of 220 V when driving a load with torque–speed characteristic given by

$$T = k_\ell \omega_r^2$$

where $k_\ell = 6 \times 10^{-3}$. Find the input frequency and rotor slip when the motor's speed is 1500 r/min. Assume that the output torque is maximum for this frequency.

Solution

The required torque output is obtained as

$$T = (6 \times 10^{-3}) \left[\frac{2\pi}{60} (1500) \right]^2 = 148 \; N \cdot m$$

This torque should be the maximum value for the voltage and frequency supply. Thus Eq. (11.29) is used:

$$148 = \frac{3(4)|V_1|^2}{4(1.6 \times 10^{-3})\omega_i^2}$$

As a result, we have

$$\frac{|V_1|}{\omega_i} = 0.28$$

Given that $|V_1| = 220/\sqrt{3}$, we get

$$\omega_i = 452 \; rad/s$$

The required frequency input is thus

$$f_i = \frac{452}{2\pi} = 71.94 \; Hz$$

The synchronous speed is obtained as

$$n_s = \frac{120f_i}{P} = 2158.3 \text{ r/min}$$

As a result, the slip is found as

$$s = \frac{n_s - n_r}{n_s} = 0.305$$

Problem 11.A.3

A separately excited dc motor is supplied by a three-phase full-wave bridge rectifier circuit whose line-to-line voltage is 230 V. The motor's armature resistance is 0.12 Ω. The motor supplies a load at a speed of 900 r/min with an output torque of 210 N · m. The armature current is 100 A under these conditions. Find the firing angle of the rectifier bridge. If the firing angle is adjusted to 65°, find the motor's output torque and its speed.

Solution

The motor's back EMF is found using

$$E_c = \frac{T\omega}{I_a} = \frac{2\pi}{60} (900) \frac{210}{100} = 197.92 \text{ V}$$

The armature voltage is obtained from

$$\begin{aligned} V_a &= E_c + I_a R_a \\ &= 197.92 + (100)(0.12) = 209.92 \text{ V} \end{aligned}$$

Thus

$$209.92 = \frac{3\sqrt{2}}{\pi} (230) \cos \alpha$$

As a result, we get

$$\alpha = 47.48°$$

Now with $\alpha_2 = 65°$,

$$V_{a2} = \frac{3\sqrt{2}}{\pi}(230)\cos 65 = 131.27 \text{ V}$$

$$E_{c2} = V_{a2} - I_a R_a = 131.27 - 12 = 119.27 \text{ V}$$

The new speed is obtained from

$$n_2 = n_1 \frac{E_{c2}}{E_{c1}} = 542.35 \text{ rpm}$$

We can thus obtain the required torque using

$$T = \frac{E_c I_a}{\omega} = 210.0 \text{ N} \cdot \text{m}$$

The torque remains the same because the armature current is constant.

Problem 11.A.4

To examine the effect of the inductance in the armature circuit of a separately excited dc motor, repeat Example 11.5 for an armature circuit time constant of 5×10^{-3} s, with all other specifications unchanged.

Solution

Equation (11.55) gives us the new critical time as

$$t^{\star}_{on} = 5 \times 10^{-3} \ln\left[1 + \frac{227.9}{300}(e^{5/5} - 1)\right]$$
$$= 4.18 \times 10^{-3} \text{ s}$$

The effect of a decreased time constant is to increase $t^{\star}_{on}$. In the present case with $t_{on} = 4 \times 10^{-3}$ s, the output current is discontinuous.

We find t_x, the time at which current ceases to flow using Eq. (11.56), as

$$t_x = 5 \times 10^{-3} \ln\left[1 + \frac{300}{227.9}(e^{4/5} - 1)\right]$$
$$= 4.8 \times 10^{-3} \text{ s}$$

The average output current is obtained, using Eq. (11.57), as

$$I_a = \left(\frac{4 \times 300}{5} - \frac{4.8 \times 227.9}{5}\right)\frac{1}{0.08}$$
$$= 265.2 \text{ A}$$

Using Eq. (11.58), with $K_1\phi_f = 2.72$, we get

$$T_0 = 2.72(265.2) = 721.3 \text{ N} \cdot \text{m}$$

The output power in this case is given by

$$P_o = 60.43 \times 10^3 \text{ W}$$

Clearly, the reduction in the armature circuit inductance gives higher armature current, torque, and power.

PROBLEMS

Problem 11.B.1

A three-phase 60-Hz eight-pole induction motor drives a load with torque–speed characteristic given by

$$T = 0.11\omega^2$$

where ω is in rad/s and T is in N $\cdot$ m. The stator applied voltage is 440 V (line to line). The rotor resistance is 0.1 Ω and its reactance is 0.8 Ω. Neglect stator impedance and the magnetizing circuit. Find the rotor slip, output torque, output power, and stator current.

Problem 11.B.2

Repeat Problem 11.B.1 if the stator applied voltage is reduced to 330 V (line to line).

Problem 11.B.3

Assume that the motor of Problem 11.A.1 drives a load with the torque–speed characteristic

$$T_\ell = 36 \times 10^{-3}\omega^2$$

where ω is in rad/s. The supply voltage is 440 V. Find the rotor slip, speed, stator current, output torque, and power. Find the input power and losses and comment on the results in relation to values obtained in Problem 11.A.1.

Problem 11.B.4

Repeat Problem 11.B.3 with the supply voltage decreased to 330 V.

Problem 11.B.5

A four-pole 60-Hz three-phase induction motor has a rotor resistance of 0.1 Ω. The motor drives a load with a torque–speed characteristic given by

$$T_\ell = 12 \times 10^{-3}\omega^2$$

where ω is in rad/s.

    a. If the motor runs at a slip of 3%, find the stator current and the developed torque.

    b. If the stator current is 100 A, find the slip and the developed torque.

Problem 11.B.6

A four-pole 60-Hz three-phase induction motor drives a load with torque–speed characteristic given by

$$T_\ell = 8 \times 10^{-3}\omega^2$$

    a. If the torque is 260 N · m, find the motor's slip and rotor resistance given that the stator current is 80 A.

    b. If the slip is 0.05, find the output torque and the armature current.

Problem 11.B.7

A four-pole 60-Hz three-phase induction motor runs at a slip of 8% and draws a stator current of 130 A when driving a square-law load. Assume that the rotor resistance is 0.08 Ω. Find the parameter k_ℓ of the load's torque–speed characteristic. Find the motor's speed, slip, and stator current for a torque output of 280 N · m.

Problem 11.B.8

An eight-pole 60-Hz 440-V three-phase induction motor has a negligible stator resistance and a rotor resistance of 0.12 Ω and a total leakage inductance of 2.2 $\times$ 10^{-3} H. The motor's supply is of the constant-volts/Hz category. Find the minimum supply frequency and the corresponding voltage. Find the maximum torque corresponding to constant-volts/Hz operation. Find the supply voltage and rotor speed corresponding to operation at 50 Hz.

Problem 11.B.9

Repeat Problem 11.B.8 for $P = 6$, $R_2 = 0.08 \ \Omega$, and $L_T = 2 \times 10^{-3}$ H. All other information is unchanged.

Problem 11.B.10

A three-phase induction motor has a rotor resistance of 0.1 Ω. The minimum frequency for operation in the constant-volts/Hz region is 10 Hz. Find the total leakage inductance of the motor.

Problem 11.B.11

An eight-pole 440-V 60-Hz three-phase induction motor is controlled so that its maximum torque is 1200 N · m under rated conditions. Find the leakage inductance of the motor's equivalent circuit. Find the rotor resistance if the minimum frequency for proper operation in the constant-volts/Hz region is 9 Hz.

Problem 11.B.12

A three-phase four-pole induction motor has the parameters

$$R_2 = 0.1 \ \Omega \qquad L_T = 1.8 \times 10^{-3} \text{ H}$$

Neglect stator resistance. The motor's supply provides a constant-volts/Hz input to the stator with a line-to-line voltage of 220 V. When driving a load with $k_\ell = 6.5 \times 10^{-3}$, the speed is 1600 r/min. Find the input frequency and rotor slip assuming that the output torque is the maximum for the frequency chosen.

Problem 11.B.13

A three-phase eight-pole induction motor has the following parameters

$$R_2 = 0.09 \ \Omega \qquad L_T = 1.7 \times 10^{-3} \text{ H}$$

Neglect stator resistance. The line-to-line voltage is 220 V in a constant-volts/Hz supply. The load has a $k_\ell = 5 \times 10^{-3}$ and is driven at a speed of 1400 r/min. Find the input frequency and rotor slip assuming that the output torque is the maximum for the frequency chosen.

Problem 11.B.14

A three-phase four-pole induction motor has a total leakage inductance of 1.6×10^{-3} H and is operating from a constant-volts/Hz supply at 220 V (line to line). The motor delivers an output torque of 160 N · m at a speed of 1500 r/min. The load has a quadratic torque-speed characteristic. Find k_ℓ and the stator's input frequency.

Problem 11.B.15

A three-phase four-pole induction motor has a total leakage inductance of 1.8×10^{-3} H and is operating from a constant-volts/Hz supply at a frequency of 55 Hz. The motor delivers an output torque of 170 N · m at a speed of 1450 r/min. Find k_ℓ of the load's quadratic torque–speed characteristic and the stator's applied voltage.

Problem 11.B.16

A 220-V three-phase supply is available to drive a separately excited dc motor through a three-phase full-wave bridge rectifier circuit connected to the armature terminals. The armature resistance is 0.15 Ω and the motor draws a current of 190 A when running at 1600 r/min with an armature voltage of 200 V dc.

    a. Find the firing angle α under the specified conditions.
    b. Find the firing angle α required for the motor to run at 1800 r/min.
    c. Find the motor's speed for a firing angle of 60°.

Problem 11.B.17

A 230-V three-phase supply is available to drive a separately excited dc motor through a three-phase full-wave bridge rectifier circuit connected to the armature terminals. The armature resistance is 0.18 Ω and the motor draws a current of 200 A when running at 1800 r/min with an armature voltage of 240 V dc.

    a. Find the firing angle α under the specified conditions.
    b. Find the firing angle α required for the motor to run at 1200 r/min.
    c. Find the motor's speed for a firing angle of 75°.

Problem 11.B.18

A 220-V three-phase supply is available to drive a separately excited dc motor through a three-phase full-wave bridge rectifier circuit connected to the armature terminals. The armature resistance is 0.12 Ω and the motor draws a current of 180 A when running at 1500 r/min with an armature voltage of 180 V dc.

    a. Find the firing angle α under the specified conditions.

    b. Find the firing angle α required for the motor to run at 875 r/min.

    c. Find the motor's speed for a firing angle of 80°.

Problem 11.B.19

A separately excited dc motor is supplied by a three-phase full-wave bridge rectifier circuit whose line-to-line voltage is 220 V. The motor's armature resistance is 0.15 Ω. The motor supplies a load at a speed of 800 r/min with an output torque of 300 N $\cdot$ m. The armature current is 90 A under these conditions. Find the firing angle of the rectifier bridge. If the firing angle is adjusted to 30°, find the motor's speed.

Problem 11.B.20

A separately excited dc motor is supplied by a three-phase full-wave bridge rectifier circuit whose line-to-line voltage is 230 V. The motor's armature resistance is 0.08 Ω. The motor supplies a load at a speed of 1000 r/min with an output torque of 220 N $\cdot$ m. The armature current is 130 A under these conditions. Find the firing angle of the rectifier bridge. If the firing angle is adjusted to 40°, find the motor's speed.

Problem 11.B.21

A separately excited dc motor is supplied by a three-phase full-wave bridge rectifier circuit whose line-to-line voltage is 120 V. The motor's armature resistance is 0.05 Ω. The motor supplies a load at a speed of 600 r/min with an output torque of 140 N $\cdot$ m. The armature current is 60 A under these conditions. Find the firing angle of the rectifier bridge. If the firing angle is adjusted to 30°, find the motor's speed.

Problem 11.B.22

For the motor of Problem 11.B.16, assume that the speed corresponding to $\alpha = 0$ is the base speed. Find the firing angle of the field rectifier circuit corresponding to a speed of 2800 r/min.

Problem 11.B.23

For the motor of Problem 11.B.17, assume that the speed corresponding to $\alpha = 0$ is the base speed. Find the firing angle of the field rectifier circuit corresponding to a speed of 2800 r/min.

Problem 11.B.24

For the motor of Problem 11.B.18, assume that the speed corresponding to $\alpha = 0$ is the base speed. Find the firing angle of the field rectifier circuit corresponding to a speed of 3000 r/min.

Problem 11.B.25

Find the required chopper on-time for the motor of Example 11.5 to deliver a torque of 400 N · m at a speed of 600 r/min. Repeat for a speed of 900 r/min. Verify that the chopper operates in the continuous current mode in both cases.

Problem 11.B.26

The armature resistance of a separately excited dc motor is 0.5 Ω and its inductance is 7 mH. The motor develops a torque of 1.34 N · m per ampere of armature current and is controlled by a one-quadrant chopper with chopping period of 5×10^{-3} s and an on-time of 4×10^{-3} s. Assume that the dc supply is 220 V. Find the average armature current, torque, and power output.

Problem 11.B.27

For the motor of Problem 11.B.26, find the required chopper on-time at a speed of 1000 r/min while delivering the same torque.

Index

E, Y.

$$\frac{12/06/96}{Sun.}$$

h=15.76